Western Hemisphere

Multiple images from satellites *Terra*, *Aqua*, *Radarsat*, and *Defense Meteorological Satellite*, and from Space Shuttle *Endeavor's* radar data of topography, all merge in a dramatic composite to show the Western Hemisphere and Eastern Hemisphere of Earth. What indications do you see on these images that tell you the time of year? These are part of NASA's Blue Marble Next Generation image collection.

Eastern Hemisphere

A Brief, Visual Approach to Physical Geography

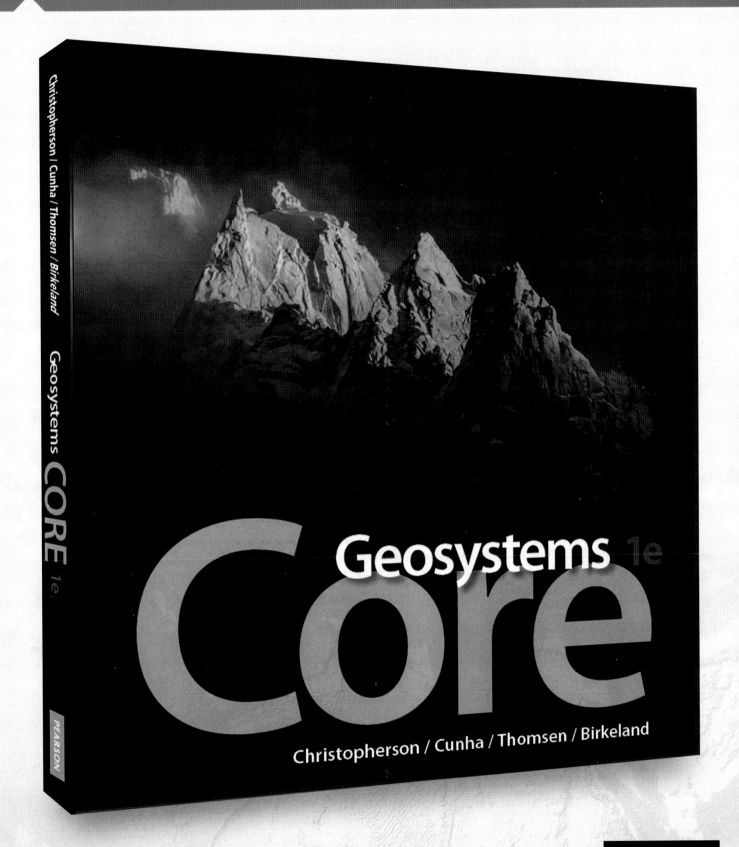

Christopherson | Cunha | Thomsen | Birkeland

Geosystems

Geosystems
Core 1e

Christopherson / Cunha / Thomsen / Birkeland

PEARSON

Brief, Modular, & Flexible

Two-page modules present the core concepts of physical geography. *Geosystems Core* focuses on a clear, concise, and highly-visual presentation of the essential science. Instructors can assign these flexible modules in whatever sequence best suits their course and teaching style. The consistent, focused, and engaging presentation prevents students from becoming lost in unnecessary detail.

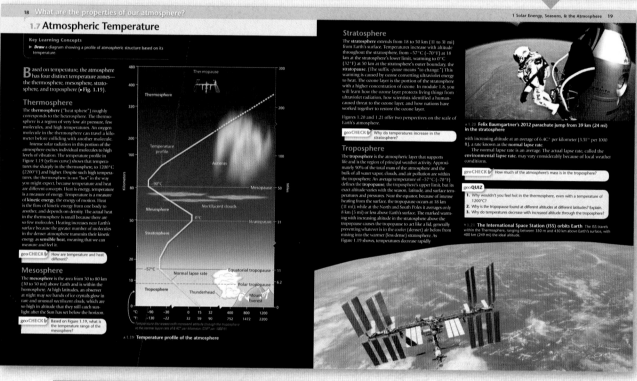

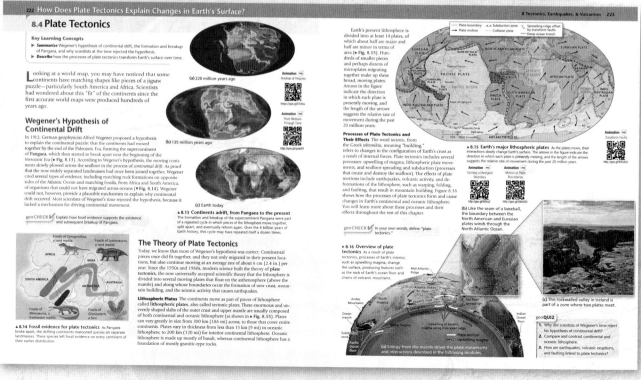

Mobile-Ready Media Bring Geography to Life

Over 130 videos & animations integrated within the chapters give readers instant access to visualizations of key physical processes, as well as applied case studies & virtual explorations of the real world. Readers use mobile devices to scan Quick Response (QR) links in the book to immediately access media as they read the chapters. These media are also available in the MasteringGeography Study Area, and can be assigned with automatically-graded assessments.

Mobile Field Trip
Videos transport students on adventures with acclaimed photographer and pilot Michael Collier, in the air and on the ground, exploring iconic landscapes of North America and the natural and human forces that have shaped them.

Project Condor
Quadcopter Videos take students out into the field through narrated & annotated quadcopter video footage, exploring the physical processes that have helped shape North American landscapes.

GeoLabs: An Integrated Lab Experience

GeoLab modules integrate the lab experience directly into the book, enabling students to get hands-on with data & the applied tools of physical geography without the need for a separate lab manual. Perfect for lab work, homework, or group work, each *GeoLab* presents a context-rich & data-driven lab activity, and includes a QR-linked *Pre-Lab* Review Video that reviews the chapter concepts needed for the activity. Associated auto-gradable assessments in MasteringGeography can be assigned for credit.

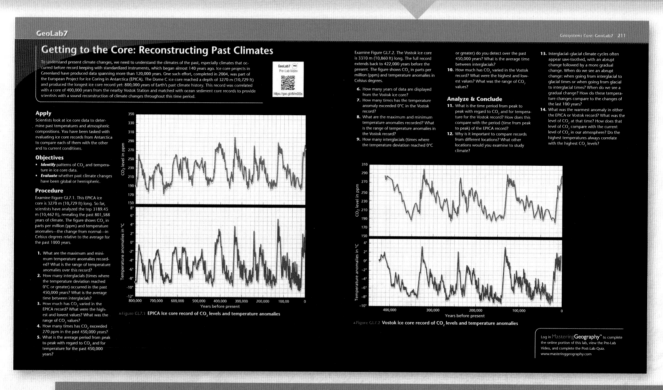

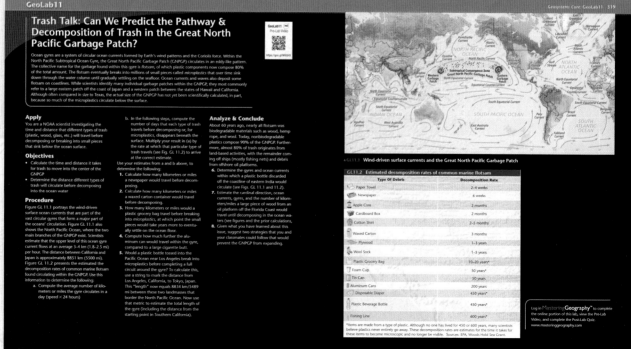

The Human Denominator of Earth Systems

The Human Denominator concludes each chapter, explicitly focusing on human-Earth relationships in physical geography & Earth systems science. These highly-visual features include maps, aerial imagery, photos of real-world applications, and a brief overview of current & potential future issues.

THE**human**DENOMINATOR 11   Oceans, Coasts, & Dunes

COASTAL SYSTEMS IMPACT HUMANS
- Rising sea level has the potential to inundate coastal communities.
- Tsunami cause damage and loss of life along vulnerable coastlines.
- Coastal erosion changes coastal landscapes, affecting developed areas; human development on depositional features such as barrier island chains is at risk from storms, especially hurricanes.

HUMANS IMPACT COASTAL SYSTEMS
- Rising ocean temperatures, pollution, and ocean acidification impact corals and reef ecosystems.
- Human development drains and fills coastal wetlands and mangrove swamps, thereby removing their buffering effect during storms.

11a

A cargo vessel ran aground on Nightingale Island, Tristan da Cunha, in the South Atlantic in 2011, spilling an estimated 1500 tons of fuel, spilling tons of soybeans, and coating endangered Northern Rockhopper penguins with oil.

11b

Dredgers pump sand through a hose to replenish beaches on Spain's Mediterranean coast, a popular tourist destination. Near Barcelona, pictured here, sand is frequently eroded during storms; natural replenishment is limited by structures that block longshore currents.

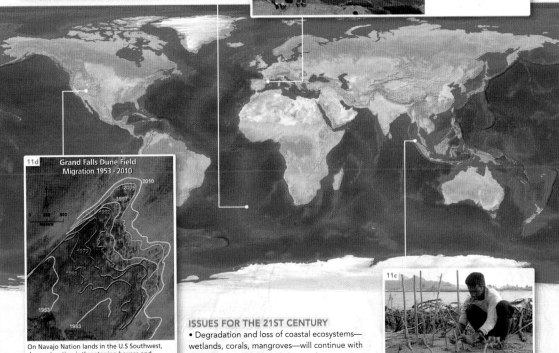

11d

Grand Falls Dune Field Migration 1953 - 2010

On Navajo Nation lands in the U.S Southwest, dune migration is threatening houses and transportation, and affecting human health. The Grand Falls dune field in northeast Arizona grew 70% in areal extent from 1997 to 2007. The increasingly dry climate of this region has accelerated dune migration and reactivated inactive dunes.

ISSUES FOR THE 21ST CENTURY
- Degradation and loss of coastal ecosystems—wetlands, corals, mangroves—will continue with coastal development and climate change.
- Continued building on vulnerable coastal landforms will necessitate expensive recovery efforts, especially as storm systems become more intense with climate change.

11c

Mangrove planting: In Aceh, Indonesia, near the site of the 2004 Indian Ocean tsunami, authorities encourage local people to plant mangroves for protection against future tsunami.

Looking Ahead
In the next chapter we examine glacial and periglacial landscapes. We will investigate how glacial formation and movement sculpts the land and leaves behind many distinctive landforms. Changes in the Earth's total mass of glacial ice is also important evidence used to monitor our changing climate.

Tools for Structured Learning

Key Concepts organize chapter modules around the big picture questions of physical geography.

Key Learning Concepts present the key information and skills that students need to master in each module, and also provide the organizing structure for the MasteringGeography item library of assessments.

GeoChecks in each module enable students to check their understanding as they read the module sections, for a "read a little, do a little" approach that fosters active critical thinking.

GeoQuizzes conclude each module, giving students a chance to check their understanding before moving on to the next module.

I-20 What tools do geographers use?

I.6 Modern Geoscience Tools

Key Learning Concepts

▶ **Explain** how geographers use the Global Positioning System, remote sensing, geographic information systems, and geovisualizations.

geoCHECK ✔ Why are at least three satellites needed to find a location using GPS?

geoCHECK ✔ Compare and contrast the two types of remote sensing.

geoCHECK ✔ Describe the two types of information that a GIS combines.

geoQUIZ

1. Explain at least two ways you have benefited from the GPS.
2. What types of remote-sensing data have you seen today? in the past week?
3. Describe the criteria for a GIS used to find a parcel of land to build a new subdivision using the following data layers: property parcels, zoning layer, floodplain layer, protected wetlands layer.

Critical Thinking, Review, & Spatial Analysis

Chapter Review includes a module-by-module summary with integrated *Review* questions, *Critical Thinking* exercises, *Visual Analysis* activities, *Interactive Mapping* activities using *MapMaster*, and *Explore* activities using *Google Earth*.

Visual Analysis

Glaciers in Alaska have been retreating dramatically due to warming temperatures. The Muir Glacier is a good example of this.

1. Examine the two photographs and describe the changes observed.

2. What are two examples visible in the photographs that show how much conditions have changed from 1941 to 2004?

Muir Glacier 1941 **(a)**

Muir Glacier 2004 **(b)**

▲R7.1 **Muir Glacier** (a) 1941 and (b) 2004.

(MG) Interactive Mapping | Login to the **MasteringGeography** Study Area to access **MapMaster.**

Climate Change

Earth's climate is changing, but not all locations will change equally. Some locations will change much more than others.

- Open MapMaster in MasteringGeography™.
- Select Global Surface Warming Worst Case Projections from the Physical Environment menu. Explore the sublayers of different temperature change projections.

1. What is the largest projected change for the land in the Northern Hemisphere? What is the largest projected change

for the land in the Southern Hemisphere? What is the projected change for the Hawaiian Islands? For your home town?

2. Describe the pattern of projected change, as a function of latitude and continentality. What are the characteristics of the locations with the highest amount of projected change? Locations with the lowest amount of projected change?

Explore Earth | Use **Google Earth** to explore the **Glaciers of Alaska.**

Over 95% of glaciers are in retreat worldwide. Glaciers in Alaska are no exception. Search for the *Columbia Glacier, Alaska*. Zoom in until you can see where the end of the glacier meets the sea. Use the *Add Path* tool to trace the outline of the end of the glacier. Turn on *Historical Imagery* (the clock button), and go back to 11/27/2007. Use the *Add Path* tool again to draw the outline of the end of the glacier.

1. Use the *Show Ruler* tool to measure the retreat from 2007 to 2013 at several places. What is the maximum and minimum retreat?

2. How many miles or kilometers per year has the glacier been retreating?

3. If the glacier continues to retreat at this rate, how long until the retreat equals your daily commute to school?

▶R7.2

Continuous Learning
Before, During, and After Class

BEFORE CLASS

Mobile Media & Reading Assignments Ensure Students Come to Class Prepared.

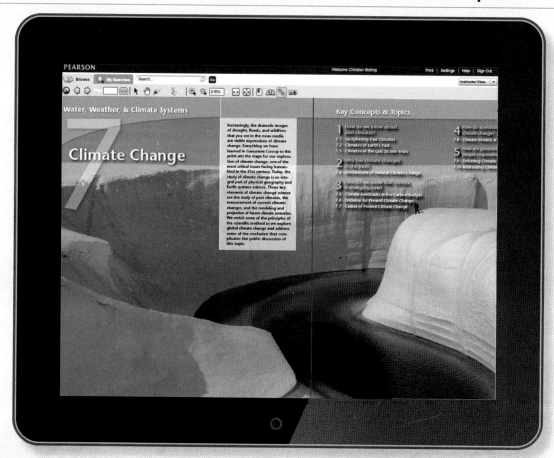

Dynamic Study Modules personalize each student's learning experience. Created to allow students to acquire knowledge on their own and be better prepared for class discussions and assessments, this mobile app is available for iOS and Android devices.

Pearson eText in MasteringGeography gives students access to the text whenever and wherever they can access the internet. eText features include:

- Now available on smartphones and tablets.
- Seamlessly integrated videos and other rich media.
- Fully accessible (screen-reader ready).
- Configurable reading settings, including resizable type and night reading mode.
- Instructor and student note-taking, highlighting, bookmarking, and search.

Pre-Lecture Reading Quizzes are easy to customize & assign

Reading Questions ensure that students complete the assigned reading before class and stay on track with reading assignments. Reading Questions are 100% mobile ready and can be completed by students on mobile devices.

with MasteringGeography™

DURING CLASS

Learning Catalytics™ & Engaging Media

What has Teachers and Students excited? Learning Catalyics, a 'bring your own device' student engagement, assessment, and classroom intelligence system, allows students to use their smartphone, tablet, or laptop to respond to questions in class. With Learning Cataltyics, you can:

- Assess students in real-time using open ended question formats to uncover student misconceptions and adjust lecture accordingly.

- Automatically create groups for peer instruction based on student response patterns, to optimize discussion productivity.

> *"My students are so busy and engaged answering Learning Catalytics questions during lecture that they don't have time for Facebook."*
>
> Declan De Paor, *Old Dominion University*

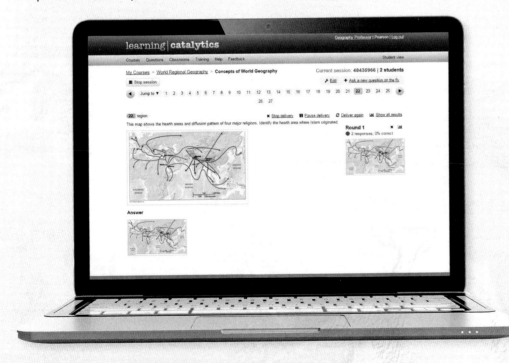

Enrich Lecture with Dynamic Media

Teachers can incorporate dynamic media into lecture, such as Videos, Mobile Field Trips Videos, MapMaster Interactive Maps, Project Condor Quadcopter videos, and Geoscience Animations.

Mastering Geography™

MasteringGeography delivers engaging, dynamic learning opportunities—focusing on course objectives and responsive to each student's progress—that are proven to help students absorb physical geography course material and understand challenging geography processes and concepts.

AFTER CLASS

Easy to Assign, Customizable, Media-Rich, and Automatically Graded Assignments

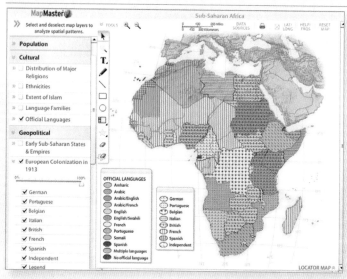

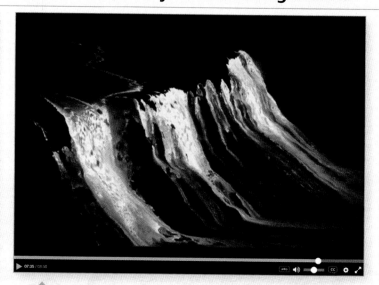

MapMaster Interactive Map Activities are inspired by GIS, allowing students to layer various thematic maps to analyze spatial patterns and data at regional and global scales. This tool includes zoom and annotation functionality, with hundreds of map layers leveraging recent data from sources such as NOAA, NASA, USGS, United Nations, and the CIA.

Geography Videos from such sources as the BBC and *The Financial Times* are now included in addition to the videos from Television for the Environment's Life and Earth Report series in **MasteringGeography**. Approximately 200 video clips for over 30 hours of footage are available to students and teachers and **MasteringGeography**.

Mobile Field Trip **Videos** have students accompany acclaimed photographer and pilot Michael Collier in the air and on the ground to explore iconic landscapes of North America and beyond. Readers scan Quick Response (QR) links in the book to access the 20 videos as they read. Also available within **MasteringGeography**.

www.masteringgeography.com

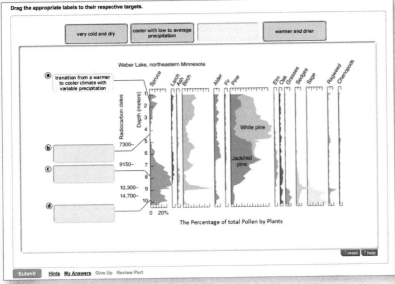

Principle of original horizontality

GeoTutors are highly visual and data-rich coaching items with hints and specific wrong answer feedback that help students master the toughest topics in geography.

Project Condor **Quadcopter Videos** take students out into the field through narrated & annotated quadcopter video footage, exploring the physical processes that have helped shape North American landscapes.

Encounter (Google Earth) activities provide rich, interactive explorations of physical geography concepts, allowing students to visualize spatial data and tour distant places on the virtual globe.

Geoscience Animations help students visualize the most challenging physical processes in the physical geosciences with schematic animations that include audio narration. Animations include assignable multiple-choice quizzes with specific wrong answer feedback to help guide students toward mastery of these core physical process concepts.

Geosystems ^{1e}

Core

Christopherson / Cunha / Thomsen / Birkeland

PEARSON

Senior Geography Editor: Christian Botting
Project Manager: Connie Long
Program Manager: Anton Yakovlev
Executive Development Editor: Jonathan Cheney
Art Development Editor: Jay McElroy
Development Manager: Jennifer Hart
Program Management Team Lead: Kristen Flathman
Project Management Team Lead: David Zielonka
Production Management: Jeanine Furino, Cenveo Publisher Services
Copyeditor: Jane Loftus
Compositor: Cenveo Publisher Services
Design Manager: Mark Ong
Interior/Cover Designer: Gary Hespenheide

Illustrators: Lachina and International Mapping Associates
Rights & Permissions Senior Project Managers: Timothy Nicholls and Maya Gomez
Photo Researcher: Lauren McFalls/Lumina
Manufacturing Buyer: Maura Zaldivar-Garcia
Executive Product Marketing Manager: Neena Bali
Senior Field Marketing Manager: Mary Salzman
Associate Media Content Producer: Mia Sullivan
Market Development Manager: Leslie Allen
Cover Photo: *Sunlight on Aiguille des Drus, Chamonix, in the European Alps, near where the borders of France, Switzerland, and Italy meet. Mount Blanc is nearby, the highest summit in the Alps;* photo by Alex Buisse–RockyNook, Photographer.

Library of Congress Cataloging-in-Publication Data

Names: Christopherson, Robert W. | Cunha, Stephen F. | Thomsen, Charles E.
Title: Geosystems core / Robert Christopherson, Stephen Cunha, Charles
  Thomsen, Ginger Birkeland.
Description: Hoboken, New Jersey : Pearson Education, [2017]
Identifiers: LCCN 2015048457| ISBN 9780321834744 (alk. paper) | ISBN
  0321834747 (alk. paper)
Subjects: LCSH: Physical geography. | Earth sciences.
Classification: LCC GB54.5 .C49 2017 | DDC 910/.02–dc23
LC record available at http://lccn.loc.gov/2015048457

About Our Sustainability Initiatives

Pearson recognizes the environmental challenges facing this planet, as well as acknowledges our responsibility in making a difference. This book is carefully crafted to minimize environmental impact. The binding, cover, and paper come from facilities that minimize waste, energy consumption, and the use of harmful chemicals. Pearson closes the loop by recycling every out-of-date text returned to our warehouse.

Along with developing and exploring digital solutions to our market's needs, Pearson has a strong commitment to achieving carbon-neutrality. As of 2009, Pearson became the first carbon- and climate-neutral publishing company, having reduced our absolute carbon footprint by 22% since then. Pearson has protected over 1,000 hectares of land in Columbia, Costa Rica, the United States, the UK and Canada. In 2015, Pearson formally adopted The Global Goals for Sustainable Development, sponsoring an event at the United Nations General Assembly and other ongoing initiatives. Pearson sources 100% of the electricity we use from green power and invests in renewable energy resources in multiple cities where we have operations, helping make them more sustainable and limiting our environmental impact for local communities.

The future holds great promise for reducing our impact on Earth's environment, and Pearson is proud to be leading the way. We strive to publish the best books with the most up-to-date and accurate content, and to do so in ways that minimize our impact on Earth. To learn more about our initiatives, please visit https://www.pearson.com/social-impact/sustainability/environment.html.

2 17

ISBN 10: **0-321-83474-7**; ISBN 13: **978-0-321-83474-4** (Student edition)
ISBN 10: **0-134-14283-7**; ISBN 13: **978- 0-134-14283-8** (Instructor's Review Copy)

www.pearsonhighered.com

Brief Contents

MasteringGeography™ Mobile-Ready Animations & Videos

Geosystems Core includes Quick Response links to over 130 mobile-ready animations and videos, which students can access using mobile devices. These media are also available in the Study Area of MasteringGeography, and can be assigned to students with quizzes.

Introduction to Physical Geography

Geoscience Animation
Map Projections

Mobile Field Trip
Introduction to Physical Geography

Video
GeoLab Pre-Lab Video

1 Solar Energy, Seasons, & the Atmosphere

Geoscience Animations
Earth Sun Relations
Formation of the Solar System
The Ozone Layer

Video
GeoLab Pre-Lab Video

2 Energy in the Atmosphere

Geoscience Animations
Global Warming, Climate Change
Earth-Atmosphere Energy Balance
The Gulf Stream

Videos
The Ozone Hole
The Ozone Layer
GeoLab Pre-Lab Video

3 Pressure, Winds, and Currents

Geoscience Animations
Coriolis Force
Global Atmospheric Circulation
Cyclones and Anticyclones
Jet Stream, Rossby Waves
Ocean Circulation
North Atlantic Deep Water Circulation
Thermohaline Circulation
El Niño and La Niña

Mobile Field Trip
El Niño

Video
GeoLab Pre-Lab Video

4 Atmospheric Water and Weather

Geoscience Animations
Water Phase Changes
Atmospheric Stability
Warm Fronts
Cold Fronts
Midlatitude Cyclones
Tornado Wind Patterns
Hurricane Wind Patterns

Mobile Field Trip
Clouds: Earth's Dynamic Atmosphere

Videos
NSSL in the Field
Radar Research at NSSL
Hurricane Hot Towers
Superstorm Sandy
Making of a Superstorm
GeoLab Pre-Lab Video

5 Water Resources

Geoscience Animations
Earth's Water and the Hydrologic Cycle
The Water Table

Mobile Field Trips
Oil Sands: An Unconventional Oil
Moving Water Across California

Videos
Hydrological Cycle
GeoLab Pre-Lab Video

6 Global Climate Systems

Geoscience Animation
Global Patterns of Precipitation

Videos
Supercomputing the Climate
GeoLab Pre-Lab Video

7 Climate Change

Geoscience Animations
Global Warming, Climate Change
End of the Last Ice Age
Earth-Sun Relations
Orbital Variations & Climate Change
The Carbonate Buffering System
Arctic Sea Ice Decline

Mobile Field Trip
Climate Change in the Arctic

Videos
20,000 Years of Pine Pollen
Taking Earth's Temperature
Keeping Up With Carbon
Temperature & Agriculture
Supercomputing the Climate
Superstorm Sandy
GeoLab Pre-Lab Video

8 Tectonics, Earthquakes, & Volcanism

Project Condor
Quadcopter Videos
Principles of Relative Dating
Intrusive Igneous Bodies

MasteringGeography™ Mobile-Ready Animations & Videos

Contents

Introduction to Physical Geography   I-2

PART I The Energy–Atmosphere System

Solar Energy, Seasons, & the Atmosphere 2

Atmospheric Energy & Global Temperatures 32

Atmospheric & Oceanic Circulations   62

PART II Water, Weather, & Climate Systems

Atmospheric Water & Weather 86

Climate Change   174

PART III   The Geosphere: Earth's Interior & Surface

Tectonics, Earthquakes, & Volcanism   212

Weathering, Karst Landscapes, & Mass Movement 246

River Systems & Landforms 268

Coastal Systems & Wind Processes   292

Glacial & Periglacial Landscapes   320

PART IV   The Biosphere

Ecosystems & Soils   344

Terrestrial Biomes   378

Preface

Welcome to *Geosystems Core,* a new exploration of physical geography! Geography is a highly visual discipline. Images of landslides, waterfalls, shrinking glaciers, monsoon deluges, climate change impacts, weather events, and tropical rainforests fill our media. Photographs portray the human response to sudden earthquakes and floods, or to more gradual phenomena such as prolonged drought effects or soil creep. For this reason, Pearson—the world's foremost publisher of geography textbooks—invites you to explore physical geography in a new, highly visual, modular approach.

Physical Geography Surrounds Us

The main purpose of *Geosystems Core* is to introduce physical geography—a geospatial science that integrates a range of disciplines concerned with Earth's physical and living systems, including geology, meteorology, biology, and ecology, among others. It is intended for use in college-level introductory courses in physical geography, Earth science, and environmental science.

Geosystems Core teaches a holistic view of Earth's environment. Central to this approach is human-environment interaction. During the last two centuries, our expanding human population became a major force in shaping Earth's environment. Humans plant crops, plough soils, domesticate animals, clear forests, build settlements, extract precious metals, and burn fossil fuels. Human agency modifies the distribution of plant and animal species. We impound and divert most of the world's major rivers, and are altering the chemistry of the oceans. Moreover, in the last 20 years, mounting evidence from every scientific field supports the case for human-induced climate change.

As an academic discipline, the roots of geography stretch back to antiquity, yet physical geography is essential to understanding current environmental issues. For example, by 320 B.C.E., the Greek philosopher and scientist Aristotle recognized how vegetation and climate changed with elevation in the Pindos Mountains of Greece. Today, contemporary geography thrives on the cutting edge of knowledge, serving as the bridge between Earth and natural sciences. New geospatial technologies such as GPS, GIS, and Remote Sensing allow humans to view, record, and analyze the world anew.

Geographers analyze environmental problems from pole to pole, and from the ocean floor to Earth's highest summits. They use new technologies to analyze acid rain deposition in mountain lakes, trace dust storms across continents, and assess the changing distribution of plants and animals on a warming planet. Knowing where things are—the spatial arrangement of everything from deserts and rainforests, to active volcanoes and hurricanes—is key to understanding geography, and is emphasized throughout *Geosystems Core*.

Although dramatic global change is underway in physical, chemical, and biological systems that support and sustain us, the environmental future of our planet need not be bleak. Population growth rates are decreasing almost everywhere on Earth. Emerging technology is leading humanity away from dependence on fossil fuels and into an era where clean and renewable energy prevails. Important advances in soil science, water conservation, and crop management are making agriculture more productive and sustainable. Advancing scientific knowledge and possibly lowering poverty rates worldwide offer enormous potential to make this twenty-first century one of great environmental restoration. For that to occur, all of us must understand the complex and interrelated environmental systems that govern our unique planet. This study of physical geography takes us along this path.

Organization & Themes

The goal of physical geography is to explain the spatial dimension of Earth's dynamic systems—its energy, air, water, weather, climate, tectonics, landforms, rocks, soils, plants, ecosystems, and biomes. *Geosystems Core* focuses on the most essential, core concepts of physical geography. The following themes present the major organizational structure of the book.

- **Earth Systems Science:** *Geosystems Core* is organized around the natural flow of energy, materials, and information in our Earth system, presenting subjects in the same sequence in which they occur in nature (atmosphere, hydrosphere, geosphere, and biosphere)—an organic, holistic Earth systems approach that is unique in this discipline.
- **Climate Change Science:** Incorporating the latest climate change science and data throughout, *Geosystems Core* includes a dedicated chapter on climate change, covering paleoclimatology and mechanisms for past climatic change, climate feedbacks and the global carbon budget, the evidence and causes of present climate change, climate forecasts and models, and actions that we can take to moderate Earth's changing climate.
- **Human-Earth Relationships:** Each chapter concludes with *The Human Denominator,* explicitly focusing on the human-Earth dimension of physical geography within context of the chapter topic. These features include maps (spatial data), real-world examples (photos), and review of both current and potential future issues that help engage students by connecting physical geography concepts to their real-world environment.
- **Geospatial Technology:** Rapidly developing technologies pervade our everyday lives. Mapping and geospatial technologies such as GPS, GIS, and RS are high growth areas, critical tools in the twenty-first century that help us visualize, measure, and analyze Earth's natural and human-built features, and make every day decisions. *Geosystems Core* integrates geospatial technology throughout all chapters to help students visualize and critically analyze the spatial dimensions of Earth's physical geography.

Structured Learning

A structured learning path and tightly integrated pedagogy give students a reliable, consistent framework for mastering the major concepts of physical geography:

- **Two-page modules present key geographical concepts** that can stand on their own and be read in any order. Instructors can assign these flexible modules in whatever sequence best suits their course and teaching style. Each module in the text contains the essential content for each concept; this focused presentation prevents students from becoming lost in unnecessary detail.
- The chapter-opening **Key Concepts** list the learning objectives for each chapter.
- **GeoChecks** and **GeoQuizzes** are integrated into each module, enabling students to check their understanding as they read the module sections, for a "read a little, do a little" approach that is engaging, and that fosters active critical thinking.
- Chapters conclude with a **Chapter Review** that includes a module-by-module summary with various types of review activities including *Critical Thinking*, *Visual Analysis*, *Interactive Mapping*, and *Explore* activities using Google Earth.
- **GeoLabs** Unique, two-page *GeoLab* capstone modules integrate a lab experience directly into the book without the need for a separate lab manual or lab section, enabling students to get hands-on with the data and tools of physical geography. Each *GeoLab* includes an online component in MasteringGeography that can be assigned and automatically graded.

Mobile Media & MasteringGeography

- **Over 130 Animations & Videos** are **QR Linked** to provide just-in-time reinforcement to learners as they read, giving students instant mobile access to visualizations of key physical processes as well as applied case studies and virtual explorations of the real world. Sources include NASA/JPL, FEMA, and NOAA, *Mobile Field Trip* Videos by Michael Collier, and *Project Condor* Quadcopter videos.
- **MasteringGeography** is an online homework, tutorial, and assessment program designed to work with *Geosystems Core* to engage students and improve results. Interactive, self-paced coaching activities provide individualized coaching to keep students on track. With a wide range of visual and media-rich activities available, including GIS-inspired MapMaster interactive maps, Encounter *Google Earth* explorations, geoscience animations, GeoTutors on the more challenging topics in geography, and a range of videos, students can actively learn, understand, and retain even the most difficult concepts.

Acknowledgments

Geosystems Core took tremendous time, resources, and focus to develop as a first edition science textbook. This highly collaborative, multi-year effort involved authors, editors, graphic artists, media producers, and experts in page design, photo research, and logistics. Our thanks to the *Pearson* team for their expertise and enthusiasm for this project. These include Senior Geography Editor Christian Botting for his vision and expertise, Program Manager Anton Yakovlev and Project Manager Connie Long for keeping us on track and on schedule, Art Development Editor Jay McElroy for taking our visions and making them into art that teaches and is beautiful, and our wonderful Executive Development Editor Jonathan Cheney for clarifying and polishing our rough thoughts into words, and Market Development Manager Leslie Allen for arranging reviews and class tests.

Our thanks also to the production team at Cenveo, in particular Jeanine Furino and Cindy Miller, whose efforts proved essential to bringing our ideas to fruition. We also thank the art and cartography studios at International Mapping: Kevin Lear and Luchina, and Senior Permissions Project Manager, Lauren McFalls for helping us find the images we requested.

Thanks to all the teachers (and their students) who served as reviewers and class testers throughout the development of *Geosystems Core*:

- **Manuscript Development & Accuracy Reviewers**

  Miriam Helen Hill, *Jacksonville State University*
  James Kernan, *State University of New York at Geneseo*
  Lisa DeChano, *University of Michigan*
  Doug Goodin, *Kansas State University*
  Janice Hayden, *Dixie State College of Utah*
  David Holt, *The University of Southern Mississippi*
  Kara Kuvakas, *Hartnell College*
  James Vaughan, *The University of Texas at San Antonio*

- **Reviewers & Class Testers**

  Alexis Aguilar, *Pasadena City College*
  Keith Bettinger, *University of Hawaii (Kapi'olani Community College)*
  Trent Biggs, *San Diego State University*
  Michael Boester, *Monroe Community College*
  Liana Boop, *San Jacinto College*
  Carsten Braun, *Westfield State University*
  Robert Bristow, *Westfield State University*
  Joan Bunbury, *Univeristy of Wisconsin La Crosse*
  Adam Burnett, *Colgate University*
  John Conley, *Fullerton College*
  Michael Davis, *Kutztown University*
  Nicole DePue, *Salem State University*
  James Dyer, *Ohio University - Athens*
  Jonathan Fleming, *University of North Alabama*
  Tyler Fricker, *Florida State University*
  Anilkumar Gangadharan, *Kennesaw State University*

Julienne Gard, *El Camino College*
Katie Gerber, *Santa Rosa Junior College*
Lawrence Gilbert, *Virginia Western Community College*
Brett Goforth, *California State University - San Bernardino*
Dafna Golden, *Mount San Antonio College*
John Greene, *University of Oklahoma*
Roy Haggerty, *Oregon State University*
Amanda Hall, *Towson University*
Janice Hayden, *Dixie State College of Utah*
Donald (Don) Helfrich, *Central New Mexico Community College*
Joseph Hinton, *City College of Chicago*
Michael Keables, *University of Denver*
Ryan Kelly, *Kentucky Community and Technical College System*
James Kernan, *State University of New York at Geneseo*
John Keyantash, *California State University - Dominguez Hills*
Chris Krause, *Glendale Community College*
Paul Larson, *Southern Utah University*
Denyse Lemaire, *Rowan University-Glassboro*
Michael Lewis, *University of North Carolina at Greensboro*
Jonathon Little, *Monroe Community College*
Jing Liu, *Santa Monica College*
Kerry Lyste, *Everett Community College*
Joy Mast, *Carthage College*
Shannon McCarragher, *Northern Illinois University*
Benjamin McDaniel, *North Hennepin Community College*
Mystyn Mills, *California State University - Long Beach*
Monika Moore, *Delta College*
Nathan Moore, *Michigan State University*
Todd Moore, *Towson University*
Patrick Olsen, *University of Idaho*
Hal Olson, *Anne Arundel Community College*
Michael Paluzzi, *Rockland County Community College*
Jeremy Patrich, *Santa Monica College*
Mark Patterson, *Kennesaw State University*
Marius Paulikas, *Kent State University*
Nancy Perry, *Northern Virginia Community College*
Mike Pesses, *Antelope Valley College*
James Powers, *Pasadena City College*
Curtis Robinson, *Sacramento City College*
Steve Schultze, *University of South Alabama*
Anil Shrestha, *Northern Illinois University*
Jeremy Spencer, *University of Akron*
John Van Stan, *Georgia Southern University*
James Vaughan, *University of Texas at San Antonio*
Megan Walsh, *Central Washington University*
Richard Watson, *Central New Mexico Community College*
Susan White, *West Los Angeles College*
Cody Wiley, *University of New Mexico*
Erika Wise, *University of North Carolina at Chapel Hill*
Julie Wulff, *Oakton Community College*

Author Acknowledgments

- **From Robert:** I thank my family, especially my wife Bobbé, for believing in this work from the first edition of *Geosystems* in 1992. I give special gratitude to all the students during my 30 years of teaching, for it is in the classroom crucible that these books are forged. I offer special thanks to my talented coauthors: Stephen, Charlie, and Ginger, for their dedicated work in extending the *Geosystems* franchise in this dramatic, new, modular presentation. And, thanks to Pearson for supporting such a creative vision for physical geography.

- **From Stephen:** Sincere thanks to Douglas R. Powell of UC Berkeley, for introducing me to Physical Geography, and for inspiration to experience first hand, what I share with my students. I also thank my students over many years. I admire their enthusiasm to learn and their wonderful minds. Finally, deepest appreciation to my family, especially spouse Mary Beth Cunha—an accomplished geospatial scientist, university faculty, and traveler extraordinaire!

- **From Charlie:** Thanks to Robert, Stephen, and Ginger, I couldn't have asked for a better team of co-authors. Thanks to Ronald I. Dorn of Arizona State University, for formally introducing me to the wonderful world of geography. Thanks to my students and colleagues at American River College. I thank my supportive family for their patience and understanding, especially my wife Leslie, and my children Emma and Finn.

- **From Ginger:** Many thanks to Robert, for his dedication in pioneering the *Geosystems* approach, and to Stephen and Charlie, for their creative work extending *Geosystems* in this new and exciting direction. My sincere gratitude goes to my husband, Karl, for our many scientific discussions and his unwavering support, and to my daughters Erika and Kelsey. Through their eyes, I see a bright future for physical geography and the challenges ahead.

- **From all of us:** Physical geography teaches us a holistic view of the intricate supporting web that is Earth's environment and our place in it. Dramatic global change is underway in human-Earth relations as we alter physical, chemical, and biological systems. Our attention to climate change science and applied topics is in response to the impacts we are experiencing and the future we are shaping. All things considered, this is a perfect time for you to be enrolled in a physical geography course! The best to you in your studies—and *carpe diem!*

Digital & Print Resources

For Students & Teachers

MasteringGeography™ with Pearson eText. The *Mastering* platform is the most widely used and effective online homework, tutorial, and assessment system for the sciences. It delivers self-paced tutorials that provide individualized coaching, focus on course objectives, and are responsive to each student's progress. The *Mastering* system helps teachers maximize class time with customizable, easy-to-assign, and automatically graded assessments that motivate students to learn outside of class and arrive prepared for lecture. MasteringGeography™ offers:

- **Assignable activities** that include GIS-inspired MapMaster™ interactive map activities, *Encounter* Google Earth™ Explorations, video activities, *Mobile Field Trips, Project Condor* Quadcopter videos, Geoscience Animation activities, map projections activities, GeoTutor coaching activities on the toughest topics in geography, Dynamic Study Modules that provide each student with a customized learning experience, end-of-chapter questions and exercises, reading quizzes, *Test Bank* questions, and more.
- **A student Study Area** with MapMaster™ interactive maps, videos, *Mobile Field Trips, Project Condor* Quadcopter videos, Geoscience Animations, web links, glossary flashcards, "In the News" readings, chapter quizzes, PDF downloads of outline maps, an optional Pearson eText and more.

Pearson eText gives students access to the text whenever and wherever they can access the Internet. Features of Pearson eText include:

- Now available on smartphones and tablets.
- Seamlessly integrated videos and other rich media.
- Fully accessible (screen-reader ready).
- Configurable reading settings, including resizable type and night reading mode.

Instructor and student note-taking, highlighting, bookmarking, and search. www.masteringgeography.com

Television for the Environment Earth Report Geography Videos, DVD (0321662989). This three-DVD set helps students visualize how human decisions and behavior have affected the environment and how individuals are taking steps toward recovery. With topics ranging from the poor land management promoting the devastation of river systems in Central America to the struggles for electricity in China and Africa, these 13 videos from Television for the Environment's global *Earth Report* series recognize the efforts of individuals around the world to unite and protect the planet.

Geoscience Animation Library, 5th edition, DVD (0321716841). Created through a unique collaboration among Pearson's leading geoscience authors, this resource offers over 100 animations covering the most difficult-to-visualize topics in physical geography, meteorology, earth science, physical geology, and oceanography.

Practicing Geography: Careers for Enhancing Society and the Environment by Association of American Geographers (0321811151). This book examines career opportunities for geographers and geospatial professionals in the business, government, nonprofit, and education sectors. A diverse group of academic and industry professionals shares insights on career planning, networking, transitioning between employment sectors, and balancing work and home life. The book illustrates the value of geographic expertise and technologies through engaging profiles and case studies of geographers at work.

Teaching College Geography: A Practical Guide for Graduate Students and Early Career Faculty by Association of American Geographers (0136054471). This two-part resource provides a starting point for becoming an effective geography teacher from the very first day of class. Part One addresses "nuts-and-bolts" teaching issues. Part Two explores being an effective teacher in the field, supporting critical thinking with GIS and mapping technologies, engaging learners in large geography classes, and promoting awareness of international perspectives and geographic issues.

Aspiring Academics: A Resource Book for Graduate Students and Early Career Faculty by Association of American Geographers (0136048919). Drawing on several years of research, this set of essays is designed to help graduate students and early career faculty start their careers in geography and related social and environmental sciences. *Aspiring Academics* stresses the interdependence of teaching, research, and service—and the importance of achieving a healthy balance of professional and personal life—while doing faculty work. Each chapter provides accessible, forward-looking advice on topics that often cause the most stress in the first years of a college or university appointment.

For Students

Applied Physical Geography—Geosystems in the Laboratory, **Ninth Edition** by Charlie Thomsen and Robert Christopherson (0321987284). A variety of exercises provides flexibility in lab assignments. Each exercise includes key terms and learning concepts linked to *Geosystems.* The Ninth Edition includes new exercises on climate change, soils, and rock identification, a fully updated exercise on basic GIS using ArcGIS online, and more integrated media, including Google Earth™ and Quick Response (QR) codes linking to Pre-Lab videos. Supported by a website with online worksheets as well as KMZ files for all of the Google Earth™ exercises found in the lab manual. www.mygeoscienceplace.com

Goode's World Atlas, 23rd Edition (0133864642). *Goode's World Atlas* has been the world's premiere educational atlas since 1923—and for good reason. It features over 250 pages of maps, from definitive physical and political maps to important thematic maps that illustrate the spatial aspects of many important topics. The 23rd Edition includes over 160 pages of digitally produced reference maps, as well as thematic maps on global climate change, sea-level rise, CO_2 emissions, polar ice fluctuations, deforestation, extreme weather events, infectious diseases, water resources, and energy production.

Pearson's Encounter Series provides rich, interactive explorations of geoscience concepts through Google Earth™ activities, covering a range of topics in regional, human, and physical geography. For those who do not use *MasteringGeography*™, all chapter explorations are available in print workbooks, as well as in online quizzes at www.mygeoscienceplace.com, accommodating different classroom needs. Each exploration consists of a worksheet, online quizzes whose results can be emailed to teachers, and a corresponding Google Earth™ KMZ file.

- *Encounter Physical Geography* by Jess C. Porter and Stephen O'Connell (0321672526)
- *Encounter World Regional Geography* by Jess C. Porter (0321681754)
  Encounter Human Geography by Jess C. Porter (0321682203)

Dire Predictions: Understanding Global Climate Change 2nd Edition by Michael Mann, Lee R. Kump (0133909778). Periodic reports from the Intergovernmental Panel on Climate Change (IPCC) evaluate the risk of climate change brought on by humans. But the sheer volume of scientific data remains inscrutable to the general public, particularly to those who may still question the validity of climate change. In just over 200 pages, this practical text presents and expands upon the essential findings of the IPCC in a visually stunning and undeniably powerful way to the lay reader. Scientific findings that provide validity to the implications of climate change are presented in clear-cut graphic elements, striking images, and understandable analogies.

The Second Edition covers the latest climate change data and scientific consensus from the IPCC *Fifth Assessment Report* and integrates mobile media links to online media. The text is also available in various eText formats, including an eText upgrade option from MasteringGeography courses.

For Teachers

Learning Catalytics is a "bring your own device" student engagement, assessment, and classroom intelligence system. With Learning Catalytics, you can:

- Assess students in real time, using open-ended tasks to probe student understanding.
- Understand immediately where students are and adjust your lecture accordingly.
- Improve your students' critical-thinking skills.
- Access rich analytics to understand student performance.
- Add your own questions to make Learning Catalytics fit your course exactly.
- Manage student interactions with intelligent grouping and timing.

Learning Catalytics is a technology that has grown out of twenty years of cutting-edge research, innovation, and implementation of interactive teaching and peer instruction. Available integrated with *MasteringGeography*™.

Instructor Resource Manual (Download) (0134142802). The manual includes lecture outlines and key terms, additional source materials, teaching tips, and a complete annotation of chapter review questions. Available from www.pearsonhighered.com/irc and in the Instructor Resources area of *MasteringGeography*™.

TestGen® Test Bank (Download) by Todd Fagin (0134142829). TestGen® is a computerized test generator that lets you view and edit *Test Bank* questions, transfer questions to tests, and print tests in a variety of customized formats. This *Test Bank* includes around 3,000 multiple-choice and short answer/essay questions. All questions are correlated against the National Geography Standards, textbook key learning concepts, and Bloom's Taxonomy. The *Test Bank* is also available in Microsoft Word® and importable into Blackboard. Available from www.pearsonhighered.com/irc and in the Instructor Resources area of *MasteringGeography*™.

Instructor Resource DVD (0134142810). The *Instructor Resource DVD* provides a collection of resources to help teachers make efficient and effective use of their time. All digital resources can be found in one well-organized, easy-to-access place. The IRDVD includes:

- All textbook images as JPEGs, PDFs, and PowerPoint™ Presentations
- Pre-authored Lecture Outline PowerPoint® Presentations which outline the concepts of each chapter with embedded art and can be customized to fit teachers' lecture requirements
- CRS "Clicker" Questions in PowerPoint™
- The TestGen software, *Test Bank* questions, and answers Electronic files of the *Instructor Resource Manual* and *Test Bank*

This *Instructor Resource* content is also available online via the Instructor Resources section of *MasteringGeography*™ and www.pearsonhighered.com/irc.

About the Authors

Robert W. Christopherson attended California State University, Chico for his undergraduate work and received his M.A. in Geography from Miami University-Oxford, Ohio. *Geosystems* evolved out of his physical geography research in grad school and thirty years of classroom teaching notes. His wife Bobbé is his principal photographer and provided more than 300 exclusive photos for each of his books. Together they completed eleven polar expeditions (most recently in summer 2013 and 2015). Robert is the recipient of numerous awards, including the 1998 and 2005 *Textbook and Academic Authors Association (TAA)* "Textbook Award" for *Geosystems* and *Elemental Geosystems, 4/e,* respectively. He was selected by American River College students as "Teacher of the Year" and received the American River College "Patrons Award." Robert received the 1999 "Distinguished Teaching Achievement Award" from the *National Council for Geographic Education* and the "Outstanding Educator Award" from the *California Geographical Society* in 1997. In 2012, California State University, Chico, presented him their "Distinguished Alumni Award." In 2013, TAA presented him with its "Lifetime Achievement Award." Robert has been deeply involved in the development of Pearson's *Geoscience Animation Library,* and he led the editorial board of Rand McNally's *Goode's World Atlas 22nd edition.* Robert currently serves on the Advisory Board of *Biosphere 2,* Earth's largest ecosystem research facility, operated by the University of Arizona.

Stephen Cunha is professor of geography at Humboldt State University. He received his B.S. and B.A. degrees from University of California, Berkeley, and his M.A. and PhD in Geography from University of California, Davis. Stephen worked ten seasons as a park ranger in Yosemite and Alaska, and three years investigating the potential for a national park and biosphere reserve in the Pamir Mountains of Tajikistan. He is an active teacher, researcher, and mountain geographer, having co-authored geography textbooks and *The Atlas of California.* His travel experience in the Americas, Asia, Oceania, Europe, and Africa, brings new international perspective and content to *Geosystems Core.* Cunha has numerous teaching and research awards, and at press time serves as President of the *Association of Pacific Coast Geographers.*

Charlie Thomsen is professor of geography at American River College, where he teaches physical geography, human geography, field classes, and GIS. He has taught field courses in Yosemite National Park, backpacking down the Lost Coast Trail, snowshoeing in the Sierra Nevada, as well as in state and national parks throughout California. His career as an educator began in high school as a Boy Scout merit badge counselor at Camp Emerald Bay on Catalina Island, and he has been teaching ever since. Professor Thomsen received his B.A. from University of California, Los Angeles and his M.A. from California State University, Chico. He is the author of Pearson's *Encounter Geosystems* and *Applied Physical Geography: Geosystems in the Laboratory,* as well as many other assessment and media projects.

Ginger Birkeland received her B.A. from the University of Colorado, Boulder, and her M.A. and PhD in Geography from Arizona State University, with a focus in fluvial geomorphology. She taught physical geography at Montana State University and field courses at the Indiana University Geologic Field Station in Montana. Ginger worked as a professional river guide for 17 years on the Colorado River in Grand Canyon, as well as on rivers in Australia and throughout the U.S. West. She also worked as a geomorphology consultant on several government-funded projects, including the Truckee River Recovery Plan in California and Nevada. She is currently a coauthor with Robert Christopherson on *Geosystems* and *Elemental Geosystems.*

Geosystems Core

Geosystems 1e

Core

Introduction to Physical Geography

Physical geography explains the spatial dimension of Earth's dynamic systems—its energy, air, water, weather, climate, tectonics, landforms, rocks, soils, ecosystems, and biomes—terms that will become familiar to you as we progress through this book. Physical geography also investigates how humans interact with Earth systems. The discipline's spatial perspective, allows geographers to examine processes and events happening at specific locations and to follow their effects across the globe. We hope you find *Geosystems Core* an important physical geography resource as you explore our unique planet and its life-supporting Earth systems. Let the journey begin!

Key Concepts & Topics

A small glacial tarn reflects the peaks and glaciers of Canada's Purcell Mountains.

I.1 The World Around Us

Key Learning Concepts

▶ *Give examples* of the kinds of events, processes, and questions that physical geography investigates.

Welcome to *Geosystems Core* and the study of physical geography. In this text, we examine the natural processes on Earth that influence our lives—ranging from weather and climate to earthquakes and volcanoes. We also examine the many ways humans interact with these Earth systems. A **system** is any set of ordered, interrelated components and their attributes, linked by flows of energy and matter—a concept we will expand upon later in this chapter. Physical geography involves the study of Earth's environments, including the landscapes, seascapes, atmosphere, and ecosystems on which humans depend. In the second decade of the 21st century, our natural world is changing, and the scientific study of Earth and its environments is more crucial than ever.

▲I.1 **Locations of events shown in Figure I.2**

(a) Flowers blooming in the Atacama Desert, Chile

(b) Destruction in Nepal from a 2015 earthquake.

▲I.2 **Events that shape our changing planet** Every day, natural disasters and the effects of ordinary human activities, such as building a dam or using fossil fuels as an energy source, can raise questions to which geographers seek the answers.

Asking Geographic Questions

Consider as examples the following events, each of which raises questions for the study of Earth's physical geography (▲ Figs. I.1 and I.2). This text provides tools for answering these questions and addressing the underlying issues.

- In 2015, El Niño rains drenched northern Chile's Atacama Desert, one of the driest places on Earth. The unexpected deluge brought catastrophic flooding. However, all that water brought something else too. Within months, an explosion of wildflowers carpeted the normally barren ground (▲ Fig. I.2a). In some places, the seeds had been dormant in the soil for decades, until this perfect combination of rainfall and spring warmth brought them to life. Will climate change bring more frequent blooms in the future?

- In April 2015, a magnitude 7.8 earthquake stuck the Himalayan nation of Nepal. The earthquake killed more than 9000 people and injured another 23,000 (▲ Fig. I.2b). Why do earthquakes occur in particular locations across the globe? Why do earthquakes of similar magnitude and duration result in thousands of human casualties in one place, but almost none in another place?

- In 2014, the U.S. National Park Service finished dismantling two dams on the Elwha River in Washington—the largest dam removals in the world to date (▶ Fig. I.2c). The project will restore a free-flowing river for fisheries and associated ecosystems. How do dams change river environments? Can rivers be restored after dam removal?

- In 2015, Hurricane Patricia off the west coast of Mexico became the most powerful tropical storm ever measured in the Western Hemisphere (▼ Fig. I.2d). Although maximum winds over the ocean reached an unprecedented 220 kph (200 mph), the storm weakened quickly as it moved over the rugged terrain of central Mexico. Why are monster storms becoming more common, and how do they threaten human life and property?

- Rapidly evolving technologies such as Global Positioning Systems (GPS), remote sensing (RS), and geographic information systems (GIS)—terms discussed later in this chapter—increase our ability to collect and analyze the data needed to answer geographic questions (▼ Fig. I.2e). The rise of citizen science, volunteered geographic information (VGI), and participatory GIS (PGIS) provide opportunities for people to help monitor Earth's natural and human properties. Which areas interest you? This book will show you many possibilities.

Asking "Where?" & "Why?" Physical geography asks *where* and *why* questions about processes and events that occur at specific locations and then follow their effects across the globe. Why does the environment vary from equator to midlatitudes and between tropical and polar regions? What produces the patterns of wind, weather, and ocean currents? How does solar energy influence the distribution of trees, soils, climates, and human populations? In this book, we explore those questions and more through geography's unique emphasis on studying factors that affect the distribution of phenomena on Earth.

Climate Change Science & Physical Geography Climate change is now an overriding focus of the study of Earth systems. The past decade experienced the highest air temperatures over land and water in the instrumental record. In response, the extent of sea ice in the Arctic Ocean continues to decline to record lows. At the same time, melting of the Greenland and Antarctica Ice Sheets is accelerating and sea level is rising. Elsewhere, intense weather events, drought, and flooding continue to increase. In presenting the state of the planet, *Geosystems Core* surveys climate change evidence and considers its implications. Welcome to an exploration of physical geography and its impact on our daily lives!

(c) Dam removal on the Elwha River, Washington

(d) Hurricane Patricia approaching the coast of Mexico

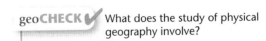 What does the study of physical geography involve?

geoQUIZ

1. Pick one of the events described above and, using your own words, list three geographic questions you would like to answer about that event.
2. Based on the examples above, would you say that humans should be considered part of the natural world? Explain your answer.
3. What is some of the evidence for climate change that scientists have observed?

(e) A student in Cambodia uses GPS to mark a location as part of a government-sponsored, land-reform effort.

I.2 The Science of Geography

Key Learning Concepts

▶ **Describe** the main perspectives of geography and distinguish physical geography from human geography.
▶ **Discuss** the use of scientific methods in geography.
▶ **Summarize** how human activities and population growth impact the environment.

The world around us is constantly changing as the events and processes described in Module I.1 transform Earth's physical environment, affecting humans and other living things. One science seeks to provide answers to our questions about these changes: **Geography** (from geo, "Earth," and graphein, "to write") studies the relationships among natural environments, geographic areas, human society, and the interdependence of all of these across Earth. For geographers, "space" is a term with a special meaning: geographic space comprises Earth's surface, but as described below, also includes much more than that.

Geographic Perspectives

As a science, geography approaches the study of Earth from a number of distinctive perspectives:

• emphasis on spatial and locational analysis
• concern with human environment–interactions (discussed below)
• adoption of an *Earth systems* perspective to analyze how the physical, biological, and human components of those systems are interconnected (discussed in Module I.3)

Given the complexity of Earth systems, it's not surprising that geography has many subfields. The field's two main divisions—human geography and physical geography—are discussed below.

Spatial & Locational Analysis The term **spatial** refers to the nature and character of physical space, its measurement, and the distribution of things within it. Geographers use **spatial analysis** as a tool to explain distributions and movement across Earth and how these processes interact with human activities.

Maps showing locations and distributions are important tools for conveying geographic data and interpreting spatial relationships. Evolving technologies such as geographic information systems and the Global Positioning System are widely used for scientific applications as millions of people access maps and locational information every day on computers and mobile devices.

Human Geography & Physical Geography Although geography integrates content from many disciplines, it splits broadly into two primary subfields: *physical geography*, which draws on the physical and life sciences, and *human geography*, which draws on the social and cultural sciences (▶Fig. I.3). The growing complexity of the human–Earth relationship in the 21st century is shifting the study of geographic processes even farther toward the synthesis of physical and human geography. This more balanced and holistic perspective is the thrust of *Geosystems Core*. Within

physical geography, research now emphasizes human influences on natural systems. For example, physical geographers monitor air pollution, examine the vulnerability of human populations to climate change, study impacts of human activities on forest health and the movement of invasive species, analyze changes in river systems caused by dam removal, and examine the response of glacial ice to changing climate.

Geography's spatial analysis method unifies the discipline more than does a specific body of knowledge. Geographers employ spatial analysis to examine how Earth's processes interact through space or over areas, and to analyze the differences and similarities between places. **Process**, a set of actions that operate in some special order, is also a central concept of geographic analysis. Therefore, **physical geography** is the spatial analysis of all the physical elements, processes, and systems that make up the environment: energy, air, water, weather, climate, landforms, soils, humans, animals, plants, microorganisms, and Earth itself.

geoCHECK ✔ Explain the two main subfields in geographical science.

The Scientific Process

The **scientific method** is the simple, organized steps leading toward concrete, objective conclusions about the natural world (▶Fig. I.4). Scientific inquiry has no single method as scientists in different fields approach their problems in different ways. However, the end result must be a conclusion that other

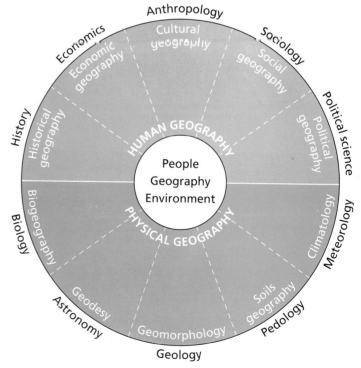

▲I.3 **The scope of geography** While physical geography focuses on processes affecting Earth systems, it shares with human geography tools, methods, and important concerns regarding the interactions among Earth's physical and human systems.

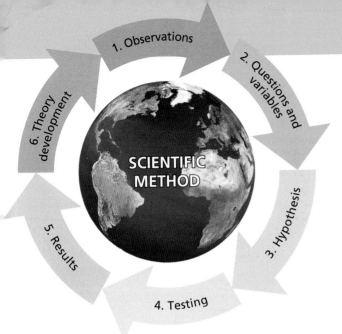

scientists can test repeatedly, either reproducing the results reached by other scientists or possibly showing that the results were false.

Using the Scientific Method Scientists who study the environment begin with clues they see in nature, followed by an exploration of the published scientific literature on their topic. Scientists then use questions and observations to form a *hypothesis*—a tentative explanation for the phenomena observed. Scientists test hypotheses using experimental studies in laboratories or natural settings (▶ **Fig. I.5**). If the results support the hypothesis, repeated testing and verification may lead to a new *theory*. A **scientific theory** is a widely accepted explanation for a phenomenon that is based on evidence and experimentation and has withstood the scrutiny of the scientific community. Reporting research results in journals and books is also part of the scientific method. Science is objective by nature and does not make value judgments. Instead, science provides people and their institutions with objective information on which to base their own value judgments. The applications of science are increasingly important as Earth's natural systems respond to the impacts of modern civilization.

▲**I.4 Scientific method continuum** Scientists continually adjust the scientific method and formulate new hypotheses based on new observations, questions and results.

geoCHECK ✔ Compare and contrast a hypothesis and a scientific theory.

a) Scientific Method Flow Chart

Real-World Observations
• Observe nature, ask questions, collect data
• Search for patterns, build conceptual or numerical models

Hypothesis and Predictions
• Formulate hypothesis (a logical explanation)
• Identify variables; determine data needed and data collection methods

Experimentation and Measurement
• Conduct experiments to test hypothesis

Result Support Hypothesis

Results Do Not Support Hypothesis
• Reject hypothesis
• Return to an earlier step of the process

Peer Review
• Communicate findings for evaluation by other scientists
• Publish scientific paper

Scientific Theory Development
• Hypothesis survives repeated testing
• Comprehensive explanation for an observation is widely accepted and supported by research

Dust darkens the surface of snowpack in the San Juan Mountains, CO, March 2009.

The dark surface on the snow is caused by a dust layer.

Other dust layers can be seen within snowpack.

Snow pit for collecting dust from snowpack, San Juan Mountains, Colorado, 13 March 2009.

(b) Using the Scientific Method Process to Study the Effects of Dust on Mountain Snowpack

1. Observations
Farmers and ranchers in southern Colorado rely on melting snow from the San Juan Mountains. Water managers have determined that the mountain snowpack now melts earlier in the spring, so water is lost before it can be used.

2. Questions and Variables
• Are air temperature increases earlier in the spring responsible for more rapid snowmelt?
• Do non-temperature factors contribute to the earlier snowmelt?

3. Hypothesis
Although the most likely explanation for earlier snowmelt is increasing temperatures, dust churned up by livestock grazing in the lowlands may also promote rapid melting, as dark dust deposited on the white snow surface absorbs heat.

4. Testing
• Review monthly temperature data on changes in air temperatures.
• Monitor and measure the deposition of dust on the surface of the mountain snowpack.

5. Results
The change in daily and seasonal air temperatures was minor. However, scientists did measure significant dust fall that darkened the snowpack, increasing the absorption of solar radiation and causing more snow to evaporate or to melt more quickly.

▲**I.5 Scientific method example application**

I.2 (cont'd) The Science of Geography

Human–Earth Interactions in the 21st Century

Throughout, *Geosystems Core* discusses issues surrounding the pervasive influence of humans on Earth systems. The global human population passed 7 billion in 2012 and is unevenly distributed among 195 countries. Virtually all population growth is in the less-developed countries that now possess 81% of the total population (▼Fig. I.6). We consider the totality of human impact on Earth to be the **human denominator**. (Each chapter in your textbook includes a Human Denominator feature that explores human impacts relevant to that chapter.) Just as the denominator in a fraction tells how many parts a whole is divided into, the growing human population and its increasing demand for resources and rising planetary impact suggest the stresses on the whole Earth system that supports us. Yet Earth's resource base—the numerator in this fraction—remains relatively fixed.

Mobile Field Trip (MG)
Introduction to
Physical Geography

https://goo.gl/B2xTBh

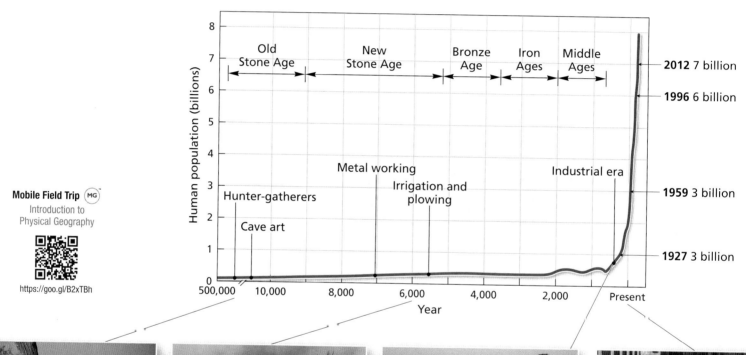

(a) Hunter-gatherers depend on wild plants and animals.

(b) Subsistence farmers use fire to clear the forest before planting crops.

(c) The plow, irrigation, and application of fertilizers enable people to produce more food on the same land year after year.

(d) Today, farmers can use new technologies to produce foods in artificial environments, as in hydroponic farming.

▲I.6 **Human population growth** Human population remained relatively low for tens of thousands of years. The shift from hunting and gathering to farming, often called the Agricultural Revolution, occurred in several different regions beginning about 10,000 years ago. A larger, more stable food supply enabled more people to live together in permanent settlements, pursue specialized occupations, and develop new technologies. Cities grew, empires emerged, and population increased at higher rates—especially after the Industrial Revolution of the late 1700s. Humans interact with and impact the environment as we obtain food. Today, people still obtain food in ways that have sustained humanity for thousands of years.

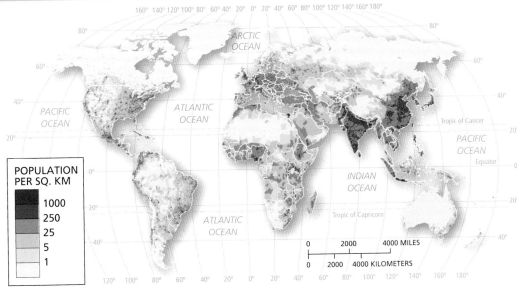

POPULATION
PER SQ. KM

	1000
	250
	25
	5
	1

(a) World population density map

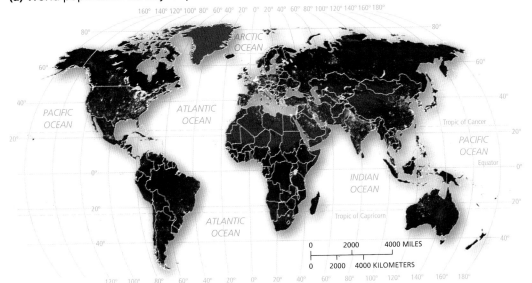

(b) Night lights around the world

▲**I.7 Population density and electric lights**

▼**I.8 Organic farming in Thailand** Organic farming is a type of sustainable agriculture that maintains soil fertility.

Approximately 38% of Earth's population lives in China and India alone (◄**Fig. I.7**). The overall planetary population is young, with 26% still under the age of 25 years. However, people in more developed countries have a greater impact on the planet per person. The United States and Canada, with about 5% of the world's population, produce about 25% of the world's gross domestic product. These two countries use more than 2 times the energy per capita of Europeans, more than 7 times that of Latin Americans, 10 times that of Asians, and 20 times that of Africans. Therefore, the impact of this 5% on the Earth systems and natural resources is enormous.

Many key issues for this century fall beneath the umbrella of geographic science, such as global food supply, energy demands, climate change, biodiversity loss, and air and water pollution. Addressing these issues in new ways is necessary to achieve **sustainability** for both human and Earth systems (◄**Fig. I.8**). The term *sustainability* refers to the ability to continue a defined activity over the long term in a way that prevents or minimizes adverse impacts on the environment. Thus physical geography is concerned with environmental sustainability measures such as the rates of natural resource harvest, the creation and release of pollutants, and the consumption of nonrenewable resources such as coal and copper (which are only sustainable if comparable and renewable substances are developed in their place). In each of these three categories, activities are not sustainable unless people can prevent or mitigate their environmental impacts. Understanding Earth's physical geography and geographic science can help to inform your thinking on these issues.

geo**CHECK** ✔ What percent of the world population is under 25 years of age?

geo**QUIZ**

1. Explain the origin of the term *geography*.
2. Describe at least two perspectives that geography uses to study Earth.
3. Identify how much more—or less—energy you might use living in Latin America, Asia, or Africa.

I.3 Earth Systems

Key Learning Concepts

▶ **Describe** systems analysis, open and closed systems.
▶ **Explain** the difference between positive and negative feedback information.
▶ **List** Earth's four spheres and classify them as biotic or abiotic.

The word *system* is used in our lives daily: "Check the car's cooling system" or "A weather system is approaching." *Systems analysis* techniques in science began with studies of energy and temperature (thermodynamics) in the 19th century. Today, systems methodology is an important analytical tool.

Systems Theory

A **system** is any set of ordered, interrelated components and their attributes, linked by flows of energy and matter, as distinct from the surrounding environment outside the system. The elements within a system may be arranged in a series or intermingled. A system comprises many interconnected subsystems. Within Earth's systems, both matter and energy are stored and retrieved, and energy is transformed from one type to another. *Matter* is mass that assumes a physical shape and occupies space. *Energy* is a capacity to change the motion and nature of matter.

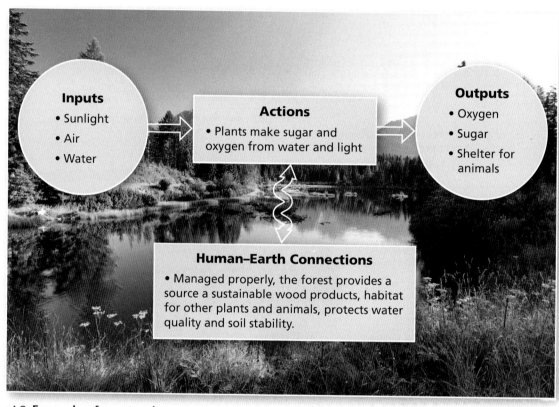

Inputs
• Sunlight
• Air
• Water

Actions
• Plants make sugar and oxygen from water and light

Outputs
• Oxygen
• Sugar
• Shelter for animals

Human–Earth Connections
• Managed properly, the forest provides a source a sustainable wood products, habitat for other plants and animals, protects water quality and soil stability.

▲I.9 **Example of a natural open system: a forest**

Earth systems may be open or closed. **Open systems** are not self-contained in that inputs of energy and matter flow into the system and outputs of energy and matter flow from the system (▲Fig. I.9). Earth is an open system in terms of energy, because solar energy enters freely and heat energy returns back into space. Within the Earth system, many subsystems interconnect. Free-flowing rivers are open systems where inputs of solar energy, precipitation, and soil particles lead to outputs of water and sediments to the ocean. A forest is another example of an open system. The input of solar energy allows trees to absorb and then store sunlight as plant materials. Forests then output oxygen that plants and animals require to survive.

In contrast, a **closed system** is self-contained and shut off from the surrounding environment. Although rare in nature, Earth itself is a closed system in terms of physical matter and resources—air, water, and natural resources. The only exceptions are the slow escape of lightweight gases from the atmosphere into space and the input of tiny meteors and cosmic dust.

System Feedback As a system operates, it often generates outputs that influence its own operations. These outputs function as "information" that returns to various points in the system via pathways called **feedback loops**. Feedback information often forms a chain of cause and effect that can further influence system operations. If the feedback information discourages change in the system, it is **negative feedback**. Negative feedback loops are common in nature. For example, when a thriving forest sinks roots deep into the soil, the amount of erosion will decrease as the vegetation absorbs increasing amounts of water, leaving less water to transport soil particles downslope.

If feedback information encourages change in the system, it is **positive feedback**. Global climate change creates an example of positive feedback as summer sea ice melts in the Arctic. As arctic temperatures rise, summer sea ice and glacial melting accelerate. This causes light-colored snow and sea-ice surfaces, which reflect sunlight and so remain cooler, to be replaced by darker-colored open ocean surfaces, which absorb sunlight and become warmer. As a result, the ocean absorbs more solar energy, which raises the temperature, which in turn melts more ice, and so forth (▶ Fig. I.10). This is a positive feedback loop, further enhancing the effects of higher temperatures and warming trends.

System Equilibrium Most systems maintain structure and character over time. A system that remains balanced over time, in which conditions are constant or recur, is in a *steady-state* **equilibrium**. For example, river channels commonly adjust their form in response to inputs of water and sediment (particles of rock or soil). These inputs may change in amount from year to year, but the channel form represents a stable average—a steady-state condition.

However, a steady-state system may demonstrate a changing trend over time, a condition described as **dynamic equilibrium**. The same river may become wider as it adjusts to greater inputs of sediment over some time scale, but the overall system will adjust to this new condition and thus maintain a dynamic equilibrium.

Systems in equilibrium tend to remain in equilibrium and resist abrupt change. However, a system may reach a **threshold**, or *tipping point*, where it can no longer maintain its character, so it lurches to a new operational level. A large flood in a river system may push the river channel to a threshold where it abruptly shifts, carving a new channel. Plant and animal communities also reach thresholds. For example, scientists identify climate change as one factor triggering a sudden decline in aspen trees in the southern Rocky Mountains.

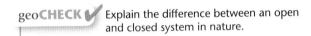 geoCHECK ✔ Explain the difference between an open and closed system in nature.

Earth Spheres & Systems Organization in *Geosystems Core*

Earth's surface is a vast area where four immense open systems interact. The three **abiotic**, or nonliving, systems overlap as the framework for the realm of the **biotic**, or living, system. The abiotic spheres are the *atmosphere* (Chapters 1–3), *hydrosphere* (Chapters 4–7), and *lithosphere* (Chapters 8–12). The biosphere is the lone biotic sphere, where all living matter on Earth is found. The living matter of Earth and everything with which it interacts is the *biosphere* (Chapters 13–14). Together, these spheres form a simplified model of Earth systems (▶ Fig. I.11).

From general layout to presentation of specific topics, *Geosystems Core* follows a systems flow. The book's structure is designed around Earth's four "spheres." Within each part, chapters and topics are arranged according to systems thinking, focusing on inputs, actions, and outputs, with an emphasis on human–Earth interactions and on interrelations among the other parts and chapters.

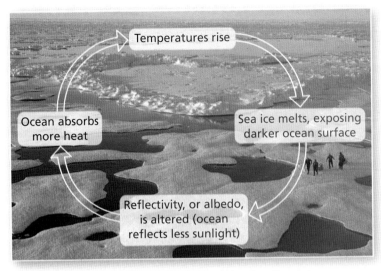
▲I.10 **The Arctic sea ice-albedo positive feedback loop**

▲I.11 **The four major Earth spheres** Of these, three are abiotic and one is biotic.

geoCHECK ✔ Describe the relationship between Earth spheres and the content organization in *Geosystems Core.*

geo**QUIZ**
1. Identify the role a "threshold" plays in an environmental system.
2. Describe an example of a "feedback" loop in nature.
3. Explain the difference between abiotic and biotic systems.

I.4 Earth Locations & Times

Key Learning Concepts

▶ *Summarize* progress in geographical knowledge about Earth's size and shape.
▶ *Explain* Earth's reference grid, including latitude and longitude and latitudinal geographic zones.
▶ *Interpret* a map of Earth's time zones.

As geographers study the physical features and processes on Earth's surface, they need to accurately locate these phenomena in space and time. You have probably noticed the network of lines that crisscrosses most globes and world maps. This "geographic grid" allows us to locate places and regions on Earth. The size and rotational velocity of Earth combine to make a 24-hour day, and Earth's annual revolution around the Sun determines the length of a year.

Earth's Dimensions & Shape

Humans have known that Earth is round since the first ship sailed over the horizon and viewers on shore saw the top sails disappear last. Our scientific understanding of Earth's size and shape began slowly, but has grown rapidly over the past 300 years. Over 2000 years ago, the Greek mathematician Pythagoras (ca. 580–500 BCE) determined that Earth is round, or *spherical*. Eratosthenes (ca. 276 BC –194 BCE) calculated the circumference of Earth in 247 BCE by comparing the angle of the Sun at noon at two different locations (▶ Fig. I.12). By the first century CE, educated people generally accepted the idea of a spherical Earth. In 1687, Sir Isaac Newton reasoned that Earth's rapid rotation produced an equatorial bulge as centrifugal force pulled Earth's surface outward. As a result, Earth's equatorial circumference is 67 km (42 mi) greater than its polar circumference. Earth is indeed slightly misshapen into an *oblate spheroid* (oblate means "flattened"), with the flatness occurring at the poles.

Today, satellite observations have confirmed with tremendous precision Earth's equatorial bulge and polar "flatness." The irregular shape of Earth's surface, coinciding with mean sea level and perpendicular to the direction of gravity, is called the **geoid**. **Figure** I.13 shows Earth's polar and equatorial circumferences and diameters.

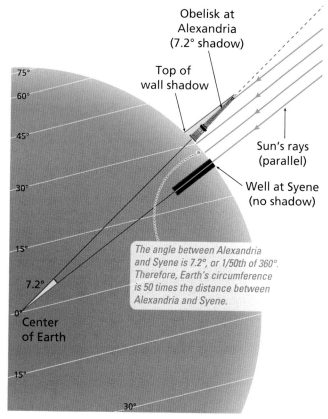

▲I.12 **Eratosthenes method for calculating Earth's circumference** Although Eratosthenes calculated the circumference of Earth over 2000 years ago, his answer, based on scientific and mathematical reasoning, was surprisingly accurate.

The angle between Alexandria and Syene is 7.2°, or 1/50th of 360°. Therefore, Earth's circumference is 50 times the distance between Alexandria and Syene.

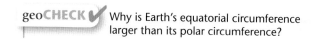

geoCHECK Why is Earth's equatorial circumference larger than its polar circumference?

Earth's Reference Grid

Fundamental to geography is an internationally accepted grid coordinate system to determine location. Geographers use pairs of numbers, or "coordinates," to locate specific points on the grid. Eratosthenes created the first world map with a rectangular grid to locate places around 200 BCE. The use of a geographic grid made it possible to accurately measure distances between locations. The terms **latitude** and **longitude** were used on maps in the first century CE to refer to distances measured in relation to standard lines on the grid. These distances are measured in degrees—units based on the division of a perfect circle into 360 equal parts (▶ Fig. I.14).

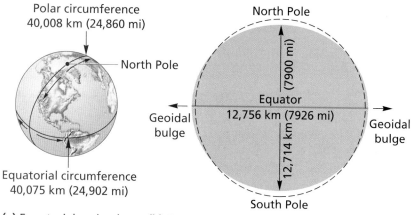

(a) Equatorial and polar circumferences

(b) Equatorial and polar diameters

Polar circumference 40,008 km (24,860 mi)

Equatorial circumference 40,075 km (24,902 mi)

North Pole

Equator 12,756 km (7926 mi)

(7900 mi)

12,714 km

Geoidal bulge

Geoidal bulge

South Pole

▲I.13 **Earth's dimensions** The dashed line is a perfect circle for comparison to Earth's geoid.

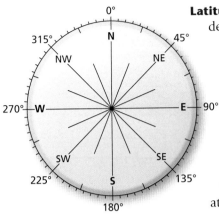

▲I.14 **360° in a circle, with the cardinal directions**

Latitude The angular distance in degrees north or south of the equator, measured from the center of Earth is latitude (▶**Fig. I.15a**). (The equator is the line that divides the spherical Earth into northern and southern hemispheres). Lines of latitude run east-west, parallel to the equator (▶**Fig. I.15b**). Latitude increases from the equator at 0° latitude, to the poles, at 90° north and south.

A line of latitude is called a **parallel**. In **Figure** I.15b, an angle of 49° is shown, and by connecting all points at 49° N, we can draw the 49th parallel. When writing the latitude of location, it is not necessary to include the word latitude, since the suffix of N or S indicates that you are giving the latitude, giving 40° N is sufficient. *Latitude* is the name of the angle (49° N), *parallel* names the line (49th parallel), and both indicate distance north of the equator.

Throughout this book, you will read references to latitudinal zones as a way of generalizing the location of different phenomena, from weather patterns to plant and animal communities. Lower latitudes are toward the equator, higher latitudes are toward the poles. The terms "the tropics" and "the Arctic" refer to environments created by different amounts of solar energy received at different latitudes. **Figure I.16** displays the names and locations of the *latitudinal geographic zones* used by geographers: *equatorial* and *tropical*, *subtropical*, *midlatitude*, *subarctic* or *subantarctic*, and *arctic* or *antarctic*. These latitudinal zones are useful for reference, but they do not have rigid boundaries. We discuss specific lines of latitude, such as the Tropic of Cancer and the Arctic Circle, in Chapter 1 as we learn about the seasons.

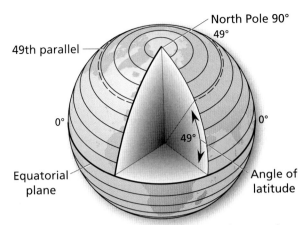

(a) Latitude is measured in degrees north or south of the Equator (0°). Earth's poles are at 90°. Note the measurement of 49° latitude.

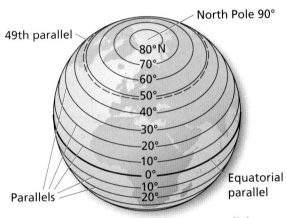

(b) These angles of latitude determine parallels along Earth's surface.

▲I.15 **Parallels of latitude** Do you know your present latitude?

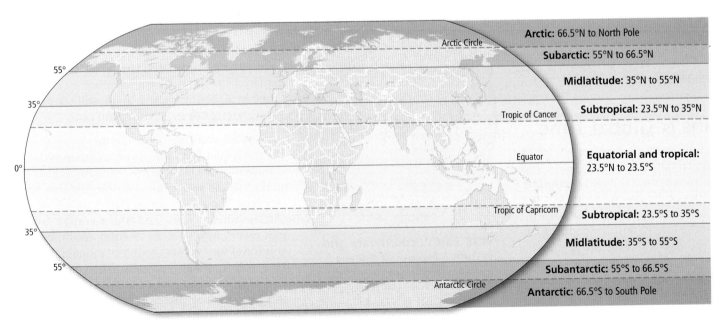

▲I.16 **Latitudinal geographic zones** Geographic zones are generalizations that characterize various regions by latitude.

I.4 (cont'd) Earth Locations & Times

Longitude The angular distance east or west of a point on Earth's surface, measured from the center of Earth is longitude (▶Fig. I.17a). On a map or globe, the lines designating these angles of longitude run north and south (Fig. I.17a). A line connecting all points along the same longitude is a **meridian**. In the figure, a longitudinal angle of 60° is shown. These meridians run at right angles (90°) to all parallels. *Longitude* is the name of the angle, *meridian* names the line, and both indicate distance in degrees east or west of the **prime meridian**, designated as 0° (▶Fig. I.17b). Earth's prime meridian—also called the *Greenwich meridian*—passes through the old Royal Observatory at Greenwich, England, as set by an 1884 treaty.

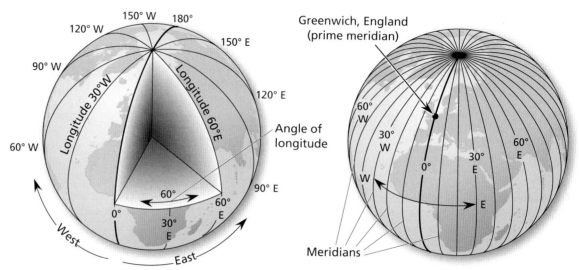

(a) Longitude is measured in degrees east or west of a 0° starting line, the prime meridian. Note the measurement of 60° E longitude.

(b) Angles of longitude measured from the prime meridian determine other meridians. North America is west of Greenwich; therefore, it is in the Western Hemisphere.

▲I.17 **Meridians of longitude** Do you know your present longitude?

Because meridians of longitude converge at the poles, the length on the ground of 1° of longitude is greatest at the equator and shrinks to zero at the poles. Longitude increases east and west from 0° at the prime meridian to 180°. Just as with latitude, it is not necessary to include the word *longitude* when writing a location's longitude. The suffix E or W indicates longitude.

Figure I.18 combines latitude and parallels with longitude and meridians to illustrate Earth's complete coordinate grid system. Note the red dot that marks 49° N and 60° E, a location in western Kazakhstan. Next time you look at a world globe, follow the parallel and meridian that converge on your location.

geoCHECK Which latitudinal zone do you live in? Why aren't lines of longitude parallel?

Meridians & Global Time

A worldwide time system is necessary to coordinate international trade, airline schedules, and daily life. Our time system is based on the fact that Earth rotates on its axis, rotating 360° every 24 hours, or 15° per hour (360° ÷ 24 = 15°).

In 1884 at the International Meridian Conference in Washington, DC, the prime meridian was set as the official standard for the world time zone system—Greenwich Mean Time (GMT). This standard time system established

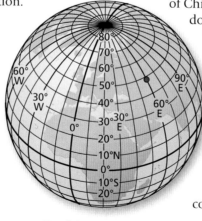

▲I.18 **Earth's coordinate grid system** Parallels of latitude and meridians of longitude allow us to locate all places on Earth precisely. The red dot is at 49° N and 60° E.

24 equally spaced standard meridians around the globe, with a time zone of 1 hour spanning 7.5° on either side of these central meridians (▶Fig. I.19). Before this universal system, time zones were not consistently defined, especially in large countries. In 1870, if you were traveling from Maine to San Francisco by railroad, you would have made 22 adjustments to keep on local time!

As you can see in **Figure I.19**, national or state boundaries and political considerations can distort time boundaries. For example, China spans four time zones, but its government decided to keep the entire country operating at the same time. Thus, in some parts of China clocks are several hours off from what the Sun is doing. In the United States, parts of Florida and west Texas are in the same time zone.

In 1972, **Coordinated Universal Time (UTC)** replaced GMT as the legal reference for official time in all countries. You might still see official UTC referred to as GMT or Zulu time.

International Date Line On the opposite side of the planet from the prime meridian is the **International Date Line** (▶Fig. I.20), which marks the line where one day officially changes to another. The International Date Line does not completely coincide with the 180th meridian, but jogs east or west to avoid dividing countries. If you travel west across the International Date Line, you would immediately gain a day, and if you travel east you immediately lose a day. From this line, the new day moves westward as Earth

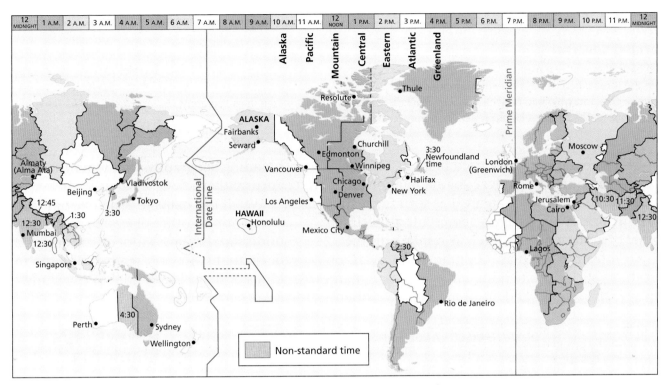

▲I.19 **Modern international standard time zones** If it is 7 p.m. in Greenwich, determine the present time in Moscow, London, Halifax, Chicago, Winnipeg, Denver, Los Angeles, Fairbanks, Honolulu, Tokyo, and Singapore.

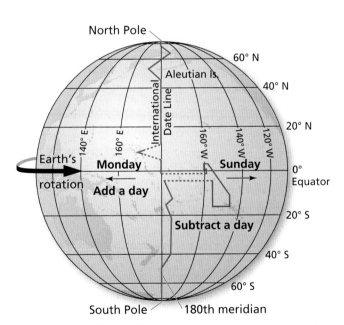

▲I.20 **International Date Line** The International Date Line (IDL) location is approximately along the 180th meridian (see the IDL location on Figure I.19). The dotted lines on the map show where island countries have set their own time zones, but their political control extends only 3.5 nautical miles (4 mi) offshore. Officially, you gain 1 day crossing the IDL from east to west.

turns eastward on its axis. At the International Date Line, the west side of the line is always 1 day ahead of the east side of the line. No matter what the time of day when the line is crossed, the calendar changes a day.

Daylight Saving Time In 70 countries, mainly in the midlatitudes, time is set ahead 1 hour in the spring and set behind 1 hour in the fall—a practice known as daylight saving time. The idea to extend daylight for early evening activities at the expense of daylight in the morning, first proposed by Benjamin Franklin, was not adopted until World War I and again in World War II to save energy by reducing the use of electric light. In 1986 and again in 2007, the United States and Canada extended the number of weeks of daylight saving time. Currently, time "springs forward" 1 hour on the second Sunday in March and "falls back" 1 hour on the first Sunday in November, except in a few places that do not use daylight saving time (Hawaii, Arizona, and Saskatchewan).

geoCHECK ✔ How many degrees apart are time zones?

geoQUIZ

1. Compare the geoid with a hypothetical Earth-like planet of the same size that is a perfect sphere. How are they similar? How are they different?
2. Why is it important to have a standard prime meridian?
3. Determine your longitude using an online map or an atlas. How many degrees are you away from a time zone central meridian (75°, 90°, 105°, 120°, 135°)? Given that Earth rotates through 1° in 4 minutes, how many minutes apart are the Sun and your watch?

1.5 Maps & Cartography

Key Learning Concepts

▶ **List** the basic elements of a map.
▶ **Explain** the three different ways of expressing map scale.
▶ **Summarize** how and why map projections were developed and how they are used in cartography.
▶ **Give examples** of the different kinds of maps and how each is used.

For centuries, geographers have used maps as tools to display information and analyze spatial relationships. A **map** is a generalized view of an area, as seen from above and reduced in size. A map usually represents a specific characteristic of a place or area, such as rainfall, airline routes, physical features such as mountains and rivers, or political features such as state boundaries and place names. **Cartography** is the science and art of mapmaking, often blending geography, mathematics, computer science, and art.

We all use maps to visualize our location in relation to other places, to plan trips, or to understand a news story or current event. Understanding how to "read" or interpret different kinds of maps is essential to our study of physical geography.

Basic Map Elements

Most maps share the same elements:

- **title**—gives the subject of the map and may also include information about who made the map, the source of map data, and the date when the map was produced
- **north arrow**—tells the reader which direction is north on the map
- **symbols**—represent features on the map using lines, patterns, areas of color, icons, and other graphic elements
- **legend**—tells the map reader what each symbol means
- **map scale**—states the mathematical relationship between the size of the map and the size of the portion of Earth the map represents (discussed below)
- **map projection**—enables showing the round Earth as a flat map (discussed below)

geoCHECK ✔ What are the basic map elements?

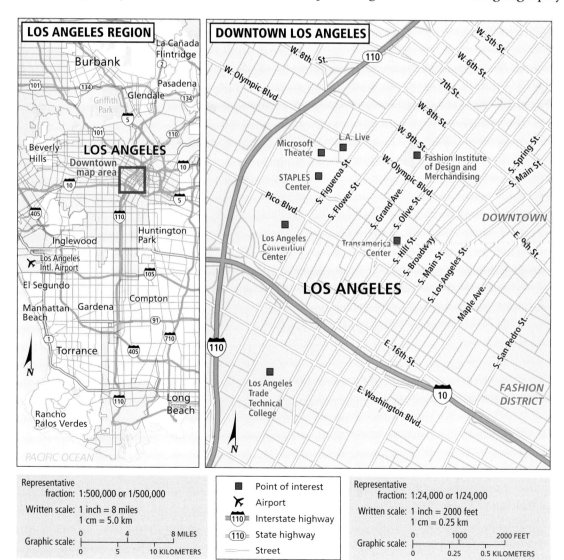

Representative
fraction: 1:500,000 or 1/500,000

Written scale: 1 inch = 8 miles
1 cm = 5.0 km

Graphic scale:
0   4   8 MILES
0   5   10 KILOMETERS

■ Point of interest
✈ Airport
(110) Interstate highway
(110) State highway
——— Street

Representative
fraction: 1:24,000 or 1/24,000

Written scale: 1 inch = 2000 feet
1 cm = 0.25 km

Graphic scale:
0   1000   2000 FEET
0   0.25   0.5 KILOMETERS

(a) Relatively small scale map of Los Angeles area shows less detail.

(b) Relatively large scale map of the same area shows a higher level of detail.

◀1.21 **Map scale** Examples of maps at different scales, with three common expressions of map scale—representative fraction, written scale, and graphic scale. Both maps are enlarged, so only the graphic scale is accurate.

The Scale of Maps

Architects, toy designers, and mapmakers all represent real things and places with models that are smaller than the thing they represent. Examples include the floorplan of a building; a diagram of a toy car, train, or plane; or a map. Each of these models has a particular *scale*, or relationship between the size of the model and the size of the actual thing it depicts. For example, an architect draws a blueprint for builders so that 0.25 inch on the drawing represents 1 foot on the building.

Cartographers do the same thing in making maps. The ratio of the size of a map to that area in the real world is the map's **scale**. Scale can be represented as a ratio (also called representative fraction), a graphic scale, or a written scale (◄**Fig. I.21**). For example, a useful scale for a local map is 1:24,000, a ratio in which 1 unit on the map represents 24,000 units on the ground. Geographers refer to as *small-*, *medium-*, or *large-scale* maps, depending upon the map's scale. A map with a scale of 1:24,000 is a large-scale map, while a scale of 1:50,000,000 is a small-scale map. The larger the number on the right, the smaller the scale. Small-scale maps have less detail for a larger area, while large-scale maps show more detail for a smaller area (**Fig. I.21**). Scale is represented as a representative fraction, a graphic scale, or a written scale (**Fig. I.21**).

Ratio Scale & Representative Fraction A ratio scale, or *representative fraction*, can be expressed with either a colon (for a ratio) or a slash (for a fraction), as in 1:24,000 or 1/24,000. No actual units of measurement are mentioned because both parts of the fraction are in the same unit: 1 cm to 24,000 cm or 1 in. to 24,000 in.

Graphic Scale A *graphic scale*, or *bar scale*, is a graphic with units to allow measurement of distances on the map. An advantage of a graphic scale is that if the map is enlarged or reduced, the scale is enlarged or reduced by the same amount, unlike written and fractional scales that become incorrect when map size changes.

Written Scale A *written scale* usually has differing, but common, units such as 1 inch equals 1 mile. For example, the ratio scale 1:24,000 conveniently converts to "1 inch equals 2000 feet" when expressed as a written scale (by dividing 24,000 by 12 in./ft).

geoCHECK ✔ Which map has more detail, a large-scale or small-scale map?

Map Projections

A globe is a small-scale, three-dimensional representation of Earth. Globes can provide an accurate representation of *area* and *shape* on Earth. However, if you wanted to go hiking or explore a new city, you need more information than a globe can provide. To provide more detail, cartographers make large-scale maps, which are two-dimensional representations of Earth. However, converting a three-dimensional sphere to a two-dimensional map causes some degree of distortion of areas and shapes. To control distortion on a flat map, cartographers use a **map projection**. By manipulating the grid coordinate system that is common to both globes and flat maps, a map projection enables cartographers to transfer data about points and lines on a globe accurately to a flat surface. Centuries ago, cartographers actually projected the shadow of a wire frame globe onto a geometric surface, such as a cylinder, plane, or cone. The wires represented parallels, meridians, and the outlines of continents. Modern cartography uses mathematical formulas to generate the many different kinds of map projections. Some are better at showing shape accurately, while others are better for showing area accurately. Cartographers must decide which characteristic to preserve, which to distort, and how much distortion is acceptable.

If you imagine taking a globe apart and trying to lay it flat on a table, that illustrates some of the problems with map projections (►**Fig. I.22**). Although large-scale maps have less distortion than small-scale maps, all maps, regardless of the projection used, have some degree of distortion.

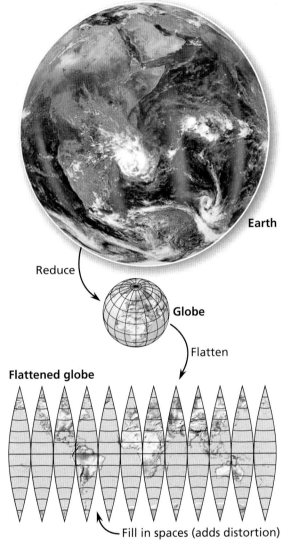

Earth

Reduce

Globe

Flatten

Flattened globe

Fill in spaces (adds distortion)

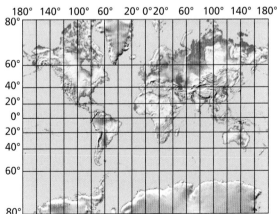

Map projection (Mercator projection–cylindrical)

▲**I.22 From globe to flat map** Conversion of the globe to a flat map projection requires a decision about which properties to preserve and the amount of distortion that is acceptable.

I.5 (cont'd) Maps & Cartography

Equal Area or True Shape? One major decision a cartographer must make when beginning a map involves choosing between projections with the properties of *equal area* and *true shape*. Cartographers designed different kinds of **equal-area** projections so that areas are correct on the map regardless of their latitude and longitude (▼**Fig. I.23a**). For example, areas measuring 10° of latitude by 10° of longitude are equal whether they are near the equator or near the poles—although the two areas differ greatly in shape. In contrast, a **true-shape** projection (also called a *conformal* projection) can correctly represent the shapes of geographic features such as coastlines and islands, but the sizes of those features can be greatly distorted (**Fig. I.22b**). The commonly used **Mercator projection** seen in **Figure I.22a** is a true-shape projection. Gerardus Mercator developed the projection in 1569 to simplify navigation. Unfortunately, as we saw in **Figure I.23b**, Mercator maps present a false view of the size of midlatitude and high-latitude regions.

If a cartographer selects an equal-area projection for a map—for example, to show the distribution of world climates—then the map will sacrifice true shape, especially where areas are stretched along the edges of the map. If a cartographer selects a true-shape projection, such as for a map used for navigation, then the map will sacrifice the property of equal area, and different regions of the map will actually have different scales.

Geosystems Core uses equal-area and compromise map projections. *Goode's homolosine projection* is an interrupted equal-area

projection and is excellent for mapping features when breaks in the map over oceans or continents is not a problem. Goode's homolosine projection is used in Geosystems Core for the world climate map in Chapter 6 (Fig. 6.), the world soil orders map (Fig. 14.8), and the terrestrial biomes map in Chapters 14 (Fig. 14.24).

The text also uses the *Robinson projection*, designed by Arthur Robinson in 1963. This is a compromise projection that is neither equal area nor true shape, but a compromise between the two. Examples of the Robinson projection in *Geosystems Core* include the latitudinal geographic zones map (**Fig. I.16**), the distribution of insolation map and the temperature ranges map in Chapter 2 (Figs. 2.5 and 2.31), the maps of lithospheric plates and volcanoes and earthquakes in Chapter 8 (Figs. 8.15 and 8.21).

The *Miller cylindrical projection* is another compromise projection used in this text. This projection was first developed by Osborn Miller and presented by The American Geographical Society in 1942. This projection is neither true shape nor true area, but is a compromise that avoids the severe scale distortion of the Mercator. Examples of the Miller cylindrical projection in *Geosystems Core* include the world time zone map in **Figure I.19**, global temperature maps in Chapter 2 (Figs. 2.29 and 2.30), and the two global pressure maps in Chapter 3 (Figs. 3.9 and 3.10).

geoCHECK ✔ Which projection described above would be best for comparing the amounts of rain forest in Latin America, Africa, and Southeast Asia? Explain.

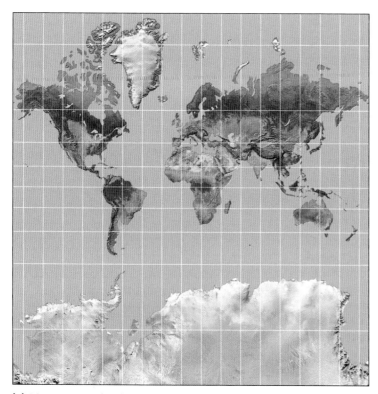

(a) Mercator projection

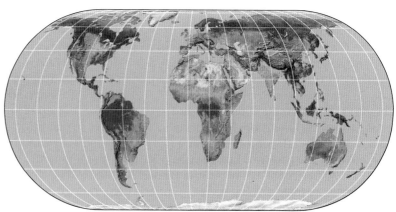

(b) Equal-area projection (Eckert IV)

Animation (MG)
Map Projections

http://goo.gl/3wii0g

▲I.23 **True-shape projections vs. equal-area**

Types of Maps

There are many kinds of maps for a vast number of purposes. Maps portray everything from Earth's physical features to political boundaries to the demographic and economic data that are important to human geographers. Physical geographers often create physical maps that show information about a physical theme such as elevation or temperature. Physical maps often use *isolines*, which are lines that represent a given value: Contour lines show elevation, isotherms show temperature, isobars show air pressure. **Topographic maps** are physical maps that can give us a sense of the terrain, or the lay of the land (▶Fig. I.24). They use different colors to represent different features, blue for water, black for human-made objects, green for vegetation, brown for contour lines. A contour line connects all points at the same elevation. Contour lines show the slope of the land as well as elevation: widely spaced contour lines indicate gentle slopes, and closely spaced contour lines indicate steep slopes. You can also use contour lines to calculate **relief**, which is the difference in elevation between two locations. Figure i.24 uses shaded relief, an artistic technique of simulated shadows that conveys a sense of what the landscape looks like. Figure i.25 shows slopes derived from digital elevation models. Other important types of physical maps are geologic maps, which show rock formations and faults (▼Fig. I.26); weather maps, which show present or future forecasts of weather; and climate maps, which show long term averages of different weather elements such as temperature or rainfall.

geoCHECK ✔ What are the two main types of maps?

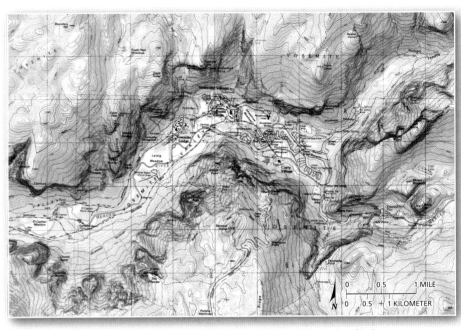

▲I.24 **Topographic map of Yosemite Valley with shaded relief**

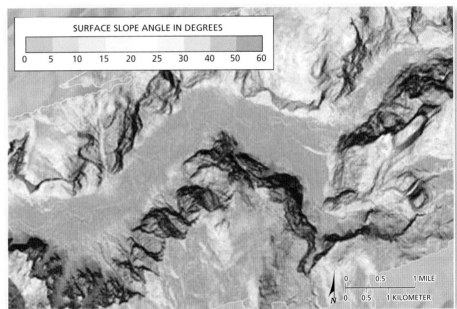

SURFACE SLOPE ANGLE IN DEGREES

0 5 10 15 20 25 30 40 50 60

▲I.25 **Surface slope map for Yosemite Valley** Very steep valley walls (red) are easily distinguished from the nearly flat valley floor (blue).

▼I.26 **Geologic map of Yosemite Valley and surrounding areas**

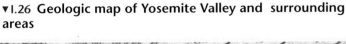

SURFICIAL DEPOSITS
- Qal — Alluvium
- Qtl — Talus
- Qti — Tioga Till
- Qpt — Pre-Tahoe Till

PLUTONIC ROCKS
- Kec — El Capitan Granite
- Khd — Half Dome Granodiorite
- Kic — Granodiorite of Illilouette Creek
- Kid — Quartz diorite
- KJdg — Diorite and gabbro
- Ks — Sentinel Granodiorite

METAMORPHIC ROCKS
- Jms — Metasedimentary rock

geoQUIZ

1. For viewing maps on a smartphone, which type of map scale would be most helpful? Explain.
2. What are the advantages of a globe over a map? Of a map over a globe?
3. Describe the two main types of distortion in map projections.
4. As a cartographer, you are asked to produce a highly accurate topographic map of the county where you live. Would you choose a large-scale or small-scale for the map? An equal area or true shape projection? Explain your answer.

I.6 Modern Geoscience Tools

Key Learning Concepts

▶ **Explain** how geographers use the Global Positioning System, remote sensing, geographic information systems, and geovisualizations.

Geographers use a number of new and evolving technologies to analyze and map Earth—the Global Positioning System (GPS), remote sensing, and geographic information systems (GIS). GPS uses satellites to provide precise locations. Remote sensing uses satellites, aircraft, and other sensors to provide visual data that enhances our understanding of Earth. GIS is a means for storing and analyzing large amounts of spatial data as separate layers of geographic information.

Global Positioning System

Using radio signals from a global network of satellites, the **Global Positioning System (GPS)** accurately determines location anywhere on or near the surface of Earth. A GPS receiver receives radio signals from the satellites and calculates the distance between the receiver and each satellite. By using signals from at least four satellites, precise locations are possible (▼Fig. I.27). GPS units also report the time, accurate to 100 billionths of a second, which is used to synchronize communications systems, electrical power grids, and financial networks.

▼I.27 Using satellites to determine location through GPS

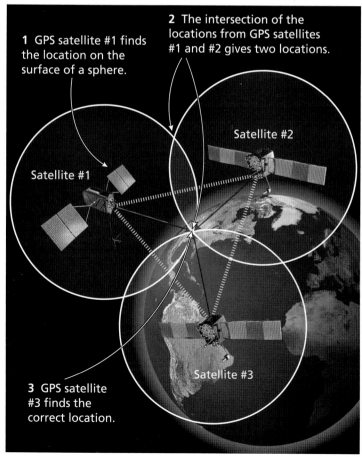

1 GPS satellite #1 finds the location on the surface of a sphere.

2 The intersection of the locations from GPS satellites #1 and #2 gives two locations.

Satellite #2

Satellite #1

3 GPS satellite #3 finds the correct location.

Satellite #3

GPS receivers are built into many smartphones and motor vehicles. The GPS is useful for many commercial and scientific applications. GPS receivers have been attached to sharks and whales to track them in real time to study their migration patterns. Airlines and shipping companies use GPS to track their vehicles, improving fuel efficiency and on-time performance.

 Why are at least three satellites needed to find a location using GPS?

Remote Sensing

Technological systems of **remote sensing** obtain information about objects without physically touching them. We do remote sensing with our eyes as we scan the environment, sensing the shape, size, and color of objects from a distance. Taking a picture with your phone is another example of remote sensing. Geographers use images captured by satellites and airborne sensors. During the last 50 years, satellite imagery has transformed Earth observation. Today, you have free access to high-quality remote-sensing imagery, through services such as Google Maps, that in the past would have been unavailable, extremely expensive, or restricted to government intelligence services. Remote sensing can be divided into passive and active remote-sensing systems.

Passive Remote Sensing Systems of passive remote-sensing record energy radiated from a surface, especially visible light and heat (▼Fig. I.28). Our eyes are passive remote sensors. Weather satellites are passive remote sensing systems with which you are probably familiar. Beginning in the 1970s, the Landsat series of satellites began recording images of Earth with sensors that captured visible light, as well as other wavelengths useful in studying agriculture, forestry, geology, regional planning, mapping, and global change research. Scientists can observe different phenomena with sensors that detect different wavelengths of energy. This allows them to compare healthy vegetation and distressed vegetation or a find outcroppings of a particular rock formation.

▼I.28 Passive remote sensing Image from October 15, 2015, showing muddy stream runoff from heavy rains in South Carolina interacting with ocean currents.

Storm runoff entering ocean

Today, sites such as Google Maps and Bing Maps show us detailed imagery, often in simulated three-dimensions, of any location in the world. Urthecast (www.urthecast.com) is now broadcasting near real-time views of Earth from cameras on the International Space Station.

Active Remote Sensing A system that directs energy at a surface and analyzes the energy returned from the surface is referred to as active remote sensing. Taking pictures with a flash in a darkened room is an example of active remote sensing. Another example is sonar, which has been used to map the ocean floor. A sonar unit emits bursts of sound and measures their return. Another technology is **LIDAR** (*li*ght and ra*dar*), which uses pulses of visible light. LIDAR units can be mounted in aircraft and on cars. LIDAR can differentiate between the first pulses returned, usually off the highest vegetation, and later returns, which are usually from the actual ground surface. This capability allows scientists to measure tree canopy heights or to virtually strip away vegetation to create a three-dimensional model of the surface (▶ **Fig. I.29**). Archaeologists have used LIDAR to discover several "lost" ancient cities in Central America. Detailed three-dimensional, LIDAR models of modern cities already exist, and LIDAR models of roads will be critical in the development of self-driving cars (▼ **Fig. I.30**).

geoCHECK ✔ Compare and contrast the two types of remote sensing.

▼**I.29 Active remote sensing** LIDAR is used to produce canopy or bare ground maps.

LIDAR survey

Conventional aerial photography cannot penetrate dense jungle canopies

(a) LIDAR uses pulses of light to form a 3D image of elevated and ground level objects.

▼**I.30 Comparison of first-return and bare ground images of the Oso landslide, WA**

Head of Oso slide

Forest

Individual trees

Vegetation on surface of slide

(a) First return shows top of vegetation

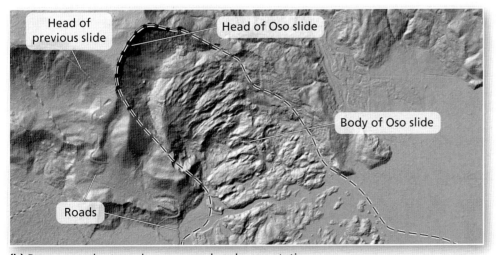

Head of previous slide

Head of Oso slide

Body of Oso slide

Roads

(b) Bare ground return shows ground under vegetation

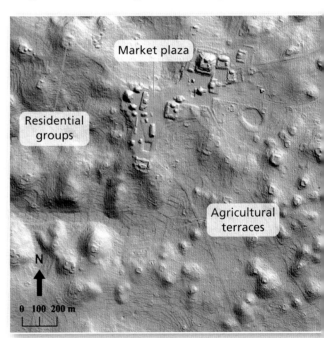

Market plaza

Residential groups

Agricultural terraces

N

θ 100 200 m

(b) LIDAR mapping of the lost city of Caracol hidden below the rain forest canopy in Central America.

I.6 (cont'd) Modern Geoscience Tools

Geographic Information Systems & Geovisualization

Techniques such as remote sensing generate large volumes of spatial data to be stored, processed, and analyzed in useful ways. A powerful tool for manipulating and analyzing this spatial data is a **geographic information system (GIS)**. A GIS is a computer-based data-processing tool that combines spatial data (where is it? what is its latitude/longitude? is it a point? a line? a polygon?) with attribute data (what is it?). In a GIS, spatial data can be organized in layers containing different kinds of data (▶ **Fig. I.31**). When you ask your phone to find the nearest coffee shop, you are using a GIS, probably without realizing it. A GIS program and a database work together to ask spatial analysis questions such as Where are you? Where are the coffee shops? Which shops are closest to you? How do you get to the nearest coffee shop? GIS systems perform these queries across multiple data layers. In the coffee shop example, three layers are required: one with your location, one with the locations of the coffee shops, and one with the layout of the streets. **Figures** I.32 and I.33 show examples of GIS analysis used to predict natural hazards and map epidemics.

▶ **I.31 Geographic information system (GIS)** Wildfires can change the response of hillsides to rainfall so that even modest rainstorms can result in dangerous flash floods and debris flows. The USGS uses a hazard assessment model that incorporates the shape of hillsides, the amount of land that is heavily burned, the steepness of hill slopes, the clay content of the soil, and the projected amount of rainfall on specific slopes to assess the probability and volume of debris flows in burned areas.

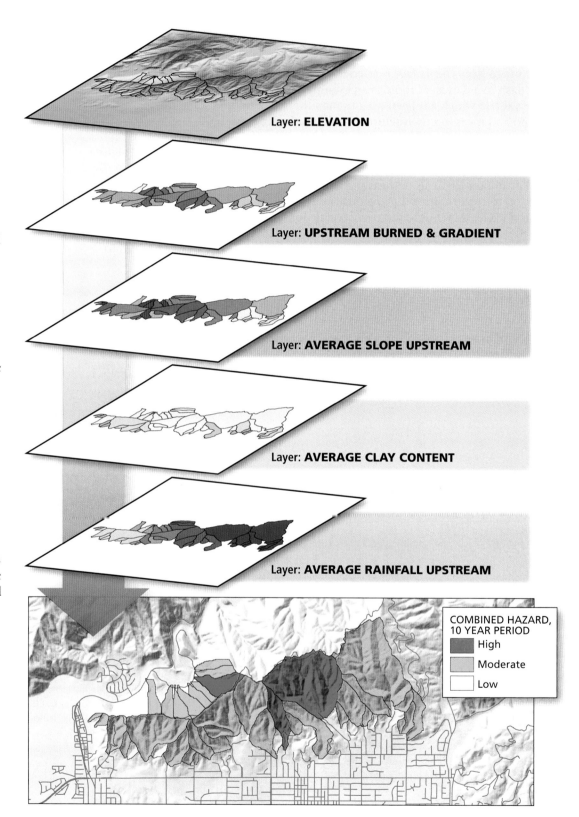

Layer: **ELEVATION**

Layer: **UPSTREAM BURNED & GRADIENT**

Layer: **AVERAGE SLOPE UPSTREAM**

Layer: **AVERAGE CLAY CONTENT**

Layer: **AVERAGE RAINFALL UPSTREAM**

COMBINED HAZARD, 10 YEAR PERIOD
- High
- Moderate
- Low

Geovisualization *Geovisualization* refers to the display of geographic information, often remote-sensing data combined with other data. Google Maps and Google Earth are two examples of geovisualization programs with which you might be familiar. Geovisualization programs often have limited GIS abilities, such as the ability to search for locations and add data layers. Many geovisualization programs allow users to upload their own data sets to combine with other user-generated data and the built-in data from the program.

geoCHECK ✔ Describe the two types of information that a GIS combines.

▲I.32 **Lahar hazard zones and arrival times for Mt. Hood**

geoQUIZ

1. Explain at least two ways you have benefited from the GPS.
2. What types of remote-sensing data have you seen today? in the past week?
3. Describe the criteria for a GIS used to find a parcel of land to build a new subdivision using the following data layers: property parcels, zoning layer, floodplain layer, protected wetlands layer.

▼I.33 **Google Earth used to track the retreat of the Jacobshavn glacier, Greenland**

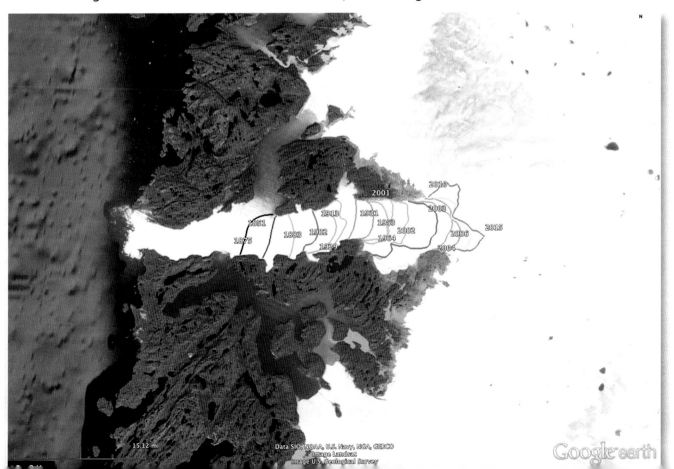

MAPS IMPACT HUMAN UNDERSTANDING

• Maps are much older than photographs.

• While maps appear in media and books everywhere, few appreciate the dynamic and vital applications they now offer.

HUMANS USE MAPS TO CHANGE THE WORLD

• Today as in the past, maps delineate empires, guide explorers, and inspire travelers to go beyond the next horizon.

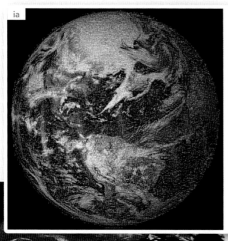

ia

On Earth Day 2014, NASA broadcast a question on social media: "Where are you on Earth Right Now?" People from 113 different countries, representing every continent, submitted over 50,000 georeferenced images. This participatory mapping created our first global selfie.

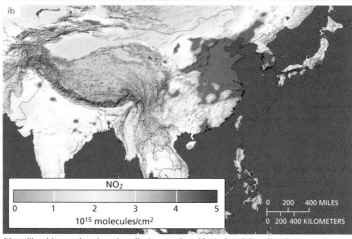

ib

NO₂

10^{15} molecules/cm²

0 | 1 | 2 | 3 | 4 | 5

0 200 400 MILES
0 200 400 KILOMETERS

Maps like this one showing air pollution produced by industrial regions in East Asia help scientists monitor changes in air quality worldwide. This is part of a world map NASA compiled based on satellite-based data on nitrogen dioxide gas, a pollutant that can form ground-level ozone, a component of smog.

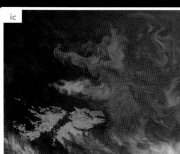

ic

Scientists around the world use remotely sensed images to measure and analyze changing vegetation cover, water resources, wildlife migration, advancing urban development, and scores of other purposes. In this example, remotely sensed images from on NASA's Terra satellite portray the waters near the Falkland Islands off the coast of southern Argentina awash in greens and blues from concentrated phytoplankton. These microscopic, plant-like organisms grow on the ocean surface, and are the foundation of a thriving ocean food chain.

ISSUES FOR THE 21ST CENTURY

• Mapping of natural and human phenomena such as earthquakes, flooding, food insecurity, and terrorist movements, will play an important role in how governments respond to the challenges each event presents.

• Rapidly evolving technological advances in geovizualization, GPS, GIS, and cartography will make geospatial science an essential tool for monitoring and analyzing human-environmental change in the 21st century.

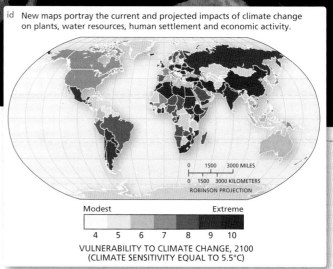

id New maps portray the current and projected impacts of climate change on plants, water resources, human settlement and economic activity.

0 1500 3000 MILES
0 1500 3000 KILOMETERS
ROBINSON PROJECTION

Modest   Extreme

4 | 5 | 6 | 7 | 8 | 9 | 10

VULNERABILITY TO CLIMATE CHANGE, 2100
(CLIMATE SENSITIVITY EQUAL TO 5.5°C)

Looking Ahead

We now embark on a journey through Earth's four spheres from the atmosphere in *Part I, Energy and Earth Systems,* to the atmosphere and hydrosphere in part *Part II, Water, Weather & Climate Systems. Part III,* The *Geosphere: Earth's Interior and Surface,* explores the processes that shape Earth's varied topography. *Part IV, The Biosphere,* analyzes the structure and function of the ecosystems and soils that sustain Earth's ecosystems, soils, and biomes.

Chapter 1 begins with the Sun, including seasonal changes in the distribution of its energy flow to Earth.

Each *Core* chapter ends with a Looking Ahead to act as a bridge from one chapter to the next.

What is physical geography?

I.1 The World Around Us

Give examples of the kinds of events, processes, and questions that physical geography investigates.

- Geography combines disciplines from the physical and life sciences with disciplines from the human and cultural sciences to attain a holistic view of Earth. Physical geography explains the spatial dimension of Earth's dynamic systems—its energy, air, water, weather, climate, tectonics, landforms, rocks, soils, plants, ecosystems, and biomes. It also asks *where* and *why* questions about processes and events that occur at specific locations and then follow their effects across the globe. The analysis of process—a set of actions or mechanisms that operate in some special order—is also central to geographic understanding. The science of physical geography is uniquely qualified to synthesize the spatial, environmental, and human aspects of our increasingly complex relationship with our home planet—Earth.

1. On the basis of information in this chapter, define physical geography and review the approach that characterizes the geographic sciences.

I.2 The Science of Geography

Describe the main perspectives of geography and distinguish physical geography from human geography.

Discuss the use of scientific methods in geography.

Summarize how human activities and population growth impact the environment.

- This spatial viewpoint examines the nature and character of Earth and the distribution of phenomena within it. Physical geography applies spatial analysis to all the physical components and process systems that make up the environment: energy, air, water, weather, climate, landforms, soils, animals, plants, microorganisms, and Earth itself. Understanding the complex relations between Earth's physical systems and human society is important to human survival. Hypotheses and theories about the Universe, Earth, and life are developed through the scientific process, which relies on a general series of steps that make up the scientific method. Results and conclusions from scientific experiments can lead to basic theories as well as applied uses for the general public. Awareness of the human denominator, the role of humans on Earth, has led to physical geography's increasing emphasis on human–environment interactions. The concept of sustainability—the ability to continue activities indefinitely while minimizing their environmental impacts—and functioning Earth systems, is important to physical geography.

2. Sketch a flow diagram of the scientific process and method, beginning with observations and ending the development of a theory.

3. Which of the following economic activities—gold mining, salmon fishing, burning fossil fuels, and wheat farming—is sustainable? Explain your answer.

I.3 Earth Systems

Describe systems analysis, open and closed systems.

Explain the difference between positive and negative feedback information.

List Earth's four spheres and classify them as biotic or abiotic.

- A system is any ordered set of interacting components and their attributes, as distinct from their surrounding environment. Earth is an open system in terms of energy, receiving energy from the Sun, but it is essentially a closed system in terms of matter and physical resources. As a system operates, information is returned to various points in the operational process via pathways of feedback loops. If the feedback discourages change in the system, it is negative feedback that opposes system changes. If feedback information encourages change in the system, it is positive feedback that encourages system changes. When the rates of inputs and outputs in the system are equal and the amounts of energy and matter in storage within the system are constant (or when they fluctuate around a stable average), the system is in dynamic equilibrium. A threshold, or tipping point, is the moment at which a system can no longer maintain its character and lurches to a new operational level. Four immense open systems powerfully interact at Earth's surface. Three of these are abiotic (nonliving)—the atmosphere, hydrosphere, and lithosphere. The fourth is the biotic (living) biosphere.

4. Identify the main difference between an open system and a closed system.

5. Identify a major difference between the four large systems, or spheres, that comprise Earth. Would life on Earth be possible if one of these four spheres did not exist? Explain your answer.

How are locations on Earth located, mapped, & divided into time zones?

I.4 Determining Earth Locations & Times

Explain Earth's reference grid: latitude and longitude and latitudinal geographic zones and time.

- Earth's equatorial circumference is 40,075 km (24,902 mi), while its polar circumference is 40,008 km (24,860 mi). Latitude is the angular distance north or south of the equator. Lines of latitude are called parallels and run east-west. Longitude is the angular distance east or west of the prime meridian. Lines of longitude are called meridians, and they converge at the poles. The prime meridian is the basis for our system of global time. There are 24 time zones, each 15° wide, but they are distorted by political boundaries. On the opposite side of the planet from the prime meridian is the International Date

Line, which marks the place where each day officially begins. No matter what the time of day when the line is crossed, the calendar changes a day. Seventy countries use daylight saving time, setting clocks 1 hour ahead in the spring and 1 hour behind in the fall.

6. Draw a simple sketch describing Earth's shape and size.

7. Define latitude and parallel and define longitude and meridian using a simple sketch with labels.

8. What and where is the prime meridian? How was the location originally selected? Describe the meridian that is opposite the prime meridian on Earth's surface.

I.5   Maps & Cartography

Define cartography and mapping basics: map scale and map projections.

- A map is a generalized view of an area, as seen from above and reduced in size. Cartography is the science and art of mapmaking, often blending geography, mathematics, computer science, and art. The ratio of the size of a map to that area in the real world is the map's scale. Scale is represented as a representative fraction, a graphic scale, or a written scale. Graphic scales are used when the map may be enlarged or reduced in size. The basic map elements are a title, the scale, a guide to the map symbols, and a north arrow. Maps can be divided into physical and political maps. Topographic maps are physical maps that can give us a sense of the terrain. Relief is the difference in elevation between two locations. The conversion of a representation of the spherical Earth to a flat map is a map projection. All projections create distortion in size or shape or both.

9. What is map scale? What are three ways it can be shown on a map?

10. Describe the differences between the characteristics of a globe and those of a flat map.

What tools do geographers use?

I.6   Modern Geoscience Tools

Describe modern geographic tools—the Global Positioning System (GPS), remote sensing, and geographic information systems (GIS).

Explain how these tools are used in geographic analysis.

- Geographers use a number of new and evolving technologies to analyze and map Earth—the Global Positioning System (GPS), remote sensing, and geographic information systems. GPS uses radio signals from satellites to accurately determine location anywhere on or near the surface of Earth. Remote sensing refers to obtaining information about objects without physically touching them. Passive remote-sensing systems record energy radiated from a surface, especially visible light and infrared energy. Active remote sensing directs energy at a surface and analyzes the energy returned from the surface. LIDAR (*li*ght and ra*dar*), is an active remote-sensing technology that uses pulses of visible light, rather than radio waves to create a three-dimensional model. A GIS is a computer-based data-processing tool that combines spatial data with attribute data. A GIS program and a database work together to ask spatial analysis questions, often across several layers of data.

11. What is a GPS and how does it assist you in finding location and elevation on Earth?

12. What is remote sensing? What are you viewing when you observe a weather satellite image on TV or in the newspaper? Explain.

13. If you were planning the development of a large tract of land, how would a GIS help you? How might planning and zoning be affected if a portion of the tract in the GIS was a floodplain or prime agricultural land?

Key Terms

abiotic, p. I-11
biotic, p. I-11
cartography, p. I-16
closed system, p. I-10
Coordinated Universal Time (UTC) , p. I-14
equilibrium, p. I-11
dynamic equilibrium, p. I-11
equal area, p. I-18

feedback loop, p. I-10
geographic information system (GIS), p. I-22
geography, p. I-6
geoid, p. I-12
Global Positioning System (GPS), p. I-20
human denominator, p. I-8

International Date Line, p. I-14
latitude, p. I-12
LIDAR, p. I-21
longitude, p. I-12
map, p. I-16
map projection, p. I-17
Mercator projection, p. I-18
meridian, p. I-14

negative feedback, p. I-10
open system, p. I-10
parallel, p. I-13
physical geography, p. I-6
positive feedback, p. I-10
prime meridian, p. I-14
process, p. I-6
relief, p. I-19
remote sensing, p. I-20

scale, p. I-17
scientific method, p. I-6
scientific theory, p. I-7
spatial, p. I-6
spatial analysis, p. I-6
sustainability, p. I-9
system, p. I-4
threshold, p. I-11
topographic maps, p. I-19
true shape, p. I-18

Critical Thinking

1. Identify the various latitudinal geography zones that roughly subdivide Earth's surface. In which zones are a) Los Angeles, b) Moscow, and c) Quito?

2. In general terms, using the scientific method as a guide, how might a physical geographer analyze water pollution in the Great Lakes?

3. What and where is the prime meridian? How was the location originally selected? Describe the meridian that is opposite the prime meridian on Earth's surface.

4. Summarize how world population growth and environmental sustainability are related.

5. Is cartography an art or a science? Explain your answer.

Visual Analysis

Figure RI.1 looks across a valley toward the Karakoram Range in Pakistan. The Indus River flows across the center portion of the image.

1. Identify evidence of each of Earth's four *spheres* in the image, and classify each of your examples as biotic or abiotic.

2. Does this picture portray and "open" or "closed" Earth system? Explain your answer.

3. Identify and describe any examples of human influences on this landscape.

▲RI.1

Explore | Use **Google Earth** to explore the **geographic grid**.

Viewing Earth from space is to see the world anew! Open Google Earth, and uncheck (or turn off) all *Borders and Labels*. On the upper right, there are three tools to navigate around Earth. Place your cursor on each tool to learn how they enable one to *Look Around*, *Move Around*, and *Zoom*. Once you are comfortable with zooming about Earth, take the following journey.

Identify and zoom in on each of the continents: Africa, Europe, Asia, North America, South America, Australia, and Antarctica. Which continent is larger: Africa or South America? Next, select the *View* menu and scroll down to and check *Grid*. The geographic grid of latitude and longitude lines will appear. Then trace the following imaginary lines around Earth: Equator, Prime Meridian, Tropic of Cancer, and the Tropic of Capricorn. Then zoom in to North America, and slowly trace a route from San Francisco to New York. Finally, enter your present location in the *Search* window, click "search," and then answer the following questions.

1. What are the latitude and longitude of your location? (It's O.K. to give the answer in whole degrees).

▲RI.2

2. Notice the geographic data displayed across the bottom of the Google Earth screen and how the data change as you move the cursor. What is the elevation of the ground surface? What is your "eye altitude"? What is the scale of your current view of the area?

3. Describe the physical features visible in your view. What effects of human activity can you see in the landscape?

Interactive Mapping | Login to the **MasteringGeography** Study Area to access **MapMaster**.

Comparing the Spatial Distribution of World Population
- Open: MapMaster in MasteringGeography
- Select: *World*. Next, turn on the *Population* categories, and select *Population Growth Rates*.

1. Which regions of Earth currently have the highest natural rate of population increase, and which areas have the lowest rate of increase?

- Next, select *Literacy Rate* from the *Population* category.

2. Identify the relationship between literacy and population growth rates Europe and Africa.

Mastering Geography™

Looking for additional review and test prep materials? Visit the Study Area in MasteringGeography™ to enhance your geographic literacy, spatial reasoning skills, and understanding of this chapter's content by accessing a variety of resources, including MapMaster™ interactive maps, videos, *Mobile Field Trips*, *Project Condor* Quadcopter videos, *In the News* RSS feeds, flashcards, web links, self-study quizzes, and an eText version of *Geosystems Core*.

Mapping for Sustainability: How Eco-Friendly is Your Campus?

GeoLab Intro (MG)
Pre-Lab Video

https://goo.gl/zH6kly

Human-environment relationships are one of the key themes of geography. One aspect of this relationship is sustainability, the idea that our impact on Earth's key systems should be minimized. College campuses across the country are taking action to become more sustainable (Figs. GLI.2 and GLI.3). Table GLI.1 lists aspects of sustainability that are relevant to your college campus. In general, buildings are more sustainable if they use less energy and water and if they produce less pollution and solid waste than buildings not designed or modified for sustainability (Fig. GLI.2).

The process of becoming more sustainable often begins with an inventory of existing conditions. In this exercise you will evaluate how sustainable your campus is by mapping sustainable features of your student center.

Apply

You are the newly elected president of the Environment Club. You ran on a platform of increasing campus sustainability and your first step is to evaluate your campus's student center in terms of sustainability. You will map the student center building and all of its sustainability features, or lack thereof, and create a plan to enhance the center's sustainability.

Objectives

- Analyze your campus's student center in terms of sustainability.
- Evaluate changes that could be made to the student center to improve its sustainability.
- Create a map, using basic map elements, to portray your campus's student center and its sustainability features.

Procedure, Part I

1. Using Table GLI.1 as a checklist, make in inventory of your student center's sustainable features, and also note the sustainability features it lacks.
2. What other sustainability features could you add to Table GLI.1? Add them to the checklist and note whether your student center has (or lacks) them.
3. What Earth systems do these sustainability efforts and features impact the most? Explain your answer.

Procedure, Part II

4. Before you can map your student center, there are some mapping decisions to be made. First, what will the scale of your map be? How large is your student center? How much of the area around the student center will you show on your map? Map scale is the ratio of the size of objects on your map to objects on the ground. The size of your map will be dictated by the size of your paper. Your campus may have a detailed downloadable map with building footprints.
5. You'll also have to decide how to use symbols to represent the sustainability features you're mapping (Fig. GLI.1). Make a list of the features and their symbols that you can use for your map's legend.
6. Draw your map of your student center and the sustainability features you've selected.
7. What features did you map? Were there new features that you weren't aware of until you started mapping?
8. What scale is your map? Write the scale as both a representative fraction (such as 1:600) and as a written scale (such as one inch equals fifty feet).

Analyze & Conclude

9. Some campuses have offices of sustainability. If you were going to make a GIS map to give to the office of sustainability, how would you organize the data? Would you group the features by geometry, with one layer for the polygons, another layer for the lines, and a third layer for the point features, or would you group them into thematic layers? Discuss your choice.
10. Were there sustainability features did you expect to find in the student center, but didn't? Were there features that you were surprised to find?
11. Overall, how sustainable is your student center? What were the most sustainable aspects? The least sustainable? Make a list of changes needed to make the center more sustainable.
12. You want to submit your map as part of a sustainability plan for your campus that will appear in the student newspaper. Write a short summary of the plan's recommendations to improve the sustainability of your campus. Work with other students in your class to assemble a plan combining everyone's recommendations and send the class plan to your campus administrator, dean, or student paper.

Table GLI.1 Sustainability Inventory

Energy
- Solar photovoltaic panels?
- Other renewables? wind turbines? solar hot water?

Buildings & Facilities
- Is the building Leadership in Energy and Environmental Design (LEED) certified?
- Sustainable materials such as hemp or sustainably harvest forest products?
- Waterless urinals?
- Innovative architecture such as straw bale, or windows and overhangs that block summer sun but let in winter sun?
- How energy efficient is the building's heating and cooling system?

Food (I f the center serves food)
- Organic food?
- Is the food sourced from local farms?

Transportation
- Public transportation: Where are the closest bus stops, light rail stops, or other public transportation facilities?
- Where are the bike racks? How many bicycles can they hold?
- Where is the Electronic Vehicle (EV) parking?
- Is there special parking for carpools?
- Other transportation features such as horse or ski parking?

Waste Reduction
- Where are the recycling containers?
- Are there compost containers in dining facility?
- Is there composting by food services?
- Are the paper towels recycled paper?
- Are the paper napkins recycled paper?

Water
- Where are the water bottle stations?
- Does the landscaping outside use drought resistant, native vegetation?
- If your campus is in an arid region, is the landscaping water saving?

▲GLI.1 **Symbols of sustainability** (Clockwise from top left): transportation (bicycle and electric vehicle); recycling; public transportation; and energy-efficient lighting.

▲GLI.2 **Energy efficiency** Solar panels on the roof Yale's School of Forestry and Environmental Studies at Kroon Hall make this building a model of sustainable practices.

▲GLI.3 **Sustainable transportation** Over 50 percent of students at the University of California, Davis travel to campus using a bike or skateboard.

1

Solar Energy, Seasons, & the Atmosphere

We follow incoming solar energy from the Sun to the top of Earth's atmosphere, and then follow that energy as it drives Earth's physical systems and influences our daily lives. This solar energy, along with Earth's tilt and rotation, produce the seasonal patterns of changing day length and Sun angle. The Sun is the ultimate energy source for most life processes in our biosphere.

We examine the atmosphere's composition, temperature, and function. Our look at the atmosphere also includes the spatial aspects of both natural and human-produced air pollution. We all interact with the atmosphere with each breath we take and in many other ways. For example, burning fossil fuels to generate electricity and power our cars and planes releases vast quantities of carbon dioxide gas into the atmosphere. These topics are essential to physical geography, for we are influencing the atmospheric composition for the future.

Key Concepts & Topics

Sunrise over the Pacific Ocean, as photographed from an altitude of 419 km (260 mi.) by Astronaut Reid Wiseman aboard the International Space Station on September 1, 2014 . We are seeing Earth's atmosphere from the side, as well as Earth's curvature. The different colors correspond to different layers in the atmosphere. This image highlights how thin our atmosphere is.

1.1 **Our Galaxy & Solar System**

Key Learning Concepts

▶ **Distinguish** among galaxies, stars, and planets.
▶ **Differentiate** between the key distances of our solar system, and locate Earth.

Physical geography focuses on planet Earth. But Earth is not alone in space, and its setting within the vastness of the universe affects conditions and processes in Earth systems.

Earth's Place in Space

Our solar system is located on one of the outer "arms" of the **Milky Way Galaxy**, a disk-shaped spiral of stars (▶Fig. 1.1). From our Earth-bound perspective in the Milky Way, the galaxy appears to stretch across the night sky like a narrow band of hazy light, although our eyes can see only a few thousand of these billions of stars. Our Sun is just one of 300 billion stars in the Milky Way Galaxy. The universe has at least 125 billion galaxies, each of which contains hundreds of billions of stars.

Our solar system condensed from a huge, slowly rotating and collapsing cloud of dust and gas, a *nebula*. **Gravity**, the mutual attraction exerted by every object upon all other objects in proportion to their mass, was the key force in this condensing solar nebula. The beginnings of the formation of the Sun and its solar system are estimated to have occurred more than 4.6 billion years ago. Within a nebula, stars condense from clouds of gas and dust, with planets gradually forming in orbits about the central mass. Astronomers study this process in other parts of the galaxy, where more than 1800 planets have been observed orbiting distant stars. An initial estimate of the number of planets in the Milky Way is 100 billion. Astronomers think there could be 500 million of these planets in the habitable zones with moderate temperatures where water can exist as a liquid.

> **geoCHECK** ✔ What is the key difference between a galaxy and a solar system?

Animation (MG)
Earth Sun Relations

http://goo.gl/XVJd3y

Our Solar Systems is in the Orion Spur of the Sagittarius Arm.

|← ————————— 100,000 light-years ————————— →|

▲**1.1 The Milky Way Galaxy** (a) The overal structure of our galaxy. (b) The Milky Way as seen from Earth.

Dimensions, Distances, & Earth's Orbit

One way to think about Earth's place in relation to the solar system and galaxy is to use the speed of light as a yardstick to measure the distances involved. The **speed of light** is 300,000 kmps (kilometers per second), or 186,000 mps (miles per second). The tremendous distance that light travels in a year, 9.5 trillion kilometers (6 trillion miles), is known as a *light-year*, and it is used as a unit of measurement for the vast universe.

The Milky Way is about 100,000 light-years from side to side, and the known universe stretches approximately 12 billion light-years in all directions. In contrast, our entire solar system is approximately 11 hours in diameter, if measured by the speed of light (◄**Fig. 1.2a**). The Moon is an average of 384,400 km (238,866 mi) from Earth, or about 1.28 seconds away at the speed of light.

Earth's orbit around the Sun is an oval or elliptical path. The average distance from Earth to the Sun is approximately 150 million kilometers (93 million miles), so light from the Sun reaches Earth in about 8 minutes and 20 seconds. While Earth is closer to the Sun in January (147,100,000 km, 91,400,000 mi) than in July (152,100,000 km, 94,500,000 mi), causing a slight variation in the amount of solar energy received, it is not enough to cause seasonal changes (◄**Fig. 1.2b**).

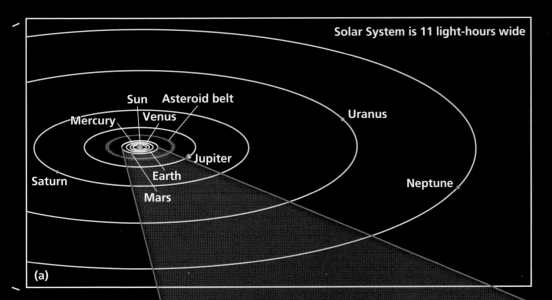

Solar System is 11 light-hours wide

Sun Asteroid belt
Mercury Venus
Uranus
Jupiter
Saturn Earth
Mars
Neptune

(a)

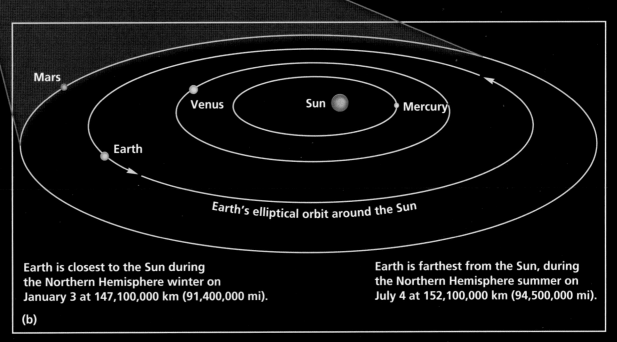

Mars
Venus Sun Mercury
Earth
Earth's elliptical orbit around the Sun

Earth is closest to the Sun during the Northern Hemisphere winter on January 3 at 147,100,000 km (91,400,000 mi).

Earth is farthest from the Sun, during the Northern Hemisphere summer on July 4 at 152,100,000 km (94,500,000 mi).

(b)

▲1.2 **Our solar system and Earth's orbit**

geoQUIZ

1. How did the solar system form from a nebula?
2. What is the diameter of our solar system? The distance across the known universe?
3. How much does the distance from Earth to the Sun vary over a year? How does this distance compare to the average distance from Earth to the Sun?

geoCHECK ✔ What is the speed of light?

1.2 Energy from the Sun

Key Learning Concepts

▶ **Describe** the Sun's operation, including solar wind.

The massive, glowing ball of gases we call the Sun continuously gives off radiant energy in all directions. The portion of that energy that reaches Earth provides energy for processes involving the planet's atmosphere, oceans, and land surface.

Solar Activity & Solar Wind

How does the Sun generate all of this energy? To generate enormous energy, you need matter—lots of it. The Sun captured about 99.9% of the matter from the original solar nebula. All the planets, their satellites, asteroids, comets, and debris are made up of the remaining 0.1%. In our entire solar system, the Sun is the only object having the enormous mass needed to sustain a nuclear reaction in its core and produce radiant energy.

The huge mass of the Sun produces tremendous pressure and high temperatures deep in its dense interior. Here, the Sun's abundant hydrogen atoms are forced together and enormous quantities of energy are liberated in the process of **fusion**. The Sun converts 4.26 million metric tons of mass to heat and light energy every second!

In addition to light and heat, the Sun constantly emits clouds of electrically charged particles. This stream of **solar wind** travels at about 50 million kilometers (31 million miles) a day, taking approximately 3 days to reach Earth.

The Sun's most conspicuous features are large **sunspots**, caused by magnetic storms on the Sun (▶ **Fig. 1.3**). The number of sunspots is related to the level of activity of the Sun—slightly more energy is radiated when there are more sunspots than when there are fewer sunspots. Individual sunspots may range in diameter from 10,000 to 160,000 km (6200 to 100,000 mi), more than 12 times Earth's diameter. The eruption in Figure 1.3a was 20 times the diameter of Earth. These produce flares and outbursts of charged material, referred to as coronal mass ejections, that affect radio and satellite communications.

A regular cycle exists for sunspot occurrences, as the number of sunspots increases and decreases, averaging 11 years from maximum to maximum. However, the cycle may vary from 7 to 17 years, with a minimum in 2009 and a maximum in 2014. (For more on the sunspot cycle, see http://solarscience.msfc.nasa.gov/SunspotCycle.shtml; for the latest space weather, see http://www.spaceweather.com/.)

geoCHECK ✔ Why is the Sun hot?

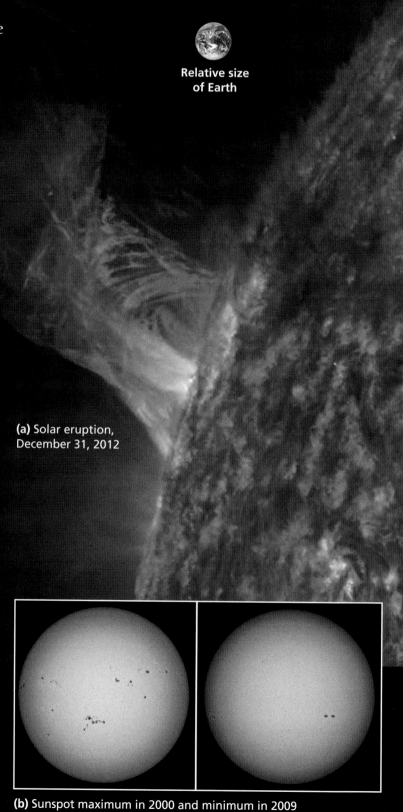

Relative size
of Earth

(a) Solar eruption,
December 31, 2012

(b) Sunspot maximum in 2000 and minimum in 2009

▲1.3 **Solar coronal mass eruption and sunspots**

Auroras

The charged particles of the solar wind first interact with the **magneto-sphere**, Earth's magnetic field, as they approach Earth. The magnetosphere is generated by the motions of Earth's molten iron outer core. The magnetosphere deflects the solar wind toward both of Earth's poles so that only a small portion of it enters the upper atmosphere.

This interaction of the solar wind and the upper layers of Earth's atmosphere produces the remarkable **auroras** that occur toward both poles (▼Fig. 1.4). These lighting effects are the *aurora borealis* (northern lights) and *aurora australis* (southern lights) in the upper atmosphere, 80–500 km (50–300 mi) above Earth's surface. They appear as folded sheets of green, yellow, blue, and red light that ripple across the skies of higher latitudes, especially poleward of 50°. The different colors of the aurora are due to different molecules in the atmosphere being excited—oxygen produces green or brownish-red light and nitrogen produces blue or red light. During a period in 2001 when the solar wind was stronger than usual auroras were visible as far south as Jamaica, Texas, and California.

(a) *Aurora australis* as seen from orbit.

(b) *Aurora borealis* over Whitehorse, Yukon Canada. On August 31, 2012, a coronal mass ejection erupted from the Sun into space, traveling at over 900 miles per second, causing this aurora four days later.

Oxygen produces green or brownish-red light.

Nitrogen produces blue or red light.

Animation (MG)
Formation of the Solar System

http://goo.gl/alti7U

▲1.4 **Auroras from orbit and from the ground**

geoCHECK ✔ What causes the auroras?

geoQUIZ
1. How much of the mass of our solar system is the Sun?
2. Why is the Sun the only object in our solar system producing heat and light?
3. Why are auroras different colors?

1.3 Electromagnetic Spectrum

Key Learning Concepts

▶ *Explain* the characteristics of the electromagnetic spectrum of radiant energy.

The essential input of energy to Earth is radiant energy from the Sun. This radiant energy travels to Earth at the speed of light. Solar radiation occupies a portion of the **electromagnetic spectrum**, which is the spectrum of all possible wavelengths of electromagnetic energy (▼Fig. 1.5). Light can be measured by both its wavelength and frequency. A **wavelength** is the distance between corresponding points on any two successive waves. The number of waves passing a fixed point in 1 second is the frequency, thus the shorter the wavelength, the higher the frequency.

Sun radiates shorter-wavelength energy than does Earth, with the Sun's emissions concentrated around 0.5 μm (micrometer) (▶Fig. 1.6). The Sun emits radiant energy composed of 8% ultraviolet, X-ray, and gamma-ray wavelengths; 47% visible light wavelengths; and 45% infrared wavelengths. Ultraviolet energy, because of its shorter wavelength, is higher energy than longer wavelength infrared energy. This is why ultraviolet energy that reaches the surface causes sunburn, skin cancer, and damages the eyesight of human beings, while we perceive infrared energy as heat.

Comparing Earth & Sun as Radiating Bodies

All objects radiate energy in wavelengths related to their surface temperatures: A hotter object emits more energy, with shorter wavelengths, than a cooler object. This holds true for the Sun and Earth. The hotter

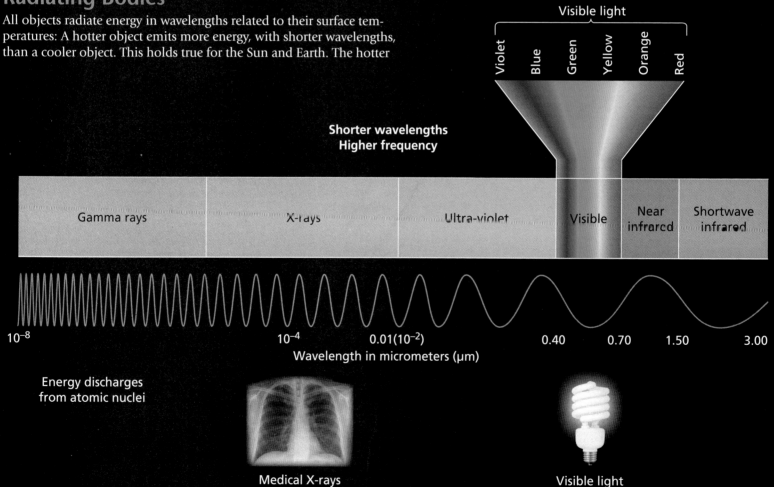

▲1.5 **Electromagnetic spectrum**

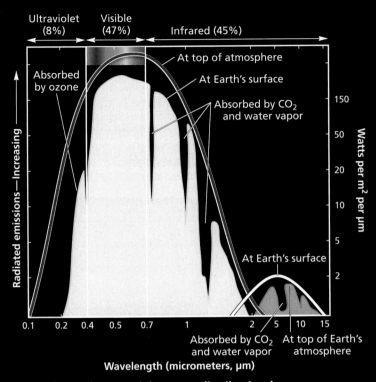

Ultraviolet (8%)   Visible (47%)   Infrared (45%)

Radiated emissions—Increasing

Absorbed by ozone

At top of atmosphere

At Earth's surface

Absorbed by CO_2 and water vapor

At Earth's surface

150
50
20
10
5
2

Watts per m² per µm

0.1   0.2   0.4   0.5   0.7   1   2   5   10   15

Absorbed by CO_2 and water vapor

At top of Earth's atmosphere

Wavelength (micrometers, µm)

▲1.6 **Solar and terrestrial energy distribution by wavelength** The left hand side of the figure shows the distribution of energy from the Sun that reaches the top of our atmosphere (purple line) and Earth's surface (yellow curve). The right hand side of the figure shows the distribution of energy from Earth out to space as measured at the surface (white line) and at the top of the atmosphere (orange curve).

The Sun's surface temperature is about 6000 K (6273°C or 11,459°F). Shorter wavelength emissions are dominant at these higher temperatures.

Earth radiates nearly all of the energy that it absorbs. Because Earth is cooler, it emits less energy and at longer wavelengths, mostly in the infrared portion of the spectrum, centered around 10.0 µm. The smooth curves on the graph in Figure 1.6 represent the radiation emitted by the Sun and Earth. While the curves are smooth, notice that there are gaps at certain wavelengths for actual outgoing radiation. These are due to water vapor, water, carbon dioxide, oxygen, ozone (O_3), and other gases in Earth's atmosphere absorbing these wavelengths.

geoCHECK Why does the Sun emit shorter wavelengths than Earth?

geoQUIZ
1. What is the relationship between wavelength and frequency?
2. How does the range of visible wavelengths compare to the range of energy emitted by the Sun?
3. Which range of wavelengths accounts for most of the energy we receive from the Sun?

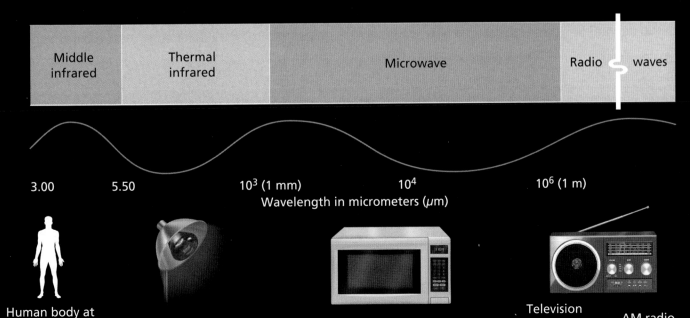

**Longer wavelengths
Lower frequency**

Middle infrared   Thermal infrared   Microwave   Radio   waves

3.00   5.50   10³ (1 mm)   10⁴   10⁶ (1 m)

Wavelength in micrometers (µm)

Human body at

Television

AM radio

1.4 Incoming Energy & Net Radiation

Key Learning Concepts
▶ **Describe** the global pattern of net radiation.

How much solar energy reaches Earth and how is it distributed? The answer to these questions involves a number of factors, including Earth's atmosphere, curved shape, and tilted axis.

The region at the top of the atmosphere, approximately 480 km (300 mi) above Earth's surface, is the outer boundary of Earth's energy system. This is where we study arriving solar radiation before it interacts with the atmosphere.

Because of its distance from the Sun, Earth intercepts only one 1/2,000,000,000 of the Sun's total energy output. This tiny fraction of the Sun's energy is still an enormous amount of energy that flows into Earth's systems.

Insolation & the Solar Constant

Solar radiation arriving at Earth's atmosphere and surface is **insolation**, derived from the words "incoming solar radiation" (▶**Fig. 1.7**). The **solar constant** is the average insolation received at the thermopause when Earth is at its average distance from the Sun, a value of 1362 watts per square meter (W/m²). Less than half of the insolation

that passes through the top of the atmosphere reaches the surface. The rest is reflected, scattered, and absorbed by the atmosphere.

geoCHECK ✔ How much insolation passes through the top of the atmosphere, but doesn't reach the surface?

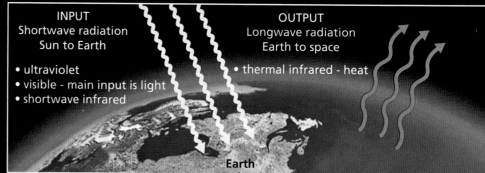

INPUT Shortwave radiation Sun to Earth	OUTPUT Longwave radiation Earth to space
• ultraviolet • visible - main input is light • shortwave infrared	• thermal infrared - heat

Earth

▲1.7 **Earth's energy budget** Light comes in to Earth from the Sun, Earth radiates heat back to space.

Uneven Distribution of Insolation

Because Earth's surface is curved, the angle of incoming sunlight is different at each latitude. Differences in the angle of sunlight by latitude result in an uneven distribution of insolation and heating. The Sun is directly over the **subsolar point**, the only location that receives "direct" sunlight, meaning rays of sunlight that strike the surface vertically, at a 90° angle. The subsolar point moves back and forth between the Tropic of Cancer at 23.5° north to the Tropic of Cancer at 23.5° south during the year. Incoming energy is more concentrated in this region, the tropics. All other places away from the subsolar point receive insolation at an angle less than 90° and receive less concentrated energy (◀**Fig. 1.8**). This effect is greater at higher latitudes. These higher latitude regions receive less insolation, mainly because of the lower angle of the Sun, but the insolation also travels through a greater thickness of the atmosphere. This results in less energy reaching the ground due to scattering, absorption, and reflection in the atmosphere. Because of these factors, the tropics receive 2.5 times more energy than the poles.

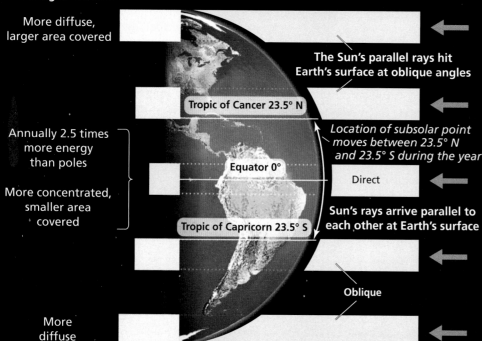

Surface area receiving insolation

More diffuse, larger area covered

Annually 2.5 times more energy than poles

More concentrated, smaller area covered

More diffuse

The Sun's parallel rays hit Earth's surface at oblique angles

Tropic of Cancer 23.5° N

Location of subsolar point moves between 23.5° N and 23.5° S during the year

Equator 0°

Direct

Sun's rays arrive parallel to each other at Earth's surface

Tropic of Capricorn 23.5° S

Oblique

▲1.8 **Insolation and Earth's curved surface**

geoCHECK ✔ Why does insolation intensity vary with latitude?

Global Net Radiation

The difference between incoming shortwave and outgoing long-wave radiation—energy inputs minus energy outputs—is called *net radiation*. The pattern of net radiation changes with latitude. Satellites measured shortwave and longwave flows of energy at the top of the atmosphere to produce the map in Figure 1.9, which shows a latitudinal imbalance in net radiation.

The tropics are a region of positive net radiation values, and the poles are regions of negative values. In middle and high latitudes, approximately poleward of 36° north and south, net radiation is negative. Negative values occur in higher latitudes because Earth's climate system loses more energy to space than it gains from the Sun. In the lower atmosphere, energy flows as wind and warm ocean currents from tropical surpluses to the polar energy deficits. The highest net radiation values, averaging 80 W/m², are above the tropical oceans along a narrow equatorial zone. Net radiation minimums are lowest over Antarctica.

The Sahara region of North Africa is surprisingly a region of net energy loss. Light-colored reflective surfaces, such as fresh snow, and desert sand reduce the absorption of incoming energy, and clear skies allow great amounts of longwave radiation to escape to space. In other regions, clouds and atmospheric pollution in the lower atmosphere affect net radiation patterns at the top of the atmosphere by reflecting more shortwave energy to space.

Having examined the flow of solar energy to Earth and the top of the atmosphere, let us now look at how seasonal changes affect the distribution of insolation as Earth orbits the Sun during the year.

geoCHECK ✔ Which region has the highest energy surplus?

geoQUIZ

1. What is the subsolar point? Where does it migrate during the year?
2. How does the amount of insolation in the tropics compare to the amount of insolation received at the poles?
3. How do the poles maintain a negative energy balance? More energy flows out from the poles than comes in from the Sun. Where does that extra energy come from?

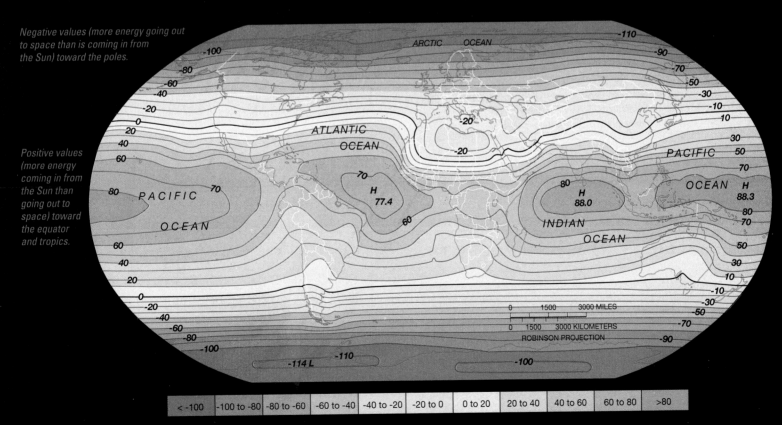

Negative values (more energy going out to space than is coming in from the Sun) toward the poles.

Positive values (more energy coming in from the Sun than going out to space) toward the equator and tropics.

< -100	-100 to -80	-80 to -60	-60 to -40	-40 to -20	-20 to 0	0 to 20	20 to 40	40 to 60	60 to 80	>80

▲1.9 **Daily net radiation patterns** Average daily net radiation flows at the top of the atmosphere. Units are W/m².

1.5 The Seasons

Key Learning Concepts

▶ **Define** solar altitude, solar declination, and day length.
▶ **Describe** the annual variability that produces Earth's seasonality.

If you have lived in one place for a full year, you may have noticed the changes in weather from summer to winter. You might have also noticed the changes in the length of the day and the angle of the Sun above the horizon. Physical geographers define the seasons in terms of Earth's changing relationship to the Sun.

Seasonality

The changes in the angle of the Sun above the horizon and changes in the hours of daylight are referred to as **seasonality**. The angle of the Sun above the horizon is the Sun's **altitude**. If the Sun is at the horizon, its altitude is 0°. If the Sun is halfway between the horizon and directly overhead, it is at 45° altitude. If the Sun is directly overhead, it is at 90° altitude. The Sun is directly overhead only at the subsolar point, where insolation is at a maximum.

The Sun's **declination** is the latitude of the subsolar point (▼**Fig. 1.10**). Declination annually migrates between the Tropic of Cancer and Tropic of Capricorn, for a total of 47° of movement. If you marked the location of the Sun in the sky at noon each day throughout the year, you would find that the Sun takes a figure-8 shaped path called an *analemma*. On the analemma chart in Figure 1.11 you can locate any date, then trace horizontally to the *y*-axis and find the subsolar point. Along the Tropic of Capricorn, the subsolar point occurs on December 21–22, at the lower end of the analemma. Following the chart, you see that by March 20–21, the subsolar point reaches the equator, and then moves on to the Tropic

of Cancer in June. As an example, use the chart to calculate the subsolar point location on your birthday. Other than Hawaii, the subsolar point does not reach the continental United States or Canada, because these locations are too far north.

Seasonality also produces changes in **day length**. Day length varies during the year, and the higher the latitude, the larger the difference between summer and winter day lengths. The equator always receives equal hours of day and night, while the poles experience a 24-hour difference in day length. Along the equator every day and night is 12 hours long, year-round. People living along 40° N (Philadelphia, Denver) or 40° S (Buenos Aires, Melbourne) experience about 6 hours' difference in daylight between winter (9 hours) and summer (15 hours). At 50° N latitude (Winnipeg, Paris) or 50° S (Falkland Islands), people experience almost 8 hours of annual day length variation.

geoCHECK ✔   How much does day length vary at the equator over a year? At latitude 40° N?

The angle of the Sun above the horizon is the Sun's altitude.	The latitude of the subsolar point is the Sun's declination.

On the June solstice, the angle of the Sun above the horizon at 23.5° N is 90°

On the June solstice, the subsolar point is at 23.5° N

On the solstices, the Sun's angle above the horizon at 0° is 66.5°

On the equinoxes, the subsolar point is at 0°

On the June solstice, the angle of the Sun above the horizon at 23.5° S is 43°

On the December solstice, the subsolar point is at 23.5°

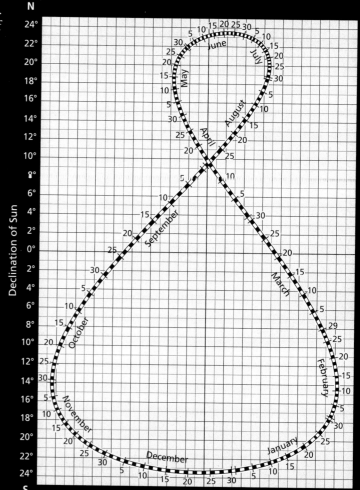

Reasons for Seasons

As Earth revolves around the Sun, a voyage that takes 1 year, it rotates daily around its **axis**, an imaginary line that goes from the North Pole to the South Pole. Seasonality is created by Earth's **revolution** around the Sun, its daily rotation, the *tilt* of its axis, the *orientation* of its axis, and its sphericity.

Revolution and Rotation Earth travels at an average speed of 107,280 kmph (66,660 mph) as it **revolves** around the Sun. Earth orbits the Sun every 365.24 days (▼**Fig. 1.12**).

Earth's **rotation**, or turning on its axis, averages slightly less than 24 hours in duration. Rotation determines day length, creates the apparent deflection of winds and ocean currents due to the Coriolis effect (see Chapter 4). Rotation

also produces the twice-daily rise and fall of the ocean tides with the gravitational pull of the Sun and the Moon. Viewed from above the equator, Earth rotates west to east, or eastward. This eastward rotation creates the Sun's apparent westward daily journey from sunrise in the east to sunset in the west. Of course, the Sun actually remains a relatively fixed position in the center of the solar system as our solar system travels through space.

Earth's rotation produces the daily pattern of day and night. The dividing line between day and night is the **circle of illumination** (▼**Fig. 1.13**).

Tilt, Orientation, and Sphericity Figure 1.13 also shows Earth's axial tilt. The plane of Earth's orbit around the Sun is the **plane of the ecliptic**. Earth's axis is tilted about 23.5° from perpendicular, but it remains fixed relative to this plane as Earth revolves around the Sun. The northern end of the axis points to Polaris, also called the North Star. If Earth orbited the Sun without axial tilt, the equator would always be the subsolar point and all other locations would receive insolation at less than a 90° angle. Earth's axial tilt of 23.5° determines the locations of the tropics, and the Arctic and Antarctic Circles. The tropics are located 23.5° away from the Equator, at the northern and southern limits of the subsolar point's annual migration. The Arctic and Antarctic Circles are located 23.5° away from poles at 90° at 66.5°. Every location between the tropics receives sunlight at a 90° angle at some point during the year, while locations north of the Arctic Circle and south of the Antarctic Circle experience 24 hours of daylight and darkness. Throughout our annual journey around the Sun, Earth's axis *points to Polaris and also keeps the same tilt* relative to the plane of the ecliptic. If we compared the axis in different months, it would always appear parallel to itself, a condition known as **axial parallelism** (▼Fig. 1.13).

Sphericity Because Earth's surface is curved, different latitudes receive sunlight at different angles on the same day (Fig. 1.8). If Earth wasn't spherical, sunlight would shine on all locations at the same angle.

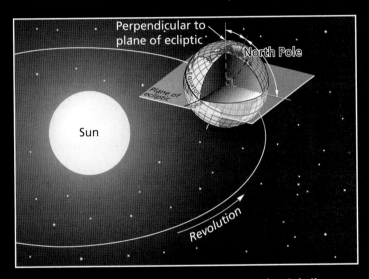

▲1.12 **Earth's revolution, rotation, and axial tilt**

geoCHECK ✔ What are the five reasons for the seasons?

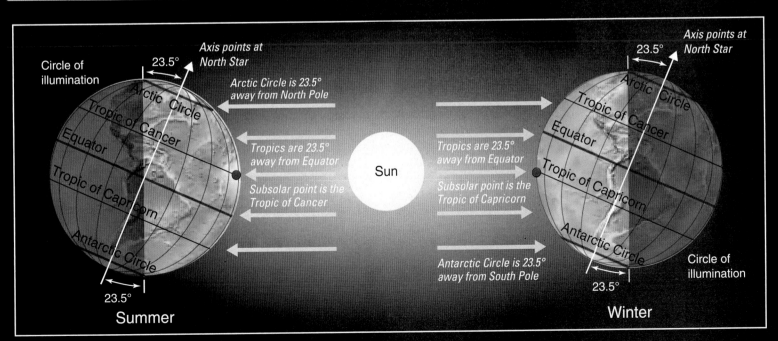

▲1.13 **Earth's circle of illumination**

1.5 The Seasons (cont'd)

March of the Seasons

During the annual cycle, or "march," of the seasons on Earth, day length is the most obvious change for latitudes away from the equator (▶Fig. 1.14). The extremes of day length occur on the *solstices* in December and June. The solstices occur when the subsolar point is at its position farthest north at the Tropic of Cancer or farthest south at the Tropic of Capricorn. During the **December solstice**, the subsolar point is the Tropic of Capricorn and locations above the **Arctic Circle** experience a 24-hour day, and locations south of the **Antarctic Circle** a 24-hour night. These day lengths are reversed during the **June solstice**, when the subsolar point is the Tropic of Cancer. In the Northern Hemisphere, the December solstice marks the beginning of winter, and is called the **winter solstice**. The winter solstice is also the midpoint of the low sun season. The June solstice marks the beginning of summer, and is called the **summer solstice**. The summer solstice is also the midpoint of the high sun season. In the Southern Hemisphere the June solstice is the winter solstice and the December solstice is the Summer solstice.

Between the solstices are the *equinoxes*, when the subsolar point is directly over the equator and day length is 12 hours at all latitudes. The **September equinox, or autumnal equinox,** marks the beginning of autumn in the Northern Hemisphere. The **March equinox, or vernal equinox,** marks the beginning of spring in the Northern Hemisphere.

Animation MG
Earth Sun Relations

http://goo.gl/XVJd3y

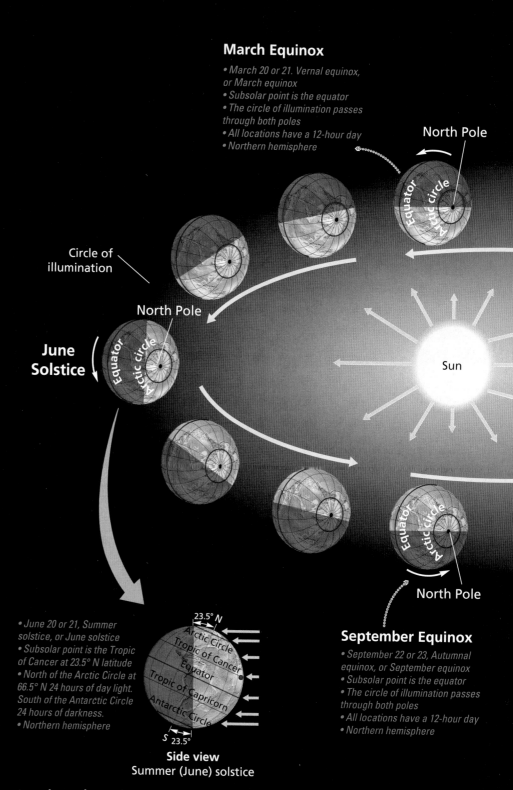

March Equinox

• *March 20 or 21. Vernal equinox, or March equinox*
• *Subsolar point is the equator*
• *The circle of illumination passes through both poles*
• *All locations have a 12-hour day*
• *Northern hemisphere*

North Pole

Equator
Arctic circle

Circle of illumination

North Pole

June Solstice

Equator
Arctic circle

Sun

Equator
Arctic Circle

North Pole

• *June 20 or 21, Summer solstice, or June solstice*
• *Subsolar point is the Tropic of Cancer at 23.5° N latitude*
• *North of the Arctic Circle at 66.5° N 24 hours of day light. South of the Antarctic Circle 24 hours of darkness.*
• *Northern hemisphere*

23.5° N
Arctic Circle
Tropic of Cancer
Equator
Tropic of Capricorn
Antarctic Circle
S 23.5°

Side view
Summer (June) solstice

September Equinox

• *September 22 or 23, Autumnal equinox, or September equinox*
• *Subsolar point is the equator*
• *The circle of illumination passes through both poles*
• *All locations have a 12-hour day*
• *Northern hemisphere*

▲1.14 **Annual march of the seasons**

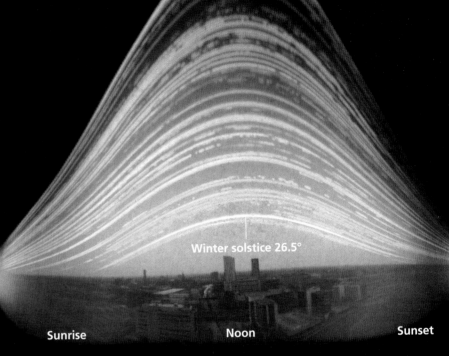

Summer solstice 73.5°

Winter solstice 26.5°

Sunrise   Noon   Sunset

▲1.15 **Seasonal observations of the sun—sunrise, noon, and sunset through the year** The pale lines in this image are time-lapse photographs of the Sun's path each day, combined to show how the path changes with the seasons. In the Northern Hemisphere, the Sun's daily path gradually rises from its lowest on the December solstice to its highest on the June solstice.

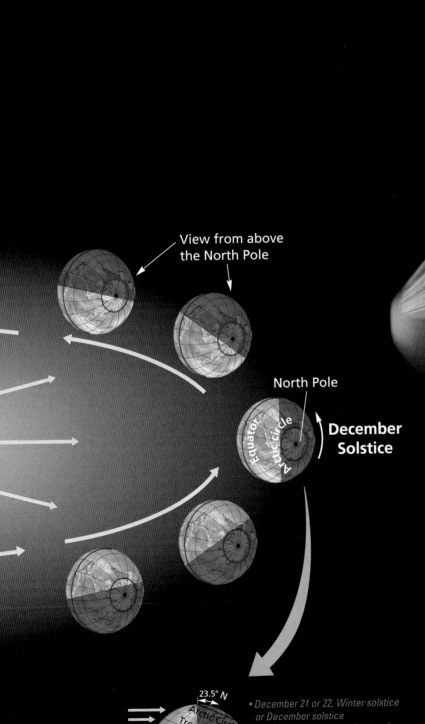

View from above the North Pole

North Pole

December Solstice

23.5° N
Arctic Circle
Tropic of Cancer
Equator
Tropic of Capricorn
Antarctic Circle
S 23.5°

• December 21 or 22, Winter solstice or December solstice
• Subsolar point is the Tropic of Capricorn at 23.5° S
• North of the Arctic Circle at 66.5° N 24 hours of darkness. South of the Antarctic Circle 24 hours of day light

Side view
Winter (December) solstice

Seasonal Observations The altitude of the Sun changes continuously through the year. In summer, the Sun is higher above the horizon at noon, and in winter, it is closer to the horizon. For example, the Sun's altitude at local noon at 40° N increases from a 26.5° angle above the horizon at the winter (December) solstice to a 73.5° angle above the horizon at the summer (June) solstice—a range of 47° (▲Fig. 1.15).

geoCHECK ✔   Where is the subsolar point on each of the solstices and equinoxes?

geoQUIZ

1. Why is axial parallelism a crucial part of seasonal changes?
2. How much does the Sun's altitude vary from the summer solstice to the winter solstice?
3. Where would the Tropics and Arctic and Antarctic Circles be if Earth's axial tilt was 30°?

1.6 Atmospheric Composition

Key Learning Concepts

▶ **Draw** a diagram showing a profile of atmospheric structure based on composition.

Earth's atmosphere is a unique mixture of gases that is the product of 4.6 billion years of development, including the life processes of living organisms. The atmosphere protects us by filtering out harmful radiation and particles from the Sun and beyond. The atmosphere is mainly composed of *air*. Air is a mixture of gases that is odorless, colorless, tasteless, and blended so thoroughly that it behaves as if it were a single gas.

The top of our atmosphere is around 480 km (300 mi) above Earth's surface. Beyond that altitude is the **exosphere**, which means "outer sphere," where the atmosphere is nearly a vacuum. The exosphere contains scarce lightweight hydrogen and helium atoms, weakly bound by gravity as far as 32,000 km (20,000 mi) from Earth.

Atmospheric Profile

If you look at a profile, or cross section, of the atmosphere, you will see that its characteristics change between the surface and the upper atmosphere. Some of these changes reflect the properties of the gases that compose the atmosphere. Other changes reflect the physical properties of the atmosphere itself, such as temperature, pressure, density, and the effects of the force of gravity. The atmosphere's chemical and physical characteristics produce its layered structure, shown in Figure 1.16.

Gravity compresses air, making it denser near Earth's surface. As you can see in Figure 1.16, the atmosphere thins rapidly with increasing altitude. **Air pressure** is approximately 1 kg/cm² (14.7 lb/in.²) at sea level.

At sea level, the atmosphere exerts a pressure of 1013.2 mb (millibar, or mb; a measure of force per square meter of surface area), or 29.92 in. of mercury (symbol, Hg), as measured by a barometer. In Canada and other countries, normal air pressure is expressed as 101.32 kPa (kilopascal; 1 kPa = 10 mb). As you will see, the atmosphere's composition and temperature, and the functions of different layers, also vary with altitude.

geoCHECK ✔ Why is air pressure higher closer to the surface?

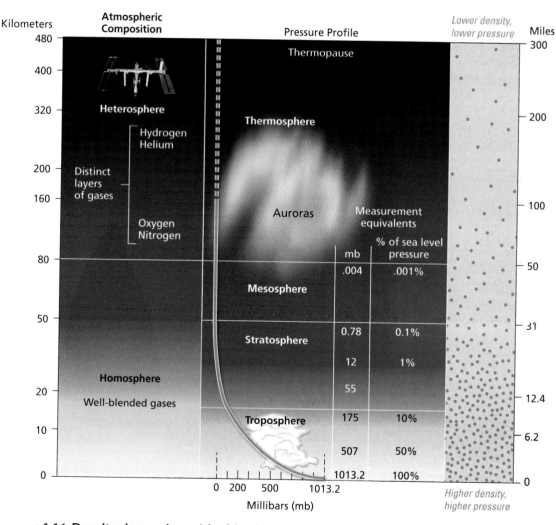

▲**1.16 Density decreasing with altitude** Have you experienced pressure changes that you could feel on your eardrums? How high above sea level were you at the time? The pressure profile plots the decrease in pressure with increasing altitude. Pressure is in millibars and as a percentage of sea level pressure. Note that the troposphere holds about 90% of the atmospheric mass (far-right % column).

Atmospheric Gases

In terms of chemical composition, the atmosphere has two broad regions, the heterosphere (80–480 km altitude) and the homosphere (Earth's surface to 80 km altitude). Gases in the homosphere are well mixed, while gases in the heterosphere are layered by density.

Heterosphere The layer of the atmosphere from about 80 km (50 mi) altitude to the exosphere is the **heterosphere**. Gases in the heterosphere are sorted by gravity into distinct layers based on their density. Oxygen and nitrogen are dominant in the lower heterosphere, while the upper portions are made of hydrogen and helium.

Homosphere The layer of the atmosphere from Earth's surface to an altitude of 80 km (50 mi) is called the **homosphere**. While air pressure decreases rapidly as you go up through the homosphere, the gases are still well mixed. The ozone layer is the region with a higher concentration of ozone, from 19 to 40 km (12 to 24 mi). Figure 1.17 shows the amounts of gases in dry, clean air in the homosphere. These gases are considered stable gases, because the proportion of gases does not vary much. CO_2 is classified as a stable gas, even though it is increasing, because its concentration varies over time periods of years to tens of thousands of years. Additionally, its concentration does not vary much globally. There are also water vapor, pollut-

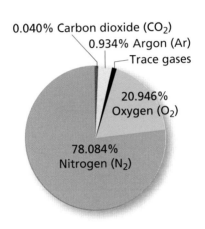

▲**1.17 Stable components of the atmosphere**

- 0.040% Carbon dioxide (CO_2)
- 0.934% Argon (Ar)
- Trace gases
- 20.946% Oxygen (O_2)
- 78.084% Nitrogen (N_2)

ants, and some trace chemicals in the lowest portion of the atmosphere. These are considered variable gases because their amounts vary over short time periods, from hours to weeks. In terms of just the stable gases, the atmosphere is made up of the following:

- 78% relatively inert nitrogen
- 21% **oxygen**. (Oxygen is crucial in life processes on Earth. Although it forms about one-fifth of the atmosphere, oxygen compounds compose about half of Earth's crust.)
- Less than 1% argon, which is inert and unusable in life processes
- 0.04% (400 parts per million [ppm]) **carbon dioxide** (CO_2)

Carbon dioxide is a natural by-product of life processes. However, atmospheric CO_2 has been increasing rapidly, as shown in Figure 1.18. Scientists agree that the increase is **anthropogenic**, or human caused. You will learn about the role of rising CO_2 concentrations in anthropogenic climate change in Chapter 7.

Carbon Dioxide and Climate Change Carbon dioxide is a *greenhouse gas*—that is, a gas that absorbs and emits infrared radiation as heat. While it makes up less than 1% of the gases in the homosphere, it is very important in maintaining Earth's temperature because it delays heat exiting the atmosphere, much like a blanket. Without CO_2 in our atmosphere, global temperatures would be below freezing. Nitrogen, oxygen, and the other 99% of the atmosphere is made up of gases that are not greenhouse gases. Over the past 200 years, the CO_2 percentage increased as a result of human activities, principally the burning of fossil fuels and deforestation. According to ice core records, CO_2 is higher now than at any time in the past 800,000 years. Since 1959, the global concentration of CO_2 in our atmosphere has increased more than 27%, reaching a peak of over 404 ppm in May, 2015 (▼**Fig. 1.18**). Today's CO_2 level far exceeds the natural range of 180–300 ppm over the last 800,000 years. If this rate of increase continues for the rest of the century, then then CO_2 levels could be 1000 ppm CO_2 in 2100.

The rate of increase in CO_2 is accelerating. From 1990 to 1999, CO_2 emissions rose at an average of 1.1% per year. Compare this to the average emissions increase since 2000 of 3.1% per year. A distinct climatic threshold is approaching at 450 ppm, sometime in the decade of the 2020s. Beyond this tipping point, many scientists believe there would be irreversible consequences, discussed in Chapter 7.

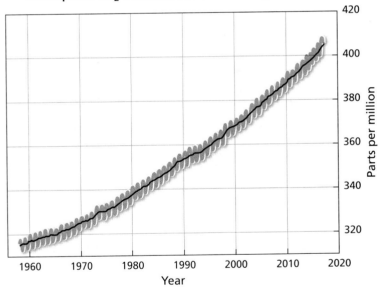

Atmospheric CO_2 at Mauna Loa Observatory, Hawai'i

▲**1.18 Recent changes in carbon dioxide levels** Atmospheric CO_2 concentrations from 1959 to the present.

geo**CHECK** ✔ What are the main gases in the atmosphere?

geo**QUIZ**

1. How high would you have to be to have half the atmosphere beneath you?
2. How do current levels of CO_2 compare to levels over the last 800,000 years?

1.7 Atmospheric Temperature

Key Learning Concepts

▶ **Draw** a diagram showing a profile of atmospheric structure based on its temperature.

Based on temperature, the atmosphere has four distinct temperature zones—the thermosphere, mesosphere, stratosphere, and troposphere (▶Fig. 1.19).

Thermosphere

The **thermosphere** ("heat sphere") roughly corresponds to the heterosphere. The thermosphere is a region of very low air pressure, few molecules, and high temperatures. An oxygen molecule in the thermosphere can travel a kilometer before colliding with another molecule.

Intense solar radiation in this portion of the atmosphere excites individual molecules to high levels of vibration. The temperature profile in Figure 1.19 (yellow curve) shows that temperatures rise sharply in the thermosphere, to 1200°C (2200°F) and higher. Despite such high temperatures, the thermosphere is not "hot" in the way you might expect, because temperature and heat are different concepts. Heat is energy, temperature is a measure of energy. Temperature is a measure of **kinetic energy**, the energy of motion. Heat is the flow of kinetic energy from one body to another, and depends on density. The actual heat in the thermosphere is small because there are so few molecules. Heating increases near Earth's surface because the greater number of molecules in the denser atmosphere transmits their kinetic energy as **sensible heat**, meaning that we can measure and feel it.

geoCHECK ✔ How are temperature and heat different?

Mesosphere

The **mesosphere** is the area from 50 to 80 km (30 to 50 mi) above Earth and is within the homosphere. At high latitudes, an observer at night may see bands of ice crystals glow in rare and unusual *noctilucent clouds*, which are so high in altitude that they still catch sunlight after the Sun has set below the horizon.

geoCHECK ✔ Based on Figure 1.19, what is the temperature range of the mesosphere?

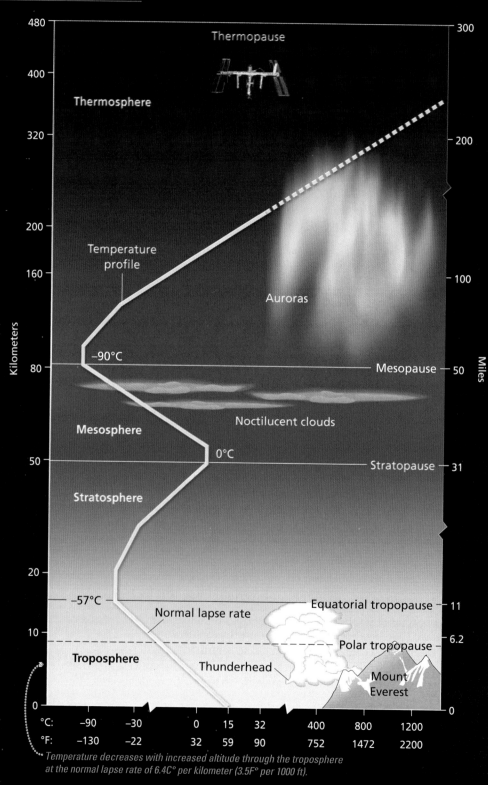

▲1.19 **Temperature profile of the atmosphere**

Temperature decreases with increased altitude through the troposphere at the normal lapse rate of 6.4C° per kilometer (3.5F° per 1000 ft).

Stratosphere

The **stratosphere** extends from 18 to 50 km (11 to 31 mi) from Earth's surface. Temperatures increase with altitude throughout the stratosphere, from –57°C (–70°F) at 18 km at the stratosphere's lower limit, warming to 0°C (32°F) at 50 km at the stratosphere's outer boundary, the **stratopause**. (The suffix *–pause* means "to change.") This warming is caused by ozone converting ultraviolet energy to heat. The ozone layer is the portion of the stratosphere with a higher concentration of ozone. In module 1.8, you will learn how the ozone layer protects living things from ultraviolet radiation, how scientists identified a human-caused threat to the ozone layer, and how nations have worked together to restore the ozone layer.

Figures 1.20 and 1.21 offer two perspectives on the scale of Earth's atmosphere.

> geo**CHECK** ✔ Why do temperatures increase in the stratosphere?

Troposphere

The **troposphere** is the atmospheric layer that supports life and is the region of principal weather activity. Approximately 90% of the total mass of the atmosphere and the bulk of all water vapor, clouds, and air pollution are within the troposphere. An average temperature of –57°C (–70°F) defines the **tropopause**, the troposphere's upper limit, but its exact altitude varies with the season, latitude, and surface temperatures and pressures. Near the equator, because of intense heating from the surface, the tropopause occurs at 18 km (11 mi); while at the North and South Poles it averages only 8 km (5 mi) or less above Earth's surface. The marked warming with increasing altitude in the stratosphere above the tropopause causes the tropopause to act like a lid, generally preventing whatever is in the cooler (denser) air below from mixing into the warmer (less dense) stratosphere. As Figure 1.19 shows, temperatures decrease rapidly

▲ 1.20 **Felix Baumgartner's 2012 parachute jump from 39 km (24 mi) in the stratosphere**

with increasing altitude at an average of 6.4C° per kilometer (3.5F° per 1000 ft), a rate known as the **normal lapse rate**.

The normal lapse rate is an average. The actual lapse rate, called the **environmental lapse rate**, may vary considerably because of local weather conditions.

> geo**CHECK** ✔ How much of the atmosphere's mass is in the troposphere?

> geo**QUIZ**
> 1. Why wouldn't you feel hot in the thermosphere, even with a temperature of 1200°C?
> 2. Why is the tropopause found at different altitudes at different latitudes? Explain.
> 3. Why do temperatures decrease with increased altitude through the troposphere?

▼ 1.21 **The International Space Station (ISS) orbits Earth** The ISS travels within the Thermosphere, ranging between 350 m and 430 km above Earth's surface, with 400 km (249 mi) the ideal altitude.

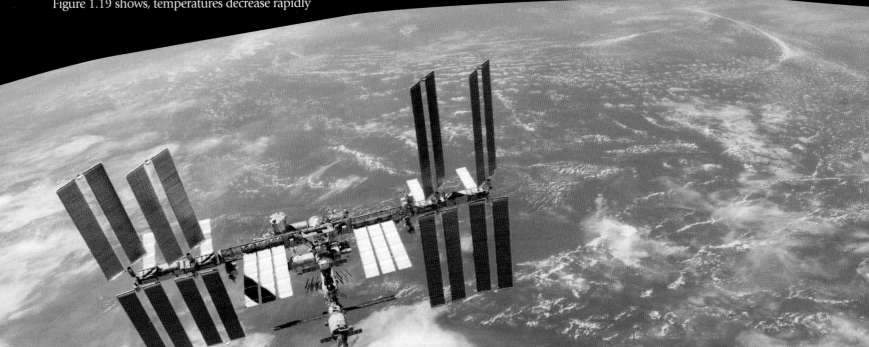

1.8 The Atmosphere's Functional Layers

Key Learning Concepts

▶ **Draw** a diagram showing a profile of atmospheric structure based on function.
▶ **Describe** conditions within the stratosphere.
▶ **Summarize** the function and status of the ozone layer.

The atmosphere has two specific functional zones, the ionosphere and the ozone layer (also called the ozonosphere), which remove most of the harmful wavelengths of incoming solar radiation and charged particles. Figure 1.22 shows where radiation is absorbed by the functional layers of the atmosphere.

Ionosphere

The outer functional layer, the **ionosphere**, extends throughout the thermosphere and the mesosphere. The ionosphere absorbs cosmic rays, gamma rays, X-rays, and shorter wavelengths of ultraviolet radiation, changing atoms to positively charged ions and giving the ionosphere its name. The glowing auroras occur principally within the ionosphere.

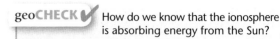
geoCHECK How do we know that the ionosphere is absorbing energy from the Sun?

Ozone layer

That portion of the stratosphere that contains an increased level of ozone is the **ozone layer**. Ozone is a highly reactive oxygen molecule made up of three oxygen atoms (O_3) instead of the usual two atoms (O_2) that make up oxygen gas. Ozone absorbs ultraviolet energy and radiates longer wavelengths of infrared radiation as heat (▼**Fig. 1.23**). This process converts most harmful ultraviolet radiation, effectively "filtering" it and safeguarding life at Earth's surface. At its densest, the ozone layer contains only 1 part ozone per 4 million parts of air. The ozone layer would be only 3 mm thick if it were compressed to surface pressure.

The ozone layer has been relatively stable over the past several hundred million years. Today, however, it is in a state of continuous change because of anthropogenic, or human-caused, pollutants.

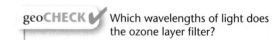
geoCHECK Which wavelengths of light does the ozone layer filter?

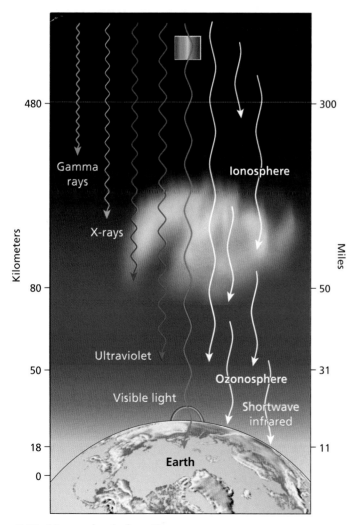
▲1.22 **Atmospheric function**

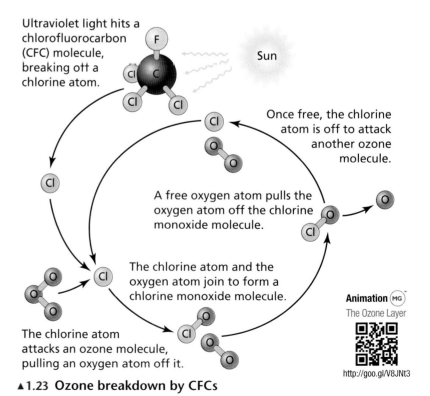

Ultraviolet light hits a chlorofluorocarbon (CFC) molecule, breaking off a chlorine atom.

Once free, the chlorine atom is off to attack another ozone molecule.

A free oxygen atom pulls the oxygen atom off the chlorine monoxide molecule.

The chlorine atom and the oxygen atom join to form a chlorine monoxide molecule.

The chlorine atom attacks an ozone molecule, pulling an oxygen atom off it.

Animation (MG)
The Ozone Layer

http://goo.gl/V8JNt3

▲1.23 **Ozone breakdown by CFCs**

Stratospheric Ozone Losses: A Continuing Health Hazard

The area of stratospheric ozone depletion above Antarctica has grown since 1979, with the record depletion in 2006 covering about 30 million square kilometers (▼Fig. 1.24). Ozone is thinning over the rest of the Southern Hemisphere, as well. In Ushuaia, Argentina, UVB intensity is 225% higher than normal.

Increased ultraviolet radiation is affecting atmospheric chemistry, biological systems, phytoplankton (small photosynthetic organisms that are largely responsible for the ocean's primary food production), fisheries, crop yields, and human skin, eye tissues, and immunity. An international scientific consensus confirmed the anthropogenic disruption of the ozone layer.

Ozone Losses Explained

Because chlorofluorocarbons (CFCs) are stable at Earth's surface and possess remarkable heat properties, they were used in aerosol sprays, as refrigerants, and in the electronics industry. They are inert—they do not dissolve in water and do not break down in biological processes. Tens of millions of tons of CFCs were sold worldwide since 1950 and released into the atmosphere. Chlorine compounds from volcanic eruptions and ocean sprays are water soluble and rarely reach the stratosphere.

In 1974, F. Sherwood Rowland and Mario Molina correctly hypothesized that stable molecules move into the stratosphere, where they are split apart by ultraviolet radiation, freeing chlorine (Cl) atoms. This process breaks up ozone molecules (O_3) and leaves oxygen molecules (O_2) in their place (Fig 1.21). Each chlorine atom can break down more than 100,000 ozone molecules, partially because they can stay there for 40 to 100 years.

Ozone loss is highest over Antarctica because of the combination of thin, icy clouds in the stratosphere and Antarctic wind patterns. Depletion of ozone develops in the Antarctic spring and usually peaks in September. Over the North Pole, conditions are more changeable, so the hole is smaller, although growing each year, with a new record set in 2011.

Video (MG)
The Ozone Hole

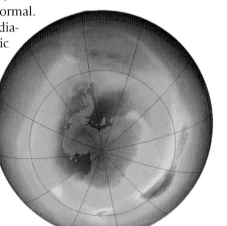

September 1980   September 2011

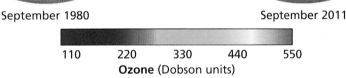

110   220   330   440   550
Ozone (Dobson units)

▲1.24 **The Antarctic ozone hole** Images show the size of the ozone "hole" in 1980 and 2011. Blues and purples show low ozone; greens, yellows, and reds show high ozone.

An International Response As the science around depletion of the ozone layer became established, sales and production of CFCs declined. During this period, however, chemical manufacturers claimed that no evidence existed to prove the ozone-depletion model, and they successfully delayed remedial action from 1974 until 1987. To make matters worse, a March 1981 presidential order permitted the export and sales of banned products. Sales increased and hit a new peak in 1987 at 1.2 million metric tons (1.32 million tons).

Finally, in 1987, an international agreement went into effect halting further sales growth. The *Montreal Protocol on Substances That Deplete the Ozone Layer* aims to reduce and eliminate CFC damage to the ozone layer. With 189 signatory countries, the protocol is regarded as probably the most successful international agreement in history.

Today, with extensive scientific evidence and verification of ozone losses, even the CFC manufacturers admit that the problem of ozone depletion is serious. Thanks to the Montreal Protocol, CFC sales continue their decline, and scientists estimate that the stratosphere will return to more normal conditions in a century if the protocol is enforced. Developing countries have been allowed to delay the end of production and use of CFCs. Recovery is proceeding slower than previously thought, despite the slowing rate of accumulation of offending chemicals.

Without this treaty the ozone layer could have disappeared by 2100, causing several million more cases of skin cancer. For their work, Rowland, Molina, and another colleague, Paul Crutzen, received the 1995 Nobel Prize for Chemistry. The Royal Swedish Academy of Sciences said "the three researchers have contributed to our salvation from a global environmental problem that could have catastrophic consequences." See http://ozonewatch.gsfc.nasa.gov/, and http://www.ec.gc.ca/ozone/ to see how our stratosphere is doing.

  geoCHECK When and where did the ozone hole reach maximum size? Explain.

geoQUIZ

1. How do CFCs break down ozone?
2. Would the ozone layer have repaired itself without the Montreal Protocol? Explain.

1.9 Variable Atmospheric Components

Key Learning Concepts

▶ **Distinguish** between natural and anthropogenic pollutants in the lower atmosphere.

Natural and anthropogenic pollutants in the troposphere are important topics of research in physical geography because they have serious human-health implications. **Pollutants** are gases, particles, and other chemicals in the atmosphere that are harmful to human health or cause environmental damage. Air pollution is not a new problem. Romans complained more than 2000 years ago about the stench of open sewers, smoke from fires, and fumes from kilns and furnaces that converted ores into metals. Air pollution has historically occurred around cities and is closely linked to human production and consumption of energy and resources.

(a)

(b)

▲**1.25 Downwind from a volcano** (a) Ash from the April 2010 eruption of Eyjafjallajökull was blown south by the jet stream, (b) disrupting international air travel.

These connections occur across the globe, as in drought-plagued Australia or the 2015 record wildfires in Canada, Alaska, and California. Each year, worldwide, wildfire acreage breaks the record of the previous year.

Because pollution moves across political borders and oceans, solutions require regional, national, and international strategies. Regulations to curb human-caused air pollution have achieved great success, although much remains to be done. Before we discuss these topics, let's examine some natural pollution sources (▲Fig. 1.25).

geoCHECK What are some of the reasons for increased annual wildfires in the American West?

▲**1.26 Smoke from California wildfires** The main fire in this image is the Rim Fire, which burned more than 257,000 acres (104,000 hectares) from August 17, 2013 to October 24, 2013.

Natural Sources of Air Pollution

Natural air pollution sources, such as wildfires and volcanoes, produce a greater volume of pollutants than do human-made sources. However, any attempt to dismiss the impact of human-made air pollution by comparing it with natural sources is misguided, for we did not evolve with anthropogenic (human-caused) contaminants.

A dramatic natural source of pollution was the 1991 eruption of Mount Pinatubo in the Philippines, perhaps the 20th century's second largest eruption. This event injected nearly 20 million tons of sulfur dioxide (SO_2) into the stratosphere.

Devastating annual wildfires on several continents produce natural air pollution (▶Fig. 1.26). Wildfire smoke contains particulate matter (dust, smoke, soot, ash), nitrogen oxides, carbon monoxide, and volatile organic compounds that darken skies and damage health. Scientists related increasing wildfires in the western United States to climate change: Higher spring and summer temperatures and earlier snowmelt are extending the wildfire season and increasing the intensity of wildfires in the western United States: "large wildfire activity increased suddenly and markedly in the mid-1980s, with higher large-wildfire frequency, longer wildfire durations, and longer wildfire seasons."[*]

[*]A. L. Westerling et al., "Warming and earlier spring increase western U.S. forest wildfire activity," *Science* 313, 5789 (Aug. 18, 2006): 940.

Natural Factors That Affect Air Pollution

Wind, local and regional landscape characteristics, and temperature inversions in the troposphere can increase the problems of air pollution.

Winds Winds transport air pollution from the United States to Canada and Europe. In Europe, the cross-boundary drift of pollution is an issue because of the proximity of countries. Dust from Africa contributes to the soils of South America and Europe, and Texas dust ends up across the Atlantic (▶ Fig. 1.27).

Pollution is carried from industry and fires in the midlatitudes to the Arctic, causing Arctic haze. This was first observed by pilots in the 1950s. Haze is a concentration of microscopic particles and air pollution that diminishes air clarity. Since these high latitudes have few people and lack industry, this haze is the product of pollutants transported to the Arctic.

Local and Regional Landscapes Local and regional landscapes are another important factor in air pollution. Surrounding mountains and hills can trap and concentrate air pollution or direct pollutants from one area to another.

Volcanic landscapes such as Iceland and Hawaii have their own natural pollution. During periods of sustained volcanic activity at Kilauea, some 2000 metric tons (2200 tons) of sulfur dioxide are produced a day, sometimes meriting public health announcements. The resulting acid rain and volcanic smog, called *vog* by Hawaiians (for volcanic smog), cause losses to agriculture as well as other economic impacts.

Temperature Inversions A temperature inversion occurs when the temperature of the atmosphere increases with altitude, rather than decreasing with altitude at the normal lapse rate (▼ Fig. 1.28). Inversions can result from certain weather conditions, such as cool air blowing in off the ocean, or when the air near the ground is radiatively cooled on clear nights, or from topographic situations that produce cold-air drainage into valleys. A normal temperature profile permits warmer and less dense air at the surface to rise, moderating surface pollution. When an inversion occurs, warm air from the surface rises, but it is trapped by the warmer air of the inversion layer.

Atlantic Ocean

Western Sahara

Mauritania

N

100 KILOMETERS   N Cape Verde

▲1.27 **Winds carrying dust in the atmosphere**

Animation (MG)
The Ozone Layer

http://goo.gl/V8JNt3

geoCHECK ✔ Why do we need to look beyond local control of pollution?

geoQUIZ

1. What are some sources of natural air pollution?
2. What are two natural factors that affect air pollution levels?

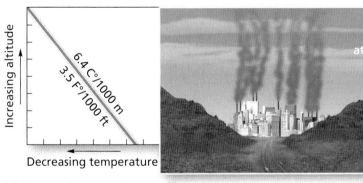

6.4 C°/1000 m
3.5 F°/1000 ft

Increasing altitude

Decreasing temperature

Mixing in the atmosphere

(a) A normal temperature profile.

Increasing altitude

Decreasing temperature

Warmer air

Mixing blocked—pollution trapped beneath inversion

Inversion layer

(b) A temperature inversion in the lower atmosphere prevents the cooler air below the inversion layer from mixing with air above. Pollution is trapped near the ground.

Inversion layer

(c) The top of an inversion layer is visible in the morning hours over a landscape.

▲1.28 **Normal and inverted temperature profiles**

1.10 Anthropogenic Pollution

Key Learning Concepts

▶ **Construct** a simple diagram illustrating the pollution from photochemical reactions in motor vehicle exhaust.
▶ **Describe** the sources and effects of industrial smog, acid deposition, and particulates.
▶ **Discuss** the Clean Air Act's impact on air pollution.

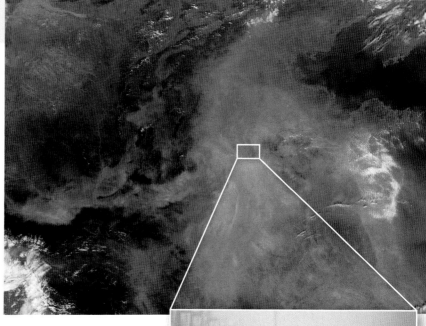

▲1.29 Air pollution in China

Anthropogenic air pollution is more common in urbanized regions and is responsible for roughly 2% of annual deaths in the United States, China, India, as well as Canada, Europe, and Mexico (▶Fig. 1.29). As urban populations grow, human exposure to air pollution increases. In 2012, for the first time, more than 50% of world population lived in metropolitan regions, some one-third with unhealthful levels of air pollution. This is a potentially massive public health issue in this century. For example, asthma rates have nearly doubled since 1980 in the United States due to motor vehicle–related air pollution. Adverse health impacts can result from several types of pollution, including photochemical smog, peroxyacetyl nitrates, industrial smog and sulfur oxides, and particulates.

Photochemical Smog

Photochemical smog is formed by sunlight interacting with chemicals in automobile exhaust (▼Fig. 1.30). Smog is the major component of anthropogenic air pollution and is responsible for the hazy sky and reduced sunlight in many of our cities.

North American urban areas may have from 10 to 100 times higher **nitrogen dioxide (NO_2)** concentrations than nonurban areas. Nitrogen dioxide and water vapor form nitric acid (HNO_3).

▼1.30 Photochemical smog

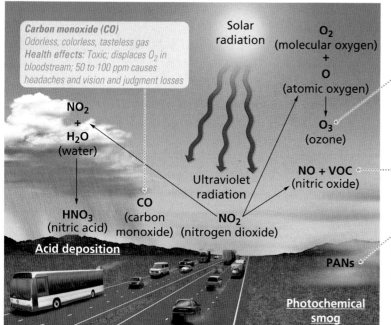

Carbon monoxide (CO)
*Odorless, colorless, tasteless gas
Health effects: Toxic; displaces O_2 in bloodstream; 50 to 100 ppm causes headaches and vision and judgment losses*

Solar radiation

O_2 (molecular oxygen)
+
O (atomic oxygen)

NO_2
+
H_2O
(water)

Ultraviolet radiation

O_3 (ozone)

NO + VOC (nitric oxide)

HNO_3 (nitric acid)

CO (carbon monoxide)

NO_2 (nitrogen dioxide)

Acid deposition

PANs

Photochemical smog

*Ozone (O_3)
Highly reactive, unstable gas
In the environment:
Damages plants
Health effects: Irritates human eyes, nose, and throat*

*Volatile organic compounds (VOCs)
In the environment: Prime agents of surface ozone formation*

*Peroxyacetyl nitrates (PANs)
In the environment: Major damage to plants, forests, crops
Health effects: No human health effects*

Nitric oxide (NO) molecules react with volatile organic compounds (VOCs) such as the hydrocarbons in gasoline. This reaction produces peroxyacetyl nitrates (PAN), which are particularly damaging to plants, including agricultural crops and forests.

Ozone is the primary ingredient in photochemical smog. The same gas that is beneficial to us in the stratosphere damages biological tissues in the lower atmosphere. Children are at greatest risk from ozone pollution—one in four children in U.S. cities is at risk of developing health problems from ozone pollution. For more information and city rankings, see www.lungusa.org/.

geoCHECK ✔ What is the major component of urban air pollution?

Industrial Smog & Sulfur Oxides

Air pollution associated with coal-burning industries is known as industrial smog (▶Fig. 1.31). **Sulfur dioxide (SO_2)** reacts with oxygen and water to make sulfate aerosols and sulfuric acid (H_2SO_4). Coal-burning electric utilities and steel manufacturing are the main sources of sulfur dioxide. Sulfur dioxide–laden air is dangerous to health, corrodes metals, and deteriorates stone building materials. Sulfuric acid and nitric acid deposition have increased since the 1970s.

Acid Deposition Sulfuric and nitric acid deposition is a major environmental issue in some areas of the United States, Canada, Europe, and Asia. Such deposition is most familiar as "acid rain," but it also occurs as "acid snow" and as dust. Normal precipitation is slightly acidic, but acid deposition has been measured at 10 times the acidity of lemon juice! Government estimates of damage from acid deposition in the United States, Canada, and Europe exceed $50 billion annually.

Acid deposition is causally linked to serious environmental problems: declining fish populations and fish kills, widespread forest damage, changes in soil chemistry, and damage to buildings and sculptures. Regions that have suffered most are the northeastern United States, southeastern Canada, Sweden, Norway, Germany, much of eastern Europe, and China. Amendments to the Clean Air Act have resulted in a 40% drop in SO_2 emissions and a 65% drop in acid rain levels since 1976.

geoCHECK ✔ What are the main sources of sulfur dioxide?

Particulates

A mixture of fine particles, both solid and liquid, that impact human health is referred to as **particulate matter (PM)**. Small particles in the air are called aerosols. The smaller the particle, the greater the health risk. These fine particles, such as combustion particles, organics, and metallic aerosols, can get into the lungs and bloodstream. Coarse particle (PM_{10}) particulates smaller than 10 microns (10 μm or less) in diameter, are of less concern, although they can irritate a person's eyes, nose, and throat. Ultrafines, which are $PM_{0.1}$, can get into smaller channels in lung tissue and cause scarring, abnormal thickening, and damage called *fibrosis*.

Sulfur oxides (SO_x, SO_2, SO_3)
Colorless gas with irritating smell produced by combustion of sulfur-containing fuels
In the environment: Leads to acid deposition
Health effects: Impairs breathing, causes human asthma, bronchitis, emphysema

Nitrogen oxides NO_x (NO, NO_2)
Reddish-brown, choking gas given off by agricultural activities, fertilizers, and gasoline-powered vehicles
In the environment: Leads to acid deposition
Health effects: Inflames respiratory system, destroys lung tissue

Particulate matter (PM)
Complex mixture of solids and aerosols, including dust, soot, salt, metals, and organic chemicals
In the environment: Dust, smoke, and haze affect visibility
Health effects: Causes bronchitis, impairs pulmonary function

SO_2 (sulfur dioxide) + O_2 (oxygen) + H_2O (water)

NO_2 ← NO_2 (nitrogen dioxide from combustion)

+ H_2O (water)

Particulates CO_2 SO_2
Industrial smog

H_2SO_4 (sulfuric acid)
Acid deposition

HNO_3 (nitric acid)
Acid deposition

Nitrogen from fertilizers

Particular matter (PM)

▲1.31 **Industrial smog**

geoCHECK ✔ Which is more harmful—fine aerosols or coarse aerosols? Explain.

Effects of the Clean Air Act

The Clean Air Act is a U.S. federal law enacted in 1970 and amended in 1977 and 1990. It established national air quality standards and air quality monitoring, automobile emission and fuel economy standards, power plant emission standards, and ozone protection standards. As a result of the Clean Air Act, our air now contains 98% less lead, 45% less carbon monoxide, 22% less nitrogen oxides, 48% less volatile organic compounds, 75% less PM_{10} particulates, 52% less sulfur oxides.

The health benefits of the Clean Air Act have far outweighed the costs of reducing air pollution. An exhaustive cost–benefit analysis found a 42-to-1 benefit-over-cost ratio! The Environmental Protection Agency (EPA) report calculated that the total direct cost to implement all the Clean Air Act rules from 1970 to 1990 was $523 billion (in 1990 dollars). The estimated direct economic benefits had a central mean of $21.2 trillion, making the estimated net financial benefit of the Clean Air Act $21.7 trillion!

In 2010, the Clean Air Act saved society $110 billion, and cost $27 billion. More than 13 million lost workdays, 3 million lost school days, 160,000 adult deaths, 1,700,000 asthma attacks, 130,000 heart attacks, and 86,000 emergency room visits were avoided in 2010 alone. Despite this, air pollution regulations have been subjected to continued political attacks.

geoCHECK ✔ How has the Clean Air Act improved air quality and saved money?

geoQUIZ

1. How does PAN form from automobile exhaust?
2. How is PM dangerous to human health?

SEASONS/ATMOSPHERE IMPACTS HUMANS

- Solar energy drives Earth systems.
- Seasonal change is the foundation of many human societies; it determines rhythm of life and food resources.
- Earth's atmosphere protects humans by filtering harmful wavelengths of light, such as ultraviolet radiation.
- Natural pollution from wildfires, volcanoes, and wind-blown dust are detrimental to human health.

HUMANS IMPACT SEASONS/ATMOSPHERE

- Climate change affects timing of the seasons. Changing temperature and rainfall patterns mean spring is coming earlier and fall is starting later. Prolonged summer temperatures heat water bodies and affect seasonal ice cover, alter animal migrations, and shift vegetation patterns to higher latitudes.
- Human-made chemicals deplete the ozone layer. Winds concentrate these pollutants over Antarctica, where the ozone hole is largest.
- Anthropogenic air pollution collects over urban areas, reaching dangerous levels in some regions, such as northern India and eastern China; other regions have improved air quality, as in the Los Angeles metropolitan area.

1a As summers get longer in Alaska, moose migrations no longer coincide with the hunting seasons of native people, who depend on the meat. Shifting animal migrations and vegetation patterns will affect ecosystems across the globe.

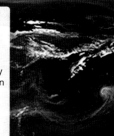

1c London implemented new low emissions standards for diesel vehicles in 2012. Owners must comply or face a daily penalty fee. Stricter regulation is one strategy to control increasing air pollution from the transportation sector.

NASA's 2012 portrait of global aerosols shows dust lifted from the surface in red, sea salt in blue, smoke from fires in green, and sulfate particles from volcanoes and fossil fuel emissions in white.

Clean burning cooking stoves will reduce the amount of fine particulates such as black carbon in developing countries. Several international initiatives are working toward this goal (see http://www.projectsurya.org/).

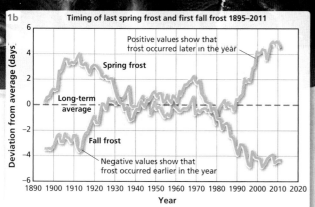

1b Timing of last spring frost and first fall frost 1895–2011

Positive values show that frost occurred later in the year

Spring frost

Long-term average

Fall frost

Negative values show that frost occurred earlier in the year

Deviation from average (days)

Year

Data for the contiguous United States show an overall trend toward a longer growing season, with a longer fall and an earlier spring.

ISSUES FOR THE 21ST CENTURY

- Ongoing climate change is altering Earth systems. Societies will need to adapt their resource base as timing of seasonal patterns changes.
- Human-made emissions must be reduced to improve air quality in Asia. Air pollution will continue to improve in regions where emissions are regulated, such as in Europe, the United States, and Canada.
- Alternative, clean energy sources are vital for reducing industrial pollution worldwide.
- Fuel efficiency, vehicle-emissions regulations, and alternative and public transportation are crucial for reducing urban pollution and CO_2 emissions that drive climate change.

Looking Ahead

In the next chapter, we follow the flow of energy through the lower portions of the atmosphere as insolation makes its way to the surface. We establish the Earth–atmosphere energy balance and examine how surface energy budgets are powered by this arrival of energy. We also begin exploring the outputs of the energy–atmosphere system, focusing on temperature concepts, temperature controls, and global temperature patterns.

What is the origin & structure of our solar system?

1.1 Our Galaxy & Solar System

Distinguish among galaxies, stars, and planets.

Differentiate between the key distances of our solar system and locate Earth.

- Our solar system—the Sun and eight planets—is located on a trailing edge of the Milky Way Galaxy. Gravity, the mutual attracting force exerted by all objects upon all other objects in proportion to their mass, is an organizing force in the universe. Solar systems form when stars (like our Sun) condense from nebular dust and gas, with planets forming in orbits around these central masses.

1. Describe the Sun's status among stars in the Milky Way Galaxy. Describe the Sun's location, size, and relationship to its planets.

2. Diagram in a simple sketch Earth's orbit about the Sun. How much does it vary during the course of a year?

1.2 Solar Energy from Sun to Earth

Describe the Sun's operation, including solar wind.

- The fusion process within the Sun generates incredible quantities of energy. Solar wind is deflected by Earth's magnetosphere, producing various effects in the upper atmosphere, including the northern and southern lights. Solar wind may also influence weather.

3. How does the Sun produce such tremendous quantities of energy?

1.3 Electromagnetic Spectrum of Radiant Energy

Explain the characteristics of the electromagnetic spectrum of radiant energy.

- Radiant energy travels outward from the Sun in all directions, representing a portion of the total electromagnetic spectrum made up of different energy wavelengths. A wavelength is the distance between corresponding points on any two successive waves.

4. What are the main wavelengths produced by the Sun? Which wavelengths does Earth radiate to space?

1.4 Incoming Energy & Net Radiation

Describe the global pattern of net radiation.

- Incoming electromagnetic radiation from the Sun is insolation. The subsolar point is where solar rays are perpendicular to the Earth's surface (radiating from directly overhead). All other locations away from the subsolar point receive slanting rays and more diffuse energy.

- The angle of incoming sunlight is different at each latitude because Earth's surface is curved. This causes an uneven distribution of insolation and heating. Higher latitude regions receive less insolation than the tropics mainly because of the lower angle of the Sun.

5. Study Figure 1.9. How does net radiation in the tropics compare with net radiation at the poles?

Why do we have seasons?

1.5 The Seasons

Define solar altitude, solar declination, and day length.

Describe the annual variability that produces Earth's seasonality.

- The angle between the Sun and the horizon is the Sun's altitude. The Sun's declination is the latitude of the subsolar point

- Earth's seasons are produced by Earth's axial tilt of about 23.5° from a perpendicular to the plane of the ecliptic, which affects the amount of solar radiation different parts of Earth receive at different times of year.

- December 21 or 22 is the winter solstice in the Northern Hemisphere, or December solstice. The subsolar point is at the Tropic of Capricorn (23.5° S).

- March 20 or 21 is the vernal equinox in the Northern Hemisphere, or March equinox. September 22 or 23 is the autumnal, equinox in the Northern Hemisphere or September equinox. On the equinoxes, the subsolar point is over the equator. All locations on Earth experience a 12-hour day and night.

- June 20 or 21 is the summer solstice in the Northern Hemisphere, or June solstice. The subsolar point is at the Tropic of Cancer (23.5° N). North of the Arctic Circle there are 24 hours of daylight, while the area from the Antarctic Circle to the South Pole is in darkness.

6. List the five physical factors that operate together to produce seasons.

7. What are the solstices and equinoxes, and what is the Sun's declination at these times?

What are the properties of our atmosphere?

1.6 Atmospheric Composition

Draw a diagram showing a profile of atmospheric structure based on composition.

- Air is a mix of gases so evenly blended it behaves as if it were a single gas.

- The weight of the atmosphere is air pressure. It decreases rapidly with altitude.

- By *composition*, we divide the atmosphere into the homosphere, from Earth's surface to 80 km, and the heterosphere, from 80 km (50 mi) to 480 km (300 mi). Gases in the homosphere are well mixed, while gases in the heterosphere are sorted by gravity.

- Carbon dioxide levels are higher now than at any other time in the past 800,000 years. Not only is the level of CO_2 increasing, but the rate of increase is also increasing.

8. Name the four most prevalent stable gases in the homosphere. Is the amount of any of these changing at this time?

1.7 Atmospheric Temperature

Draw a diagram showing a profile of atmospheric structure based on temperature.

- Using *temperature* as a criterion, we identify the thermosphere as the outermost layer, corresponding roughly to the heterosphere in location. Its upper limit is at an altitude of approximately 480 km.
- In the homosphere, temperature criteria define the mesosphere, stratosphere, and troposphere.

9. Draw and label a diagram showing the structure of the atmosphere based on its temperature.

1.8 Atmospheric Function

Draw a diagram showing a profile of atmospheric structure based on function.

Describe conditions within the stratosphere.

Summarize the function and status of the ozonosphere, or ozone layer.

- The outermost region by function is the ionosphere, extending through the heterosphere and partway into the homosphere. It absorbs cosmic rays, gamma rays, X-rays, and shorter wavelengths of ultraviolet radiation and converts them into kinetic energy. Within the stratosphere, the ozonosphere, or ozone layer, absorbs ultraviolet radiation, raising the temperature of the stratosphere.
- The reduction of stratospheric ozone, the ozone layer, by CFCs during the past several decades is a hazard for society and many natural systems.

10. What are the two primary functional layers of the atmosphere and what does each do?

11. Why is stratospheric ozone so important? Describe the effects created by increases in ultraviolet light reaching the surface.

1.9 Variable Atmospheric Components

Distinguish between natural and anthropogenic pollutants in the lower atmosphere.

- The spatial aspects of natural and human-caused pollutants are important topics in physical geography and have serious human health implications
- Wildfires and volcanoes are natural sources of air pollution. Wildfires burn larger areas, in part due to climate change.
- Winds, local and regional landscapes, and temperature inversions all can concentrate pollution.

12. Why does a temperature inversion worsen an air pollution episode?

1.10 Anthropogenic Pollution

Construct a simple diagram illustrating the pollution from photochemical reactions in motor vehicle exhaust.

Describe the sources and effects of industrial smog, acid deposition, and particulates.

Discuss the Clean Air Act's impact on air pollution.

- Anthropogenic air pollution is responsible for roughly 2% of annual deaths in the United States, China, India, as well as Canada, Europe, and Mexico.
- Photochemical smog results from the interaction of sunlight and the products of automobile exhaust.
- Human-produced industrial smog over North America, Europe, and Asia is related to coal-burning power plants.

13. Summarize the cost–benefit results from the first 20 years under Clean Air Act regulations.

Critical Thinking

1. If the Sun were hotter, how would that change the curve of energy distribution by wavelength (Fig. 1.6)?

2. Given that Earth is closest to the Sun in January and farthest from the Sun in July, why is July warmer than January in the Northern Hemisphere?

3. How does Earth's axial tilt relate to seasonality? How would a greater axial tilt affect seasonality?

4. How could you evaluate the relative harm of anthropogenic air pollution versus natural air pollution?

5. How is the percentage of people suffering from asthma related to the percentage living in urban areas? What is one likely explanation of this correlation?

Visual Analysis

1. Describe what is happening in this photograph of Salt Lake City, Utah, in winter. What is this phenomenon called?

2. If you could measure changes in air temperature from near the ground up to about the same altitude as the mountaintops, how would the air temperature change? Explain.

▶R1.1

Interactive Mapping | Login to the **MasteringGeography** Study Area to access **MapMaster**.

Insolation

- Open: MapMaster in MasteringGeography
- Select: *Insolation,* then select *Vegetation* from the *Physical Environment* menu.

The Insolation layer shows annual mean insolation in watts per square meter received at the surface, and the Vegetation layer shows world plant regions.

1. What are the insolation values associated with desert shrub vegetation?

2. What are the insolation values associated with tropical forest vegetation?

3. What is the insolation value for where you live? What type of vegetation is found there?

Explore | Use **Google Earth** to explore seasonal changes in the **Sun's declination**.

▲R1.2

Seasonal Declination Changes

Google Earth allows you to model sunlight for any day and time. We can use this to see how the position of the circle of illumination moves during the year. Open Google Earth, zoom out so you can see the entire Earth. Click the Sun button, found between the Sky view button and the Historic imagery button. Move the earth until you can see the dividing line between day and night. Click on the wrench button in the Date and Time Options pop-up window. Click on the month in the End date/time box. Click through the entire year by using the up arrow in the End date/time box. Although the circle of illumination is just an artistic depiction of the actual circle of illumination, we can still use it for some general analysis.

1. Which areas would have 24 hours of darkness on 12/21? (You may have to tilt the Earth so you can see each of the poles.)

2. Which areas would have 24 hours of light on 12/21?

3. The Sun is perpendicular to the circle of illumination. Click through the weeks until it appears that the circle of illumination passes through both poles and the Sun's declination is 0°. What day does this occur on in Google Earth? What seasonal event is this? When is the Sun directly over the equator?

4. Continue to click through the year until the circle of illumination is illuminating the largest area north of the Arctic Circle. You might want to tilt Earth so you can see all of the Arctic Circle. On what day does it appear that there is the most illumination north of the Arctic Circle? What seasonal event is this?

MasteringGeography™

Looking for additional review and test prep materials? Visit the Study Area in MasteringGeography™ to enhance your geographic literacy, spatial reasoning skills, and understanding of this chapter's content by accessing a variety of resources, including MapMaster™ interactive maps, videos, *Mobile Field Trips, Project Condor* Quadcopter videos, *In the News* RSS feeds, flashcards, web links, self-study quizzes, and an eText version of *Geosystems Core.*

Key Terms

air (p. 16)
air pressure (p. 16)
altitude (p. 12)
Antarctic Circle (p. 14)
Arctic Circle (p. 14)
auroras (p. 7)
autumnal equinox (p. 27)
axial parallelism (p. 13)
axial tilt (p. 13)
axis (p. 13)
circle of illumination (p. 13)
day length (p. 12)

December solstice (p. 14)
declination (p. 12)
electromagnetic spectrum (p. 8)
environmental lapse rate (p. 19)
exosphere (p. 16)
fusion (p. 6)
gravity (p. 4)
heterosphere (p. 17)
homosphere (p. 17)
industrial smog (p. 24)
insolation (p. 10)
ionosphere (p. 20)

June solstice (p. 14)
kinetic energy (p. 18)
magnetosphere (p. 7)
March equinox (p. 14)
mesosphere (p. 18)
Milky Way Galaxy (p. 4)
nitrogen dioxide (NO_2) (p. 24)
normal lapse rate (p. 19)
oxygen (p. 17)
ozone layer (p. 20)
particulate matter (PM) (p. 25)

photochemical smog (p. 24)
plane of the ecliptic (p. 13)
pollutants (p. 22)
revolution (p. 13)
rotation (p. 13)
sensible heat (p. 18)
September equinox (p. 14)
solar constant (p. 10)
solar wind (p. 6)
speed of light (p. 5)
stratosphere (p. 19)
subsolar point (p. 10)

sulfur dioxide (SO_2) (p. 24)
summer solstice (p. 14)
sunspots (p. 6)
temperature inversion (p. 23)
thermosphere (p. 18)
Tropic of Cancer (p. 12)
Tropic of Capricorn (p. 14)
tropopause (p. 19)
troposphere (p. 19)
wavelength (p. 8)
winter solstice (p. 14)
vernal equinox (p. 27)

Seasonal Changes: What is the role of latitude in changing insolation & day length?

GeoLab1 (MG)
Pre-Lab Video

https://goo.gl/pHJn0k

Over the course of a year, the area where we live experiences seasonal changes in the amount of insolation it receives, the angle of the Sun, and the length of daylight. Seasonal changes in insolation for a particular location are largely driven by the angle of the Sun above the ground, and the length of daylight--both of which depend on the latitude of that location.

▲ GL1.1 **Spain's Gemasolar array**

Apply

In this lab, you will be working as an analyst for an electrical power company that is planning to build a number of new solar power generation arrays at different locations. To help assess the energy potential of each location, you will use charts and graphs to determine insolation amounts and the Sun's altitude, estimate day length during the year, and track the path of the Sun. This information has applications in areas ranging from agriculture to architecture and urban design.

Objectives

- Analyze seasonal changes in insolation
- Estimate day length
- Calculate hourly insolation values

Procedure I: Tracking Daily Insolation

Figure GL 1.1 shows daily insolation as a function of time of year and latitude. Refer to the chart to answer the questions below.

1. How much insolation is received at 0° N on the summer solstice? On the winter solstice? How much does insolation vary from summer to winter at this location?

2. How much insolation is received at 40° N on the summer solstice? On the winter solstice? How much does insolation vary from summer to winter at this location?

3. How much insolation is received at 60° N on the summer solstice? On the winter solstice? How much does insolation vary from summer to winter at this location?

4. How much insolation is received at 90° N on the summer solstice? On the winter solstice? How much does insolation vary from summer to winter at this location?

5. What is your latitude? How much insolation is received at that location on the summer solstice? On the winter solstice? How much does insolation vary from summer to winter at your location?

6. How does the difference between summer and winter insolation vary with latitude? Which latitudes have the most consistent insolation?

Log in MasteringGeography™ to complete the online portion of this lab, view the Pre-Lab Video, and complete the Post-Lab Quiz.
www.masteringgeography.com

Procedure II: Tracking Seasonal Variations

Table GL.1.1 shows the hours of daylight at different latitudes in the Northern Hemisphere.

7. Calculate day length for 30° N, 40° N, 50° N, 60° N, and your latitude. Enter the values in Table GL1.1.

8. What is the average hourly insolation at the Equator on the summer solstice? To find this value, divide the daily insolation value by the number of hours in the day.

9. What is the average hourly insolation at 40° N on the summer solstice? On the winter solstice?

10. What is the average hourly insolation at 90° N on the summer solstice?

11. What is the average hourly insolation at your latitude on the summer solstice? On the winter solstice?

12. Which of these locations receives the most intense (highest watts per hour per square meter) insolation? How many watts per hour per square meter does this location receive, and when does this occur?

Analyze & Conclude

13. Figure GL.1.1 shows changes in insolation related to the angle of the Sun and day length. However, the amount of insolation over the poles varies. How does the maximum insolation at the North Pole's summer solstice compare to the maximum insolation at the South Pole's summer solstice? What other Earth–Sun relationship could explain this difference?

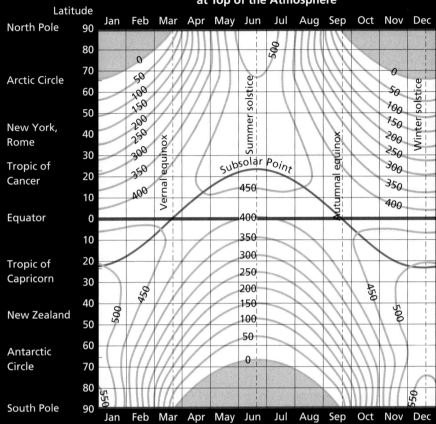

Daily Receipt of Insolation (W/m²) at Top of the Atmosphere

▲GL1.2 **Daily insolation received at the top of the atmosphere.**
The total daily insolation received at the top of the atmosphere is charted in watts per square meter per day by latitude and month. (1 W/m²/day = 2.064 cal/cm²/day).

14. Do you think that latitude or day length is more important in insolation? Compare the North Pole with 40° N and with the equator in your analysis and explain your answer.

15. Your company plans to build three solar power arrays, at 30° N, 40° N, and 50° N. What is one main way in which the design of each array would need to differ, assuming each is intended to generate the same amount of power? Explain your answer.

16. Your company is considering building solar power arrays near the Arctic Circle in Alaska and Canada. What are the advantages and disadvantages of these locations for solar power?

Table GL 1.1 Day length—the time between sunrise and sunset—at selected latitudes for the Northern Hemisphere.

	Winter Solstice (December Solstice) December 21–22			Vernal Equinox (March Equinox) March 20–21			Summer Solstice (June Solstice) June 20–21			Autumnal Equinox (September Equinox) September 22–23		
	A.M.	P.M.	Day length	A.M.	P.M.	Day length	A.M.	P.M.	Day length	A.M.	P.M.	Day length
0°	6:00	6:00	12	6:00	6:00	12	6:00	6:00	12	6:00	6:00	12
30°	6:58	5:02		6:00	6:00	12	5:02	6:58		6:00	6:00	12
40°	7:26	4:34		6:00	6:00	12	4:34	7:26		6:00	6:00	12
50°	8:05	3:55		6:00	6:00	12	3:55	8:05		6:00	6:00	12
60°	9:15	2:45		6:00	6:00	12	2:45	9:15		6:00	6:00	12
90°	No sunlight			Rising Sun			Continuous sunlight			Setting Sun		

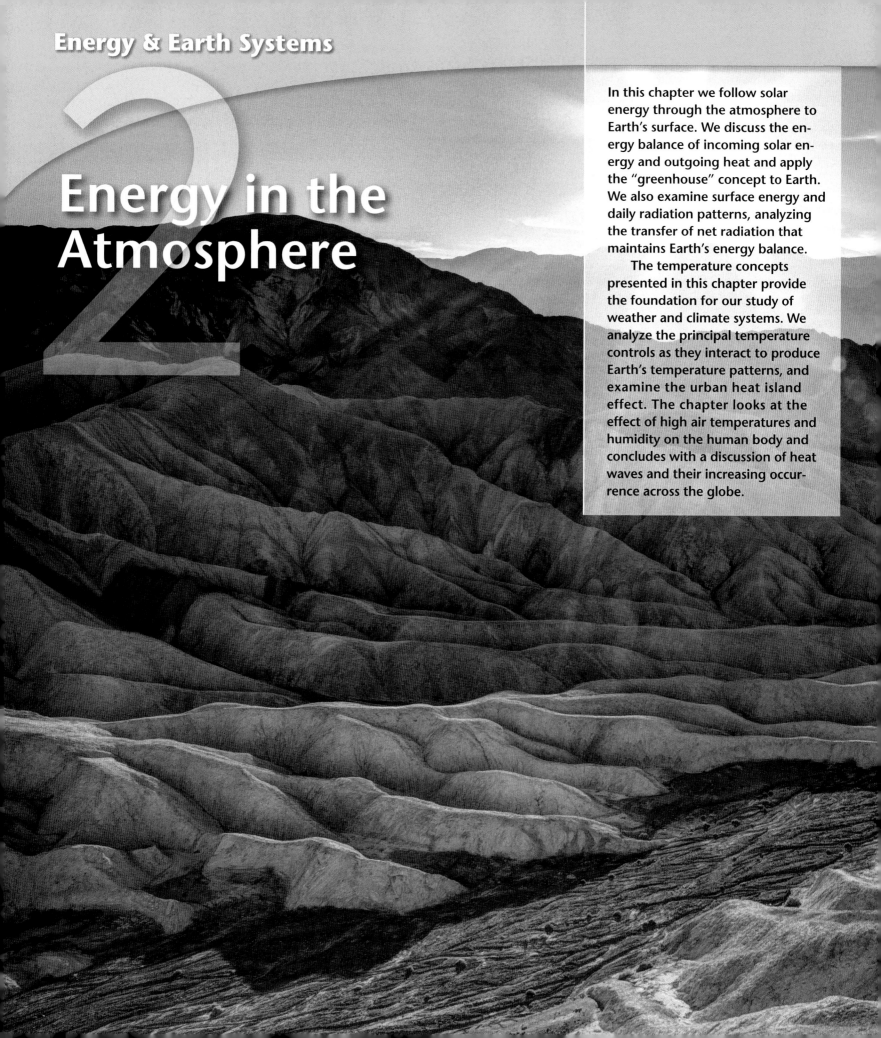

Energy in the Atmosphere

In this chapter we follow solar energy through the atmosphere to Earth's surface. We discuss the energy balance of incoming solar energy and outgoing heat and apply the "greenhouse" concept to Earth. We also examine surface energy and daily radiation patterns, analyzing the transfer of net radiation that maintains Earth's energy balance.

The temperature concepts presented in this chapter provide the foundation for our study of weather and climate systems. We analyze the principal temperature controls as they interact to produce Earth's temperature patterns, and examine the urban heat island effect. The chapter looks at the effect of high air temperatures and humidity on the human body and concludes with a discussion of heat waves and their increasing occurrence across the globe.

Key Concepts & Topics

A hiker approaches Zabriskie Point in Death Valley, California. The highest air temperature ever recorded occurred in Death Valley on July 10, 1913, when the temperature reached 56.7°C (134°F).

2.1 Energy Balance Essentials

Key Learning Concepts

▶ **Define** energy and heat.

▶ **Explain** four types of heat transfer: radiation, conduction, convection, and advection.

Animation (MG)
Global Warning,
Climate Change

http://goo.gl/cTHCHK

E nergy from the Sun enters Earth's atmosphere, warms Earth's surface, evaporates water, creates weather, and travels back out into space. Figure 2.1 is a simplified diagram of this process. Our budget of atmospheric energy includes shortwave radiation inputs (ultraviolet light, visible light, and near-infrared wavelengths) that are converted into longwave radiation outputs (thermal infrared).

Energy & Heat

Animation (MG)
Earth–
Atmosphere
Energy Balance

http://goo.gl/7UYgTM

Energy is the capacity to do *work*, or to move material. *Kinetic energy* is the energy of motion, produced by the vibrations of molecules that we measure as *temperature*. *Potential energy* is stored energy that has the capacity to do work under the right conditions. Gasoline has potential energy that is released when it is burned in a car's engine. The water in the

reservoir above a hydropower dam has potential energy that is released when gravity pulls the water through the turbines and into the river downstream. In both of these examples, potential energy is converted into kinetic energy.

Types of Heat Heat is the flow of kinetic energy between molecules because of a temperature difference between them. Heat always flows from an area of higher temperature to an area of lower temperature. An example is the transfer of heat when you place your hand on a cold desk. Heat flows from your hand into the desk, warming it.

Two types of heat energy are important for understanding Earth–atmosphere energy budgets. **Sensible heat** can be "sensed" by humans as temperature because it comes from the kinetic energy of molecular motion. **Latent heat** ("hidden" heat) is the energy gained or lost when a substance changes from one state to another, such as from water vapor to liquid water (gas to liquid) or from water to ice (liquid to solid). In latent heat transfer, as long as a change in state is taking place, the substance itself does not change temperature.

 **geoCHECK** ✔ What is heat?

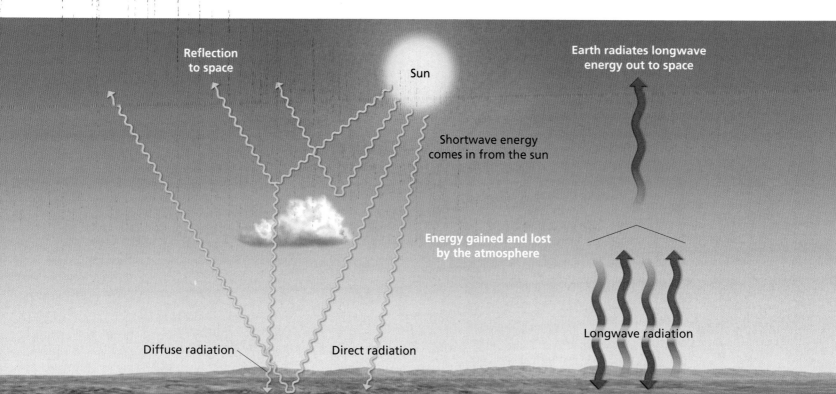

▲2.1 Simplified Earth–atmosphere energy system

Methods of Heat Transfer

Heat energy can be transferred in a number of ways throughout Earth's atmosphere, land, and water bodies. *Radiation* is the transfer of heat in electromagnetic waves, such as radiation that travels from the Sun to Earth, or as from a fire or a burner on the stove (▶ Fig. 2.2). Waves of radiation do not need to travel through a medium, such as air or water, in order to transfer heat.

Conduction and Convection Conduction is the molecule-to-molecule transfer of heat energy as it flows from areas of higher temperature to those of lower temperature. As molecules warm, the rate of their vibration increases, causing collisions that produce motion in neighboring molecules, transferring heat from warmer to cooler materials.

Gases and liquids also transfer energy by **convection**, which is movement caused by differences in temperature and density within a fluid. In the atmosphere or in bodies of water, warmer (less dense) masses tend to rise and cooler (more dense) masses tend to sink, establishing a pattern of flow called a convection current. *Advection* refers to mainly horizontal movement, as opposed to the mainly vertical movement of convection.

You have probably experienced these energy flows in your kitchen: Heat is conducted through the handle of a pan, and boiling water bubbles in convective motions.

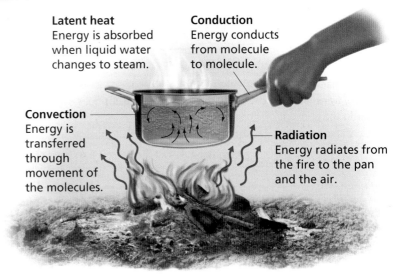

Latent heat
Energy is absorbed when liquid water changes to steam.

Conduction
Energy conducts from molecule to molecule.

Convection
Energy is transferred through movement of the molecules.

Radiation
Energy radiates from the fire to the pan and the air.

▲ 2.2 Heat transfer mechanisms

geoCHECK ✔ How is energy transferred through space from the Sun to Earth?

Also in the kitchen, you may use a convection oven that has a fan to circulate heated air to uniformly cook food.

Shortwave & Longwave Radiation

The flow of energy into Earth's atmosphere is mainly shortwave energy, while the flow of energy from Earth out to space is mainly longwave energy. Figure 2.3 shows global monthly values for energy flow in watts per square meter (W/m^2) of both reflected and emitted radiation. In Figure 2.3a, lighter regions indicate where more sunlight is reflected into space, for example, by lighter-colored land surfaces such as deserts or by cloud cover such as that over tropical lands. Green and blue areas illustrate where less light is reflected because more shortwave energy is absorbed at the surface.

In Figure 2.3b, orange and red regions show where more longwave radiation was absorbed and then emitted to space. Less longwave energy is escaping in the blue and purple regions. Those blue regions of lower longwave emissions over tropical lands are due to tall, thick clouds along the equatorial region (the Amazon, equatorial Africa, and Indonesia). These clouds also cause higher shortwave reflection. Subtropical desert regions exhibit greater longwave radiation emissions, owing to the presence of little cloud cover and greater radiative energy losses from surfaces that absorb a lot of energy (▲ Fig 2.4).

▲ 2.4 Cloud-free skies over desert dunes in Morocco.

geoCHECK ✔ Where are the highest values of outgoing longwave radiation?

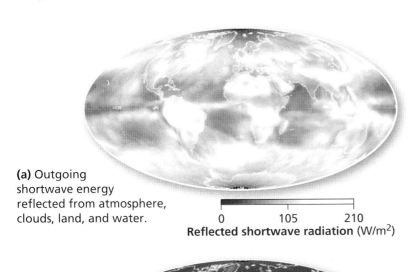

(a) Outgoing shortwave energy reflected from atmosphere, clouds, land, and water.

| 0 | 105 | 210 |

Reflected shortwave radiation (W/m^2)

(b) Longwave energy emitted by land, water, atmosphere, and cloud surfaces back to space.

| 100 | 210 | 320 |

Outgoing longwave radiation (W/m^2)

▲2.3 Shortwave and longwave radiation

geoQUIZ

1. Compare and contrast kinetic and potential energy.
2. How does latent heat differ from sensible heat?
3. Describe how energy is transferred by conduction and convection.

2.2 Insolation Input & Albedo

Key Learning Concepts

▶ *Identify* pathways for solar energy through the troposphere to Earth's surface: transmission, scattering, refraction, and absorption.

▶ *Explain* the concept of albedo (reflectivity).

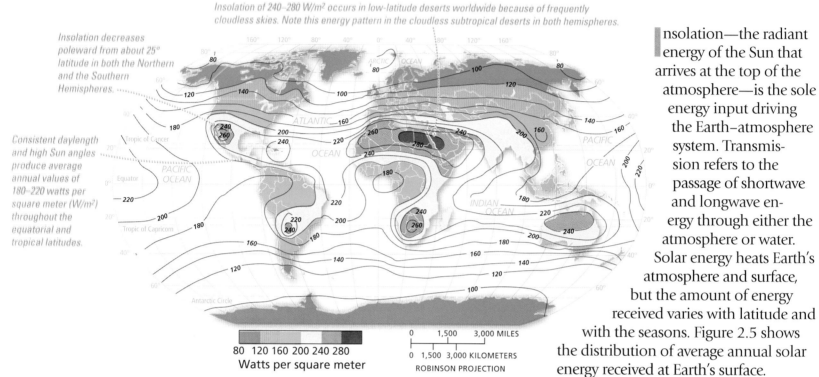

Insolation of 240–280 W/m² occurs in low-latitude deserts worldwide because of frequently cloudless skies. Note this energy pattern in the cloudless subtropical deserts in both hemispheres.

Insolation decreases poleward from about 25° latitude in both the Northern and the Southern Hemispheres.

Consistent daylength and high Sun angles produce average annual values of 180–220 watts per square meter (W/m²) throughout the equatorial and tropical latitudes.

80 120 160 200 240 280
Watts per square meter

0   1,500   3,000 MILES
0 1,500 3,000 KILOMETERS
ROBINSON PROJECTION

▲ 2.5 **Distribution of insolation**

nsolation—the radiant energy of the Sun that arrives at the top of the atmosphere—is the sole energy input driving the Earth–atmosphere system. Transmission refers to the passage of shortwave and longwave energy through either the atmosphere or water. Solar energy heats Earth's atmosphere and surface, but the amount of energy received varies with latitude and with the seasons. Figure 2.5 shows the distribution of average annual solar energy received at Earth's surface.

▼ 2.6 **Rayleigh scattering makes the sky blue**

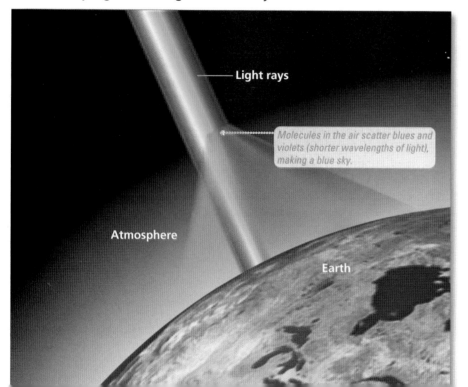

Light rays

Molecules in the air scatter blues and violets (shorter wavelengths of light), making a blue sky.

Atmosphere

Earth

Scattering (Diffuse Radiation)

When atmospheric gases, dust, cloud droplets, water vapor, and pollutants change the direction of light's movement without altering its wavelengths, **scattering** occurs. For example, the blue color of the sky and the red colors at sunrise and sunset are caused by *Rayleigh scattering* (◀ **Fig. 2.6**). In this process, small gas molecules in the air scatter shorter wavelengths of light—the blues and violets—making a blue sky. A sky filled with smog and haze appears almost white because the larger particles from air pollution scatter all wavelengths of visible light.

The position of the Sun in the sky affects the degree to which light is scattered. When the Sun is overhead, rays travel through a shorter distance through the atmosphere than when the Sun is closer to the horizon. Light at sunrise and sunset undergoes more scattering of shorter wavelengths, leaving oranges and reds to reach the observer at sunset or sunrise.

Some incoming insolation is diffused by clouds and atmosphere and transmits to Earth as diffuse radiation, the downward component of scattered light.

geoCHECK ✔ Which latitudes receive the highest insolation?

Refraction

As insolation enters the atmosphere, it passes from virtually empty space into atmospheric gases. The transition

▲2.7 **Desert mirage**

from one medium (a near vacuum) to another (air) causes the insolation to change speed and direction in a process called **refraction**. Refraction creates mirages and twinkling stars when light travels through layers of air with different temperatures (▶ Fig. 2.7). Refraction also occurs when insolation passes from air into water. A rainbow is created when visible light passes through many raindrops and is refracted and reflected toward the observer.

An interesting function of refraction is that it adds approximately 8 minutes of daylight because the atmosphere bends the rays of light. We see the Sun's image about 4 minutes before the Sun actually peeks over the horizon. Similarly, as the Sun sets, its refracted image is visible from over the horizon for about 4 minutes afterward.

geoCHECK ✔ Why is the sky blue?

Reflection & Albedo

A portion of arriving energy bounces directly back into space without being absorbed: This is **reflection**. It is an important control over the amount of insolation that is available for absorption by a surface. **Albedo** refers to the percentage of insolation that is reflected (0% is total absorption; 100% is total reflection). In the visible wavelengths, darker colors have lower albedos, and lighter colors have higher albedos. For example, brilliant white sand dunes (▼Fig. 2.8a) have a much higher albedo than a rainforest canopy (▼Fig. 2.8b).

Earth and its atmosphere reflect 31% of all insolation when averaged over a year. The full Moon has only a 6%–8% albedo value. Earthshine is four times brighter than moonlight. It is not surprising that astronauts report that our planet looks startlingly beautiful from space.

geoCHECK ✔ What does it mean to say that a surface has a low albedo?

Absorption

The assimilation of radiation by molecules of matter and its conversion from one form of energy to another is **absorption**. Insolation that is not part of the 31% reflected from Earth's surface and atmosphere is absorbed and converted into either infrared radiation or chemical energy by plants in photosynthesis.

Land and water surfaces absorb about 45% of incoming insolation, and atmospheric gases, dust, clouds, and stratospheric ozone absorb about 24% of incoming insolation. We will look at these numbers in greater detail in Figure 2.13.

geoCHECK ✔ How does Earth's albedo compare with the Moon's?

geoQUIZ

1. Compare insolation at the equator with insolation at the tropics. Which latitude receives greater amounts? What accounts for the difference?
2. How is scattering different from albedo?
3. What are three things that can happen to light as it passes through the atmosphere?

(a) High albedo surface, White Sands, New Mexico.

▲2.8 **High and low albedo surfaces**

(b) Lower albedo surface, rainforest canopy in Indonesia.

2.3 The Greenhouse Effect, Clouds, & Atmospheric Warming

As you have learned, clouds are an important factor in the energy balance in Earth's atmosphere. The composition of the atmosphere—especially greenhouse gases—also affects the way the atmosphere absorbs and emits radiant energy.

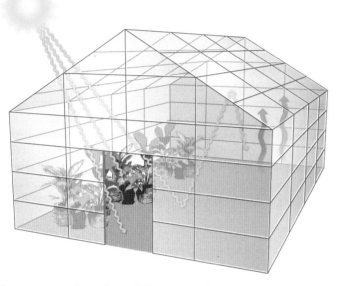

(a) Transparent glass allows light to pass through to the inside of the greenhouse, where it is absorbed by soil, plants, and other materials. The glass acts as a one-way filter by trapping the longwave heat energy.

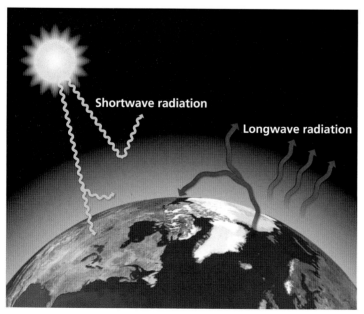

(b) Our atmosphere does not trap longwave radiation, but rather absorbs and then reradiates heat to Earth's surface and out to space. Today's increasing amount of carbon dioxide is absorbing more heat, warming the lower atmosphere.

The Greenhouse Effect & Atmospheric Warming

Although Earth's temperature changes from place to place throughout the year, the planet's average temperature, while increasing, tends to remain relatively stable over time. Part of the reason for this is that Earth's atmosphere moderates temperatures. Earth emits longwave energy from its surface and atmosphere out to space. However, some of this longwave radiation is absorbed by gases in the lower atmosphere and then emitted back toward Earth. Some of these important gases are carbon dioxide, water vapor, methane, nitrous oxide, and chlorofluorocarbons (CFCs): These are the **greenhouse gases**. The absorption and emission of energy by greenhouse gases is an important factor in warming the troposphere. The rough similarity between this process and the way a greenhouse operates gives the process its name—the **greenhouse effect** (◀Fig. 2.9).

geoCHECK ✔ Which wavelengths of energy are absorbed by greenhouse gases in the atmosphere?

Clouds & Earth's "Greenhouse"

Clouds come in many different shapes and sizes. They also affect energy in the atmosphere in complex ways (▶Fig. 2.10). Clouds can reduce insolation by 75%, which cools Earth's surface. They can also act as insulation by absorbing and radiating longwave radiation from Earth, warming the surface.

Effects of Different Cloud Types Clouds affect the heating of the lower atmosphere differently, depending on cloud type. The percentage of cloud cover is important, and the cloud type, height, and thickness (water content and density) are also important. High-altitude, ice-crystal clouds reflect insolation with albedos of about 50%, whereas thick, lower cloud cover reflects about 90% of incoming insolation.

◀2.9 The greenhouse effect

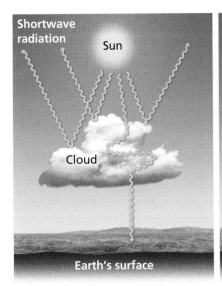
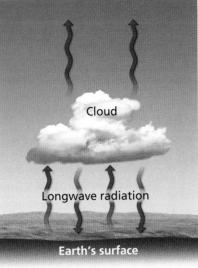

(a) Clouds reflect and scatter shortwave radiation, returning a high percentage to space.

(b) Clouds absorb and reradiate longwave radiation emitted by Earth; some longwave energy is lost to space and some moves back toward Earth's surface.

▲2.10 **The effect of clouds on shortwave and longwave radiation**

To understand the actual effects of clouds on the atmosphere's energy budget, we must consider the transmission of shortwave and longwave radiation and cloud type (▲Fig. 2.10).

Effects of Jet Contrails Jet contrails (condensation trails) produce high cirrus clouds stimulated by aircraft exhaust—sometimes called false cirrus clouds (▼Fig. 2.11). High-altitude ice-crystal clouds reduce outgoing longwave radiation and therefore contribute to global heating. However, these contrail-seeded false cirrus clouds reduce the amount of insolation reaching the surface because they are more dense and have a higher albedo than normal cirrus clouds.

The 3-day grounding of commercial air traffic following the September 11, 2001, terrorist attacks on the World Trade Center helped scientists study contrail effects on daily temperature range—the difference between the daytime maximum and the nighttime minimum temperatures. Scientists compared weather data from 4000 stations for the 3-day shutdown with data from the past 30 years. Their research suggests that contrails reduce the daily temperature range in regions with a high density of aircraft.

geoCHECK ✔ How do cirrus clouds affect surface temperatures, compared with thicker, lower clouds?

Aerosols & Global Temperatures

The 1991 eruption of Mount Pinatubo in the Philippines injected 15–20 million tons of sulfur dioxide droplets into the stratosphere. Winds then spread the aerosols worldwide. The increase in atmospheric albedo produced a temporary average cooling of 0.5°C (0.9°F).

Industrialization is producing a haze of pollution, including sulfate aerosols, that is changing the reflectivity of the atmosphere. Pollution causes both an atmospheric warming through absorption by pollutants such as black carbon particles produced by burning forests and diesel fuel and a surface cooling by sulfate aerosols reducing insolation reaching the surface. Black carbon produced by ships in the Arctic is of special concern. When black carbon, or soot, falls to the ground, it dramatically reduces the albedo of snow and ice and accelerates melting. The decline in insolation reaching Earth's surface because of pollution is called **global dimming**. One recent study estimates that aerosols reduced surface insolation by 20% during the first decade of this century.

geoCHECK ✔ How would reducing air pollution affect surface temperatures?

geoQUIZ

1. How does Earth's atmosphere act like a greenhouse? How does it differ from a greenhouse?
2. How do clouds both warm and cool Earth's surface?
3. Describe how more ships in the Arctic could lead to more melting of sea ice.

(a) Low, thick clouds reflect most shortwave insolation to space, leading to atmospheric cooling.

(b) High, thin clouds transmit most shortwave insolation to the surface, leading to atmospheric warming.

▲2.11 **Energy effects of cloud types**

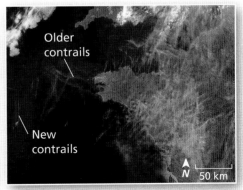

(c) Contrails over Brittany, France, in 2004. Newer contrails are thin; older contrails have widened and formed thin, high cirrus clouds, with an overall warming effect on Earth.

2.4 Earth–Atmosphere Energy Balance

Key Learning Concepts

▶ ***Describe*** the Earth–atmosphere energy balance and the patterns of global net radiation.

Earth's energy balance has inputs from the Sun and outputs to space. Because of increasing levels of greenhouse gases, Earth currently returns slightly less energy to space than it receives from the Sun. This imbalance of approximately 0.6 watts per square meter seems small, but it is roughly equal to the energy of detonating four small nuclear bombs per second.

Distribution of Energy by Latitude

Figure 2.12 shows how the Earth–atmosphere energy balance varies by latitude. Between the tropics, the angle of incoming insolation is high and day length is consistent, with little seasonal variation, so more energy is gained than lost: Energy surpluses dominate. In the polar regions, the Sun is low in the sky, surfaces are light in color (ice and snow) and reflective, and for up to 6 months during the year no insolation is received, so more energy is lost than gained: Energy deficits prevail. At around 36° latitude, a balance exists between energy gains and losses for the Earth–atmosphere system.

The imbalance of net radiation from the tropical surpluses to the polar deficits drives global circulation patterns because the huge amounts of heat energy available in the tropics tend to flow toward cooler areas. Winds, ocean currents, and weather systems move this energy from the tropics to the poles. Dramatic examples of such energy and mass transfers are tropical cyclones—hurricanes and typhoons. Forming in the tropics, these powerful storms mature and migrate to higher latitudes, carrying with them enormous quantities of energy, water, and water vapor.

geoCHECK ✔ Explain the cause of the transport of heat energy from the tropics toward the pole.

Inputs & Outputs

Figure 2.13 summarizes the Earth–atmosphere radiation balance. It brings together all the elements discussed to this point in the chapter by following 100% of arriving insolation through the troposphere. The shortwave portion of the budget is on the left in the illustration; the longwave part of the budget is on the right.

Out of 100% of the solar energy arriving:

- Earth's average albedo accounts for the 31% of insolation that is reflected.
- Absorption by atmospheric clouds, dust, and gases involves another 21% and accounts for the atmospheric heat input.
- Stratospheric ozone absorption and radiation account for another 3% of the atmospheric budget.
- About 45% of the incoming insolation transmits through to Earth's surface as direct and diffuse short-wave radiation that is absorbed and heats the surface, where it is stored in the ground as heat energy.

Earth's atmosphere and surface eventually emit into space the longwave radiation part of the budget: 21% (atmosphere heating) + 45% (surface heating) + 3% (ozone emission) = 69%.

geoCHECK ✔ At what latitudes can you find low levels of incoming energy? Where can you find high levels of incoming energy?

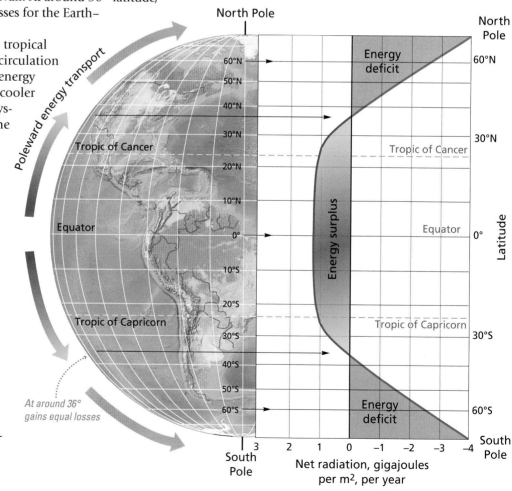

▲2.12 Energy budget by latitude

Incoming Shortwave Radiation
Solar energy cascades through the lower atmosphere where it is absorbed, reflected, and scattered. Clouds, atmosphere, and the surface reflect 31% of this isolation back to space.

Outgoing Longwave Radiation and Albedo
Over time, Earth emits, on average, 69% of incoming energy to space. When added to the amount of energy reflected (31%), this equals the total energy input from the Sun (100%).

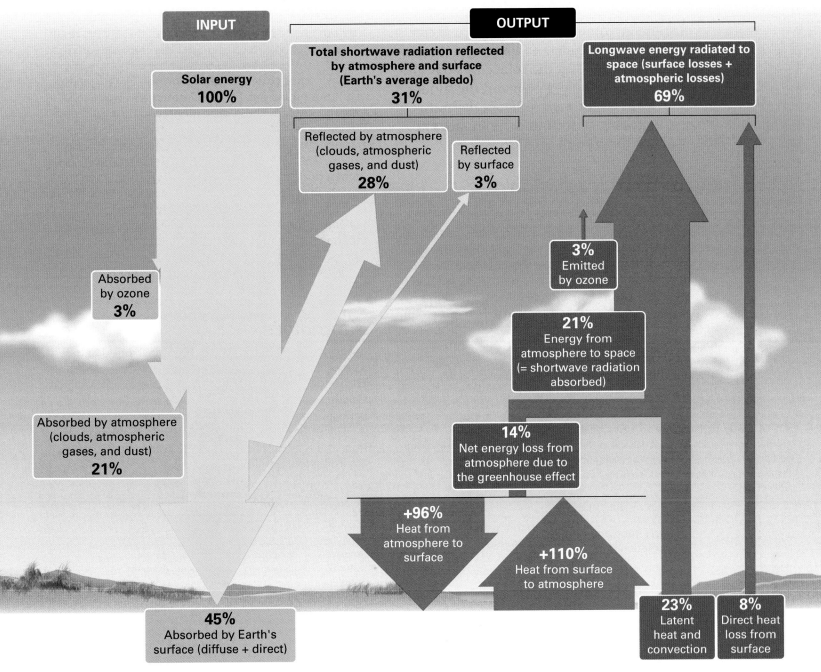

INPUT

OUTPUT

Solar energy
100%

Total shortwave radiation reflected by atmosphere and surface (Earth's average albedo)
31%

Longwave energy radiated to space (surface losses + atmospheric losses)
69%

Reflected by atmosphere (clouds, atmospheric gases, and dust)
28%

Reflected by surface
3%

Absorbed by ozone
3%

Absorbed by atmosphere (clouds, atmospheric gases, and dust)
21%

3%
Emitted by ozone

21%
Energy from atmosphere to space (= shortwave radiation absorbed)

14%
Net energy loss from atmosphere due to the greenhouse effect

+96%
Heat from atmosphere to surface

+110%
Heat from surface to atmosphere

45%
Absorbed by Earth's surface (diffuse + direct)

23%
Latent heat and convection

8%
Direct heat loss from surface

▲2.13 **Earth–atmosphere energy balance**

Animation (MG)
Earth–Atmosphere Energy Balance

http://goo.gl/7UYgTM

geoQUIZ
1. Why is Earth's energy balance currently experiencing an energy surplus?
2. Explain why the tropics receive high levels of incoming energy and the poles receive low levels of incoming energy.
3. Referring to Figure 2.13, summarize the inputs and outputs that maintain Earth's energy balance.

2.5 Energy Balance at Earth's Surface

Key Learning Concepts

▶ **Explain** daily temperature patterns and surface energy flows.

After energy comes to Earth's surface, it creates patterns of high and low temperatures. These patterns vary over the course of a day and over the course of the year. Once at the surface heat energy flows through different pathways—evaporating water or warming the surface and atmosphere, for example.

Patterns of Insolation & Temperature

Temperatures vary according to a typical daily pattern (▼Fig. 2.14). Times of peak air temperature lag after times of peak insolation. Incoming energy starts to arrive at sunrise, peaks at noon, and ends at sunset. The warmest time of day occurs when the maximum insolation is absorbed and emitted to the atmosphere from the ground, not at the moment of maximum insolation. When incoming energy exceeds the outgoing energy, air temperature increases; temperature peaks when incoming energy begins to diminish as the afternoon insolation decreases. In contrast, the coldest time of day is at sunrise. The delay

between noon and the warmest part of the day is a result of the time it takes for heat to transfer from the surface to the air.

The annual pattern of insolation and air temperature exhibits a similar lag. For the Northern Hemisphere, January is usually the coldest month, occurring after the December solstice and the shortest days. Similarly, the warmest months of July and August occur after the June solstice and the longest days. The insolation curve has the highest peak at the time of the summer solstice (around June 21 in the Northern Hemisphere and December 21 in the Southern Hemisphere). The delay between the date of maximum insolation and the warmest day is a result of the time it takes to warm bodies of water.

geoCHECK ✔ When is the warmest time of day?

Simplified Surface Energy Balance

Energy and moisture are continually exchanged at the surface, creating climates of great variety at different scales. The following discussion might be easier to follow if you visualize an actual surface at a small scale—perhaps a park, a front yard, or a place on campus.

The surface receives visible light and longwave radiation, and it reflects light and emits longwave radiation according to the following scheme:

+SW ↓	−SW ↑	+LW ↓	−LW ↑	= NET R
Energy input from insolation— some is absorbed, some is reflected	Energy lost to space by reflection	Infrared energy emitted by atmospheric gases and absorbed by the ground	Infrared energy lost to space from the surface or absorbed by atmospheric gases	

We use SW for shortwave and LW for longwave for simplicity.

Adding and subtracting the energy flow at the surface completes the calculation of **net radiation** (NET R), or the balance of all radiation at Earth's surface. As the components of this equation vary with day length through the seasons, cloudiness, and latitude, NET R varies.

▼**2.14 Daily radiation and temperature curves** This radiation and temperature plot for a typical day shows the changes in insolation (solid line) and air temperature (dashed line). Comparing the curves reveals a lag between the insolation peak at local noon (the insolation peak for the day) and the warmest time of day.

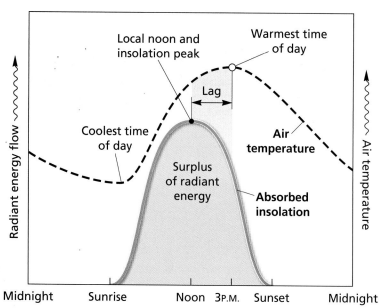

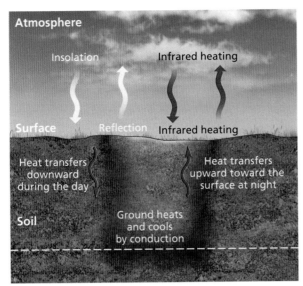

▲2.15 **Net radiation** Idealized input and output of energy at the surface and within a column of soil.

Energy that flows from the ground to the atmosphere is a loss from the surface

Sensible heat transfer normally occurs within the top meter of the ground

Net Radiation The net radiation available at Earth's surface is the final outcome of the energy balance process (▲Fig. 2.15). Net radiation flows from a surface through three pathways:

1. *Sensible heat* is the transfer between air and surface through convection and conduction within materials.
2. *Latent heat of evaporation* is the energy that is stored in water vapor as water evaporates. Water absorbs large quantities of latent heat energy as it changes state to water vapor, removing this heat energy from the surface. This heat energy is later released to the environment when water vapor changes state back to a liquid.
3. *Ground heating and cooling* is the energy that flows into and out of the surface (land or water) by conduction.

Variation in the flow of NET R among sensible heat (energy we can feel), latent heat (energy for evaporation), and ground heating and cooling produces the variety of environments we experience in nature. On the MasteringGeography website, you can find a comparison of daily radiation budgets for a station in the southern California desert and one in an agricultural valley in British Columbia.

Water managers in hot, dry environments such as southern California apply surface energy balance principles when they make decisions about water supply. Imagine you are the director of an irrigation district. You need to be able to provide water to your farmers for their crops, especially on the hottest days when latent heat transfer by evaporation from soil and transpiration from plants is high. To help you plan water deliveries, researchers use remote sensing data and a detailed surface energy balance equation to calculate the amount of water used by crops in your district. Their method measures insolation, albedo, sensible heat, and ground heating. These values are plugged into the SEBAL (surface energy balance algorithm for land) equation to calculate the actual amount of water lost to the atmosphere on a particular day. In the SEBAL equation, insolation equals latent heat plus albedo, sensible heat, and ground heating. Once the amounts of energy used in albedo, sensible heat, and ground heating are known, the remaining energy represents the latent heat used to evaporate water. An example of this method is shown for the Imperial Valley in southern California below in (▼Figure 2.16). The method allows irrigation district managers to better plan for water deliveries, as well as allowing farmers to avoid underwatering or buying unneeded water.

geoCHECK ✔ What are the three main pathways for surface heat?

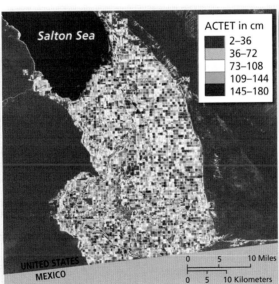

(a) Analysis of actual evaporation.

ACTET in cm
2–36
36–72
73–108
109–144
145–180

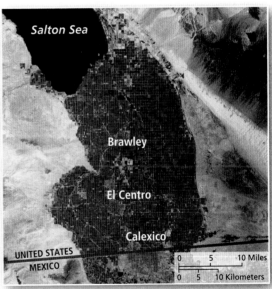

(b) Satellite image showing fields and surrounding desert.

▲2.16 **Water evaporation values, Imperial Valley, California** ACTET in the legend represents the actual amount of water that evaporated.

geoQUIZ

1. What are the sources of heating that create the warmest time of day?
2. What are the components of calculating NET R?
3. How would the heat flows for a site in a desert differ from a site in a region that recieves the same amount of insolation, but has lush vegetation?

2.6 Temperature Concepts & Measurement

Key Learning Concepts

▶ **Define** the concept of temperature.
▶ **Distinguish** between Kelvin, Celsius, and Fahrenheit temperature scales and how they are measured.

Heat is the flow of kinetic energy between molecules, and **temperature** is a measure of the average kinetic energy (motion) of individual molecules. The effect of temperature we feel is the sensible heat transfer from warmer objects to cooler objects. Temperature and heat are related because changes in temperature are caused by the absorption or emission (gain or loss) of heat energy.

Temperature Scales

Scientists have developed three scales used in measuring temperature. The Fahrenheit, Celsius, and Kelvin scales (▶Fig. 2.17) are named after the scientists who developed them.

The Fahrenheit (F) scale places the melting point of ice at 32°F, with an additional 180° to the boiling point of water at 212°F. Ice has only one melting point, but water has several freezing points. The scale is named for its developer, Daniel G. Fahrenheit, a German physicist (1686–1736).

About a year after the adoption of the Fahrenheit scale, Swedish astronomer Anders Celsius (1701–1744) developed the Celsius (C) scale. He placed the melting point of ice at 0°C and the boiling temperature of water at sea level at 100°C, dividing his scale into 100 degrees.

British physicist Lord Kelvin (born William Thomson, 1824–1907) proposed the Kelvin scale (K) in 1848. The Kelvin scale starts at *absolute zero*, the temperature at which atomic and molecular motion completely stop. Kelvin scale readings are proportional to the actual kinetic energy in a material. Zero on the Kelvin scale equals –459.67°F and –273°C. The Kelvin scale's melting point for ice is 273 K, and its boiling point of water is 373 K.

The United States remains the only major country still using the Fahrenheit scale. The continuing pressure from the scientific community, and the rest of the world, makes adoption of Celsius and SI (Système International) units inevitable in the United States. This textbook presents Celsius (with Fahrenheit equivalents in parentheses) throughout.

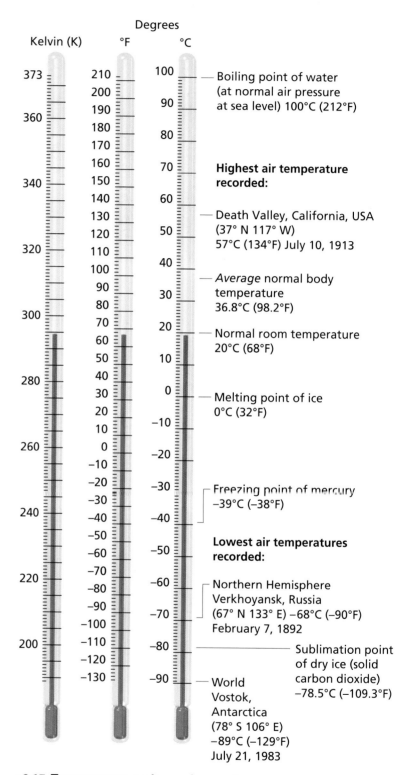

▲2.17 **Temperature scales and records**

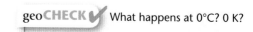 geoCHECK ✔ What happens at 0°C? 0 K?

and other heat flows; LSTs tend to be highest in dry environments with clear skies and surfaces with low albedo that absorb solar radiation.

Temperature readings are taken daily at more than 16,000 weather stations worldwide. Some stations also report the duration of temperatures, rates of rise or fall, and temperature variation over time throughout the day and night.

Meteorologists use temperature data to compile statistics on daily, monthly, and annual high, low, and mean temperatures.

- The *daily mean temperature* is an average of daily minimum–maximum readings.
- The *monthly mean temperature* is the total of daily mean temperatures for the month divided by the number of days in the month.
- An *annual temperature range* expresses the difference between the lowest and highest monthly mean temperatures for a given year.

In order to better track global climate change, the World Meteorological Organization (WMO) wants to establish a reference network with one station per 250,000 km² (95,800 mi²). NASA's Goddard Institute for Space Studies constantly refines the methodology of collecting average temperature measurements. In order to maintain a quality data set, scientists assess each station for the possibility of artificially high readings caused by human activities.

Web Site (MG)
World Meteorological Organization

http://goo.gl/elxgoq

geoCHECK ✔ Why are weather stations painted white?

geoQUIZ

1. How does heat differ from temperature?
2. How is air temperature different from land surface temperature?
3. Why does the WMO want to increase the number of weather stations worldwide?

▲ 2.18 **Thermometer shelter and weather station on sea ice in Antarctica**

Measuring Temperature

Temperatures are directly measured with either liquid-in-glass or electronic thermometers. Liquid-in-glass thermometers are sealed glass tubes filled with either mercury or alcohol.

Thermometers for official readings are placed outdoors in small shelters with high albedo and good ventilation (▲ **Fig. 2.18**). They are placed 1.2–1.8 m (4–6 ft) above a vegetated, natural surface. Official temperature measurements are made in the shade to prevent the Sun from heating the thermometer directly.

Satellites measure land surface temperature (LST) or land "skin" temperature, which is often much hotter than air temperature (▶ **Fig. 2.19**). You have felt this difference when walking barefoot across hot sand or pavement: The surface under your feet is much hotter than the air around your body. Land surface temperatures record the heating of the ground from insolation

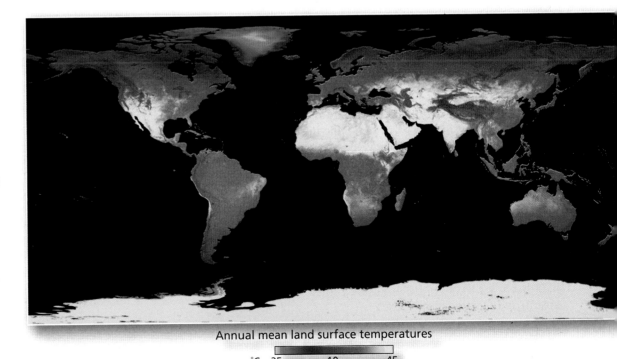

Annual mean land surface temperatures

°C −25 ▬▬▬ 10 ▬▬▬ 45

▲ 2.19 **Annual mean land surface temperatures**

2.7 Principal Temperature Controls

Key Learning Concepts

▶ *Explain* the effects of latitude, altitude and elevation, and cloud cover on global temperature patterns.

▶ *Describe* the land–water heating differences that produce continentality and marine effects on temperature ranges.

The interaction of several physical factors, called *controls*, produces Earth's temperature patterns. These principal temperature controls are latitude, altitude and elevation, cloud cover, land–water heating differences and the influence of ocean currents and sea-surface temperatures.

Latitude

Latitude is the most important control on annual average temperature and temperature range. The subsolar point moves between the tropics of Cancer and Capricorn during the year. As one moves away from the tropics and the subsolar point, annual insolation decreases and the difference between summer and winter insolation increases. The four cities graphed in Figure 2.20 demonstrate the effects of latitudinal position on temperature. From equator to poles, Earth ranges from continually warm in the low-latitude tropics, to seasonally variable in the temperate midlatitudes, to continually cold in the polar high latitudes.

geoCHECK ✔ What aspects of temperature are related to latitude?

Altitude & Elevation

Two terms, altitude and elevation, are commonly used to refer to heights on or above Earth's surface. *Altitude* refers to airborne objects or heights above Earth's surface. *Elevation* usually refers to the height of a point on Earth's surface above sea level (▶ **Fig. 2.21**). Therefore, the height of a flying jet is expressed as altitude, whereas the height of a mountain ski resort is expressed as elevation.

In the thinner atmosphere at high elevations in mountainous regions or on high plateaus, surfaces gain and lose energy rapidly to the atmosphere. Within the troposphere, temperatures decrease with increasing altitude above Earth's surface at the normal lapse rate of 6.4 °C/1000 m (3.5 °F/1000 ft). The density of the atmosphere also diminishes with increasing altitude. The density of the atmosphere at an elevation of 5500 m (18,000 ft) is about half that at sea level. As the atmosphere thins, the air loses its ability to absorb and radiate sensible heat.

The snow line is the elevation where temperatures are cold enough for snow to remain through the year. The snow line's location is mainly a function of elevation and latitude. Permanent ice fields and glaciers exist

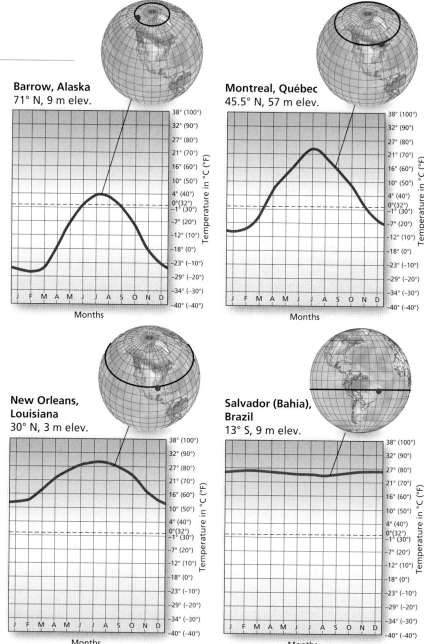

▲2.20 **Temperature patterns by latitude**

on equatorial mountain summits in the Andes and East Africa at approximately 5000 m (16,400 ft) and at sea level in Antarctica.

geoCHECK ✔ How does altitude and elevation affect temperatures?

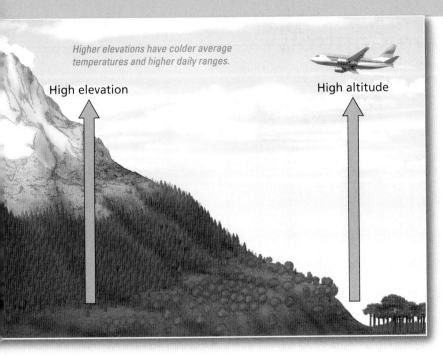

Higher elevations have colder average temperatures and higher daily ranges.

High elevation

High altitude

▲2.21 **Altitude and elevation**

Cloud Cover

Approximately 50% of Earth is cloud-covered at any given moment. Clouds are the most variable factor influencing Earth's radiation budget and are the subject of extensive research (see http://isccp.giss.nasa.gov/ and http://science.larc.nasa.gov/ceres/ for examples).

Clouds lower daily maximum temperatures and raise nighttime minimum temperatures, and their effect varies with cloud type, height, and density. Cloud moisture

reflects, absorbs, and liberates large amounts of energy released upon condensation and cloud formation.

 How do clouds affect temperatures?

Land–Water Heating Differences

Water bodies tend to moderate temperature patterns, and more extreme temperatures occur in continental interiors. The physical nature of land (rock and soil) and water (oceans, seas, and lakes) is the reason for land–water heating differences: Land heats and cools faster than water, producing temperature contrasts between continental interiors and areas near bodies of water. These contrasts are the result of differences between land and water in evaporation, transparency, specific heat, and movement (▼Fig. 2.22).

Evaporation When water evaporates and changes to water vapor, heat energy is absorbed and stored in the water vapor as latent heat, cooling the surface from which the water evaporated. Approximately 86% of all evaporation on Earth is from the oceans. As surface water evaporates, it absorbs energy from the immediate environment, resulting in a lowering of temperatures. Temperatures over land, with far less water to evaporate, experience less evaporative cooling than do marine locations.

Transparency Solid ground is opaque; water is transparent. Because of water's **transparency,** light passes through water to an average depth of 60 m (200 ft) in the ocean. This illuminated zone can reach to depths of 300 m (1000 ft). Available heat energy is distributed over a much greater volume, forming a larger energy reservoir than that of land surfaces.

Light striking soil is absorbed, heating the ground surface. Maximum and minimum temperatures generally are experienced right at ground level. Below the surface, temperatures remain about the same throughout the day.

Specific Heat Water requires far more energy to change its temperature than does land, and therefore water has a higher **specific heat,** the heat capacity of a substance. On average, the specific heat of water is about four times that of soil.

Movement Land is a rigid, solid material, whereas water is a fluid and is capable of movement. Differing temperatures and currents result in a mixing of cooler and warmer waters, and that mixing spreads the available energy over an even greater volume than if the water were still.

 Why does land heat and cool faster than water?

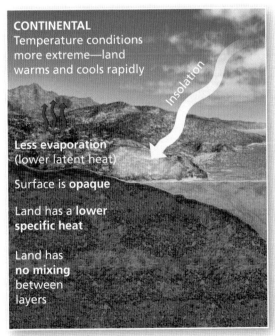

CONTINENTAL
Temperature conditions more extreme—land warms and cools rapidly

Insolation

Less evaporation (lower latent heat)

Surface is **opaque**

Land has a **lower specific heat**

Land has **no mixing** between layers

(a) Continental conditions

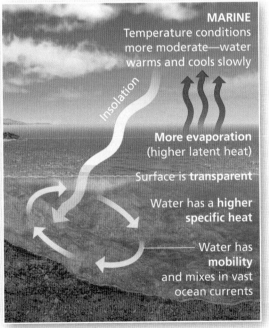

MARINE
Temperature conditions more moderate—water warms and cools slowly

Insolation

More evaporation (higher latent heat)

Surface is **transparent**

Water has a **higher specific heat**

Water has **mobility** and mixes in vast ocean currents

(b) Marine conditions

▲2.22 **Land–water heating differences**

2.7 (cont'd) Principal Temperature Controls

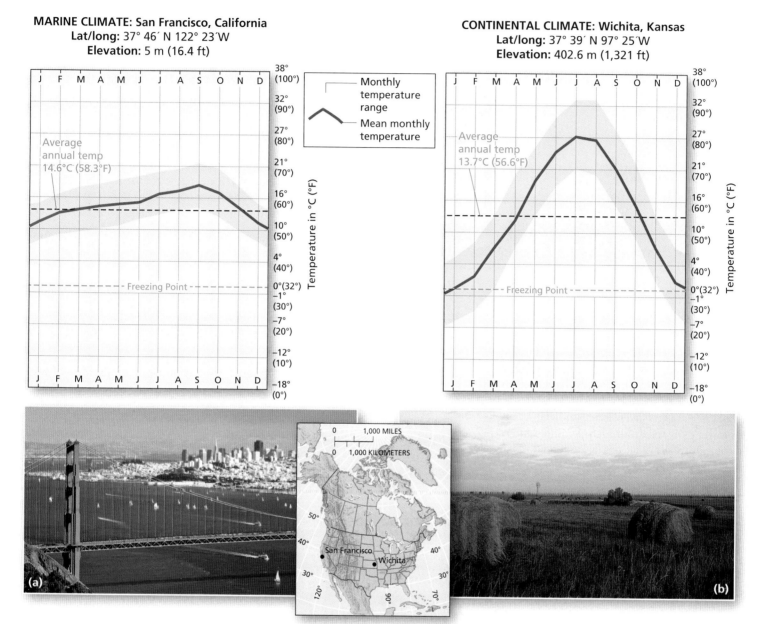

MARINE CLIMATE: San Francisco, California
Lat/long: 37° 46′ N 122° 23′W
Elevation: 5 m (16.4 ft)

CONTINENTAL CLIMATE: Wichita, Kansas
Lat/long: 37° 39′ N 97° 25′W
Elevation: 402.6 m (1,321 ft)

Monthly temperature range
Mean monthly temperature

Average annual temp 14.6°C (58.3°F)

Average annual temp 13.7°C (56.6°F)

Freezing Point

▲2.23 Marine and continental cities—(a) San Francisco, California, and (b) Wichita, Kansas

Marine & Continental Climates

The land-water heating differences discussed above have different effects on climate depending on geographic location. The **marine effect** (or *maritime* effect) describes the moderating influence of the ocean along coastlines or on islands. **Continentality** refers to the effect of location in a continental interior less affected by the sea and having a greater daily and yearly temperature range.

San Francisco, California, and Wichita, Kansas, provide us examples of the marine effect and continentality (▲Fig. 2.23). Both cities are at approximately the same latitude. Because of the marine effect, San Francisco has moderate temperatures within a narrow range all year. Wichita, in contrast, has a much wider temperature range because of continentality.

In Eurasia, coastal Trondheim, Norway and inland Verkhoyansk, Russia also illustrate this pattern. They are both at similar latitudes, however coastal Trondheim has a 17 C° (30.6 F°) annual temperature range, while Verkhoyansk has a continental location and a 63 C° (113 F°) range. The range between the January record low and the July record high for Verkhoyansk is an incredible 105 C° (189 F°)!

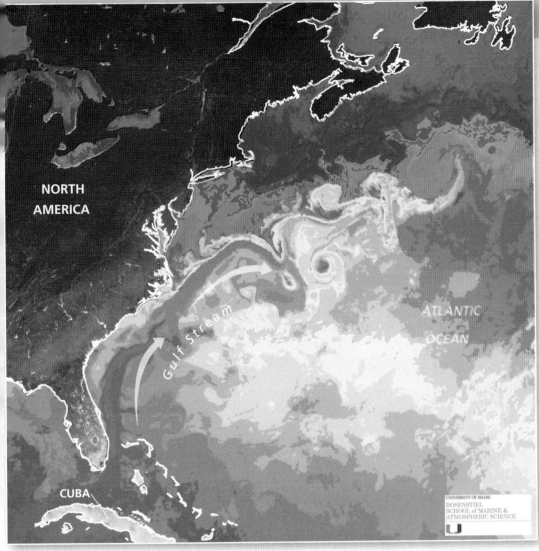

NORTH AMERICA

Gulf Stream

ATLANTIC OCEAN

CUBA

UNIVERSITY OF MIAMI
ROSENSTIEL
SCHOOL of MARINE &
ATMOSPHERIC SCIENCE

Temperature

| 2° - 9°C (36°-49°F) | 10° - 16°C (50°-62°F) | 17° - 24°C (63°-75°F) | 25° - 29°C (76°-84°F) |

▲2.24 **The Gulf Stream**

Animation (MG)
The Gulf Stream

http://goo.gl/i2IGHN

Ocean Currents & Sea-Surface Temperatures

The general pattern of ocean circulation also affects temperature patterns both in the oceans and above adjacent landmasses. Warm currents occur off the east coasts, while cold currents occur off the west coasts. For example, the Gulf Stream moves northward off the east coast of North America, carrying warm water far into the North Atlantic (◀Fig. 2.24). Iceland, coastal Scandinavia, and northwestern Europe experience much milder temperatures than would be expected just south of the Arctic Circle (66.5°). In the western Pacific Ocean, the Kuroshio, or Japan Current, functions much the same as the Gulf Stream, having a warming effect on Japan, the Aleutians, and the northwestern margin of North America.

Figure 2.25 displays sea-surface temperature satellite data for July 2013. The Western Pacific Warm Pool is the deep red color in the southwestern Pacific Ocean, with temperatures above 30°C (86°F). This region has the highest average ocean temperatures in the world. Note the cooler ocean temperatures associated with cold currents off the west coasts of North and South America, Europe, and Africa.

Warm currents affect more than just temperatures. Warm water evaporates more easily than colder water, so places located by warm currents will have higher humidity and rainfall than places located by cold currents.

geoCHECK ✔ How do west coast sea-surface temperatures compare with those on the east coast?

geoQUIZ

1. What is the most important control for average annual temperature?
2. Describe three ways that land and water heat differently.
3. How does continentality affect temperature range?

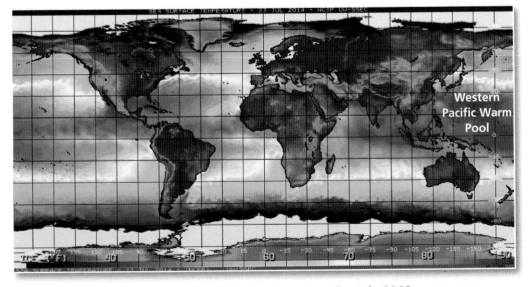

Western Pacific Warm Pool

▲2.25 **Average monthly sea-surface temperatures for July 2013**

2.8 The Urban Environment

Key Learning Concepts

▶ **List** typical urban heat island conditions and their causes.
▶ **Contrast** the microclimate of urban areas with that of surrounding rural environments.
▶ **Describe** ways that cities can reduce the urban heat island effect.

Microclimates are small areas, ranging from square meters to square kilometers, with climates that are different from the region around them. Urban microclimates generally differ from nearby nonurban areas by regularly experiencing temperatures as much as 6°C (10°F) hotter than surrounding suburban and rural areas. The surface energy characteristics of urban areas are similar to those of desert locations because of a lack of vegetation in both environments.

Urban Heat Islands

The physical characteristics of urbanized regions produce an **urban heat island** (UHI) that has both average maximum and minimum temperatures higher than nearby rural settings (▼ Fig. 2.26). A UHI is hotter toward the downtown central business district and cooler over areas of trees and parks.

UHI effects are greater in bigger cities than smaller ones. The difference between urban and rural heating is more pronounced in cities surrounded by forest rather than by dry sparsely vegetated environments. UHI effects also tend to be highest in cities with dense population and slightly lower in cities with more urban sprawl. Go to www.nasa.gov/topics/earth/features/heat-island-sprawl.html for an interesting study on UHIs in the U.S. Northeast.

In the average city in North America, heating is increased by modified urban surfaces such as asphalt and glass, flat surfaces, pollution, and human activity such as industry and transportation. For example, driving an average car (25 miles per gallon or 10 km/l) 1 mile produces enough heat to melt a 14-lb bag of ice (4.5 kg ice per kilometer driven). The removal of vegetation and the increase in human-made materials that retain heat are two of the most significant UHI causes.

Sensible heat is less in urban forests than in other parts of the city because of shading from tree canopies and plant processes such as transpiration that add moisture into the air. In New York City, daytime temperatures average 5–10°C (9–18°F) cooler in Central Park than in the greater metropolitan area. The temperature differences between parks and paved surfaces are clear in the infrared image of Sacramento, California, shown in Figure 2.27.

Most major cities also produce a **dust dome** of airborne pollution trapped by certain characteristics of air circulation in UHIs: The pollutants collect with a decrease in wind speed in urban centers; they then rise as the surface heats and remain in the air above the city, affecting urban energy budgets. Table 2.1 lists some of the factors that cause UHIs and compares selected climatic elements of rural and urban environments.

 geoCHECK How does a tree-covered park compare with a parking lot as a factor in the UHI effect?

Mitigating UHI Effects

City planners and architects can reduce UHI effects with strategies such as planting of vegetation in parks and open space (urban forests) and encouraging "green" roofs like rooftop gardens, high-albedo "cool" roofs, and lighter-color "cool" pavements. In addition to lowering urban outdoor temperatures, such strategies keep buildings' interiors cooler, thereby reducing energy consumption and greenhouse gases caused by fossil-fuel emissions. During the

▼ **2.26 Typical urban heat island profile**

Urban heat island typically 1–3 C° (2–5 F°) warmer than rural regions

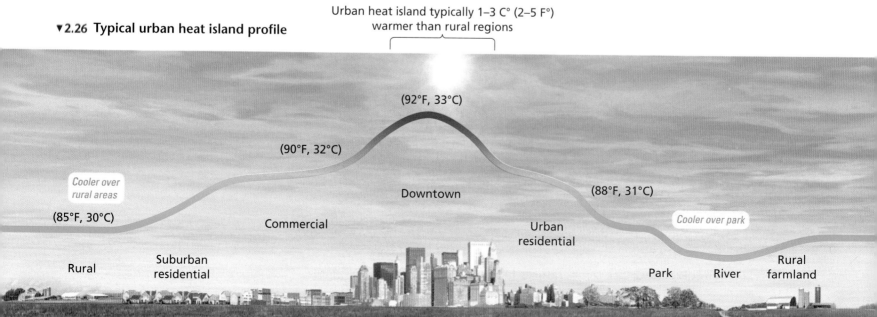

(92°F, 33°C)

(90°F, 32°C)

Cooler over rural areas

(85°F, 30°C)

Downtown

(88°F, 31°C)

Cooler over park

Commercial

Urban residential

Rural

Suburban residential

Park

River

Rural farmland

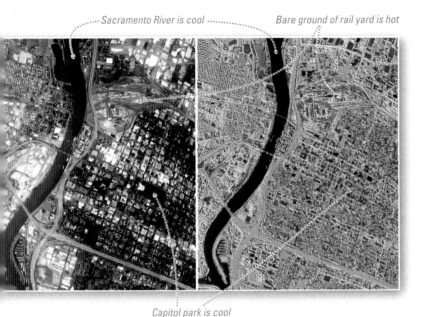

Sacramento River is cool ·········   ········· Bare ground of rail yard is hot

········ Capitol park is cool

▲2.27 **Infrared images (black & white and false color) of urban heat island Sacramento, CA** The white areas, mostly rooftops, are about 60°C (140°F) and the dark areas are approximately 29-36°C.(85-96°F). The hottest spots are buildings, seen as white rectangles of various sizes. Sacramento's rail yard is the orange area east of the Sacramento River, which flows from top to bottom.

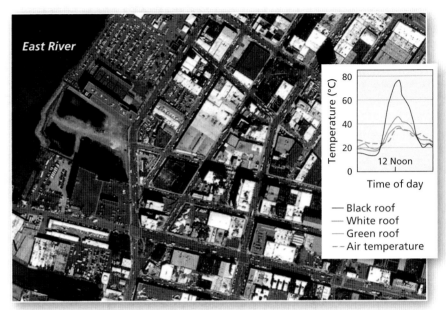

East River

— Black roof
— White roof
— Green roof
-- Air temperature

▲2.28 **Effects of rooftop materials on urban heat islands**

hottest day of the 2011 New York City summer, temperature measurements for a white roof covering were 24°C (42°F) cooler than for a traditional black roof nearby (▲**Fig. 2.28**). Solar panels also cool roofs so that temperatures under the shade of the panels can be dramatically cooler.

With 60% of the global population expected to live in cities by the year 2030, and with air and water

temperatures rising because of climate change, UHI issues are emerging as a significant concern both for physical geographers and for the public at large. Studies have found a direct correlation between peaks in UHI temperatures and heat-related illness and fatalities. More information is available from the Environmental Protection Agency, including mitigation strategies, publications, and hot topic discussions, at www.epa.gov/hiri/.

geoCHECK ✔ Why are cities warmer than rural areas?

geoQUIZ

1. What climate type is most similar to the climate of cities? Explain.
2. Where are the coolest areas in cities, usually?
3. What are two climatic effects of urban air pollution?
4. What are some actions or policies that would reduce the UHI effect?

Table 2.1 Urban Heat Island Factors

Driving Factor	Climatic Element Affected	Urban Compared to Rural
Metal, glass, asphalt, concrete, brick surfaces conduct up to three times more heat than soil.	More net radiation	Hotter
Surfaces with low albedo absorb and retain more heat.	Lower albedo	Hotter
Buildings interrupt wind flows causing an urban canyon effect.	Lower wind speed	Less windy
Heat is generated by homes, vehicles, and factories.	Higher temperatures • annual mean • winter low • summer high	Hotter
Air pollution absorbs insolation.	Higher temperatures	Hotter
Condensation nuclei from air pollution.	More cloudiness and fog More precipitation Snow	More clouds More rain Less snow in city, more snow downwind
Urban desert effect: less plant cover and more sealed surfaces mean less cooling from evaporation and plant transpiration. Water runs off pavement quickly.	Lower relative humidity Less evaporation More flooding	Lower Less More

2.9 Earth's Temperature Patterns

Key Learning Concepts

▶ **Interpret** the pattern of Earth's temperatures on January and July temperature maps and on a map of annual temperature ranges.

We now examine mean global air temperature patterns in January and July and the pattern of global temperature ranges. January and July are used instead of the solstice months of December and June because of the lag time between insolation received and maximum or minimum temperatures.

The lines on temperature maps are *isotherms*. An **isotherm** is a type of isoline—a line along which there is a constant value—that connects points of equal temperature and portrays the temperature pattern, just as a contour line on a topographic map illustrates points of equal elevation.

January & July Temperatures

January Temperatures Figure 2.29 shows January's mean temperatures for the world. The **thermal equator** is an isotherm connecting all points of highest mean temperature, roughly 27°C (80°F). As you might expect, the thermal equator shifts south or north with the changing seasons. In the Northern Hemisphere in January, isotherms shift equatorward as cold air chills the continental interiors. Warmer conditions extend farther north than over land than over the oceans at comparable latitudes (▼Fig 2.29).

July Temperatures Figure 2.30 shows average July worldwide temperatures. Notice how the thermal equator shifts northward with the summer Sun and reaches the Persian Gulf-Pakistan-Iran region. The Persian Gulf is the site of the highest recorded sea-surface temperature of 36°C (96°F). July temperatures in Verkhoyansk, Russia, average more than 13°C (56°F), which results in a 63°C (113°F) seasonal range.

The hottest places on Earth occur in Northern Hemisphere deserts during July, caused by clear skies, strong surface heating, virtually no surface water, and few plants. Prime examples are portions of the Sonoran Desert of North America, the Sahara of Africa, and the Lut Desert in Iran. The highest shade temperature ever recorded, 57°C (134°F), occurred on July 10, 1913, at Death Valley, **California**.

geoCHECK ✔ How does the movement of the thermal equator over Eurasia compare to its movement over the Pacific Ocean?

▼2.29 **January temperatures**

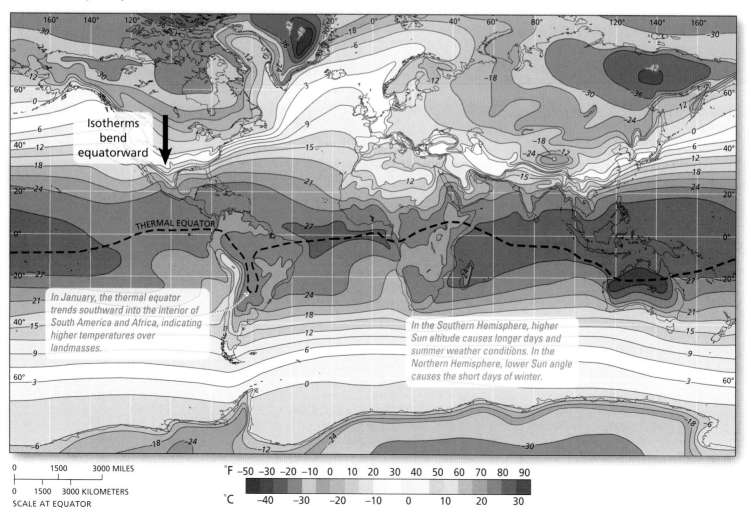

Isotherms bend equatorward

THERMAL EQUATOR

In January, the thermal equator trends southward into the interior of South America and Africa, indicating higher temperatures over landmasses.

In the Southern Hemisphere, higher Sun altitude causes longer days and summer weather conditions. In the Northern Hemisphere, lower Sun angle causes the short days of winter.

| 0 | 1500 | 3000 MILES |
| 0 | 1500 | 3000 KILOMETERS |
SCALE AT EQUATOR

°F −50 −30 −20 −10 0 10 20 30 40 50 60 70 80 90

°C −40 −30 −20 −10 0 10 20 30

Annual Temperature Ranges

The temperature range map shows regions that experience the greatest annual extremes and, in contrast, the most moderate temperature ranges, as demonstrated in Figure 2.31. The largest temperature ranges occur at subpolar locations within the continental interiors of North America and Asia, where average ranges of 64°C (115°F) are recorded (dark brown area on the map). The Southern Hemisphere, in contrast, has little seasonal variation in mean temperatures, owing to the lack of large landmasses and the vast expanses of water to moderate temperature extremes.

Southern Hemisphere temperature patterns are generally maritime, and Northern Hemisphere patterns feature continentality, although interior regions in the Southern Hemisphere experience some continentality effects.

geoCHECK ✔ Which region sees the highest annual temperature range? Why does this extreme range occur at this particular location?

geoQUIZ

1. Where are the highest surface temperatures found in January? July? Explain.
2. What is the highest temperature range in the Southern Hemisphere? Explain.

▲2.31 **Temperature ranges**

▼2.30 **July temperatures**

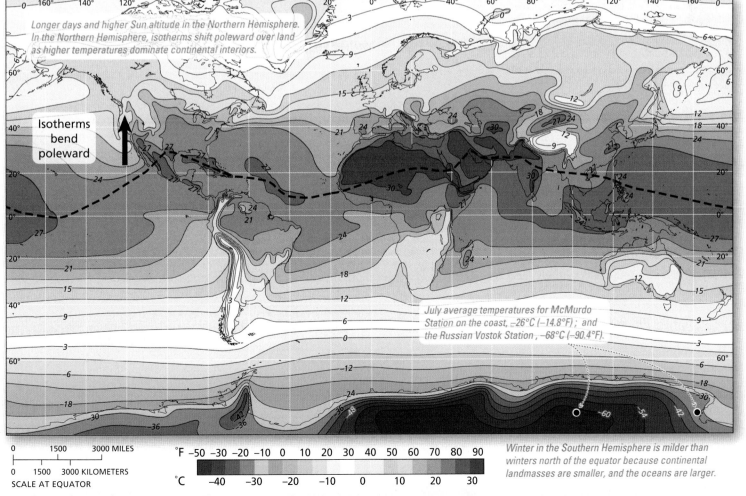

Longer days and higher Sun altitude in the Northern Hemisphere. In the Northern Hemisphere, isotherms shift poleward over land as higher temperatures dominate continental interiors.

Isotherms bend poleward

July average temperatures for McMurdo Station on the coast, −26°C (−14.8°F); and the Russian Vostok Station, −68°C (−90.4°F).

Winter in the Southern Hemisphere is milder than winters north of the equator because continental landmasses are smaller, and the oceans are larger.

0	1500	3000 MILES
0	1500	3000 KILOMETERS

SCALE AT EQUATOR

°F −50 −30 −20 −10 0 10 20 30 40 50 60 70 80 90
°C −40 −30 −20 −10 0 10 20 30

2.10 Wind Chill, the Heat Index, & Heat Waves

Key Learning Concepts

▶ *Discuss* wind chill, heat waves, and the heat index as they relate to human heat response.

Our bodies subjectively sense temperature and react to temperature changes. *Apparent temperature* is the general term for the temperature as it is perceived by humans. Both humidity and wind affect apparent temperature and our sense of comfort. Wind tends to make us feel colder—the effect known as wind chill. And high humidity can make us feel hotter. A **heat wave**—defined as 3 days in a row with temperatures above 90°F (32.2°C)—can pose health hazards to overheated humans. To rate the dangers of hot weather, scientists have developed the *heat index*.

Wind Chill

The *wind-chill index* is important to those who experience winters with freezing temperatures. As wind speeds increase, heat loss from the skin increases, and the wind-chill factor rises. To track the effects of wind on apparent temperature, the National Weather Service (NWS) uses the *wind-chill temperature index*, a chart plotting the temperature we feel as a function of actual air temperature and wind speed (▶Fig. 2.32). Clothing that prevents wind access to your skin can help protect against wind chill.

geoCHECK ✔ Why do you feel colder when wind blows over your skin?

Actual Air Temperature in °C (°F)

Wind speed, kmph (mph) ↓ / Calm	4 (40)	−1 (30)	−7 (20)	−12 (10)	−18 (0)	−23 (−10)	−29 (−20)	−34 (−30)	−40 (−40)
16 (10)	1 (34)	−6 (21)	−13 (9)	−20 (−4)	−27 (−16)	−33 (−28)	−41 (−41)	−47 (−53)	−54 (−66)
32 (20)	−1 (30)	−8 (17)	−16 (4)	−23 (−9)	−30 (−22)	−37 (−35)	−44 (−48)	−52 (−61)	−59 (−74)
48 (30)	−2 (28)	−9 (15)	−17 (−1)	−24 (−12)	−32 (−26)	−39 (−39)	−47 (−53)	−55 (−67)	−62 (−80)
64 (40)	−3 (27)	−11 (13)	−18 (−1)	−26 (−15)	−34 (−29)	−42 (−43)	−49 (−57)	−57 (−71)	−64 (−84)
80 (50)	−3 (26)	−11 (12)	−19 (−3)	−27 (−17)	−35 (−31)	−43 (−45)	−51 (−60)	−59 (−74)	−67 (−88)

Frostbite times: ☐ 30 min. ▨ 10 min. ■ 5 min.

▲2.32 **Wind-chill temperature index**

The Heat Index

Persistent high temperatures present challenges to people, especially those living in cities in humid environments. The heat index (HI) indicates the human body's reaction to air temperature and water vapor. Higher amounts of water vapor in the air (the higher the humidity) mean less evaporative cooling. The HI indicates how the air feels to an average person.

Figure 2.33 is an abbreviated version of the HI used by the NWS. The table beneath the graph describes the effects of HI categories on higher-risk groups. The combination of high temperature and high humidity severely reduces the body's natural ability to regulate internal temperature.

geoCHECK ✔ Would you feel warmer with high temperatures and high or low humidity?

Temperature in °C (°F)

Relative Humidity (%) ↓	27 (80)	28 (82)	29 (84)	30 (86)	31 (88)	32 (90)	33 (90)	34 (94)	36 (96)	37 (98)	38 (100)	39 (102)	40 (104)	41 (106)	42 (108)	43 (110)
40	27 (80)	27 (81)	28 (83)	29 (85)	31 (88)	33 (91)	34 (94)	36 (97)	38 (101)	41 (105)	43 (109)	46 (114)	48 (124)	51 (106)	54 (130)	58 (136)
50	27 (81)	28 (83)	29 (85)	31 (88)	33 (91)	35 (95)	37 (99)	39 (103)	42 (108)	45 (113)	48 (118)	51 (124)	55 (131)	58 (137)		
60	28 (82)	29 (84)	31 (88)	33 (91)	35 (95)	38 (100)	41 (105)	43 (110)	47 (116)	51 (123)	54 (129)	58 (137)				
70	28 (83)	30 (86)	32 (90)	35 (95)	38 (100)	41 (105)	44 (112)	48 (119)	52 (126)	57 (134)						
80	29 (84)	32 (89)	34 (94)	38 (100)	41 (106)	45 (113)	49 (121)	54 (129)								
90	30 (86)	33 (91)	37 (98)	41 (105)	45 (113)	50 (122)	55 (131)									
100	31 (87)	35 (95)	39 (103)	44 (112)	49 (121)	56 (132)										

Likelihood of heat disorders with prolonged exposure or strenuous activity:

Caution — Fatigue and heatcramps possible

Extreme caution — Heat cramps, heat exhaustion, or heat stroke possible

Danger — Heat cramps or heat exhaustion likely; heat stroke probable

Extreme danger — Heat stroke imminent

▲2.33 **Heat-index for various temperatures and relative humidity levels**

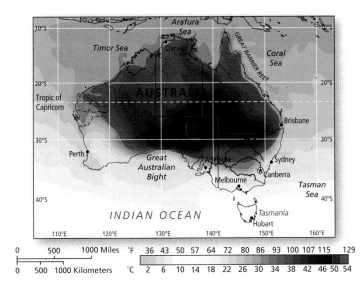

▲**2.34 Australian heat wave of 2013**   During the heat wave, temperatures were so much higher that a new color had to be added to the map to show them.

Heat Waves & Climate Change

Climate scientists think that as climate change brings higher temperatures, heat waves will become more frequent. Present global average temperatures are higher than at any time during the past 125,000 years. Global temperatures rose an average of 0.17°C (0.3°F) per decade since 1970, and this rate is accelerating. Summer temperatures in the United States and Australia rose to record levels during heat waves in 2012 and 2013. The Australian heat wave of 2013 lasted weeks, with temperatures topping 45°C (113°F) (▲**Fig. 2.34**).

Scientists study temperature patterns using temperature anomalies. An *anomaly* is the difference between a temperature and the long-term average temperature selected as the reference period or *baseline*. Figure 2.35 shows the differences between anomalies by decade compared with average temperatures from the baseline period of 1951–1980. The trend toward warmer temperatures and their orange and red colors is apparent in the 2000s, especially in the Northern Hemisphere.

The last 15 years included the warmest years in the climate record. According to NOAA's National Climatic Data Center (NCDC), the year 2012 was the hottest year in the contiguous United States since record-keeping began in 1895.

geoCHECK ✔   What is the difference between temperatures and temperature anomalies?

geoQUIZ

1. How long would it take to develop frostbite if it was −23°C with a 16 kilometer per hour (kph) breeze?
2. What is the likelihood of developing a heat disorder if the temperature is 34°C with 70% relative humidity?
3. The temperature anomalies in Figure 2.35 are compared with temperatures from 1951–1980. How would the anomalies be different if they were compared with temperatures from 1991–2010?

Decadal Surface-Temperature Anomalies (C°)

1970s

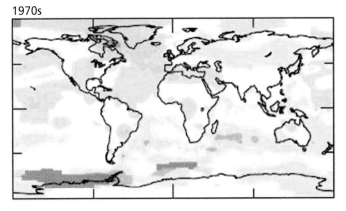

1980s

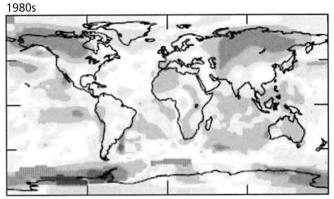

1990s

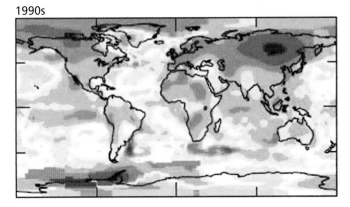

2000s

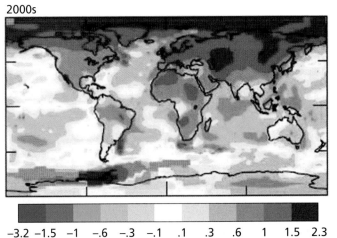

−3.2 −1.5 −1 −.6 −.3 −.1   .1   .3   .6   1   1.5 2.3

▲**2.35 Global temperature anomalies**

TEMPERATURE IMPACTS HUMANS

- The Earth–atmosphere system balances itself naturally, maintaining temperature patterns and planetary systems that support Earth, life, and human society.
- Solar energy is harnessed for power production worldwide, by technology ranging from small solar cookers to large-scale photovoltaic arrays.

2a

The International Maritime Organization, made up of 170 countries, is developing policies to improve energy efficiency and reduce diesel ship emissions, especially black carbon. In the Arctic, soot and particulates darken ice surfaces, decrease albedo, and enhance melting.

HUMANS IMPACT ENERGY BALANCE & TEMPERATURE

- Humans produce atmospheric gases and aerosols that affect clouds and the Earth–atmosphere energy budget, which, in turn, affect temperature and climate. For example, fossil-fuel burning produces carbon dioxide and other greenhouse gases that warm the lower atmosphere.
- Urban heat island effects accelerate warming in cities, which house more than half the global human population.

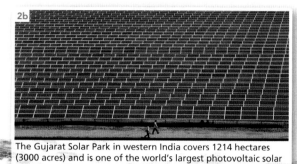

2b

The Gujarat Solar Park in western India covers 1214 hectares (3000 acres) and is one of the world's largest photovoltaic solar facilities, with a capacity of 500 MW.

This February NASA Blue Marble true-color image shows land surfaces, oceans, sea ice, and clouds.

2d
East River
Con Edison

In Queens, New York, lighter-colored rooftops help lower temperatures and the overall UHI effect. Studies confirm the higher temperatures of black rooftops as compared to white or green (vegetated) rooftops.

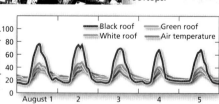

Temperature (°C) — Black roof, White roof, Green roof, Air temperature
August 1, 2, 3, 4, 5

2c

Wildfires raged across Australia during the 2013 summer heat wave, fueled by drought and record-breaking high temperatures. Near Hobart, Tasmania, fires destroyed over 80 homes.

ISSUES FOR THE 21ST CENTURY

- Improved energy efficiency and renewable energy sources, such as solar, can reduce the use of energy from fossil fuels, thus slowing the addition of anthropogenic greenhouse gases to the atmosphere.
- Strategies to reduce urban heat island effects can help lessen the dangers of heat waves in cities and slow the general atmospheric warming trend.
- Continued increases in average global air and ocean temperatures will enhance climate change effects worldwide.

Looking Ahead

In the next chapter, we consider the global circulation of winds and ocean currents and the forces that interact to produce these movements. Multiyear fluctuations in global circulation patterns, such as El Niño and La Niña in the Pacific Ocean, have far-reaching effects on world climates.

How does energy flow?

2.1 Energy Balance Essentials

Define energy and heat.

Explain four types of heat transfer: radiation, conduction, convection, and advection.

- The energy of motion is kinetic energy, measured as temperature. The flow of kinetic energy because of a temperature difference is heat. Potential energy is stored energy. Energy we can feel is sensible heat. Latent heat is the heat gained or lost in phase changes, such as from solid to liquid, while the substance's temperature remains unchanged. Four mechanisms of heat transfer are radiation, conduction, convection, and advection.

1. Define temperature, heat, sensible heat, and latent heat. Give an example of each type of heat transfer.

2.2 Insolation Input & Albedo

Identify pathways for solar energy through the troposphere to Earth's surface—transmission, scattering, refraction, and absorption.

Explain the concept of albedo (reflectivity).

- Transmission is the passage of radiation through either the atmosphere or water. Scattering occurs when light reflects in all directions, without changing its wavelength. Refraction bends light, changing its wavelength. Albedo is the reflective quality of a surface, reported as the percentage of insolation that is reflected.

2. Define *refraction*. How is it related to day length? To a rainbow?

3. List several types of surfaces and their albedo values. What determines the reflectivity of a surface?

2.3 The Greenhouse Effect, Clouds, & Atmospheric Warming

Analyze the effect of clouds and aerosols on atmospheric heating and cooling.

Explain the greenhouse concept as it applies to Earth.

Describe the effect of aerosols on atmospheric temperatures.

- Clouds, aerosols, and atmospheric pollutants can cool and heat the atmosphere. The greenhouse effect refers to gases in the lower atmosphere absorbing and reradiating infrared heat. Carbon dioxide, water vapor, and methane are the main greenhouse gases.

4. How do high, thin ice clouds affect temperatures differently than low, thick clouds?

5. What are the similarities and differences between an actual greenhouse and the gaseous atmospheric greenhouse? Why is Earth's greenhouse effect changing?

Where does energy flow?

2.4 Earth–Atmosphere Energy Balance

Describe the Earth–atmosphere energy balance and the patterns of global net radiation.

- The atmospheric energy budget has shortwave radiation inputs (UV, visible light, and near-infrared) and shortwave and longwave (thermal infrared) radiation outputs. In the tropical latitudes, high Sun angle and consistent day length produce energy surpluses. In polar regions, a low Sun angle, high albedo, and up to 6 months of darkness produce energy deficits. This imbalance of radiation drives global circulations of both energy and mass through wind, ocean currents, and weather. Net radiation (NET R) is the sum of all shortwave (SW) and longwave (LW) radiation gains and losses.

6. Sketch a simple energy balance diagram for the troposphere. Label each shortwave and longwave component and the directional aspects of related flows.

2.5 Energy Balance at Earth's Surface

Explain daily temperature patterns and surface energy flows.

- During an average day, air temperature peaks between 3:00 and 4:00 P.M. and dips to its lowest point right at or slightly after sunrise. Air temperature lags behind each day's peak insolation. The warmest time of day occurs not at the moment of maximum insolation, but at the moment when a maximum of insolation is absorbed.

7. Why is there a temperature lag between the highest Sun altitude and the warmest time of day? Relate your answer to the insolation and temperature patterns during the day.

2.6 Temperature Concepts & Measurement

Define the concept of temperature.

Distinguish between Kelvin, Celsius, and Fahrenheit temperature scales and how they are measured.

- Temperature is a measure of the average kinetic energy, or molecular motion, of individual molecules in matter. The Kelvin scale uses 100 units between the melting point of ice (273 K) and the boiling point of water (373 K). The Celsius scale has 100 degrees between the melting point of ice (0°C) and the boiling point of water (100°C). The Fahrenheit scale has 180 degrees between the melting point of ice (32°F) and the boiling point of water (212°F).

8. What is the difference between temperature and heat?

What controls Earth's temperatures?

2.7 Principal Temperature Controls

Explain the effects of latitude, altitude and elevation, and cloud cover on global temperature patterns.

Describe the land–water heating differences that produce continental and marine effects on temperature ranges.

- Principal temperature controls are latitude (the distance north or south of the equator), cloud cover, altitude, and elevation. Land–water heating differences are due to evaporation (water cools the surface from which it evaporates); transparency (light heats a much greater volume of water); specific heat (it takes much more energy to change the temperature of water than land); and movement (ocean currents transfer heat). The Gulf Stream carries warm water up the east coast of North America, far into the North Atlantic, warming Europe. Marine or maritime temperature ranges are much lower than continental temperature ranges.

9. Explain why land and water heat differently.

10. Differentiate between temperature ranges at marine versus continental locations. Give an example of each from the text discussion.

What are Earth's temperature patterns?

2.8 The Urban Environment

List typical urban heat island conditions and their causes.

Contrast the microclimate of urban areas with that of surrounding rural environments.

Describe ways that cities can reduce the urban heat island effect.

- Cities have hotter microclimates than nonurban areas. This is the urban heat island (UHI) effect, due to low-albedo surfaces, low evaporation, air pollution, and heat generated by machines.

11. Briefly describe the climatic effects of urban environments, as compared with nonurban environments.

2.9 Earth's Temperature Patterns

Interpret the pattern of Earth's temperatures on January and July temperature maps and on a map of annual temperature ranges.

- Isotherms are isolines that connect points of equal temperature and show temperature patterns. Isotherms generally trend east–west. The thermal equator moves north and south with the summer Sun and moves more over continents than over the oceans

12. Describe and explain the extreme temperature range experienced in north-central Siberia between January and July.

2.10 Wind Chill, the Heat Index, & Heat Waves

Discuss wind chill, heat waves, and the heat index as they relate to human heat response.

- The wind-chill factor shows the effect of the combination of cold temperatures and wind. Recent heat waves, prolonged periods of high temperatures lasting days or weeks, have caused fatalities and billions of dollars in economic losses. The heat index (HI) shows how the combination of high temperature and high humidity decreases the human body's ability to cool through evaporation from skin.

13. On a day when temperature reaches 37.8°C (100°F), how does a relative humidity reading of 50% affect apparent temperature?

Key Terms

absorption (p. 37)	global dimming (p. 39)	heat wave (p. 54)	net radiation (NET R)	sensible heat (p. 34)
albedo (p. 37)	greenhouse effect	isotherm (p. 52)	(p. 42)	specific heat (p. 47)
conduction (p. 35)	(p. 38)	jet contrails (p. 39)	reflection (p. 37)	temperature (p. 14)
continentality (p. 48)	greenhouse gases	latent heat (p. 34)	refraction (p. 37)	thermal equator (p. 52)
convection (p. 35)	(p. 38)	marine effect (p. 48)	scattering (diffuse	transparency (p. 47)
dust dome (p. 50)	heat (p. 34)	microclimate (p. 50)	radiation) (p. 36)	urban heat island (p. 50)

Critical Thinking

1. Would Death Valley have reached 57°C (134°F) if it were in a wetter climate? Explain your thinking using sensible heat and latent heat.

2. State a hypothesis predicting what might be the effect of lower air pollution levels on insolation and temperature? Explain.

3. Why do clouds have different effects on surface temperatures depending upon their altitude and thickness?

4. Describe how latitude affects the average temperature and the temperature range of a location?

5. If you were a landscape architect, what design choices would you use to reduce the urban heat island effect if you were designing a new shopping center?

Visual Analysis

The green roof of the California Academy of Science in San Francisco is a 2.5-acre living roof and home to almost 90 different species, and over 1.7 million native plants. The living roof also absorbs 70% of the rainwater that falls on it, which amounts to over 2 million gallons each year.

1. What in the photo could tell you about how the roof would affect the temperature of the building and its surroundings?

2. How would the flow of heat be different for this roof, compared with a roof made of composite shingles?

3. What physical factors would make this roof cooler than a normal roof?

4. What physical factors associated with climate change would increase the likelihood that more people would choose this type of roof?

(a)   (b)

▲ R2.1 Green roof on the California Academy of Science, San Francisco, CA

Interactive Mapping | Login to the **MasteringGeography** Study Area to access **MapMaster**.

Insolation

- Open: MapMaster in MasteringGeography.
- Select: *Global Mean Temperature for January* from the *Physical Environment* menu.

This layer shows mean temperature for January.

1. Which region has the highest temperature? Which region has the lowest temperatures?

2. What factors contribute to the high temperatures? What factors contribute to the low temperatures?

Deselect the *Global Mean Temperature for January* layer and select the *Global Mean Temperature for July* layer.

3. What causes the temperature patterns across South America at 20° South?

4. What causes the curved appearance of the isotherms off the west and east coasts of North and South America?

Explore | Use **Google Earth** to explore the **Lut Desert in Iran**.

Albedo Values

Places that appear bright in Google Earth have high albedo. Zoom and pan to Africa. How does central Africa compare with northern Africa? What factors are responsible for the high and low albedo values of those regions?

Search for the *Lut Desert, Iran*. This location is one of the hottest places on Earth.

Zoom into the *Lut Desert*.

1. How would the flow of heat in the Lut Desert differ from the flow of heat in central Africa, especially the latent heat of evaporation?

2. What explains this difference?

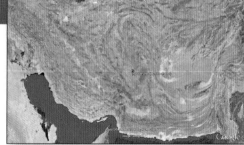

▲ R2.2

Mastering Geography™

Looking for additional review and test prep materials? Visit the Study Area in MasteringGeography™ to enhance your geographic literacy, spatial reasoning skills, and understanding of this chapter's content by accessing a variety of resources, including

MapMaster™ interactive maps, videos, *Mobile Field Trips, Project Condor* Quadcopter videos, *In the News* RSS feeds, flashcards, web links, self-study quizzes, and an eText version of *Geosystems Core*.

Some Like It Hot: Global Temperature Patterns

GeoLab2 (MG)
Pre-Lab Video

https://goo.gl/DEGwHD

Over the course of a year, seasonal changes in insolation, sun angle, and daylength help to create the patterns of annual mean temperatures and annual temperature ranges. The amount of seasonal change is largely a function of latitude, with higher latitudes having a greater variation between summer and winter temperatures. Since the oceans moderate temperatures, coastal locations have smaller temperature ranges than locations that are inland. Together latitude and continentality are the most important factors in determining temperature ranges.

Apply

As a commercial real estate broker, you know that different businesses prefer to locate in areas that can take advantage of certain climate characteristics. Because you have knowledge and understanding of temperature patterns, you can help clients find locations for their businesses that will take advantage of climate.

In this lab you will use a map and graphs to determine mean temperatures and evaluate the importance of continentality, latitude, and other factors in annual mean temperature. The second part of this GeoLab in MasteringGeography explores annual temperature range).

Objectives

- Evaluate the effects of continentality and latitude upon mean annual temperatures and temperature ranges.
- Compare the effects of continentality and latitude upon temperature ranges.
- Use your knowledge of global temperature patterns to identify locations for commercial real estate clients.

Procedure: Patterns of Annual Mean Temperatures

One of the patterns of climate is annual mean temperature. Before you start showing properties to clients, let's make sure that you understand what factors determine annual mean temperatures, and what patterns those factors create.

Fig GL 2.1 shows mean annual temperatures across the United States' 48 contiguous states, and Fig GL 2.2 shows annual mean temperatures on a global scale.

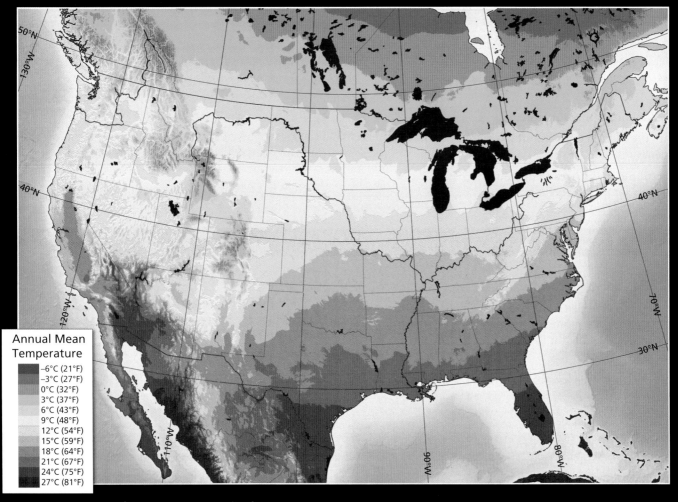

Annual Mean Temperature

	−6°C (21°F)
	−3°C (27°F)
	0°C (32°F)
	3°C (37°F)
	6°C (43°F)
	9°C (48°F)
	12°C (54°F)
	15°C (59°F)
	18°C (64°F)
	21°C (67°F)
	24°C (75°F)
	27°C (81°F)

▲GL2.1 **Mean annual temperature: United States**

1. According to the text, what are the main influences on mean temperature? What are two examples of how those influences shape the temperature patterns shown in Figure GL 2.1?
2. What accounts for the difference in temperature between West Virginia and Kentucky?
3. What accounts for the difference in temperature between Colorado and Kansas?
4. What is the main influence on the temperatures along 90°W?
5. What is the range in temperatures along 90°W, from 30°N to 70°N?
6. What is the rate of temperature change at 90°W, from 30°N to 70°N in temperature degrees per degree of longitude? Take the temperature range you found in question 5 and divide it by the number of degrees from 30°N to 70°N.

7. Refer to Fig GL2.2. What is the annual mean temperature at 50°N, 0°? What is the annual mean temperature at 50°N, 60°E? What is the annual mean temperature at 50° N, 120°E?

Analyze & Conclude

8. What factors, other than latitude, are responsible for the mean temperature patterns across North America seen in Fig GL2.1?
9. Describe the pattern of mean temperatures across California at 35°N. Explain the factors that cause the temperature to vary from west to east.
10. Which state has the lowest average annual temperatures over its entire area? The highest?
11. Referring to Fig GL2.2, compare the mean temperatures across 35°N from 120°W to 75°W with the mean temperatures across

35°N from 40°E to 120°E. What is the range across North America? What is the range across Eurasia? What explains the difference between them?
12. What annual mean temperatures would you expect to find across Australia, if it moved 20° south?
13. Based on Fig GL2.1 identify states in the United States for your client that would be the best for a ski resort (mean annual temperature below 0°C), citrus farms (mean annual temperature above 21°C, and beach resorts (mean annual temperature above 21°C and below 24°C).
14. Based on Fig GL2.2 identify for your client two global locations that would be suitable for long term cryogenic storage of important seeds. To be suitable, the annual mean temperature needs to be as cold or colder than –15°C.

Log in Mastering**Geography**™ to complete the online portion of this lab, view the Pre-Lab Video, and complete the Post-Lab Quiz.
www.masteringgeography.com

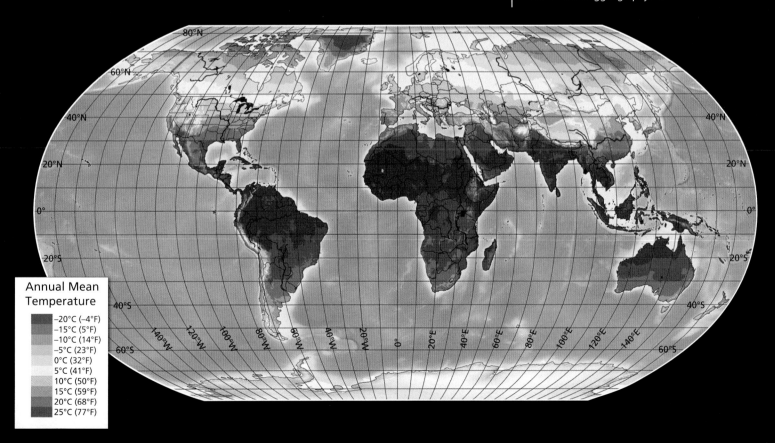

Annual Mean Temperature

- –20°C (–4°F)
- –15°C (5°F)
- –10°C (14°F)
- –5°C (23°F)
- 0°C (32°F)
- 5°C (41°F)
- 10°C (50°F)
- 15°C (59°F)
- 20°C (68°F)
- 25°C (77°F)

▲GL2.2 **Mean annual temperature: World**

Pressure, Winds, & Currents

3

The dramatic dust storm shown in the chapter opener photo is a visible expression of wind. In this chapter, we examine the temperature and pressure differences that cause wind to blow. We will also examine the forces that act on moving air to create global and local wind patterns. We then look at the circulation of Earth's atmosphere, including the principal pressure systems, patterns of global surface winds, upper atmosphere winds, monsoons, and local winds. Finally, we consider Earth's wind-driven oceanic currents and explain multiyear shifts in atmospheric and oceanic flows. The energy driving all this movement comes from one source: the Sun.

Key Concepts & Topics

A haboob, or dust storm, 600 m high (2000 ft) and 100 km (60 mi) wide, engulfs Phoenix, Arizona, in July 2012. This dust storm was driven by monsoon winds up to 80 kmph (50 mph), cut off power to over 9000, and closed the international airport for 20 minutes.

Wind Essentials

Key Learning Concepts

▶ **Define** the concept of air pressure.
▶ **Describe** instruments used to measure air pressure.
▶ **Define** wind.
▶ **Explain** how wind is measured, how wind direction is determined, and how winds are named.

A ir flows from one region to another because of differences in air pressure. These pressure differences occur because of Earth's uneven heating by the Sun. On a global scale, the imbalance between the heating of the tropics and the poles creates global patterns of circulation that include winds, ocean currents, and weather systems. At a local scale, the differences in heating between the ocean and the land create daily land–sea breeze patterns. The complex patterns of air circulation on our planet are the result of the interplay between these global and local factors. Figure 3.1 shows the normally invisible wind circulation pattern on September 1, 2015.

▲3.1 **Wind patterns** This image based on satellite data shows, from left to right, Hurricane Kilo, Hurricane Ignacio, and Hurricane Jimena on September 1, 2015. This was the first time that there were three hurricanes this powerful in the same ocean basin.

Air Pressure & Its Measurement

What causes air pressure? Air molecules in the atmosphere are pulled toward the center of Earth by gravity. The molecules bounce off each other, creating air pressure through their motion and number. Air molecules exert more force when they are closer together, as at Earth's surface, or when the temperature is higher and they are moving more quickly. Recall from Chapter 1 that air pressure decreases with altitude. In addition, air pressure rises and falls as weather systems move across Earth's surface.

Air pressure is measured using a barometer. In 1643, Evangelista Torricelli invented a device for measuring air pressure—the **mercury barometer** (◀**Fig. 3.2**). Torricelli sealed a 1 m tall glass tube at one end, filled it with mercury (Hg), and inverted it into a dish of mercury. He determined that the average height of the column of mercury in the tube was 760 mm (29.92 in.) and that it varied day to day as the weather changed. A more compact design is the **aneroid barometer**. The aneroid barometer principle is simple: A small chamber, partially emptied of air, is sealed and connected to a mechanism attached to a needle on a dial. As air pressure increases, it compresses the chamber; as air pressure decreases, the chamber expands—changes in air pressure move the needle. An aircraft altimeter is a type of aneroid barometer.

Normal sea-level pressure is 1013.2 mb (millibar, which expresses force per square meter of surface area) or 29.92 in. of mercury (Hg). In Canada and other countries, normal air pressure is expressed as 101.32 kPa (kilopascal; 1 kPa = 10 mb). As shown in Figure 3.3, the normal range of Earth's atmospheric pressure from strong high pressure to deep low pressure is about 1050 to 980 mb (31.00 to 29.00 in.).

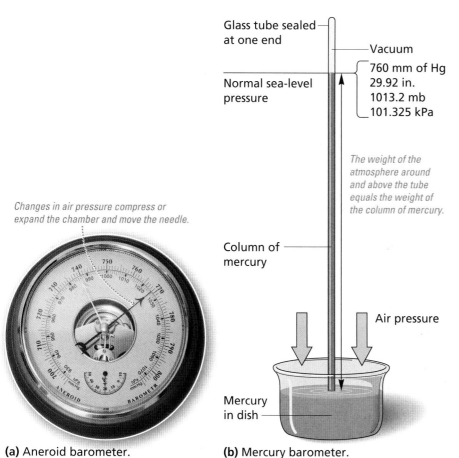

Glass tube sealed at one end

Vacuum

760 mm of Hg
29.92 in.
1013.2 mb
101.325 kPa

Normal sea-level pressure

The weight of the atmosphere around and above the tube equals the weight of the column of mercury.

Changes in air pressure compress or expand the chamber and move the needle.

Column of mercury

Air pressure

Mercury in dish

(a) Aneroid barometer.

(b) Mercury barometer.

▲3.2 **Aneroid and mercury barometers**

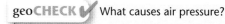 geoCHECK ✔ What causes air pressure?

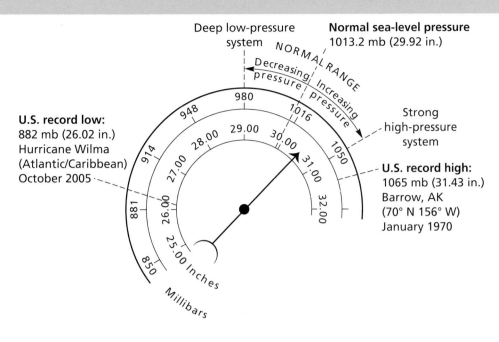

▲3.3 Air pressure readings and conversions Scales express air pressure in millibars and inches of mercury (Hg), with average air pressure values and recorded pressure extremes.

These are the 16 principal wind directions used by meteorologists. A wind from the west is a westerly wind (it blows eastward).

A wind out of the south is a southerly wind (it blows northward).

▲3.5 Wind compass Sixteen wind directions are identified on a wind compass. Winds are named for the direction from which they originate.

Wind

Temperature differences caused by uneven heating of Earth's surface cause different regions to have different air pressures. Differences in air pressure between one location and another produce **wind**, the horizontal motion of air. Wind flows from areas of higher air pressure toward areas of lower air pressure. Turbulence adds a vertical element in updrafts and downdrafts. Wind's two principal properties are speed and direction. An **anemometer**

▼3.4 Wind The same wind forces that drive the windsurfer also create the waves he's surfing on.

measures wind speed in kilometers per hour (kmph), miles per hour (mph), meters per second (mps), or knots. A *knot* is a nautical mile per hour, covering 1 minute of longitude at the equator in an hour, equivalent to 1.85 kmph, or 1.15 mph. The Beaufort wind scale is a descriptive scale useful in estimating wind speed based on simple observations. For example, at Beaufort scale 2 leaves are rustling in the breeze, and at scale 4 whole branches are swaying in the wind. The Beaufort scale is posted on our MasteringGeography website. A **wind vane** determines wind direction. The standard measurement is taken 10 m (33 ft) above the ground to reduce surface effects on wind direction.

Winds are named for the direction from which they originate (▲Fig. 3.5). For example, a north wind blows from the north to the south. The most common winds in the midlatitudes are called the westerlies, because they blow from the west toward the east.

Data (MG)
Beaufort scale

http://goo.gl/5DxNUR

geoCHECK ✔ What are the two main properties of wind that we measure?

geoQUIZ

1. Compare and contrast a mercury barometer and an aneroid barometer.
2. What was the lowest air pressure measured in the US? Where was the lowest air pressure in the US measured? What was the highest air pressure measured in the US? Where was the lowest air pressure in the US measured?.

3.2 Driving Forces within the Atmosphere

Key Learning Concepts

▶ **Explain** the four driving forces within the atmosphere: gravity, pressure gradient force, Coriolis force, and friction force.

▶ **Explain** how the pressure gradient, Coriolis, and friction forces affect winds in cyclones and anticyclones.

Four forces determine the speed and direction of winds. Gravitational force, pressure-gradient force, Coriolis force, and friction force operate on moving air and ocean currents and influence global wind circulation patterns.

Gravitational Force

Gravity pulls objects, including air molecules, toward the center of Earth. The gravitational force creates air pressure by compressing the atmosphere so that pressure is greatest at Earth's surface, and decreases with altitude. Without gravity, there would be no air pressure, and therefore no wind.

geoCHECK ✔ Why is air pressure higher closer to the surface?

Pressure-Gradient Force

A pressure gradient is the difference in air pressure between two points on Earth's surface. These pressure differences establish a **pressure-gradient force** that causes winds by driving air from areas of higher air pressure to areas of lower pressure (▲**Fig. 3.6**). Without a difference in air pressure between two locations, there would be no wind. Recall that these high- and low-pressure areas are mainly formed by the unequal heating of Earth's surface. Cold, dense air at the poles exerts greater pressure than warm, less dense air along the equator.

An **isobar** is an isoline that connects points of equal pressure on a weather map. Closely spaced contour lines on a topographic map indicate a steeper slope on land, and close isobars indicate steepness in the pressure gradient. Just as a ball rolls down a steep hill more quickly than a gently sloped hill, a steeper pressure gradient generates higher winds than a less-steep pressure gradient.

geoCHECK ✔ How is the pressure-gradient force related to wind speed?

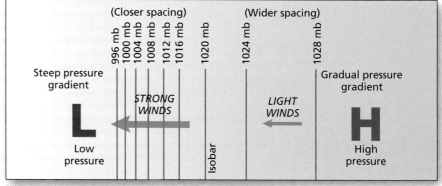

(a) Pressure gradient, isobars, and wind strength.

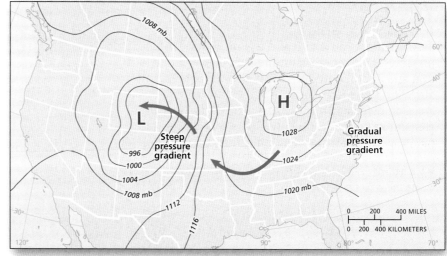

(b) Pressure gradient and wind strength portrayed on a weather map.

▲**3.6 Pressure gradient and wind speed**

Coriolis Force

Once air is moving, it is acted upon by two forces. The **Coriolis force** (also called the Coriolis effect) makes wind appear to be deflected in relation to Earth's rotating surface. Because Earth rotates eastward, objects appear to curve to the right in the Northern Hemisphere and to the left in the Southern Hemisphere (▶**Fig. 3.7**). The Coriolis force is an effect of Earth's rotation. Earth's rotational speed varies with latitude, increasing from 0 kmph at the poles to 1675 kmph (1041 mph) at the equator. The Coriolis force is zero along the equator, increases to half the maximum deflection at 30° N and 30° S, and reaches maximum deflection at the poles. The Coriolis force also increases as the speed of the moving object increases, so faster winds have greater apparent deflection. Without the Coriolis force, winds would move along straight paths between high- and low-pressure areas.

A common misconception about the Coriolis force is that it affects water draining out of a sink, tub, or toilet. Moving water or air must cover some distance before the Coriolis force noticeably deflects it. Water movements down a drain are too small in extent to be noticeably affected by this force.

geoCHECK ✔ What causes the Coriolis force?

Friction Force

The **friction force** slows wind speeds and reduces Coriolis force close to the surface. Surface friction extends to a height of about 500 m (around 1600 ft). In general, rougher surfaces produce more friction. Because surface friction decreases wind speed, it reduces the effect of the Coriolis force and causes winds to move across isobars at an angle.

geoCHECK ✔ How does friction affect surface winds?

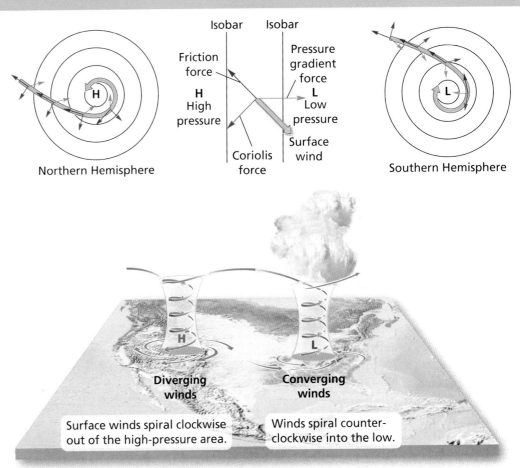

▲3.8 The pressure-gradient force, Coriolis force, and friction force that produce winds

Diverging winds — Surface winds spiral clockwise out of the high-pressure area.

Converging winds — Winds spiral counter-clockwise into the low.

Northern Hemisphere

Southern Hemisphere

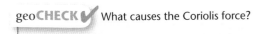

Animation (MG)
Coriolis Force

http://goo.gl/bs5jnb

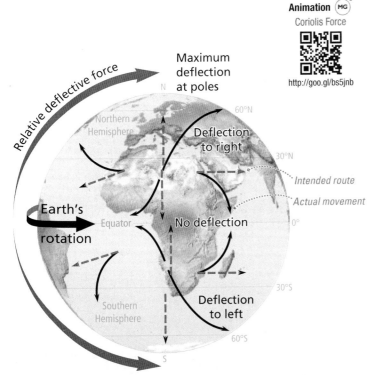

▲3.7 The Coriolis force—an apparent deflection

The Effects of Forces on Winds

Winds result from a combination of the four factors discussed above. Figure 3.8a shows the flow of wind from high to low pressure. High pressure is associated with descending and diverging air, and low pressure is associated with converging and ascending air. Surface friction decreases both the speed of winds and the Coriolis force. Winds close to the surface result from the pressure-gradient force, the Coriolis force, and friction. The net effect of these forces is wind that flows at a 20° to 45° angle to the pressure gradient (▲Fig. 3.8). Northern Hemisphere winds descend and spiral out from a high-pressure area clockwise to form an **anticyclone** and spiral into a low-pressure area counterclockwise and ascend to form a **cyclone**. In the Southern Hemisphere, these circulation patterns are reversed, with winds flowing counterclockwise out of anticyclonic high-pressure cells and clockwise into cyclonic low-pressure cells. Above 500 m (1600 ft), the two forces acting upon wind are the pressure-gradient force and the Coriolis force. Figure 3.8 shows that the result of these two forces is for wind to flow parallel to the isobars. Such winds, called **geostrophic winds**, are characteristic of upper tropospheric circulation.

geoCHECK ✔ How is the flow of geostrophic winds different from the flow of surface winds?

geoQUIZ
1. Briefly describe the four forces that cause winds.
2. Why does the Coriolis force deflect objects to their left in the Southern Hemisphere?

3.3 Global Patterns of Pressure & Motion

Key Learning Concepts

▶ *Locate* the primary high- and low-pressure areas and principal winds.

Earth's atmospheric circulation is driven by the flow of energy from the surplus at the equator toward the energy deficits at the poles (refer back to Figure 2.12). Uneven heating of Earth's surface results in global patterns of high and low pressure that in turn create Earth's patterns of surface winds, weather, and ocean currents. These movements transfer vast amounts of energy from the tropics toward the poles. The atmosphere is the main mechanism for transferring energy from about 35° latitude to the poles in each hemisphere, whereas ocean currents move more heat in the zone from the equator to 17° N and S. Atmospheric circulation also spreads natural and human-caused air pollutants worldwide.

Geographers view atmospheric circulation at three scales: primary circulation—worldwide patterns of pressure and winds; secondary circulation—continent-sized, migrating weather systems; and tertiary circulation—local winds.

Primary High-Pressure & Low-Pressure Areas

The following discussion of Earth's pressure patterns will refer often to Figures 3.9 and 3.10, which shows average surface barometric pressure in January and July. The spacing and position of the isobars also show the speed and direction of surface winds.

The primary high- and low-pressure areas appear on these maps as cells or uneven belts of similar pressure that stretch across the face of the planet, interrupted by landmasses. The primary winds flow from the primary high-pressure areas to the primary low-pressure areas.

Warmer, less-dense air along the equator rises, creating the low pressure region of the **intertropical convergence zone** or **ITCZ**; and colder, more-dense air at the poles sinks, creating the weak **polar high-pressure cells**. Air sinking over the tropics creates the **subtropical high-pressure cells**, and warm air moving poleward from the tropics rises, creating the **subpolar low-pressure cells**. In both hemispheres, winds are generally *easterly* (westward-moving) toward the equator from the tropics, and they are generally *westerly* (eastward-moving) in the middle and high latitudes.

geoCHECK ✔ Where are the four bands of pressure in each hemisphere?

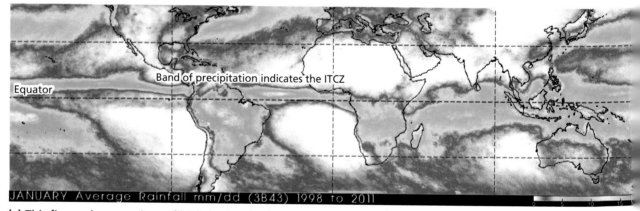

(a) This figure shows regions of high (red and yellow) precipitation along the ITCZ.

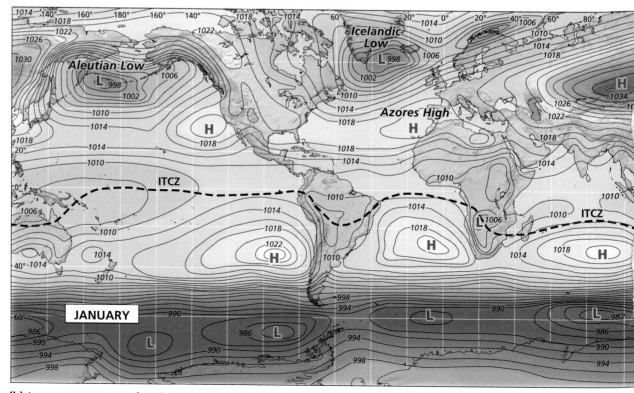

(b) January average surface barometric pressures (millibars); dashed line marks the general location of the intertropical convergence zone (ITCZ).

▲3.9 **ITCZ and global barometric pressures for January**

ITCZ: Warm & Rainy

Constant high Sun altitude and 12-hour days make large amounts of energy available in the ITCZ throughout the year. Warm, moist air rises and cools, releasing latent heat and producing condensation and rainfall. Vertical cloud columns frequently reach the top of the troposphere. The rising air causes surface winds to converge along the entire region of low pressure.

The combination of heating and convergence forces air aloft and forms the ITCZ. The ITCZ is identified by bands of clouds in Figures 3.9 and 3.10. In January, the zone crosses northern Australia and dips southward in eastern Africa and South America. In July, the ITCZ can be seen over southern Asia, central Africa, and northern South America.

Within the ITCZ, winds are calm or mildly variable because of the even pressure gradient and the vertical ascent of air. The rising air from the equatorial low-pressure area spirals upward to the north and south. Beginning at about 20° N and 20° S, air descends, forming high-pressure systems in the subtropical latitudes. This pattern of air rising in the ITCZ, moving toward the poles, and sinking in the subtropics creates the **Hadley cells**, two circulation cells on either side of the ITCZ (▼**Fig. 3.10**).

geoCHECK ✔ What is the general weather pattern caused by the ITCZ?

▼**3.10 ITCZ and global barometric pressures for July**

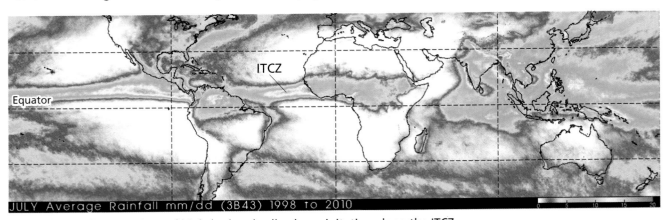

(a) This figure shows regions of high (red and yellow) precipitation along the ITCZ.

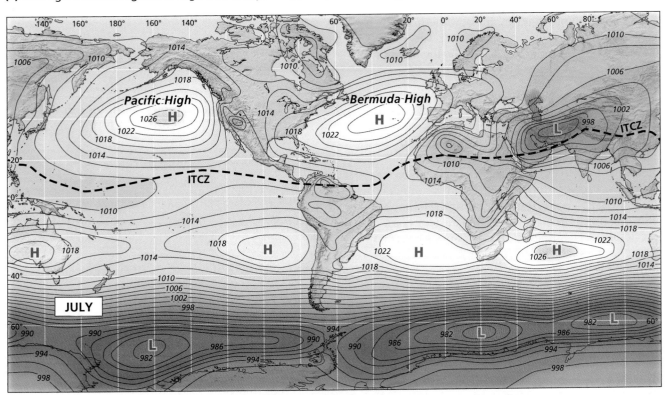

(b) July average surface barometric pressures (millibars); dashed line marks the general location of the intertropical convergence zone (ITCZ).

3.3 (cont'd) Global Patterns of Pressure & Motion

Subtropical High-Pressure Cells: Hot & Dry

Broad high-pressure zones of hot, dry air exist in both hemispheres between 20° and 35° latitude. The clear, frequently cloudless skies over the Sahara and the Arabian Deserts are typical of these zones, referred to as the subtropical high or subtropical high-pressure cells. Powered by the Hadley cells and their strong vertical movement of air in the ITCZ, these surface high-pressure cells are characterized by wind diverging at the surface.

The high-pressure centers form as air above the subtropics in the Hadley cells is pushed downward and heats by compression on its descent to the surface. Warmer air has a greater water vapor capacity than cooler air, making this descending warm air relatively dry because of its large water vapor capacity compared to its low water vapor content. The air is also dry because of the moisture that was lost as precipitation as the air initially rose in the ITCZ. One of the main reasons for the world's deserts being found in the subtropics is the warm, dry air from the subtropical high-pressure cells.

Air sinking to the surface spreads out and flows both toward the poles and toward the equator, creating two important belts of winds. In the Northern Hemisphere, the air flowing toward the equator forms the **northeast trade winds**. In the Southern Hemisphere, air flowing toward the equator forms the **southeast trade winds**. The wind flowing from the subtropical highs toward the poles in both hemispheres is deflected by the Coriolis force to create the **westerlies**, the main winds in the midlatitudes. The westerlies initially flow toward the pole from the subtropical high, but they are deflected to flow from the west as they move toward the poles.

Midlatitude Circulation Patterns The midlatitudes are the region that connects the subtropical high-pressure cells with the subpolar low-pressure cells and the polar high-pressure cells. The heat surplus in the tropics causes energy to flow toward the poles in the form of the warm, moist westerlies, which meet cooler air flowing from the poles along the polar front. The westerlies and the polar easterlies meet along the polar front, creating the midlatitude winter storms that will be explored in the next chapter.

Shifting Pressure Cells While sometimes referred to as belts of high pressure, the subtropical highs are actually distinct high pressure centers, or cells. The more important cells are named for their locations. For example, the Atlantic subtropical high-pressure cell is the *Bermuda high* (when it is in the western Atlantic in summer) or the *Azores high* (when it migrates to the eastern Atlantic in winter; Refer back to Fig. 3.9). The **Pacific High**, or **Hawaiian High**, dominates the Pacific in July, blocking storms coming from the west and moving them poleward. The high-pressure cell moves southward in winter, allowing storms to move through to the west coast of North America. In the Southern Hemisphere, large high-pressure centers over the Pacific, Atlantic, and Indian Oceans shift north and south with the subsolar point (refer back to Fig. 3.9).

The entire high-pressure system migrates 5° to 10° in latitude with the summer high Sun. The eastern sides of these anticyclonic systems are drier and more stable and feature cooler ocean currents than do the western sides. The drier eastern sides of these systems and dry summer conditions influence climate along subtropical and midlatitude west coasts. In fact, Earth's major deserts generally occur within the subtropical belt and extend to the west coast of each continent except Antarctica.

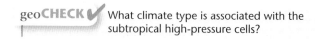 geoCHECK ✔ What climate type is associated with the subtropical high-pressure cells?

Subpolar Low-Pressure Cells along the Polar Front: Cool & Moist

Moving poleward from the subtropical high-pressure cells, we encounter the subpolar low-pressure cells. In January, two low-pressure cyclonic cells exist over the oceans around 60° N: the North Pacific *Aleutian low* and the North Atlantic *Icelandic low* (▶Fig. 3.10). Both cells dominate in winter and weaken or disappear in summer as the high-pressure systems in the subtropics strengthen. The area of contrast between cold, dry air from the polar and Arctic regions and warm, moist air brought by the westerlies forms the **polar front**, where masses of air with different characteristics battle, bringing storms and rain. This front encircles Earth, centered around these low-pressure areas. The changing weather patterns in the midlatitudes that we experience are made by secondary highs and lows along the polar front that migrate north and south with the seasons. Low-pressure cyclonic storms, hundreds to thousands of kilometers in diameter, migrate out of the Aleutian and Icelandic frontal areas, bringing precipitation to North America and Europe, respectively. Northwestern sections of North America and Europe generally are cool and moist as a result of the passage of these cyclonic systems.

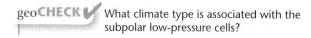 geoCHECK ✔ What climate type is associated with the subpolar low-pressure cells?

▼3.11 **Typical cool and moist conditions in the Pacific Northwest temperate rain forest**

**The Polar Front and Subpolar
Low-Pressure Cells**
Persistent lows over the North Pacific
and North Atlantic cause cool, moist
conditions. Cold, northern air masses
clash with warmer air masses to the
south, forming the polar front.

Polar High-Pressure Cells
A small atmospheric polar mass is cold and
dry with weak high pressure. LImited solar
energy results in weak, variable winds called
the polar easterlies.

*Surface air flowing from the
subtropical high-pressure cells
generates Earth's principal
surface winds: the westerlies
and the trade winds. The
westerlies are the dominant
surface winds from the
subtropics to high latitudes.*

*The trade winds, or trades,
converge on the equatorial
low-pressure trough. The
trade winds pick up large
quantities of moisture as they
return through the Hadley
circulation cells for another
cycle of uplift and
condensation. These Hadley
cells are the circuit completed
by winds rising along the ITCZ.
Air moves northward and
southward into the subtropics,
descending to the surface and
returning to the ITCZ as the
trade winds.*

**Intertropical
Convergence Zone (ITCZ)**
Lying along the equator,
the ITCZ is a trough of
low pressure and light
winds. Moist, unstable air
rises in the ITCZ, causing
heavy precipitation
year-round.

Subtropical High-Pressure Cells
Persistent highs produce regions where
air is pushed downward, compressed,
and warmed. Earth's major deserts
form beneath these cells.

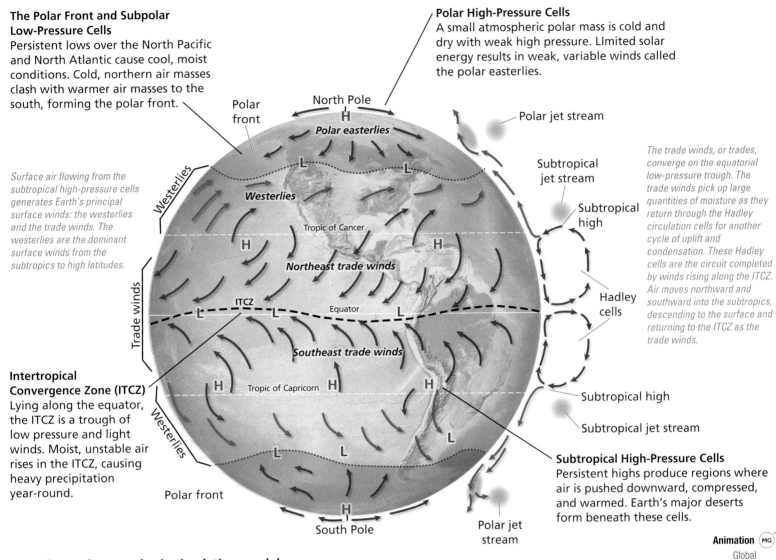

▲3.12 **General atmospheric circulation model**

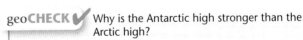
Polar high-pressure cells, and the winds they drive, are
weak because the polar atmosphere receives little energy from
the Sun (▲Fig. 3.12). Cold, dry, and weak winds move away
from the polar region in an anticyclonic direction. They descend and diverge clockwise in the Northern Hemisphere and
counterclockwise in the Southern Hemisphere and form the
weak, variable winds of the **polar easterlies**.

Of the two polar regions, the Antarctic has the stronger
and more persistent high-pressure system, the **Antarctic high**,
that forms over the Antarctic landmass. When an Arctic polar
high-pressure cell does form, it tends to locate over the colder
northern continental areas in winter (the Canadian or Siberian
high) rather than directly over the relatively warmer Arctic
Ocean.

geoCHECK ✔ Why is the Antarctic high stronger than the
Arctic high?

geo**QUIZ**

1. Explain two reasons that the subtropical highs create the dry conditions
   that are associated with desert climates.
2. Describe the two wind systems created by the subtropical high-pressure
   cells.
3. Relate the general temperature and precipitation characteristics of the
   tropics to the ITCZ and the trade winds.
4. Which two wind systems combine to create the pattern of weather
   along the polar front?

3.4 Upper Atmospheric Circulation

Key Learning Concepts

▶ **Explain** pressure patterns and winds in the middle and upper troposphere.
▶ **Describe** Rossby waves.
▶ **Define** the jet streams.

Circulation in the middle and upper troposphere is an important part of the atmosphere's general circulation. The middle and upper atmosphere is home to wind systems that guide storms and create weather patterns at the surface.

Pressure in the Upper Atmosphere

Weather maps of surface conditions are made with isobars, lines of equal air pressure. For upper-air maps, we plot the height above sea level that the pressure we are studying is found. For example, Figure 3.11a shows 500 mb plotted as a constant isobaric surface—all points have the same atmospheric pressure.

We use this 500-mb level to analyze upper-air winds and how they might affect surface weather conditions. Just as with surface maps, closer spacing of the isobars indicates faster winds; wider spacing indicates slower winds. Upper-air maps include areas of high pressure called **ridges** and areas of low pressure called **troughs**. The 500-mb height contours showing a ridge bend poleward; the contours showing a trough bend equatorward.

Ridges and troughs in the upper-air wind flow are important in sustaining surface cyclonic (low-pressure) and anticyclonic (high-pressure) circulation. Winds near the ridges slow and converge, causing air to descend and create high pressure at the surface, whereas winds near the area of maximum wind speeds along the troughs accelerate and diverge, causing low pressure air to converge and ascend from the surface. Compare the wind-speed indicators and labels in Figure 3.11a near the ridge with the wind-speed indicators around the trough, as well as the relationship between upper-air convergence and surface divergence.

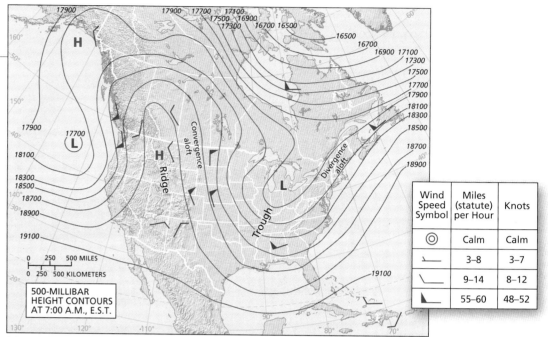

Wind Speed Symbol	Miles (statute) per Hour	Knots
◎	Calm	Calm
╰—	3–8	3–7
╰╲	9–14	8–12
╰╱	55–60	48–52

(a) Contours show height (in feet) at which 500-mb pressure occurs—a constant isobaric surface. The pattern of contours reveals geostrophic wind patterns in the troposphere ranging from 16,500 to 19,100 ft above sea level.

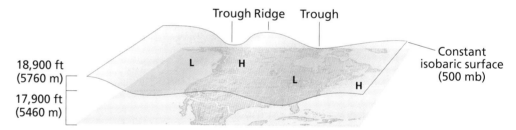

(b) Note the "ridge" of high pressure over the Intermountain West, at an altitude of 5760 m (18,900 ft), and the "trough" of low pressure over the Great Lakes region and off the Pacific Coast, at an altitude of 5460 m (17,900 ft).

Animation (MG)
Cyclones and Anticyclones

http://goo.gl/ZjE1jy

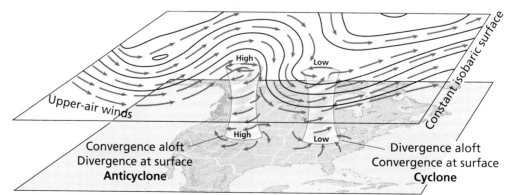

(c) Note areas of convergence aloft (corresponding to surface divergence) and divergence aloft (corresponding to surface convergence).

▲**3.13 Analysis of a constant isobaric surface for an April day**

Rossby Waves

Within the westerly flow of upper-air winds are **Rossby waves** (▼Fig. 3.14). Rossby waves flow along the polar front, the line of conflict between colder and drier air brought from the north by the polar easterlies and warmer and more moist air brought up from the south by the westerlies. The development of Rossby waves begins with ripples along the polar front that increase in amplitude to form waves. The extremely cold conditions during the "polar vortex" of 2014 were due to one of these waves growing in size and extending farther south than usual. These ripples develop into waves of ridges and troughs of higher and lower air pressure. If you live in the midlatitudes, the storms you experience during winter occur along the polar front.

geoCHECK ✔ Where do Rossby waves occur?

Animation (MG)
Jet Stream & Rossby Waves

http://goo.gl/Q2k47H

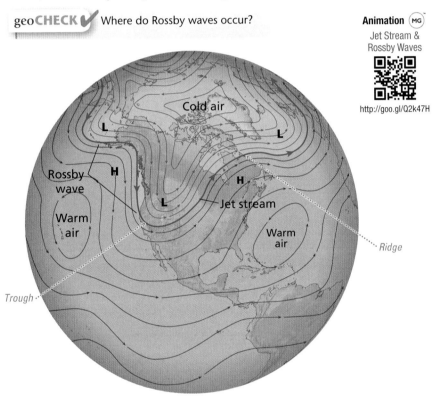

▲3.14 **Rossby waves in the upper atmosphere**

Jet Streams

The most prominent movements in these upper-level westerly wind flows are the **jet streams**, migrating rivers of wind that influence surface weather systems (▶Fig. 3.15). Their locations were shown in Figure 3.10. The jet streams are 160–480 km (100–300 mi) wide but only 900–2150 m (3000–7000 ft) thick, with core speeds that can exceed 300 kmph (190 mph). They tend to weaken during summer and strengthen during winter as the streams shift closer to the equator. The pattern of ridges and troughs causes variation in jet stream speeds. The strength of the jet stream can be seen in this satellite image of the eruption of the Eyjafjallajökull volcano in Iceland, which injected volcanic debris into the jet stream flow (▶Fig. 3.15b).

The *polar jet stream* meanders between 30° and 70° N, at the top of the troposphere, along the polar front, at altitudes between 7600 and 10,700 m (24,900 and 35,100 ft). The polar jet stream can migrate as far south as Texas, steering colder air masses into North America and influencing surface storm paths traveling eastward. In the summer, the polar jet stream exerts less influence on storms by staying over higher latitudes. The *subtropical jet stream* meanders from 20° to 50° N and may occur over North America simultaneously with the polar jet stream.

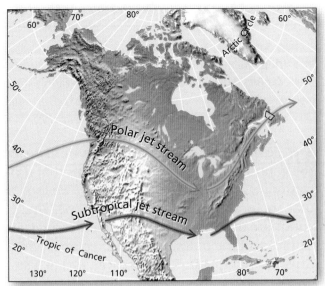

(a) Average locations of the two jet streams over North America.

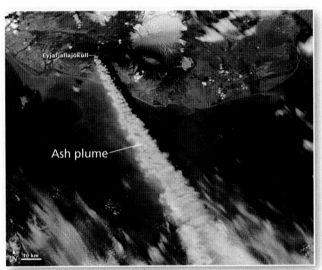

(b) Jet stream transport of volcanic ash. The jet stream blew ash from the Eyjafjallajokull volcano in Iceland south to Europe, in April, 2010. Since volcanic ash destroys jet engines; airports were shut down, and thousands of flights canceled. People's attention was focused on the guiding jet stream as it impacted their flight schedules and lives.

▲3.15 **Polar jet stream**

geoCHECK ✔ In which season is the polar jet stream strongest?

geoQUIZ

1. How does upper-level convergence affect surface air flow?
2. Are higher wind speeds associated with ridges or troughs? Explain why this occurs.
3. How could pilots use the jet stream to their advantage?

3.5 Local & Regional Winds

Key Learning Concepts

▶ **Describe** the conditions that produce several types of local winds.
▶ **Explain** regional monsoons in relation to Earth's atmospheric circulation.

We have looked at global-scale wind systems, the primary air circulation patterns that drive the westerlies and the trade winds. Now we look at winds that are local or regional in their scope. The global-scale winds are driven by heating differences between the tropics and the poles; local and regional winds are also driven by temperature differences, just at a smaller scale.

Local Winds

Local winds often arise because of differences in temperature and pressure that develop between areas of land and water or between parcels of air above land at different elevations. Cooler air is more dense (higher pressure) and tends to flow toward areas where the air is warmer (lower pressure).

Land & Sea Breezes Heating differences between land and water produce the land and sea breezes that occur on most coastlines (▶Fig. 3.16). Land warms faster than the water during the day. The warm air rises because it is less dense, which causes air pressure at the surface to drop by 1–2 mb, which triggers an onshore flow of cooler marine air to replace the rising warm air—usually strongest in the afternoon. The onshore flow of cooler air can lower the temperature by 2° to 10° C (3.6° to 18° F). At night, land cools faster than the water, cooler air over the land subsides and flows offshore over the warmer water, where the air is lifted. The land breezes are usually not as strong as the sea breezes, because land has a rougher surface than the ocean, so friction reduces the wind speed.

Mountain & Valley Breezes These breezes result from mountain air cooling rapidly at night and valley air heating rapidly during the day (▼Fig. 3.17). Warm air rises upslope during the day, and at night, cooler air subsides downslope into the valleys.

▼3.17 **Valley and mountain breezes**

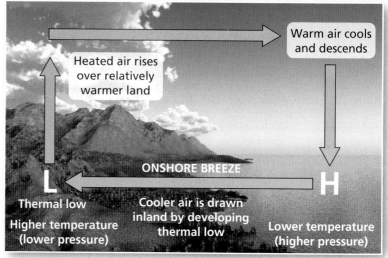

(a) Daytime sea-breeze conditions

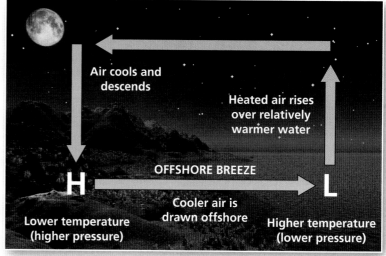

(b) Nighttime land-breeze conditions

▲3.16 **Sea and land breezes** Temperature and pressure patterns vary for (a) daytime sea breezes and (b) nighttime land breezes.

(a) Daytime valley-breeze conditions

(b) Nighttime mountain-breeze conditions

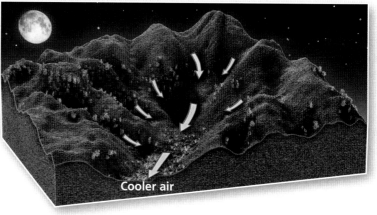

California
Great Basin
Santa Ana Winds
Pacific Ocean
Mexico

▲3.18 **Santa Ana winds and wildfires** The Santa Anas blow from high pressure over the Great Basin to low pressure over the Pacific coast, sending smoke plumes over the Pacific Ocean.

Santa Ana Winds Strong, dry winds sometimes flow across the desert towards Southern California coastal areas. These *Santa Ana winds* occur when high pressure builds over the Great Basin of the western United States in the fall, with low-pressure offshore. The winds are heated by compression as the air flows from higher to lower elevations, and gains speed as they move through constricting valleys. These winds bring hot, dry, dusty conditions to populated areas near the coast and create dangerous wildfire conditions (▲ Figure 3.18). Santa Ana winds encourage wildfires by drying out fire fuel, and coming at the end of the Southern California summer drought, before the fall rains begin.

Katabatic Winds Usually stronger than local winds and of larger, regional scale, *katabatic winds* occur when air at the surface cools, becomes denser, and flows downslope from an elevated plateau or highland. Gravity winds are not specifically related to the pressure gradient. Antarctica and Greenland experience ferocious katabatic winds that blow off the ice sheets.

geoCHECK ✔ Would sea-breezes be stronger during winter or summer? Explain.

Monsoon Winds

Regional winds are part of Earth's secondary atmospheric circulation. The **monsoons** are seasonally shifting wind systems caused by the annual cycle of migrating pressure belts. Monsoons occur in the tropics over Southeast Asia, Indonesia, India, northern Australia, and equatorial Africa. A mild version of such a monsoonal-type flow affects the southwestern United States. The location and size of the Asian landmass and its proximity to the Indian Ocean drive the monsoons of southern and eastern Asia (▶ Fig. 3.19). In the winter, high pressure dominates central Asia, while the equatorial low-pressure tough (ITCZ) is over the central Indian Ocean. This pressure gradient produces cold, dry winds from the Asian interior over the Himalayas and across India. These winds dry out the landscape, especially in combination with the hot temperatures from March through May. The ITCZ shifts northward during June to September, and high temperatures create

a thermal low in the Asian interior. Subtropical high pressure dominates the Indian Ocean, with a sea-surface temperature of 30°C (86°F). As a result of this reversed pressure gradient, hot winds pick up moisture from the warm ocean as they flow toward and are lifted by the Himalayas.

When the monsoonal rains arrive from June to September, they bring welcome relief from the dust, heat, and parched land of Asia's springtime. World-record rainfalls drench India. Cherrapunji, India, shown on the map, received both the second highest average annual rainfall (1218 cm, or 479.5 in.) and the highest single-year rainfall (2647 cm, or 1042 in.) on Earth. Because the monsoon rains provide critical water resources for the south Asian region, scientists are looking at how climate change affects monsoon strength and timing. Examples of the changes that may be underway include the 94.2 cm (37.1 in.) of rain that fell in a few hours on July 27, 2005, in Mumbai, India, causing widespread flooding; the extensive flooding across India in August 2007; the record-breaking monsoon rains and flooding in 2015 in Pakistan, India, and Myanmar.

geoCHECK ✔ When is the rainy season in the Asian monsoon?

geoQUIZ
1. Which local wind would be best suited to power the electrical demand of air conditioning for California?
2. Why are katabatic winds not necessarily related to the pressure gradient?
3. Which two pressure systems create the wet and dry seasons of the Asian monsoon?

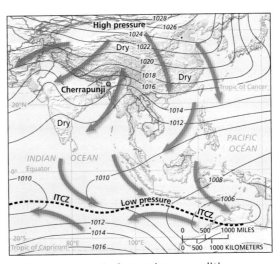

◀3.19 **The Asian monsoons**

(a) Northern Hemisphere winter conditions.

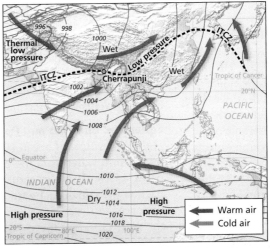

(b) Northern Hemisphere summer conditions.

Warm air
Cold air

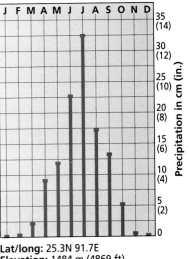

Lat/long: 25.3N 91.7E
Elevation: 1484 m (4869 ft)
Total ann. precip.: 1218.0 cm (479.5 in.)

(c) Precipitation at Cherrapunji, India.

3.6 Ocean Currents

Key Learning Concepts

▶ **Describe** the basic pattern of Earth's major surface ocean currents.
▶ **Explain** the causes and effects of upwelling and downwelling.
▶ **Describe** the deep thermohaline circulation of Earth's oceans.

Along with the pattern of global winds discussed earlier in this chapter, ocean currents play an important role in the transfer of heat energy from the tropics toward the poles. The circulation systems of the atmosphere and the oceans are intimately connected, because the major driving force for ocean currents is the frictional drag of the winds on the surface of the ocean. As wind moves over the rough ocean surface, the resulting friction transfers energy from the wind to the water, causing the water to move. This main force is modified by the Coriolis force, density differences caused by temperature and salinity, the configuration of the continents and ocean floor, and astronomical forces that cause the tides.

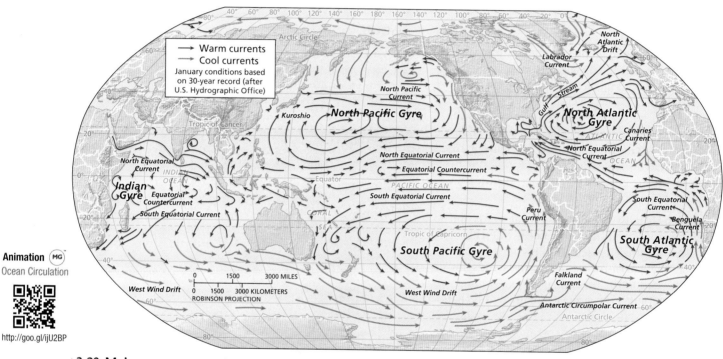

Animation (MG)
Ocean Circulation

http://goo.gl/ijU2BP

▲3.20 **Major ocean currents**

Surface Currents

Ocean currents are driven by the winds flowing out of the subtropical high-pressure cells in both hemispheres (▲Fig. 3.20). The large circular currents in the oceans are called **gyres**. Because ocean currents flow over long distances, the Coriolis force deflects them.

Some scientists study the details of the flow of these gyres by tracking objects washed overboard by storms. Examples of objects carried by these gyres include the cargo spill of nearly 30,000 rubber ducks, turtles, and frogs that washed overboard in 1994 in the North Pacific (▶Fig. 3.21).

Marine debris circulating in the Pacific gyre, including plastic in the "Pacific garbage patch," is the subject of ongoing scientific study. The Pacific garbage patch is a large region with a high concentration of tiny transparent particles of plastic, trapped by the currents of the Pacific gyre.

Equatorial Currents and Poleward Flows
The circular flow of ocean currents transports warm water from the equator to the poles, where the water cools and moves back to the equator. The trade winds drive the ocean surface waters westward along the equator. When these currents reach the eastern shores of the continents, water actually piles up to an average height of 15 cm (6 in.). This phenomenon, called the **western intensification**, forms poleward-moving currents.

The piled-up water moves poleward in tight channels along the eastern shorelines of the continents. In the Northern Hemisphere, the Gulf Stream and the Kuroshio move northward as a result of western intensification. The warm, deep, clear water of the Gulf Stream moves at 3–10 kmph (1.8–6.2 mph) and is 50–80 km (30–50 mi) wide and 1.5–2.0 km (0.9–1.2 mi) deep.

Upwelling Flows Ocean currents can move vertically as well as horizontally, because of local winds and differences in temperature, salinity, and density. Upwelling currents are formed where surface water is swept away from a coast. Cool nutrient-rich water rises from great depths to replace the water that was pushed aside. These currents exist off the Pacific coasts of North and South America and the subtropical and midlatitude west coast of Africa. These areas are some of Earth's prime fishing regions.

geoCHECK ✔ Do gyres follow the pattern of high- or low-pressure circulation?

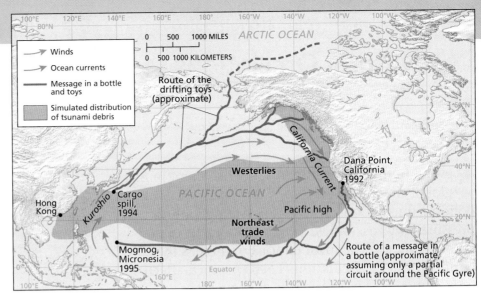

▲3.21 **Transport of marine debris by Pacific Ocean currents** The paths of a message in a bottle and toy rubber duckies show the movement of currents around the Pacific gyre. The distribution of debris from the 2011 Japan tsunami is a computer simulation based on expected winds and currents through January 7, 2012.

Thermohaline Circulation— Deep Currents

Differences in density caused by differences in temperatures and salinity produce deep currents called **thermohaline circulation** (*thermo-* refers to temperature and *-haline* refers to salinity, or the amount of salts dissolved in water). Although these currents flow more slowly than surface currents, thermohaline circulation moves larger volumes of water, and a large amount of heat as well.

While thermohaline circulation doesn't have a beginning or end, our discussion starts with the flow of the Gulf Stream toward the Arctic Ocean (▼Fig. 3.22). When this warm water mixes with the cold water of the Arctic Ocean, it cools, increases in density, and sinks. In addition, some water forms sea ice, leaving the salt behind to make the ocean water more saline, also increasing its density. As the cold temperature and increased salinity cause denser water to sink, surface water moves to replace the sinking water, forming a current and driving the thermohaline circulation. This process also occurs in the southern hemisphere as warm equatorial surface currents meet cold Antarctic waters. As this water moves northward, it warms; feeding the warm, shallow currents in the Indian Ocean and North Pacific.

A complete circuit of these surface and subsurface currents may require 1000 years.

Thermohaline circulation appears to play a profound role in global climate. Studies show that warming Arctic temperatures and melting polar ice could disrupt the deep current in the North Atlantic. Vast areas of melting sea ice and land ice, as in Greenland, are adding fresh water to the Arctic Ocean. Because the fresh water is less dense than salt water, these changes could affect the rate of sinking, and slow or stop the North Atlantic deep ocean circulation.

geoCHECK ✔ What causes surface water to sink and form the deep currents of thermohaline circulation?

geo**QUIZ**

1. Describe the forces that create surface ocean currents.
2. What is western intensification? How is it related to warm surface currents?
3. How would a warmer Arctic affect thermohaline circulation?

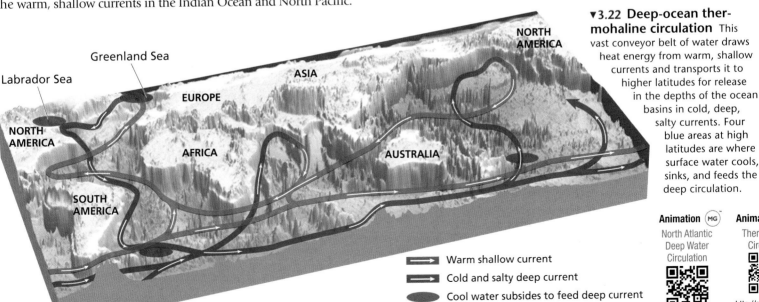

▼3.22 **Deep-ocean thermohaline circulation** This vast conveyor belt of water draws heat energy from warm, shallow currents and transports it to higher latitudes for release in the depths of the ocean basins in cold, deep, salty currents. Four blue areas at high latitudes are where surface water cools, sinks, and feeds the deep circulation.

⇨ Warm shallow current
⇨ Cold and salty deep current
⬭ Cool water subsides to feed deep current

Animation (MG)
North Atlantic
Deep Water
Circulation
http://goo.gl/tgGqUg

Animation (MG)
Thermohaline
Circulation
http://goo.gl/vJjCBX

3.7 Natural Oscillations in Global Circulation

Key Learning Concepts

▶ **Summarize** the main effects and timescales of the El Niño–Southern Oscillation.

Several multiyear oscillations are important in the global circulation picture. These oscillations are periodic fluctuations in sea-surface temperatures, air pressure patterns, and winds that ultimately affect global climates. The most famous of these is the El Niño–Southern Oscillation (ENSO), which affects temperature and precipitation on a global scale. Here we describe ENSO and its effects upon weather and climate.

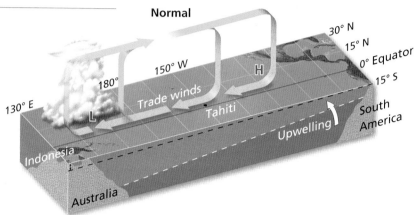

(a) Normal wind, pressure patterns, and upwelling patterns across the Pacific Ocean.

El Niño–Southern Oscillation

The El Niño–Southern Oscillation (ENSO) in the Pacific Ocean causes the greatest year-to-year variability of temperature and precipitation on a global scale. Although named El Niño ("the boy child") because these episodes seem to occur around Christmas, they can occur as early as spring and summer and persist through the year.

El Niño–ENSO's Warm Phase During an **El Niño**, air pressures and trade winds shift (▶Fig. 3.23b). Air pressure is higher than usual over the western Pacific and lower than usual over the eastern Pacific. The shift in air pressure causes the northeast trade winds to weaken and even reverse. The change in wind and ocean current direction causes cooler sea-surface temperatures in the western Pacific. At the same time, sea-surface temperatures increase up to 8 °C (14 °F) above normal in the central and eastern Pacific as warm water replaces the normally cold, nutrient-rich water along Peru's coastline. The shift in air pressure patterns and ocean temperatures create the El Niño–Southern Oscillation, or ENSO. Higher air pressure over the western Pacific causes drier conditions, while the lower air pressure over the eastern Pacific is associated with rising air and wetter conditions. The shift in air pressure also causes equatorial ocean currents to weaken or reverse. Normally, the current flowing eastward away from the west coast of South America creates an upwelling of cold water. During an ENSO, the wind and warm surface waters flow toward the South American coast slowing the upwelling currents of nutrient-rich water. This loss of nutrients affects the entire marine food chain, from plankton to birds and seals that feed on fish.

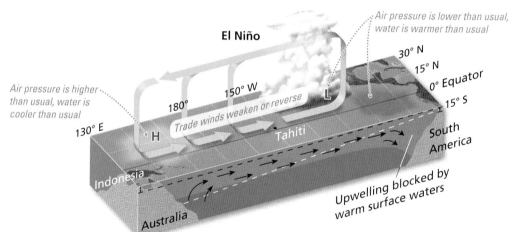

(b) El Niño wind, pressure patterns, and upwelling patterns across the Pacific Ocean.

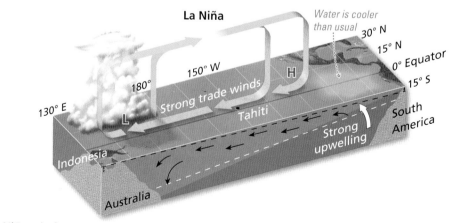

(c) La Niña wind, pressure patterns, and upwelling patterns across the Pacific Ocean.

▲**3.23 El Niño and La Niña**

Mobile Field Trip (MG)
El Niño

https://goo.gl/yQos7h

Animation (MG)
El Niño and
La Niña

http://goo.gl/1YAdt9

The expected interval for ENSO recurrence is 3 to 5 years. The frequency and intensity of ENSO events increased through the 20th century. The two strongest ENSO events in 120 years occurred in 1982–1983 and 1997–1998. An El Niño subsided in May 2010. Although the pattern began to build again in late summer 2012, it resulted in a weak El Niño that ended in early 2013. Conditions indicate that 2015 might see an ENSO more extreme than the El Nino of 1997–1998.

La Niña—ENSO's Cool Phase When surface waters in the central and eastern Pacific cool below normal, the condition is dubbed **La Niña**, Spanish for "the girl" (◄Fig. 3.23c). This condition is weaker and less consistent than El Niño. The strength or weakness of an El Niño does not correlate with the strength or weakness of La Niña, or vice versa. Following the record 1997–1998 ENSO event, the subsequent La Niña was not as strong as predicted.

In contrast, the 2010–2011 La Niña was one of the strongest on record. In Australia, this event corresponded with the wettest December in history. Heavy rainfall led to the country's worst flooding in 50 years in Queensland and across eastern Australia. Weeks of precipitation flooded an area the size of France and Germany combined and caused the evacuation of thousands of people.

Global Effects Related to ENSO and La Niña El Niño is tied to strong hurricanes in the eastern Pacific and droughts in South Africa, India, Australia, and the Philippines (▼Figs 3.24a and c). In India, every drought for more than 400 years seems

linked to this warm phase of ENSO. In the United States, the Northeast and Midwest states have warmer and drier winters, while the Southwest has cooler and wetter winters.

La Niña often brings wetter conditions throughout Indonesia, the South Pacific, and northern Brazil (▼Fig. 3.24b). During El Niño years Atlantic hurricanes are stronger, and during La Niña years they are weaker. In North America, La Niña brings higher amounts of rainfall across the upper Midwest, the northern Rockies, Northern California, and the Pacific Northwest. Canada typically experiences cooler and snowier winters. South America has drought conditions across Peru and Chile, while northern Brazil receives more rain than normal. See http://www.esrl.noaa.gov/psd/enso/ for more on ENSO, or go to NOAA's El Niño Theme Page at http://www.pmel.noaa.gov/toga-tao/el-nino/nino-home.html.

geoCHECK ✔ Explain what patterns are affected by ENSO.

geoQUIZ

1. How does the ENSO affect your weather? Explain the effects during the El Niño and La Niña phases of the ENSO.
2. How do conditions in the eastern and western Pacific Ocean differ during El Niño and La Niña?
3. Why does an El Niño reduce upwelling off the coast of South America?

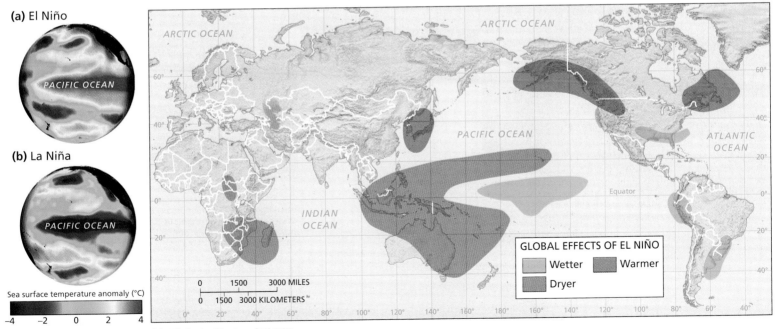

(a) El Niño

(b) La Niña

Sea surface temperature anomaly (°C)

−4   −2   0   2   4

(c) Global effects of El Niño.

GLOBAL EFFECTS OF EL NIÑO
Wetter   Warmer
Dryer

▲3.24 **El Niño and La Niña** (a) Typical sea surface temperature anomalies for an El Niño. (b) Typical sea surface temperature anomalies for a La Niña.

ATMOSPHERIC & OCEANIC CIRCULATION IMPACT HUMANS

• Wind and pressure contribute to Earth's general atmospheric circulation, which drives weather systems and spreads natural and anthropogenic pollution across the globe.
• Natural oscillations in global circulation, such as ENSO, affect global weather.
• Ocean currents carry human debris and non-native species into remote areas and spread oil spills across the globe.

HUMANS IMPACT ATMOSPHERIC & OCEANIC CIRCULATION

• Climate change may be altering patterns of atmospheric circulation, especially in relation to Arctic sea-ice melting and the jet stream, as well as possible intensification of subtropical high-pressure cells.
• Air pollution in Asia affects monsoonal wind flow; weaker flow could reduce rainfall and affect water availability.

3a

In June 2012, a dock washed ashore in Oregon, 15 months after a tsunami swept it out to sea from Misawa, Japan. The dock traveled about 7280 km (4524 mi) on ocean currents across the Pacific.

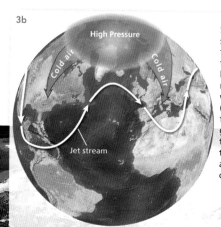

3b

High Pressure

Cold air   Cold air

Jet stream

Scientists think that melting sea ice in the Arctic owing to climate change is altering the temperature and pressure balance between polar and midlatitude regions. This weakens the jet stream (which steers weather systems from west to east around the globe), creating large meanders that bring colder conditions to the United States and Europe, and high-pressure conditions over Greenland.

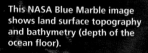

This NASA Blue Marble image shows land surface topography and bathymetry (depth of the ocean floor).

3d

Windsurfers enjoy the effects of the mistral winds off the coast of southern France. These cold, dry winds driven by pressure gradients blow southward across Europe through the Rhône River valley.

3c

In August 2010, monsoon rainfall caused flooding in Pakistan that affected 20 million people and led to over 2000 fatalities. The rains came from an unusually strong monsoonal flow, worsened by La Niña conditions. See Chapter 14 for satellite images of the Indus River during this event.

ISSUES FOR THE 21ST CENTURY

• Wind energy is a renewable resource that is expanding in use.
• Ongoing climate change may affect ocean currents, including the thermohaline circulation

Looking Ahead

In Chapter 4, you will learn how the atmosphere's water vapor content and temperature determine humidity. You will also explore how clouds and fog form, and consider the role of air masses in midlatitude cyclones. After an introduction to the basics of weather forecasting, you will see how atmospheric conditions lead to thunderstorms, tornadoes, and tropical cyclones

Chapter 3 Review

Why do we have wind?

3.1 Wind Essentials

Define the concept of air pressure, and **describe** instruments used to measure air pressure.

Define wind, and **explain** how wind is measured, how wind direction is determined, and how winds are named.

- Air pressure is created by the motion, size, and number of molecules. A barometer measures air pressure at the surface, whether it is a mercury barometer or an aneroid barometer.

- Wind is the horizontal movement of air across Earth's surface. Wind speed is measured with an anemometer, and its direction with a wind vane. Winds are named for the direction they come from.

1. How does air exert pressure?

2. What is normal sea-level pressure in millibars? In millimeters? In inches?

3.2 Driving Forces within the Atmosphere

Explain the four driving forces within the atmosphere—gravity, pressure-gradient force, Coriolis force, and friction force.

- Earth's gravitational force creates air pressure. The pressure-gradient force drives winds from high pressure to low pressure. The Coriolis force deflects objects to the right in the Northern Hemisphere and to the left in the Southern Hemisphere because of the rotation of Earth. The friction force slows winds next to the surface. The pressure gradient and Coriolis force produce geostrophic winds, which move parallel to isobars, above the surface frictional layer.

- Winds descend and diverge in a high-pressure stem, or anticyclone, spiraling outward in a clockwise direction in the Northern Hemisphere. In a low-pressure system, or cyclone, winds converge and ascend, spiraling upward in a counterclockwise direction in the Northern Hemisphere. The rotational directions are reversed for each in the Southern Hemisphere.

3. Explain how the Coriolis force appears to deflect atmospheric and oceanic circulations.

4. Describe the horizontal and vertical air motions in a high-pressure anticyclone and in a low-pressure cyclone.

What are the patterns of air & ocean circulation?

3.3 Global Patterns of Pressure & Motion

Locate the primary high- and low-pressure areas and principal winds.

- The primary pressure regions are the equatorial low, the weak polar highs (at both the North and the South Poles), the subtropical highs, and the subpolar lows.

- Along the equator, air rises in the ITCZ and descends in the subtropics in each hemisphere.

- The subtropical highs on Earth are generally between 20° and 35° latitude in each hemisphere. In the Northern Hemisphere, they include the Bermuda high, Azores High, and Pacific High. Winds flowing out of the subtropics to higher latitudes produce the westerlies in both hemispheres, while winds flowing toward the equator produce the trade winds.

- Cold polar air collides with warmer air from the subtropics to form the polar front. The Aleutian low and Icelandic low dominate the North Pacific and Atlantic, respectively. The weak and variable polar easterlies diverge from the highs at each pole, the stronger of which is the Antarctic high.

5. Construct a simple diagram of Earth's general circulation, including the four principal pressure belts and the three principal wind systems.

6. What is the relationship among the Aleutian low and migratory low-pressure cyclonic storms in North America?

3.4 Upper Atmospheric Circulation

Describe upper-air circulation, and **define** the jet streams.

- The variations in the altitude of the 500-mb surface show the ridges and troughs around high- and low-pressure systems. Areas of converging upper-air winds sustain surface highs, and areas of diverging upper-air winds sustain surface lows.

- Vast waves along the polar front are Rossby waves. Prominent streams of high-speed westerly winds in the upper-level troposphere are the polar and subtropical jet streams.

7. How is the 500-mb surface, especially the ridges and troughs, related to surface pressure systems? How are upper-level divergence and surface lows related? Convergence aloft and surface highs?

3.5 Local Winds

Explain several types of local winds and the regional monsoons.

- Land–water heating differences create land and sea breezes. Temperature differences during the day and evening between valleys and mountain summits cause mountain and valley breezes. Katabatic winds, or gravity drainage winds, need an elevated plateau or highland, where layers of air at the surface cool, become denser, and flow downslope.

- The shifting ITCZ causes intense summer rains and dry winter conditions in Southeast Asia, Indonesia, India, northern Australia, equatorial Africa, and southern Arizona.

8. Explain the factors that produce land and sea breezes.

9. Describe the seasonal pressure patterns that produce the Asian monsoonal wind and precipitation patterns. Contrast January and July conditions.

3.6 Ocean Currents

Describe the basic pattern of Earth's major surface ocean currents and deep thermohaline circulation.

- Ocean currents are caused by the frictional drag of wind, the Coriolis force, and differences in density. The circulation around subtropical highs in both hemispheres is usually offset toward the western side of each ocean basin.

- The trade winds push enormous quantities of water that pile up along the eastern shore of continents in a process known as the western intensification. Where surface water is swept away from a coast, an upwelling current of cold, nutrient-rich water occurs. These currents generate vertical mixing of heat energy and salinity.

- Density differences related to temperature and salinity create flows of deep, sometimes vertical, thermohaline circulation. A complete circuit of thermohaline circulation can take 1000 years. Increased surface temperatures in the ocean and atmosphere, coupled with climate-related changes in salinity, can alter the rate of thermohaline circulation in the oceans.

10. Where on Earth are upwelling currents experienced? How are upwelling currents related to fishing conditions?

11. What is meant by deep-ocean thermohaline circulation? At what rates do these currents flow?

What are multiyear oscillations?

3.7 Natural Oscillations in Global Circulation

Summarize the main effects and timescales of the El Niño—Southern Oscillation.

- The most famous of the several multiyear fluctuations is the El Niño–Southern Oscillation (ENSO) in the Pacific Ocean, which affects interannual variability in climate on a global scale.

- During an El Niño, air pressures and trade winds shift so that air pressure is higher than usual and sea-surface temperatures are lower than usual in the western Pacific, while air pressure is lower than usual and sea-surface temperatures are higher than usual in the eastern Pacific. During an ENSO, the wind and warm surface waters flow toward the South American coast, slowing the upwelling currents of nutrient-rich water, affecting the entire marine food chain, from plankton to birds and seals that feed on fish.

- When central and eastern Pacific sea-surface temperatures are below normal, a La Niña is occurring. The strength of a La Niña is unrelated to the strength of an El Niño.

12. Describe the changes in sea-surface temperatures and atmospheric pressure that occur during El Niño and La Niña, the warm and cool phases of the ENSO.

Key Terms

anemometer, p. 65
aneroid barometer, p. 64
Antarctic high, p. 71
anticyclone, p. 67
Coriolis force, p. 66
cyclone, p. 67
El Niño, p. 78

friction force, p. 67
geostrophic winds, p. 67
gyres, p. 76
Hadley cells, p. 69
Hawaiian high, p. 70
intertropical convergence zone (ITCZ) , p. 68
isobar, p. 66
jet streams, p. 73
La Niña, p. 79

mercury barometer, p. 64
monsoons, p. 75
northeast trade winds, p. 70
Pacific high, p. 70
polar easterlies, p. 71
polar front, p. 70
polar high-pressure cells, p. 68

pressure-gradient force, p. 66
ridges, p. 72
Rossby waves, p. 73
southeast trade winds, p. 70
subpolar low-pressure cells, p. 68
subtropical high-pressure cells, p. 68

thermohaline circulation, p. 77
trade winds, p. 70
troughs, p. 72
westerlies, p. 70
western intensification, p. 76
wind, p. 65
wind vane, p. 65

Critical Thinking

1. Describe or draw a diagram showing how the principal winds would blow if Earth rotated in the opposite direction. Include the warm and cold surface ocean currents.

2. How do the subtropical high-pressure cells affect summer rainfall in California and the Mediterranean? How would those climates be affected if the subtropical high-pressure cells became stronger throughout the year?

3. Compare how the El Niño and La Niña phases of ENSO affect weather patterns in the United States.

4. If the oceans continue to warm, how would warmer sea-surface temperatures affect the Asian monsoon?

5. Briefly explain how thermohaline circulation influences Europe's weather and how it would change if thermohaline circulation stopped or reversed.

Visual Analysis

Hundreds of millions of tons of dust are blown each year from the deserts of Africa, across the Atlantic Ocean, to become part of beaches in the Caribbean and soil in the Amazon. Some dust storms are so large that to properly appreciate their scale they need to be seen from space.

The natural-color image to the right was captured by the Suomi NPP satellite on July 31, 2013. The milky lines that run vertically across the image are actually the reflection of sunlight off the ocean. The maps to the right show the relative concentrations of aerosol particles on July 31 and August 1–2. The satellite also gathered ultraviolet radiation that revealed the presence of the airborne dust. Higher concentrations of dust are orange, lower concentrations and sun glint are yellow.

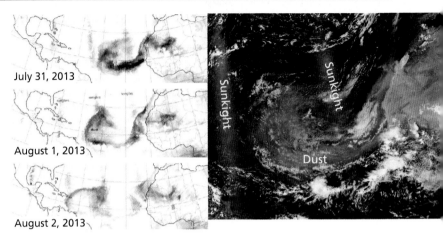

▲ **Figure R3.1 Visual Analysis** Dust from the Sahara Desert and Africa's interior in the summer of 2013.

1. How many days did it take the dust to travel from the middle of the Atlantic to Jamaica?

2. Which wind system was mainly responsible for blowing the dust toward South America?

Interactive Mapping | Login to the **MasteringGeography** Study Area to access **MapMaster**.

El Niño & La Niña

- Open: MapMaster in MasteringGeography.
- Select: *Sea Surface Temperature El Niño* from the *Physical Environment* menu.

This layer shows sea-surface temperature anomalies during the El Nino phase.

1. Which region has the highest temperature anomaly?

2. How much warmer than normal is this region?

- Select the *Sea Surface Temperature El Niño* layer and from the Physical Environment menu.

3. How much cooler is the Pacific at 160° W during La Niña than during El Niño?

4. Does El Niño or La Niña affect a larger area? Describe the regions that have higher and lower temperatures during a La Niña.

5. Given that there are better fishing conditions when there are upwelling currents off of the South American coast, describe fishing conditions during an El Niño and a La Niña.

Explore | Use **Google Earth** to explore Earth's **Winds & Clouds**.

Wind & Clouds

Click on the Information link in the *Weather* folder in the *Layers* window. You should see a pop-up window with a link to add the 24-hour clouds animation. Add the animation and set it to play on loop.

Google Earth does not have a built-in wind layer, but by using the clouds layer we can observe patterns of wind across Earth.

1. Identify the latitude and longitude of a midlatitude storm system or clouds that are an example of primary circulation. Which primary circulation is driving this system?

2. Identify the latitude and longitude of a tropical storm system or clouds that are an example of a different primary circulation. Which primary circulation is driving this system?

3. There is usually a belt of cumulonimbus clouds that are along the ITCZ. They often look like exploding popcorn. Identify the latitude of the region where you see them.

▲ **Figure R3.2**

MasteringGeography™

Looking for additional review and test prep materials? Visit the Study Area in MasteringGeography™ to enhance your geographic literacy, spatial reasoning skills, and understanding of this chapter's content by accessing a variety of resources, including MapMaster™

interactive maps, videos, *Mobile Field Trips*, *Project Condor* Quadcopter videos, *In the News* RSS feeds, flashcards, web links, self-study quizzes, and an eText version of *Geosystems Core*.

El Niño or La Niña?

The El Niño Southern Oscillation (ENSO) affects weather worldwide, influencing droughts, hurricanes, rainfall, and flooding. The shifts in sea-surface temperature and the trade winds also affect fishing conditions off the coasts of Ecuador and Peru. Scientists have a good understanding of the effects of ENSO but are currently unable to predict the occurrence of an ENSO more than one season ahead.

GeoLab3 (MG)

Pre-Lab Video

https://goo.gl/RO5EeJ

Apply

In order to understand patterns and changes in ocean temperatures, researchers use maps with isotherms. You are the lead climatologist for the Southern California Water District. You will use data from the Tropical Atmosphere Ocean Project (TAO) to create an isotherm map of subsurface temperatures in the tropical Pacific Ocean. The TAO project, operated by NOAA, has an array of 70 moored buoys that monitor surface temperature, rainfall, and wind, as well as ocean temperatures from 1 m to 500 m, ocean chemistry, and currents (Figs. GL 3.1 and GL.3.4). The Japan Agency for Marine-Earth Science and Technology operates the Triangle Trans-Ocean Buoy Network (TRITON) buoys. The TAO buoys are in the eastern and central Pacific (blue dots), and the TRITON buoys (yellow dots) are in the western Pacific.

Objectives

- Graph ocean temperatures
- Identify the location of the thermocline
- Apply your understanding of ENSO to determine whether an El Niño or a La Niña event is occurring

Procedure

Review the sections in module 3.7, Natural Oscillations in Global Circulation, that discuss the ENSO and its related phenomena, including Figure 3.22. Recall that isotherms are lines of equal temperature. You will use the buoy temperature data to draw isotherms.

In this exercise you will use data from the TAO/TRITON buoy array to determine whether El Niño or La Niña conditions were present during two different years.

1. On Figure GL 3.2, find and draw the 26°C isotherm.

To draw the isotherms:

a. Find a pair of values in each column that are greater and less than 26. For example, in the 150° E column, the temperature is 27.3° at 50 m and 25.1° at 75 m. The isotherm is between 50 m and 75 m, but it would be slightly closer to the 25.1°C because 26°C is closer to 25.1°C than to 27.3°C.

b. The 26°C isotherm has been started between 140° E and 180°. Continue the 26°C isotherm across the Pacific Ocean to 90° W.

c. Draw the 23°C, 20°C, 17°C, 14°C, 11°C, and 9°C isotherms.

2. Using colored pencils, shade the areas between the isotherms, using this key.

Table GL 3.1 Color key °C	
>26	Red
24–26	Orange
20–23	Yellow
17–19	Light green
14–16	Dark green
11–13	Light blue
09–10	Dark blue
<9	Violet

3. Repeat steps 1 and 2, using Figure GL 3.3.

4. On Figure GL 3.2, at which longitude do you find water temperatures greater than 26°C?

5. On Figure GL 3.3, at which longitude do you find water temperatures greater than 26°C?

6. How would you determine wind direction based upon the location of the highest sea-surface temperatures?

7. On Figure GL 3.2, which direction is the wind blowing? On Figure GL 3.3, which direction is the wind blowing?

8. Does Figure GL 3.2 or 3.3 show El Niño conditions? What evidence supports this?

9. Does Figure GL 3.2 or 3.3 show La Niña conditions? What evidence supports this?

10. Which longitudes on Figure GL 3.2 would support the strongest convection and highest rainfall? Explain your choice.

11. Which longitudes on Figure GL 3.3 would you expect to find the best fishing? Explain your choice.

12. On Figure GL 3.2, which longitudes and over which part of the Pacific Ocean (north, south, east, or west) would experience drought conditions? Explain your choice.

13. The thermocline is a layer of water where temperatures change rapidly with depth. To find the thermocline, look for the region where the isotherms are close together. Explain how the depth of the thermocline off of the South American coast changes between El Niño and La Niña conditions.

Analyze & Conclude

14. How would the air pressure and sea-surface temperature changes associated with El Niño potentially affect precipitation in the western United States? Explain your answer.

15. Why does the change in ocean temperature patterns you identified in question 13 have such a major impact on Earth's weather?

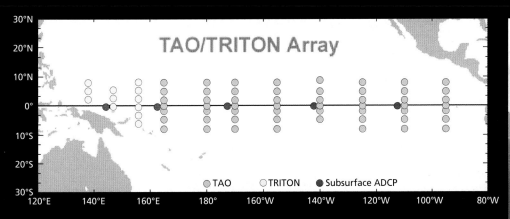

GL3.1 TAO/TRITON Array

TAO/TRITON Array

Legend: ○ TAO ○ TRITON ● Subsurface ADCP

Longitude axis: 120°E — 140°E — 160°E — 180° — 160°W — 140°W — 120°W — 100°W — 80°W
Latitude axis: 30°N, 20°N, 10°N, 0°, 10°S, 20°S, 30°S

Monthly Mean Ocean Temperatures – December 1997

Depth (m)	140°E	150°E	160°E	170°E	180	170°W		140°W	120°W	110°W	100°W
0	29.7	29.8	29.4	29.1	29.0	29.5		29.7	29.0	28.6	28.4
	29.3	29.5	29.1	29.1	29.0	29.4		29.6	29.1	28.4	28.2
50	27.3	27.9	27.7	28.5	29.0	29.4		29.8	29.0	28.4	28.0
	25.1	25.4	24.9	26.2	28.4	29.3		29.7	28.9	28.3	27.7
								29.4	28.7	28.2	27.2
100	22.7	22.5	22.9	22.5	24.8	27.8		28.3	28.1	27.4	25.1
	20.6	20.2	20.8	19.5	19.6	22.5		24.3	24.2	22.4	21.0
150	18.6	17.2	17.9	16.5	15.9	17.0		18.0	16.3	16.6	18.1
200	15.0	13.8	13.8	12.7	12.7	13.1		13.5	13.5	14.2	15.7
250	12.7	12.4	11.9	11.9	12.0	12.3					
300	11.6	11.4	11.4	11.3	11.3	11.8		12.1	12.1	12.4	12.5
350											
400											
450											
500	8.0	8.3	8.3	8.7	8.8	8.5		8.3	7.8	8.0	8.0

Longitude: 140°E 150°E 160°E 170°E 180 170°W 160°W 150°W 140°W 130°W 120°W 110°W 100°W 90°W

GL3.2 Ocean Thermal Structure Map—December 1997

Monthly Mean Ocean Temperatures – December 1998

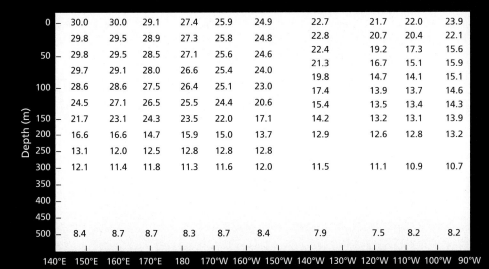

Depth (m)	140°E	150°E	160°E	170°E	180	170°W		140°W	120°W	110°W	100°W
0	30.0	30.0	29.1	27.4	25.9	24.9		22.7	21.7	22.0	23.9
	29.8	29.5	28.9	27.3	25.8	24.8		22.8	20.7	20.4	22.1
								22.4	19.2	17.3	15.6
50	29.8	29.5	28.5	27.1	25.6	24.6		21.3	16.7	15.1	15.9
	29.7	29.1	28.0	26.6	25.4	24.0		19.8	14.7	14.1	15.1
100	28.6	28.6	27.5	26.4	25.1	23.0		17.4	13.9	13.7	14.6
	24.5	27.1	26.5	25.5	24.4	20.6		15.4	13.5	13.4	14.3
150	21.7	23.1	24.3	23.5	22.0	17.1		14.2	13.2	13.1	13.9
200	16.6	16.6	14.7	15.9	15.0	13.7		12.9	12.6	12.8	13.2
250	13.1	12.0	12.5	12.8	12.8	12.8					
300	12.1	11.4	11.8	11.3	11.6	12.0		11.5	11.1	10.9	10.7
350											
400											
450											
500	8.4	8.7	8.7	8.3	8.7	8.4		7.9	7.5	8.2	8.2

Longitude: 140°E 150°E 160°E 170°E 180 170°W 160°W 150°W 140°W 130°W 120°W 110°W 100°W 90°W

GL3.3 Ocean Thermal Structure Map—December 1998

Next Generation ATLAS Current Meter Mooring

- Sea surface temperature-conductivity sensor (1 m depth)
- 3/8" Wire rope
- Conducting cable
- Subsurface sensors (temperature, conductivity current, pressure)
- Suburface Current Mooring
- ADCP
- Pressure/Temperature Sensor
- 700 m
- 3/4" Nylon line
- Acoustic Release
- Acoustic Release
- Anchor (6,000 lbs)
- Anchor (1,600 lbs)

GL3.4 Surface and Subsurface Buoys

Log in Mastering Geography™ to complete the online portion of this lab, view the Pre-Lab Video, and complete the Post-Lab Quiz.
www.masteringgeography.com

4

Atmospheric Water & Weather

In this chapter, we see how energy from the Sun combines with water to make clouds, rain, storms, tornadoes, and hurricanes—in other words, **weather**. Weather is the day-to-day condition of the atmosphere. **Climate** is the long-term average (over decades) of weather conditions and extremes in a region. Important elements that contribute to weather are temperature, air pressure, relative humidity, wind speed, wind direction, and seasonal factors such as insolation.

Meteorology is the scientific study of the atmosphere—*meteor* means "heavenly" or "of the atmosphere." Meteorologists study the atmosphere's physical characteristics and motions, the complex connections of atmospheric systems, and weather forecasting. Understanding atmospheric water is an essential part of meteorology and weather prediction.

Key Concepts & Topics

A thunderstorm passes just north of Grand Island, Nebraska. Storms such as this often produce large hail and lightning.

4.1 Water's Unique Properties

Key Learning Concepts

▶ *Describe* the heat properties of water.
▶ *Identify* the traits of water's three physical states: solid, liquid, and gas.

Water is the most common compound on Earth's surface and has several unique qualities. Water:

- is colorless, odorless, and tasteless.
- is the only substance on Earth that occurs in all three states—ice, liquid, and gas. (▶ Fig. 4.1)
- requires large amounts of energy to change its temperature.
- is the only substance that expands when it freezes.
- dissolves more substances than anything else.
- weighs 1 g/cm³ (gram per cubic centimeter), 62.3 lb/ft³, or 8.337 lb/gal.

Critical to water's role in the atmosphere are the three states of water and the energy needed to change its temperature.

Heat Properties & Phase Changes of Water

Specific heat is the amount of heat energy required to change the temperature of a substance. A calorie (cal) is the amount of energy required to raise the temperature of 1 g of water (at 15 °C) by 1 Celsius degree. To

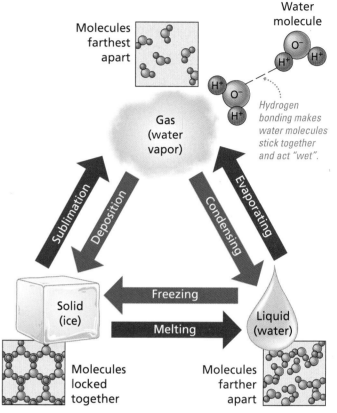

▲4.1 **Water's three physical states and the phase changes between them** Water exists in three physical states—solid, liquid, and gas. The transitions between these states are phase changes.

▼4.2 **The three physical states of water: snow and ice are solid; the ocean is liquid; water vapor in the air is a gas**

raise the temperature of water by 1 Celsius degree requires 1 cal, so the specific heat of water is 1 cal per gram. It takes approximately five times as much energy to raise or lower the temperature of water as it does to change the temperature of the same mass of granite, a common type of rock.

When water changes from one state to another, it undergoes a **phase change** (◀ Fig. 4.2). Heat energy must be added to or released from water for it to undergo a phase change. Heat energy that is absorbed or released as water changes phases is **latent heat**. Even though energy is absorbed or released by the water undergoing a phase change, the temperature of the water does not change. Heat exchanged during phase changes of water provides more than 30% of the energy that powers the general circulation of the atmosphere.

Ice, the Solid Phase As water cools, it contracts in volume and reaches its greatest density at 4 °C (39 °F). As it continues to cool, it can expand up to 9%, creating hexagonal (six-sided) structures.

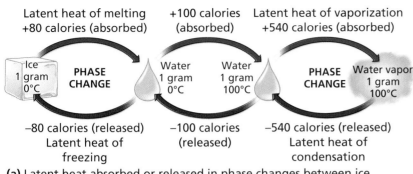

Latent heat of melting
+80 calories (absorbed)

+100 calories
(absorbed)

Latent heat of vaporization
+540 calories (absorbed)

−80 calories (released)
Latent heat of
freezing

−100 calories
(released)

−540 calories (released)
Latent heat of
condensation

(a) Latent heat absorbed or released in phase changes between ice and water and water vapor. To transform 1 g of ice at 0°C to 1 g of water vapor at 100°C requires 720 cal: 80 + 100 + 540.

▲**4.3 Water's heat energy characteristics**

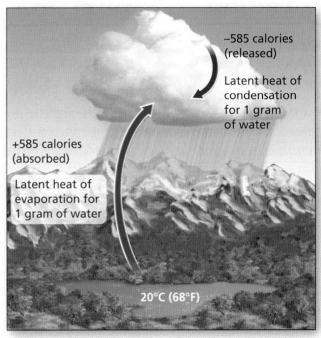

(b) The exchange of latent heat between water in a lake and the atmosphere.

Pure ice floats because it is 9% less dense than water. Without this lower density, much of Earth's freshwater would be in ice on the ocean floor.

For ice to change to water, the motion of the molecules must increase. While no temperature change occurs between ice at 0°C (32°F) and water at 0°C, 80 cal* of heat energy must be absorbed for the phase change of 1 g of ice to melt to 1 g of water. Both the latent heat of freezing and the latent heat of melting involve 80 cal.

Sublimation refers to the direct change of ice to water vapor (the change from water vapor to ice is deposition). The latent heat of sublimation is 620 cal, which is equal to the latent heat of melting combined with the latent heat of vaporization.

Water, the Liquid Phase To raise the temperature of 1 g of water at 0°C (32°F) to boiling at 100°C (212°F), we must add 100 cal, gaining an increase of 1 Celsius degree (1.8 Fahrenheit degree) for each calorie added. No phase change is involved in this temperature gain.

Water Vapor, the Gas Phase Water vapor is an invisible gas. The phase change from liquid to vapor at boiling temperature, at normal sea-level pressure, requires the absorption of 540 cal for each gram, the **latent heat of vaporization**. Vaporization refers to liquid water at boiling temperature becoming a gas, while evaporation happens at temperatures below boiling. You can feel the absorption of latent heat as evaporative cooling on your skin when it is wet. When water vapor condenses to a liquid, each gram gives up its stored 540 cal as the **latent heat of condensation**.

geoCHECK ✔ How does the amount of heat energy required to vaporize water compare to the heat energy required to melt ice?

*Remember that a calorie is the amount of energy required to raise the temperature of 1 g of water (at 15°C) by 1 Celsius degree.

Water Phase Changes in the Atmosphere

To evaporate water at 20°C (68°F), each gram of water must absorb approximately 585 cal from the environment as the **latent heat of evaporation** (▲Fig. 4.3). This is slightly more energy than the 540 cal required if the water was at boiling temperature.

The energy is released when air cools and water vapor condenses back into the liquid state, forming moisture droplets and releasing 585 cal per gram of water as the latent heat of condensation. Because of these unique heat properties, water is a major contributor of energy to the atmosphere. Latent heat absorbed by water vapor in the tropics is transported toward the poles by global wind patterns. When water vapor condenses, heat energy is released into the atmosphere, warming the atmosphere and powering storms.

geoCHECK ✔ Where does the heat energy in water vapor go when water vapor condenses to liquid water?

Animation (MG)
Water Phase
Changes

http://goo.gl/iC6lzW

geo**QUIZ**

1. How are the physical properties of water related to its physical structure?
2. What are the differences between heat absorbed by water that changes its temperature and heat absorbed by water that changes its phase?
3. Describe how latent heat is involved with the transfer of heat from the tropics to the poles.

4.2 Humidity—Water Vapor in the Atmosphere

Key Learning Concept

▶ **Define** humidity, relative humidity, and dew-point temperature.

Humidity refers to water vapor in the air. The amount of water vapor that can be present in the air is mainly a function of temperature. We are often aware of humidity when it is either too high or too low. North Americans spend billions of dollars a year to adjust humidity, either with air-conditioning (extracting water vapor and cooling) or with air-humidifying (adding water vapor).

Relative Humidity

After air temperature, relative humidity is the most common piece of information in local weather broadcasts. **Relative humidity** is the ratio (given as a percentage) of the amount of water vapor that is present in the air compared to the maximum water vapor possible in the air at a given temperature (▼Fig. 4.4).

Relative humidity changes as the temperature or the amount of water vapor in the air changes. Water molecules in the atmosphere exist as a gas, a liquid, and a solid. The molecules go back and forth between these states as they absorb or release heat energy. Warmer air increases the rate of evaporation, and cooler air tends to increase the rate of condensation. Warmer air has more energy to evaporate water than cooler air, so more water vapor can be present in warmer air than cooler air. When the amount of water vapor present is equal to the maximum amount possible, the relative humidity is 100%, and the air is **saturated**. In saturated air, an addition of water vapor or a decrease in temperature results in condensation, forming dew, clouds, fog, or precipitation.

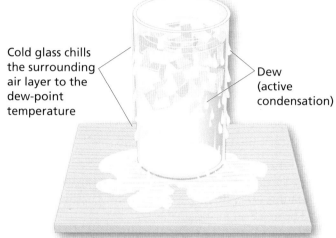

Cold glass chills the surrounding air layer to the dew-point temperature

Dew (active condensation)

(a) When the air cools to the dew-point temperature, water vapor condenses out of the air and onto the glass as dew.

(b) When temperatures are below freezing, water vapor will deposit as ice, onto a windshield for example.

▲**4.5 Dew-point and frost-point temperature examples**

The temperature at which a given mass of air becomes saturated and condensation begins is the **dew-point temperature** (▲Fig. 4.5). When temperatures are below freezing, the *frost point* occurs when air becomes saturated and ice crystals are deposited, forming frost.

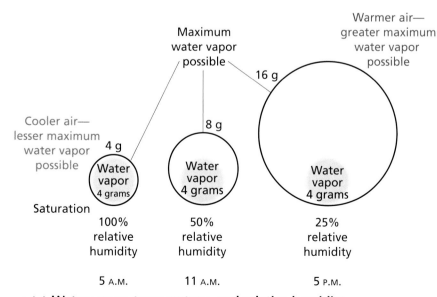

Cooler air—lesser maximum water vapor possible

Maximum water vapor possible

Warmer air—greater maximum water vapor possible

16 g

8 g

4 g

Water vapor 4 grams

Water vapor 4 grams

Water vapor 4 grams

Saturation

100% relative humidity

50% relative humidity

25% relative humidity

5 A.M.

11 A.M.

5 P.M.

▲**4.4 Water vapor, temperature, and relative humidity**

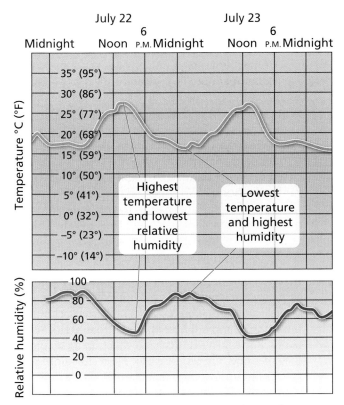

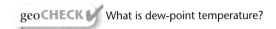

As temperature rises, there is more energy available to evaporate water than at lower temperatures. If the amount of water vapor remains constant and temperature rises, the amount of water vapor possible increases, so relative humidity will decrease. This results in the pattern during a typical day of temperature rising and relative humidity falling—and vice versa (◀Fig. 4.6). While the relative humidity varies, the actual amount of water vapor present in the air may remain the same. The relationship between relative humidity and temperature creates daily and seasonal patterns of higher and lower humidity. Relative humidity is highest at dawn, when air temperature is lowest. Seasonally, January relative humidity readings are higher than July readings because air temperatures are lower in winter.

Water vapor absorbs and releases thermal infrared energy, making it possible for satellites to measure water vapor in the troposphere (▼Fig. 4.7). Maps showing the amount of water vapor in the atmosphere are important in weather forecasting because it shows how much moisture could fall as rain or snow (shown in Fig. 4.7 as total precipitable water vapor).

geoCHECK ✔ What is dew-point temperature?

▲4.6 **Typical temperature and relative humidity patterns**
As temperature rises, relative humidity decreases. When temperature falls, relative humidity increases.

▼4.7 **Average total precipitable water vapor in mm, May 2009** Higher temperatures in the tropics cause higher amounts of water vapor in the atmosphere. Compare the total precipitable water amounts along the equator with the amounts along 40°N.

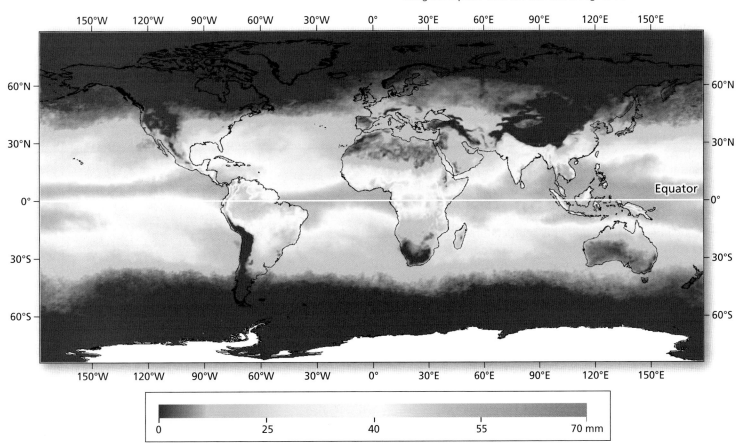

4.2 (cont'd) Humidity—Water Vapor in the Atmosphere

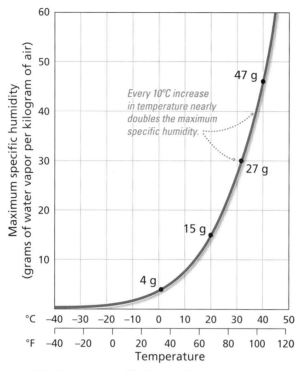

▲4.8 **Maximum specific humidity** Maximum specific humidity is the maximum possible water vapor in a mass of water vapor per unit mass of air (g/kg).

Expressions of Humidity

Two more ways to express humidity are specific humidity and vapor pressure.

Specific humidity While relative humidity changes with temperature, **specific humidity** stays the same as temperature changes (▲**Fig. 4.8**). Specific humidity is the mass of water vapor (in grams) per mass of air (in kilograms).

The maximum mass of water vapor possible in a kilogram of air at any specified temperature is the maximum specific humidity, plotted in Figure 4.8. The graph illustrates that, for every temperature increase of 10°C (18°F), the maximum specific humidity of air nearly doubles. Notice that there are larger changes at higher temperatures. While the maximum doubles with each 10°C increase, warming from −30°C (−22°F) to −20°C (−4°F) results in an increase from 0.5 g maximum to 1.0 g maximum, while an increase from 30°C (86°F) to 40°C (104°F) results in an increase from 27 g to 47 g. Specific humidity is useful in describing the moisture content of large air masses and provides information necessary for weather forecasting.

Vapor Pressure When water molecules evaporate, they become a gas and are a portion of the air pressure. The share of air pressure that is made up of water-vapor molecules is **vapor pressure**, expressed in millibars (mb) (▶**Fig. 4.9**).

Air that contains as much water vapor as possible at a given temperature is at saturation vapor pressure. Any temperature increase or decrease will change the saturation vapor pressure.

The inset in Figure 4.9 compares saturation vapor pressures over water and over ice surfaces at subfreezing temperatures. Saturation vapor pressure is lower over ice than over water because the rigid structure and strong molecular bonds in ice make it harder for water molecules to leave an ice surface than to leave a water surface, where molecules are not locked together as tightly. This is important in the process of raindrop formation in clouds that are below freezing temperature.

Distribution of Temperature & Humidity The relationship between temperature and humidity explains why warm tropical air over the ocean can contain so much water vapor, thus providing latent heat to power tropical storms. It also explains why cold air is "dry" and why extremely cold air toward the poles does not produce a lot of precipitation. You might have heard someone say that it is too cold to snow. When air temperatures are very low, the air contains little water vapor,

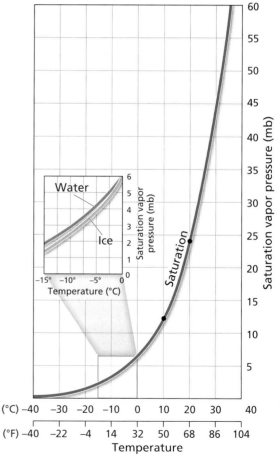

▲4.9 **Saturation vapor pressure** Saturation vapor pressure is the maximum possible water vapor, as measured by the pressure it exerts (mb). Inset compares saturation vapor pressures over water surfaces with those over surfaces at subfreezing temperatures.

Another instrument is a sling psychrometer, which consists of two thermometers mounted side by side on a holder (▼Fig. 4.11). One thermometer is the dry-bulb thermometer, and it records the ambient (surrounding) air temperature. The other thermometer is the wet-bulb thermometer, and its bulb is covered by a moistened cloth wick.

If the air is dry, water evaporates quickly; the water vapor absorbs latent heat from the wet-bulb thermometer, cooling the thermometer and causing its temperature to lower (the difference between the two temperatures is the wet-bulb depression). On the other hand, in conditions of high humidity, little water evaporates from the wick, so the wet-bulb temperature only drops a little.

geoCHECK ✔ How does relative humidity differ from specific humidity?

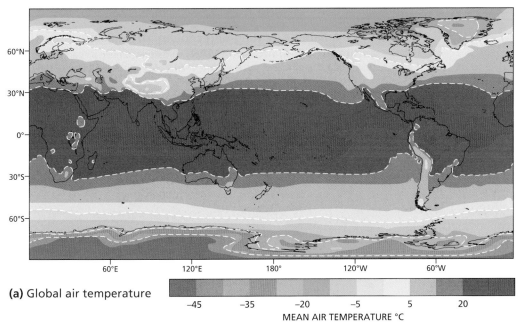

(a) Global air temperature

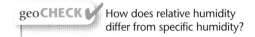

MEAN AIR TEMPERATURE °C

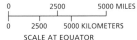

SCALE AT EQUATOR

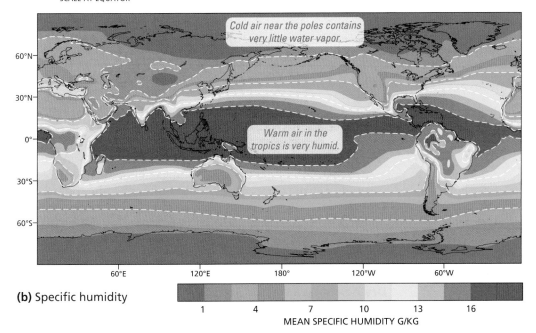

Cold air near the poles contains very little water vapor.

Warm air in the tropics is very humid.

(b) Specific humidity

1   4   7   10   13   16
MEAN SPECIFIC HUMIDITY G/KG

▲4.10 **Global air temperature and specific humidity**

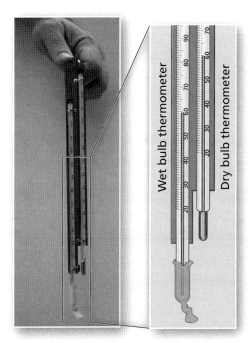

▲4.11 **The sling psychrometer is a tool for measuring humidity**

even though it is at 100% relative humidity. For example, at −20° C (−4°F), the maximum specific humidity is 1 g of water vapor per kilogram of air. (▲Fig. 4.10).

Instruments for Measuring Humidity Various instruments measure relative humidity. The hair hygrometer uses the principle that human hair changes as much as 4 percent in length between 0% and 100% relative humidity. As the hair absorbs or loses water in the air, it changes length, indicating relative humidity.

geo**QUIZ**

1. When is relative humidity highest during a typical day?
2. Why would meteorologists analyze specific humidity of the atmosphere, rather than relative humidity?
3. What is the relative humidity if the wet-bulb and dry-bulb temperatures of a sling psychrometer are the same?

4.3 Atmospheric Stability

Key Learning Concept

▶ *Define* atmospheric stability.

Meteorologists use the term *parcel* to describe a body of air. Think of an air parcel as a volume of air, perhaps 300 m (1000 ft) in diameter or more with specific temperature and humidity characteristics. Two opposing forces work on a parcel of air: an upward *buoyant force* and a downward *gravitational force* (▼Fig. 4.12). The parcel's temperature determines its density— warm air is less dense than cold air. When a parcel of air is warmer than the air around it, the parcel is less dense than the surrounding air, and the parcel's buoyancy increases.

Stability refers to the tendency of an air parcel to rise, remain in place, or fall. A parcel that is warmer than the air around it is *unstable*, and it will rise until it reaches an altitude where the air around it has a density and temperature similar to its own. A *stable* parcel will either stay where it is or sink. A hot-air balloon rises because it is warmer and therefore less dense than the air around it. A cold-air balloon wouldn't rise because it would be more dense than the air around it.

geoCHECK ✔ Why do you think most hot-air balloon launches start early in the morning?

Adiabatic Processes

The normal lapse rate and the environmental lapse rate (ELR) refer to the change in temperature with increasing altitude through the troposphere. The **normal lapse rate** is the average decrease in temperature with increase in altitude through the troposphere, measured at 6.4 Celsius degrees/1000 m (3.5 Fahrenheit degrees/1000 ft). The **environmental lapse rate (ELR)** is the actual

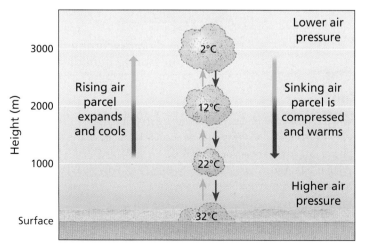

▲**4.12 Adiabatic heating and cooling** Rising air expands and cools, sinking air is compressed and heats.

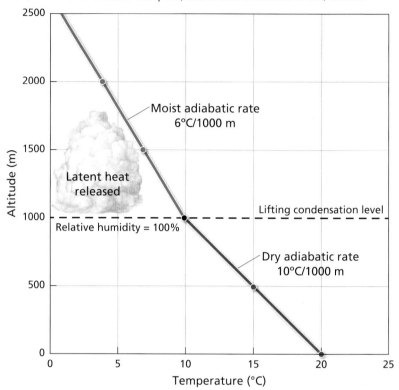

▼**4.13 Adiabatic heating and cooling: dry adiabatic rate (DAR) and moist adiabatic rate (MAR)** Unsaturated rising air cools at the DAR of 10°C/1000 m until it cools to dew point, then it cools at the MAR of 6°C/1000 m.

lapse rate at a particular place and time and varies by several degrees per thousand meters. The **Adiabatic** rates refer to the change in temperature of a moving parcel of air. Adiabatic means occurring without a loss or gain of heat—in this case, without any heat exchange between the surrounding environment and the vertically moving parcel of air. *Adiabatic* describes the cooling and warming rates for a parcel of rising or falling air. A *rising* parcel of air cools by expansion because of lower air pressure at higher altitudes. *Sinking* air heats by compression. The two adiabatic lapse rates depend on the moisture conditions in the parcel: the dry adiabatic rate (DAR) and the moist adiabatic rate (MAR).

Dry Adiabatic Rate The **dry adiabatic rate (DAR)** is the rate at which "dry" air cools by expansion (if ascending) or heats by compression (if descending) (▲Fig. 4.13). "Dry" refers to air that has less than 100% relative humidity. The average DAR is 10 Celsius degrees/1000 m (5.5 Fahrenheit degrees/1000 ft).

Moist Adiabatic Rate The **moist adiabatic rate (MAR)** is the rate at which an ascending air parcel that is moist, or saturated, cools by expansion (▶Fig. 4.13). The average MAR is 6 Celsius degrees/1000 m (3.3 Fahrenheit degrees/1000 ft). This is roughly 4 Celsius degrees (2 Fahrenheit degrees) less than the dry adiabatic rate. As a saturated parcel of air rises and cools, water vapor condenses and releases the latent heat of condensation, forming a cloud. The altitude at which condensation occurs is the **lifting condensation level**. The higher the water-vapor content is, the more latent heat is released. The MAR is much lower than the DAR in warm air, whereas the two rates are more similar in cold air. From the 6 Celsius degrees/1000 m (3.3 Fahrenheit degrees/1000 ft) average, the MAR varies with moisture content and temperature and can range from 4–10 Celsius degrees per 1000 m (2–5.5 Fahrenheit degrees per 1000 ft).

 geoCHECK Why is the DAR higher than the MAR?

Stable & Unstable Atmospheric Conditions

The rate of cooling of a parcel compared to the environmental lapse rate determines the stability of the atmosphere. Assume that a lifting mechanism such as surface heating or a mountain range is present to get the parcel started. The atmosphere is *unstable* if a rising parcel cools more slowly than the air around it (▶Fig. 4.14). The parcel will become warmer, less dense, and more buoyant than the surrounding air, and it will continue to rise. As it rises, cools, and becomes saturated, clouds may form (▼Fig. 4.15a). The atmosphere is *stable* if an air parcel cools more quickly than the air around it and is denser than the surrounding air (▼Fig 4.15b). The denser air parcel resists lifting, and the sky remains generally cloud-free. If the environmental lapse rate is between the DAR and MAR, the atmosphere is *conditionally unstable*. If the rising parcel is unsaturated and cools at the DAR, it will be stable. If the parcel becomes saturated and cools at the MAR, it can become unstable.

geoCHECK What is an example of an ELR that would create atmospheric conditions that are stable? Unstable? Conditionally unstable?

geoQUIZ

1. What two opposing forces act on parcels of air?
2. What would cause the ELR to vary?
3. Why would an air parcel with more water vapor become unstable before an air parcel with less water vapor would?

▼4.14 **Instability of an air parcel: ELR less than DAR, but greater than MAR** A parcel of air cooling at the DAR will be colder than the air around it, while a parcel of air cooling at the MAR can become warmer than the surrounding air and thus unstable.

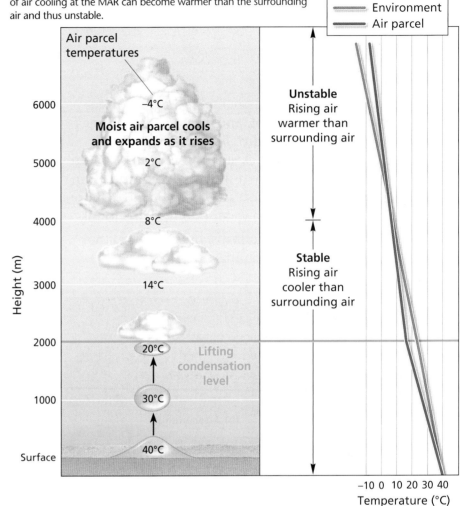

▼4.15 **Unstable and stable conditions**

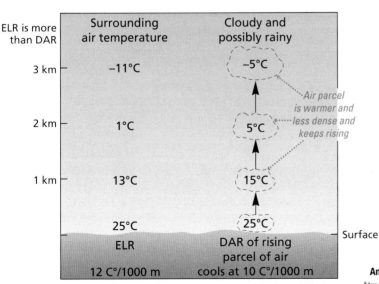

(a) **Unstable** If the ELR is greater than the DAR, conditions are unstable because the parcel will cool more slowly than the air around it. The parcel will become warmer than the air around it and continue to rise because it is less dense than the air around it.

Animation MG
Atmospheric Stability

http://goo.gl/ztSTTd

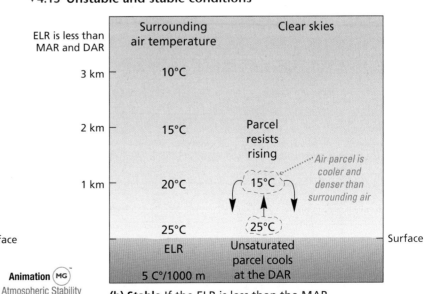

(b) **Stable** If the ELR is less than the MAR, conditions are stable because the parcel will cool more quickly than the air around it. The cooler parcel will be more dense than the air around it, and will not rise on its own.

4.4 Clouds & Fog

Key Learning Concepts

▶ **Explain** cloud formation, and identify major cloud classes and types, including fog.

▶ **Describe** the two ways that raindrops form.

Clouds and fog are fundamental indicators of stability, moisture content, and weather. A **cloud** is a visible group of moisture droplets or ice crystals suspended in air. Fog is just a cloud on the ground. Of the many cloud types, we present the most common examples in a simple classification scheme (▼Fig. 4.16). **Precipitation** is liquid or solid water falling from clouds.

Clouds may contain raindrops, but not when they first form. Clouds are a great mass of moisture drop-lets, each invisible without magnification. A **moisture droplet** is approximately 20 μm (micrometers) in diameter (0.002 cm, or 0.0008 in.) and is made of a million or more smaller droplets.

Condensation requires saturated air and **condensation nuclei**, microscopic particles that always are present in the atmosphere. Ordinary dust, soot, and ash from volcanoes and forest fires and particles from burned fuel typically provide these nuclei. Urban air has high concentrations of these nuclei, while maritime air masses have lower concentrations.

geoCHECK ✔ What two conditions are required for condensation to occur?

▼4.16 **Principal cloud types**

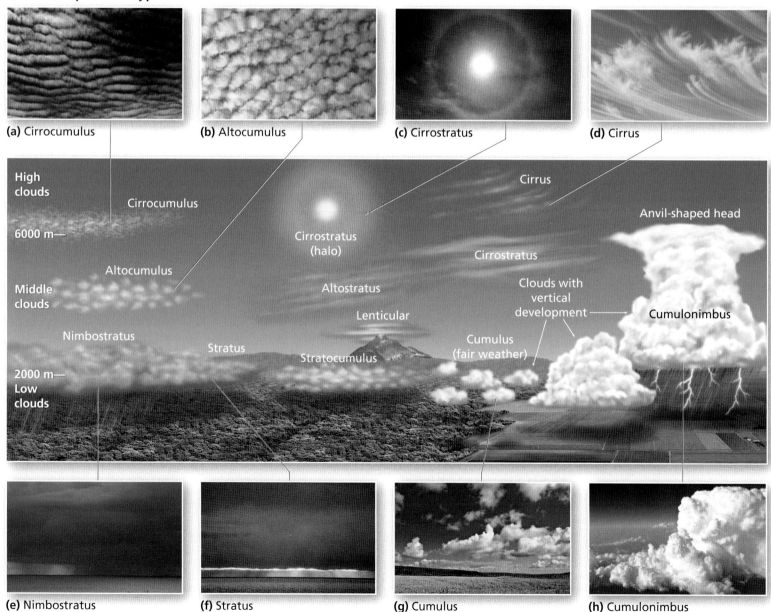

(a) Cirrocumulus

(b) Altocumulus

(c) Cirrostratus

(d) Cirrus

High clouds

Cirrocumulus

Cirrus

6000 m—

Cirrostratus (halo)

Anvil-shaped head

Altocumulus

Cirrostratus

Middle clouds

Altostratus

Clouds with vertical development

Cumulonimbus

Lenticular

Nimbostratus

Stratus

Cumulus (fair weather)

Stratocumulus

2000 m—
Low clouds

(e) Nimbostratus

(f) Stratus

(g) Cumulus

(h) Cumulonimbus

Cloud Types & Identification

In 1803, English biologist and amateur meteorologist Luke Howard created a classification system for clouds that we use today. Howard based his system on cloud shape and altitude and used Latin names for the different types. The basic cloud shapes are flat, puffy, and wispy: *Stratiform* clouds are flat and layered (or horizontally developed); *cumuliform* clouds are puffy and globular (or vertically developed); *cirriform* clouds are wispy, high-altitude clouds made of ice crystals.

The four altitudinal classes of clouds are: low, middle, high, and vertically developed.

Low Clouds Low clouds, from the surface up to 2000 m (6500 ft), are stratus or cumulus (Latin for "layer" and "heap," respectively). **Stratus** clouds are flat, dull, gray, and featureless. If they yield drizzling precipitation, they are **nimbostratus** (the prefix *nimbo-* denotes "stormy" or "rainy").

Cumulus clouds appear bright and puffy, like cotton balls. Vertically developed cumulus clouds extend beyond low altitudes into middle and high altitudes.

Stratocumulus may fill the sky in patches of lumpy, grayish, low-level clouds. Near sunset, these spreading puffy stratiform remnants may catch and filter the Sun's rays, sometimes indicating clearing weather.

Middle-Level Clouds The prefix alto- (meaning "high") denotes middle-level clouds made of water droplets, from 2000 m (6500 ft) to 6000 m (20,000 ft). **Altocumulus** clouds may appear as: patchy rows, wave patterns, a "mackerel sky," and lens-shaped clouds.

Lenticular, or lens-shaped, clouds form on the crests of mountain ranges (▼Fig. 4.17) under stable conditions. Air flowing upslope to the summit is lifted, causing cooling and condensation. Because the atmosphere is stable, the air sinks after passing over the summit, causing compression, heating, and evaporation. These clouds indicate stable atmospheric conditions—if the atmosphere were unstable, this lifting would form cumulonimbus clouds. Lenticular clouds often appear stationary, but they are not. The flow of moist air continuously resupplies the cloud on the windward side as air evaporates from the cloud on the leeward side.

High Clouds Ice crystals in thin concentrations compose **cirrus** clouds (Latin for "curl of hair") occurring above 6000

▼ 4.17 Lenticular cloud

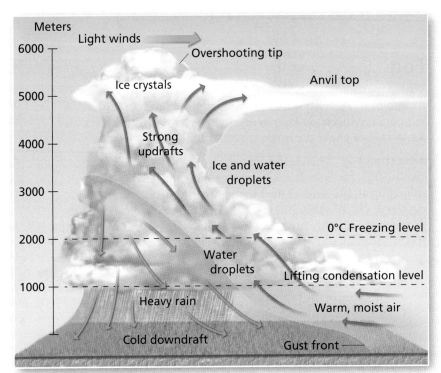

(a) Structure of a cumulonimbus cloud. Violent updrafts and downdrafts occur within the cloud. Blustery wind gusts are in front of the cloud.

(b) A dramatic cumulonimbus thunderhead over Africa at 13.5° N latitude near the Senegal–Mali border.

▲ 4.18 Cumulonimbus thunderhead

m (20,000 ft). Cirrus clouds can indicate an oncoming storm, especially if they thicken and lower in elevation. The prefix *cirro-*, as in cirrostratus and cirrocumulus, indicates other high clouds that form a thin veil or puffy appearance, respectively.

Vertically Developed Clouds A cumulus cloud can develop into a towering giant called **cumulonimbus** (▲Fig. 4.18). Such clouds are called thunderheads because of their shape and associated lightning and thunder. Note the surface wind gusts, updrafts and downdrafts, heavy rain, and ice crystals at the top of the rising cloud column. High-altitude winds or the temperature inversion just above the tropopause may shear the top of the cloud into the characteristic anvil shape of the mature thunderhead. Rising temperatures in the stratosphere act as an inversion layer, keeping the air from rising further.

geoCHECK ✔ Which types of clouds are made of ice crystals?

4.4 (cont'd) Clouds & Fog

Fog

With a maximum visibility of 1 km (3300 ft), fog is a cloud on the ground. **Fog** tells us that the air at ground level is saturated. A temperature-inversion layer generally caps a fog layer with warmer temperatures above the inversion altitude.

Advection Fog When warm, moist air moves over a cooler surface—ocean currents, lake surfaces, or snow masses—the layer of migrating air becomes chilled to the dew point, and **advection fog** develops. Off all subtropical west coasts in the world, summer fog forms in the manner just described (▶Fig. 4.19). Types of advection fog include:

- **Evaporation Fog** Another type of advection fog forms when cold air lies over the warmer water of a lake, an ocean surface, or even a swimming pool. This wispy **evaporation fog**, or steam fog, may form as water molecules evaporate from the water surface into the cold overlying air, where they condense (▼Fig. 4.20). When evaporation fog happens at sea, it is called sea smoke.
- **Upslope Fog** Advection fog also forms when moist air is cooled to dew-point temperature as it is lifted to higher elevations by a hill or mountain. The resulting adiabatic cooling forms **upslope fog** as a stratus cloud. Along the Appalachians and the eastern slopes of the Rockies, such fog is common in winter and spring.
- **Valley Fog** Another advection fog associated with topography is **valley fog**. Because cool air is denser than warm air, it settles in low-lying areas, chilling the air near the ground to its dew-point temperature and producing fog (▶Fig. 4.21).

Radiation Fog When radiative cooling of the ground chills the air layer above that surface to the dew-point temperature, **radiation fog** forms. This fog occurs over moist ground, especially on clear nights, but does not occur over water because water does not cool appreciably overnight (▶Fig. 4.22).

geoCHECK ✔ How does valley fog differ from radiation fog?

▼ **4.20 Evaporation fog**

▲ **4.19 Advection fog in San Francisco**

Precipitation

Water that falls from the sky is precipitation. This term refers to rain, snow, hail, or any of the other forms it takes. You might have noticed that rain falls from clouds, but not from all clouds. This is because cloud droplets—including the swirling droplets you can see in fog—are too small to fall from a cloud. For rain to fall, the droplets must grow larger.

Two mechanisms that form raindrops are collision–coalescence and the Bergeron–Findeisen process (▶Fig. 4.23). In warm clouds (above freezing), droplets collide and coalesce into drops—the collision-coalescence mechanism. In clouds that are cold enough to have both ice crystals and liquid water, the Bergeron–Findeisen process occurs. Because saturation vapor pressure is lower over an ice surface than over a water surface, water vapor evaporates from water droplets and deposits onto ice crystals. This process of ice deposition continues until the crystals grow large enough to fall. If the ice crystals fall through freezing air, they fall as snow. If the air is warmer, the ice crystals melt and fall as rain.

In deserts or under arid conditions with a very dry layer of air between clouds and the ground, rain will fall, but evaporate before it reaches the ground. This tantalizing phenomenon is **virga**. Virga is common in the Western United States, the Canadian Prairies, the Middle East, and North Africa. (▶Fig. 4.24).

geoCHECK ✔ Describe the two ways that raindrops form.

Mobile Field Trip MG
Clouds: Earth's
Dynamic Atmosphere

https://goo.gl/aZ53QY

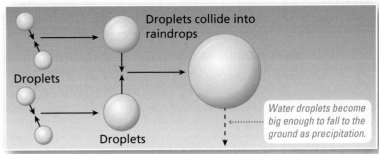

(a) The collision-coalescence mechanism

Droplets
Droplets collide into raindrops
Droplets

Water droplets become big enough to fall to the ground as precipitation.

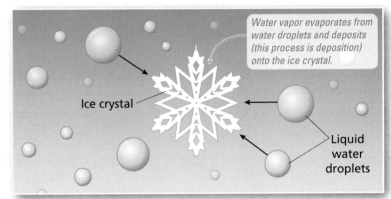

Water vapor evaporates from water droplets and deposits (this process is deposition) onto the ice crystal.

Ice crystal

Liquid water droplets

(b) The Bergeron–Findeisen process Water vapor sublimates from the ice crystal and evaporates from the water droplets, but more water deposits on the ice crystal than condenses on the water droplets.

▲ 4.23 **Mechanisms of raindrop formation and ice crystal growth**

▲ 4.21 **Valley fog during a rare temperature inversion at the Grand Canyon**

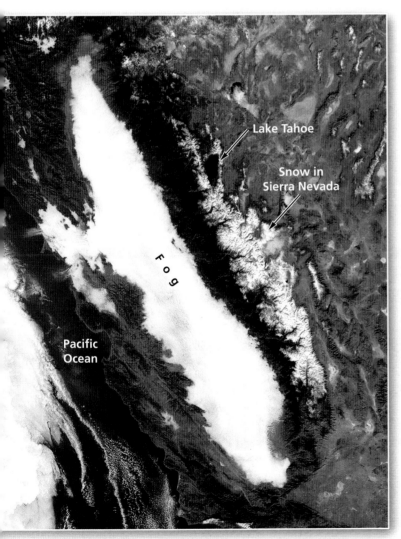

Lake Tahoe

Snow in Sierra Nevada

Fog

Pacific Ocean

▲ 4.22 **Winter radiation fog fills the Central Valley of California**

▲4.24 **Virga occurs when rain falls but evaporates before it reaches the ground**

geoQUIZ

1. Which clouds indicate stable atmospheric conditions? Which clouds form under unstable atmospheric conditions?
2. What conditions are required to form a lenticular cloud?
3. Why would radiation fog not form over water surfaces?

4.5 Air Masses

Key Learning Concept

▶ *Describe* air masses that affect North America.

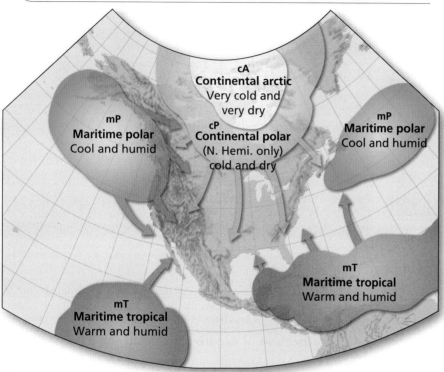

(a) Winter pattern. The polar and arctic air masses (mP, cP, and cA) shift south with the subsolar point in winter

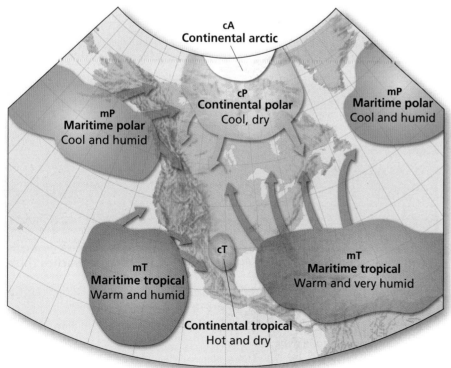

(b) Summer pattern. The tropical air masses (mT and cT) shift north during summer. The more ustable mT air mass in the southeast brings summer thunderstorms.

▲**4.25 Main North American air masses** Air masses shift position between winter and summer. The arrows on the map represent the directions in which the air masses tend to move.

The interactions between Earth's surface and its overlying air create regional air masses with uniform temperature, humidity, and stability characteristics. Each of these distinctive bodies of air is an **air mass**. They initially reflect the characteristics of their *source region*, but then move away from the source region and gradually take on the characteristics of their new location as they migrate with the winds. Air masses interact to produce weather patterns. For example, weather forecasters speak of a "cold Canadian air mass" or a "moist tropical air mass." In terms of thickness, they sometimes extend through the lower half of the troposphere.

Classifying Air Masses

We classify air masses generally according to the moisture and temperature characteristics of their source regions:

- Moisture—m for maritime (wet) and c for continental (dry).
- Temperature (latitude factor)—A for arctic, P for polar, T for tropical, E for equatorial, and AA for antarctic.

Continental polar (cP) air masses are found only in the Northern Hemisphere (◀**Fig. 4.25**). They are most developed in winter and cold-weather conditions and typically have cold, stable air; clear skies; and high pressure. The Southern Hemisphere lacks the necessary continental landmasses at high latitudes to create cP air masses.

Maritime polar (mP) air masses occur over the northern Atlantic and Pacific Oceans and feature cool, moist, unstable conditions throughout the year.

Maritime tropical (mT) air masses include the mT Gulf/Atlantic and the mT Pacific.

High humidity in the East and Midwest is from the mT Gulf/Atlantic air mass, which is particularly unstable and active from late spring to early fall. The mT Pacific air mass is stable to conditionally unstable and is lower in moisture content and available energy. Partly because of these characteristics of the mT Pacific air mass, the southwestern United States receives lower average precipitation than the rest of the country.

geoCHECK Which types of air masses are most common where you live?

Air Mass Modification

As air masses migrate from source regions, they take on the temperature and humidity characteristics of the land over which they pass. For example, an mT Gulf/Atlantic air mass may carry humidity to Chicago and on to Winnipeg, but it will gradually lose its initial characteristics of high humidity and warmth with each day's passage northward.

Similarly, below-freezing temperatures occasionally reach into southern Texas and Florida, brought by an invading winter cP air mass from the north. However, that air mass warms from the −50°C (−58°F) of its winter source region in central Canada, especially after it leaves areas covered by snow.

Modification of cP air as it moves south and east produces higher snowfall to the east of each of the Great Lakes (▶Fig. 4.26a). As below-freezing air passes over the warmer Great Lakes, it absorbs heat energy and moisture from the lake surfaces (▶Fig. 4.26b). This produces heavy lake-effect snowfall downwind into Ontario, Québec, Michigan, Ohio, Pennsylvania, and New York—with some areas receiving annual average snowfalls in excess of 325 cm (130 in.).

This moisture supply is shut off when lakes freeze over, usually in January. However, air and lake temperatures have been rising due to global warming. The Great Lakes have lost 71% of their ice cover since 1973, according to research by the Great Lakes Environmental Research Laboratory. The Great Lakes, including Lake Superior, were almost ice-free, with just 5% ice coverage, during the 2011–2012 winter season. Summer surface water temperatures on Lake Superior increased approximately 4.5 F° (2.5 C°) from 1979 to 2006, according to research by the University of Minnesota–Duluth's Large Lakes Observatory. If temperatures continue to rise, lakes will freeze over later in the season, or not at all. Ironically, global warming may increase lake-effect snows.

geoCHECK ✔ Why does Buffalo, New York, receive substantially more snow than Detroit, Michigan?

geoQUIZ

1. Which types of air masses are found only in the Northern Hemisphere?
2. Describe how an air mass would be modified as it travels from the Gulf of Mexico northward to Hudson Bay, Canada.
3. How could warmer temperatures increase lake-effect snow amounts?

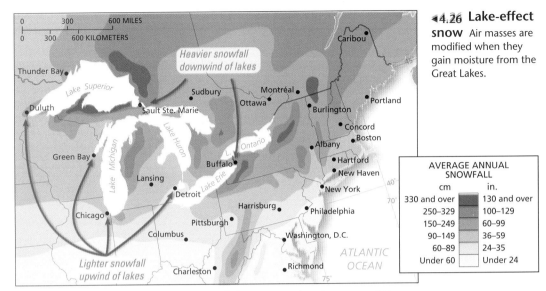

▶4.26 Lake-effect snow Air masses are modified when they gain moisture from the Great Lakes.

(a) Heavy snowfall occurs downwind of the Great Lakes, because of the prevailing winds, the westerlies.

AVERAGE ANNUAL SNOWFALL	
cm	in.
330 and over	130 and over
250–329	100–129
150–249	60–99
90–149	36–59
60–89	24–35
Under 60	Under 24

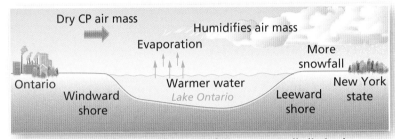

(b) Processes causing lake-effect snowfall are generally limited to about 50 km (30 mi) to 100 km (60 mi) inland.

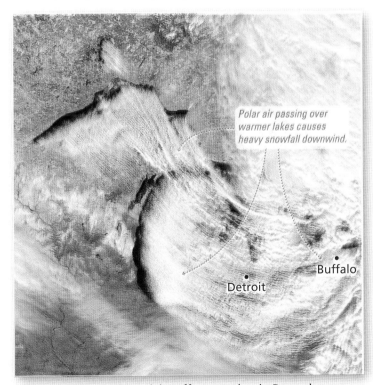

(c) Satellite image shows lake-effect weather in December.

4.6 Processes That Lift Air

Key Learning Concept

▶ *Identify* four types of atmospheric lifting mechanisms.

Air must be cooled to its dew-point temperature to condense, form clouds, and bring rain. When air is lifted, it expands and cools adiabatically. The four main lifting mechanisms are:

- Convergent lifting—air flows toward an area of low pressure.
- Convectional lifting—local surface heating causes warm air to rise.
- Orographic lifting—air is lifted upward by a barrier such as a mountain range.
- Frontal lifting—warmer air is lifted upward by cooler air.

Convergent Lifting

Air flowing into an area of low pressure converges and ascends in **convergent lifting** (▼Fig. 4.27). For example, the southeast and northeast trade winds converge along the equatorial region, forming the intertropical convergence zone (ITCZ), a region of frequent cumulonimbus clouds and high average annual precipitation (see Figs. 3.10 and 3.11).

▲**4.28 Convectional lifting** Local heating of the ground warms air parcels, causing them to rise.

Convectional Lifting

Surface heating of the land from the dark soil in a plowed field or an urban heat island can cause **convectional lifting** (▲Fig. 4.28).

Convectional lifting can be small scale, as in the field example above, but it can also affect a larger region. Florida's precipitation can be caused by both convergence and convection. Heating of the land causes convergence of onshore winds from the Atlantic Ocean and the Gulf of Mexico. Because the Sun gradually heats the land and the air above it, convectional showers tend to form in the afternoon and early evening. Florida has the highest frequency of days with thunderstorms of any state in the United States.

Orographic Lifting

Orographic lifting occurs when a moving air mass encounters a barrier, often a mountain range (▶Fig. 4.29). If the atmosphere is stable, lenticular clouds may form. Unstable conditions can form a line of cumulus and cumulonimbus clouds. Because orographic lifting happens regardless of atmospheric stability, it is the most consistent lifting mechanism. It also enhances rain produced by convectional lifting and frontal lifting.

The wetter intercepting slope is the windward slope, and the drier far-side slope is the leeward slope. Moisture

▲**4.27 Convergent lifting** Air flowing into an area of low pressure converges and ascends.

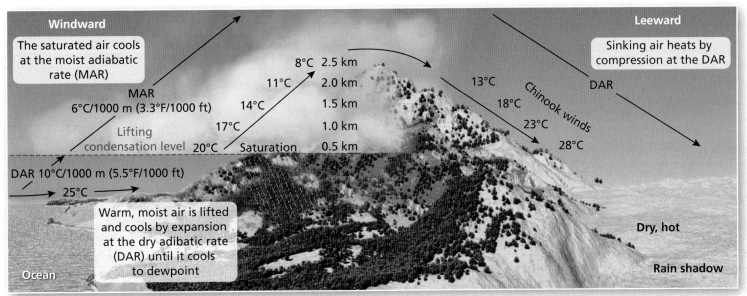

Windward

The saturated air cools at the moist adiabatic rate (MAR)

MAR
6°C/1000 m (3.3°F/1000 ft)

Lifting condensation level

8°C   2.5 km
11°C   2.0 km
14°C   1.5 km
17°C   1.0 km
20°C / Saturation   0.5 km

DAR 10°C/1000 m (5.5°F/1000 ft)

25°C

Warm, moist air is lifted and cools by expansion at the dry adibatic rate (DAR) until it cools to dewpoint

Ocean

Leeward

Sinking air heats by compression at the DAR

13°C   DAR
18°C   Chinook winds
23°C
28°C

Dry, hot

Rain shadow

(a) Winds force warm, moist air up the windward mountain slope, producing adiabatic cooling, eventual saturation and condensation, cloud formation, and precipitation. On the leeward side, air sinks and heats by compression, creating the hot, dry rain shadow.

(c) Rainier station is on the windward side.

(d) Yakima station is on the leeward side.

▲ 4.29 **Orographic lifting**

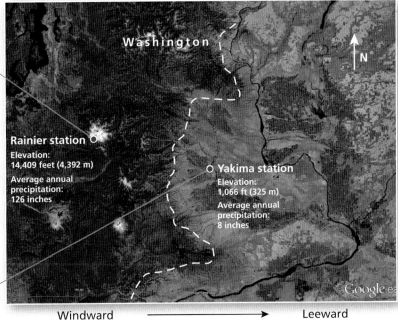

Washington

N

Rainier station ○
Elevation:
14,409 feet (4,392 m)

Average annual precipitation:
126 inches

○ **Yakima station**
Elevation:
1,066 ft (325 m)

Average annual precipitation:
8 inches

Google ea

Windward ⟶ Leeward

(b) Windward and leeward precipitation patterns across Washington.

condenses from the lifting air mass on the windward side of the mountain; on the leeward side, the sinking air heats by compression, and any remaining water in the air evaporates. Air masses that are warm and moist before lifting on the windward side often become hot and dry after descending the leeward slope.

In North America, chinook winds (called föhn or foehn winds in Europe) are the warm, downslope air flows characteristic of the leeward side of mountains. Such winds caused a jump in temperature of 27 Celsius degrees (49 Fahrenheit degrees) in only two minutes at Spearfish, South Dakota on January 22, 1943.

4.6 (cont'd) **Processes That Lift Air**

(a) Mt. Wai'ale'ale receives an average of 1234 cm (486 in.) of rain.

(b) The rainshadow side of Kaua'i receives and average of 50.8 cm (20 in.) of rain.

▲4.30 **Windward slope and rain shadow**

The term **rain shadow** is applied to dry regions leeward of mountains. East of the Cascade Range, Sierra Nevada, and Rocky Mountains, such rain-shadow patterns predominate. The precipitation pattern of windward and leeward slopes can be seen worldwide.

The wettest places on Earth experience orographic lifting (▲Fig. 4.30). Mount Wai'ale'ale, on the windward side of the island of Kaua'i, Hawai'i, received an annual average of 1234 cm (486 in., or 40.5 ft) of rain from 1941 to 1992. In contrast, the rain-shadow side of Kaua'i receives only 50 cm (20 in.) of rain annually. Cherrapunji, India, is in the Assam Hills, between the Bay of Bengal and the Himalayas. Orographic lifting of moisture from the summer monsoons resulted in the wettest year ever recorded, 2647 cm (1042 in., or 86.8 ft) of rain, as well as 930 cm (366 in., or 30.5 ft) of rainfall in 1 month.

geoCHECK ✔ Why is the rain shadow on the downwind side of a mountain range?

Frontal Lifting: Cold & Warm Fronts

When air masses with different temperatures and humidities meet, they do not immediately mix, but form a boundary layer between them called a *front*. The leading edge of a warm air mass is a **warm front**, and the leading edge of a cold air mass is a **cold front**.

Warm Fronts As shown in Figure 4.31a, a line with semicircles facing in the direction of frontal movement denotes a warm front. Because warm air is less dense, an advancing warm air mass slides up over the cooler air and tends to push the cooler air into the characteristic wedge shape of a warm front.

Warm fronts have a characteristic pattern of cloud development: High cirrus and cirrostratus clouds announce the advancing frontal system; then clouds lower and thicken to altostratus; finally, the clouds lower and thicken to stratus and nimbostratus within several hundred kilometers of the front.

Cold Fronts As shown in Figure 4.31b, a cold front is denoted by a line with triangular spikes that point in the direction of frontal movement along an advancing cP or mP air mass. The steep face of the cold front is caused by its cold, dense air rapidly lifting the less-dense air of the warmer air mass. Cold fronts travel faster than warm fronts because the cold dense air mass of the cold front easily lifts the warmer, less dense air mass. As the cold front advances, the wind direction shifts, and the temperature and air pressure both drop. As the line of most intense lifting passes, usually just ahead of the front itself, air pressure drops to a local low. Clouds may build along the cold front into cumulonimbus clouds, which may appear as an advancing wall of clouds. Precipitation usually is heavy and can be accompanied by hail, lightning, and thunder.

After the cold front passes, the winds will usually be northerly in the Northern Hemisphere (or southerly in the Southern Hemisphere). Temperatures are lower and air pressure higher in response to the cooler, more-dense air. Cloud cover breaks and clears.

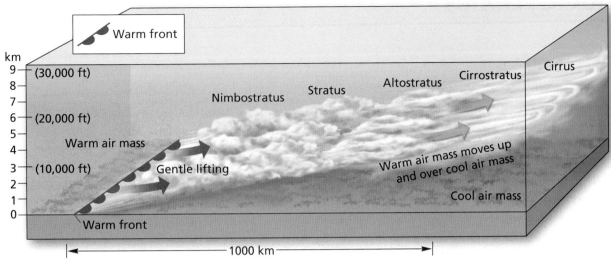

(a) Warm front Note the sequence of cloud development as the warm front approaches. Warm air slides upward over a wedge of cooler, passive air near the ground. Gentle lifting of the warm, moist air produces nimbostratus and stratus clouds and drizzly rain showers, in contrast to the more dramatic cold-front precipitation.

Animation MG
Warm Fronts

http://goo.gl/MtA6cZ

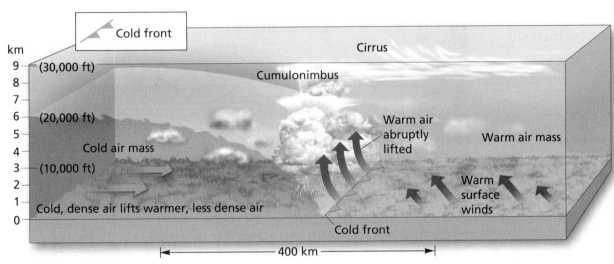

(b) Cold front Denser, advancing cold air forces warm, moist air to lift quickly. Preciptiation in a cold front is often heavier and briefer than the precipitation from a warm front.

Animation MG
Cold Fronts

http://goo.gl/TFFeBt

▲4.31 **Warm and cold fronts** The leading edge of an air mass is its front.

▼4.32 **A powerful squall line over the prairie in Glasgow, Montana**

A fast-advancing cold front can cause violent lifting and create a zone right along or slightly ahead of the front called a **squall line** (◄Fig. 4.32). Along a squall line, wind patterns are turbulent and wildly changing, and precipitation is intense. The well-defined frontal clouds in the photograph rise abruptly, with new thunderstorms forming along the front. Tornadoes also may develop along such a squall line.

geoCHECK ✔ Which lifting mechanism is most common where you live?

geo**QUIZ**

1. Which type of lifting is responsible for the most consistent rainfall?
2. What is the cloud pattern that indicates an approaching warm front?
3. Compare and contrast the precipitation that occurs along warm fronts and cold fronts in terms of how the precipitation forms and its intensity.

4.7 Midlatitude Cyclones

Key Learning Concept

▶ **Describe** the life cycle of a midlatitude cyclonic storm system.

Animation (MG)

Midlatitude Cyclones

http://goo.gl/ZWjpVS

Warm and cold fronts often are found together, spiraling around a center of low pressure and forming a storm called a **midlatitude cyclone**, or **wave cyclone**. These storms are formed by the conflict between warm, moist tropical air masses and cool, dry polar air masses. The air masses combine to form a warm front and a cold front spiraling around the converging, rising air of a low-pressure center. They are called wave cyclones because, when seen from above, the two fronts resemble a wave (▶Fig. 4.33).

Wave cyclones, which can be 1600 km (1000 mi) wide, dominate weather patterns in the middle and higher latitudes of both the Northern and the Southern Hemispheres. A midlatitude cyclone can form along the polar front, particularly in the region of the Icelandic and Aleutian low-pressure cells in the Northern Hemisphere. Other areas associated with cyclone development and intensification include the eastern slope of the Rocky Mountains, the Gulf Coast, and the eastern seaboard.

The high-speed winds of the jet streams guide cyclonic systems and their air masses along storm tracks. Typical storm tracks in North America follow the subsolar point and shift north with the subsolar point toward the poles in summer and south toward the equator in winter. As the storm tracks begin to shift northward in the spring, cP and mT air masses are in their clearest conflict. This is the time of strongest frontal activity, featuring thunderstorms and tornadoes.

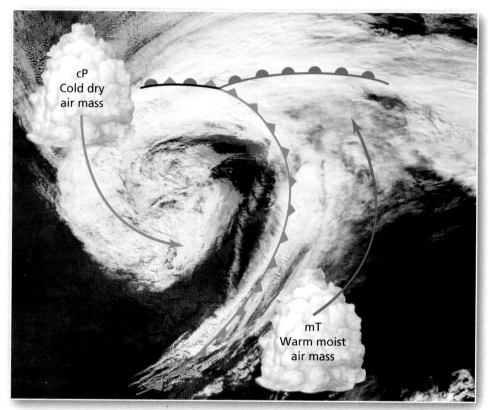

cP
Cold dry
air mass

mT
Warm moist
air mass

▲**4.33 Observing a midlatitude cyclone** Satellite images reveal the flow of moist and dry air that drives a midlatitude cyclone.

Life Cycle of a Midlatitude Cyclone

A midlatitude cyclone takes roughly 3–10 days to progress through its life cycle. Figure 4.33 shows the birth, maturity, and death of a typical midlatitude cyclone in several stages.

- **Cyclogenesis** The first stage is cyclogenesis (▶Fig. 4.34-1). This process usually begins along the polar front, where cold and warm air masses converge and are drawn into conflict. A slight disturbance along the polar front, perhaps a small change in the path of the jet stream, can start the converging, ascending flow of air and thus a surface low-pressure system. For a wave cyclone to form along the polar front, an area of divergence aloft must match an area of convergence at the surface.

- **Open Stage** In the open stage, to the east of the developing low-pressure center, warm air begins to move northward along an advancing front, while cold air advances southward to the west of the center (▶Fig. 4.34-2). As the midlatitude cyclone matures, the counterclockwise flow draws the cold air mass from the north and west and the warm air mass from the south. In the cross section, you can see the profiles of both a cold front and a warm front and each air mass segment.

- **Occluded Stage** Next is the occluded stage (▶Fig. 4.34-3). The colder cP air mass is denser than the warmer mT air mass and acts like a bulldozer blade, lifting the warmer air mass. Cold fronts can travel at an average speed of 40 kmph (25 mph),

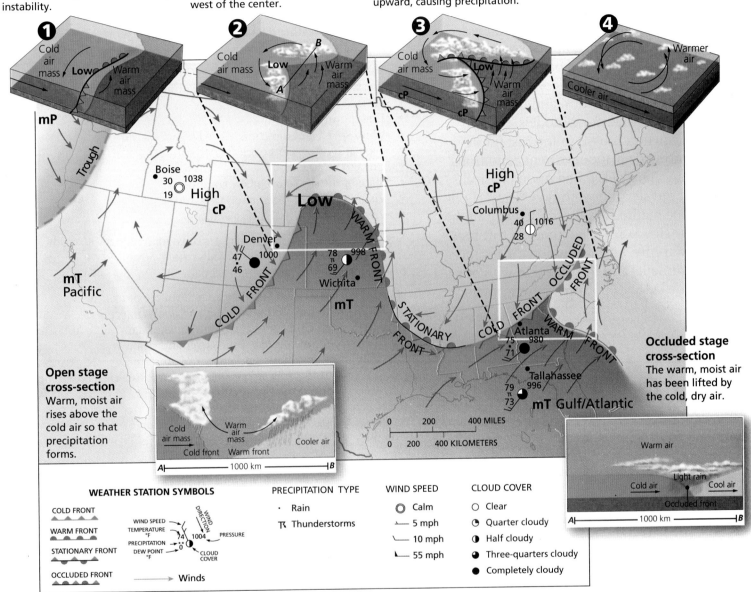

Stage 1: Cyclogenesis
A disturbance develops along the polar front or in certain other areas. Warm air converges near the surface and begins to rise, creating instability.

Stage 2: Open stage
Cyclonic, counterclockwise flow pulls warm, moist air from the south into the low-pressure center while cold air advances southward west of the center.

Stage 3: Occluded stage
The faster-moving cold front overtakes the slower warm front and wedges beneath it. This forms an occluded front, along which cold air pushes warm air upward, causing precipitation.

Stage 4: Dissolving stage
The midlatitude cyclone dissolves when the cold air mass completely cuts off the warm air mass from its source of energy and moisture.

Open stage cross-section
Warm, moist air rises above the cold air so that precipitation forms.

Occluded stage cross-section
The warm, moist air has been lifted by the cold, dry air.

WEATHER STATION SYMBOLS

COLD FRONT

WARM FRONT

STATIONARY FRONT

OCCLUDED FRONT

WIND SPEED
TEMPERATURE °F
PRECIPITATION
DEW POINT °F

WIND DIRECTION
PRESSURE
CLOUD COVER

Winds

PRECIPITATION TYPE
· Rain
ꓫ Thunderstorms

WIND SPEED
○ Calm
⌐ 5 mph
⌐ 10 mph
⌐ 55 mph

CLOUD COVER
○ Clear
◔ Quarter cloudy
◑ Half cloudy
◕ Three-quarters cloudy
● Completely cloudy

▲**4.34 Stages of a midlatitude cyclone** The life-cycle of a midlatitude cyclone has four stages that unfold at the meeting of cold and warm air masses.

whereas warm fronts average roughly half that, at 16–24 kmph (10–15 mph). Thus, a cold front often overtakes the warm front, wedging beneath it and producing an **occluded front**.

- **Dissolving Stage** Finally, the dissolving stage of the midlatitude cyclone occurs after the warm air mass is lifted and cooled while the cold air mass is gradually warmed by the surface. Remnants of the cyclonic system then dissipate in the atmosphere (▲ **Fig. 4.34-4**). Because the tropics receive more energy than the poles, this process will begin again as new warm, moist air from the tropics meets cold, dry air from the poles.

geoCHECK Why is it necessary to have two different air masses to form a midlatitude cyclone?

geoQUIZ

1. Which types of air masses form midlatitude cyclones?
2. How does the jet stream affect the movement of midlatitude cyclones?
3. Why does occlusion occur in a midlatitude cyclone?

4.8 Weather Forecasting

Key Learning Concept

▶ **List** the measurable elements that contribute to weather forecasting.

Meteorologists collect weather data and use computer models based upon past events to predict weather. Building a database of wind, pressure, temperature, and moisture conditions is key to numerical weather prediction and the development of weather-forecasting models. Numerical weather prediction is used to forecast future weather by putting data about current conditions into mathematical models based upon physical laws, like heat transfer and wind flow. These models use the same methods as those used to examine possible future climates. The accuracy of forecasts continues to improve with technological advancements in instruments and software and with our increasing knowledge of the atmospheric interactions that produce weather.

Important weather data necessary for the preparation of a weather map include the following:

- Barometric pressure
- Pressure tendency (steady, rising, falling)
- Surface air temperature
- Dew-point temperature
- Wind speed and direction
- Type and movement of clouds
- Current weather
- State of the sky (current sky conditions)
- Visibility; vision obstruction (fog, haze)
- Precipitation since last observation

Key to weather forecasting today are the data obtained by weather satellites; automated, ground-based sensors; advanced computer processing systems; and Doppler radar.

Video **MG**

NSSL In the Field

http://goo.gl/es8T0V

Weather Satellites Satellites are one of the key tools in forecasting weather and analyzing climate. Weather satellites collect vast amounts of data on air and surface temperatures, cloud cover and cloud type, atmospheric water vapor, and areas and amounts of precipitation. These data are also used for assessing climatic change. In the United States, the National Weather Service (NWS) provides weather forecasts and current satellite images (see *http://www.nws.noaa.gov/*). Internationally, the World Meteorological Organization coordinates weather information (see http://www.wmo.ch/).

Automated Surface Observing System Weather information in the United States comes mainly from the Automated Surface Observing System (ASOS), which is installed at over 900 airports across the country. An ASOS instrument array is made up of numerous sensors that supply continuous data for weather elements, helping the NWS increase the timeliness and accuracy of its forecasts using on-the-ground information (◀Fig. 4.35).

Advanced Weather Interactive Processing System The National Oceanic and Atmospheric Administration (NOAA) developed the Advanced Weather Interactive Processing System (AWIPS), which consists of high-tech computer workstations that assemble information, thereby improving forecast accuracy and severe-weather warnings (▶Fig. 4.36). AWIPS is able to display and integrate a variety of data—for example, air pressure, water vapor, humidity, Doppler radar, lightning strikes in real time, and wind profiles.

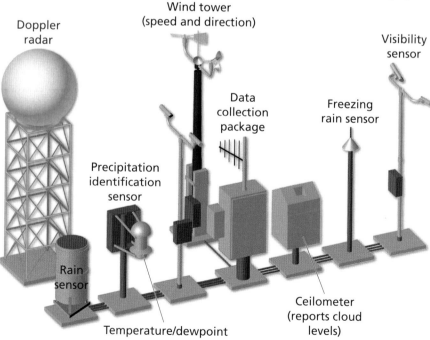

Doppler radar

Wind tower (speed and direction)

Visibility sensor

Data collection package

Freezing rain sensor

Precipitation identification sensor

Rain sensor

Temperature/dewpoint sensor

Ceilometer (reports cloud levels)

▲**4.35 Automated Surface Observing System (ASOS) instrument array and doppler radar.**

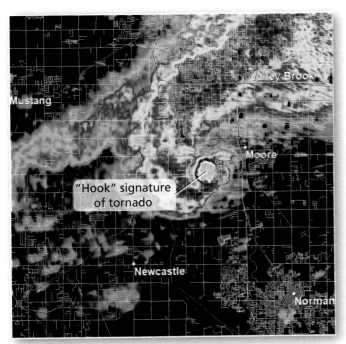

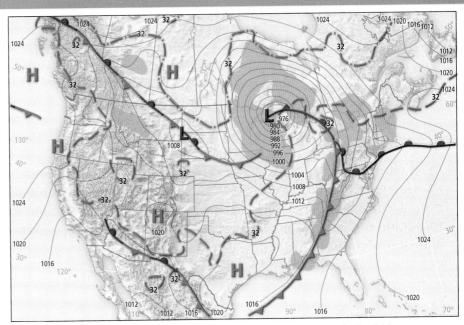

(a) Surface weather map at 7:00 A.M., E.S.T. Feb. 21, 2014. The dashed blue lines show regions with freezing temperatures, and the light blue areas indicate precipitation.

▲ **4.37** Doppler radar hook from tornado that devastated Moore, OK

Doppler Radar An essential element of weather forecasting is Doppler radar. It detects the intensity and direction of moisture droplets toward or away from the radar, indicating wind direction and speed. This information is critical to providing accurate severe-storm warnings, including tornado warnings. The rapidly rotating funnel cloud of a tornado creates a "hook" signature (▲ **Fig. 4.37**).

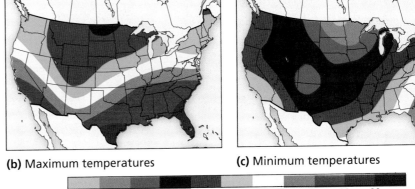

(b) Maximum temperatures   **(c)** Minimum temperatures

0   10   20   30   40   50   60   70   80   90

▲ **4.38** The daily weather map shows surface conditions, temperatures, air pressure, frontal boundaries, and precipitation

Weather Maps & Forecasting Weather maps play an important role in weather forcasting by showing current conditions. Figure 4.38 shows the web version of the NWS daily weather map. Examples of the standard weather symbols used on maps are shown in Figure 4.34, p. 107. While actual storms are widely varied in shape and duration, you can apply this general midlatitude cyclone model, along with your understanding of warm and cold fronts, to reading the daily weather map.

Video MG
Radar Research at NSSL

http://goo.gl/W3eqzY

For links to weather maps, current forecasts, satellite images, and the latest radar, please go to the Mastering Geography web site.

▲ **4.36** AWIPS computers that process complex data are critical for forecasting and monitoring real-time weather events.

geo**CHECK** ✔ What can weather satellites tell us about the atmosphere?

geo**QUIZ**

1. What are the main types of data needed to predict weather?
2. How can Doppler radar predict and identify tornadoes?

4.9 Thunderstorms

Key Learning Concepts

► **Explain** how thunderstorms form.
► **Describe** the characteristics of thunderstorms, including related phenomena such as lightning and thunder, hail, atmospheric turbulence, and derechos.

The flow of energy across Earth can set into motion destructive, violent weather. On a humid summer afternoon, the combination of warm, humid air and the Sun's energy can quickly lead to the formation of dark, threatening clouds, dangerous lightning strikes, and stormy conditions.

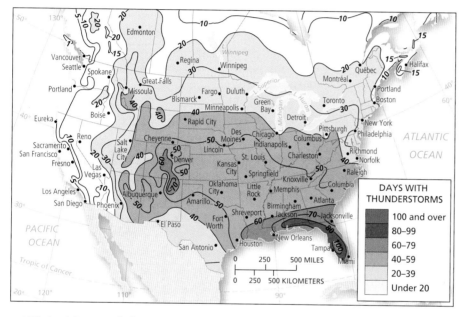

▲ 4.39 **Incidence of thunderstorms**

Formation & Characteristics of Thunderstorms

Giant cumulonimbus clouds can create dramatic weather moments—squall lines of heavy precipitation, lightning, thunder, hail, blustery winds, and tornadoes. Thunderstorms form during unstable atmospheric conditions. The life cycle of a thunderstorm has three stages of development. In the first stage, warm, moist air is lifted, often by convergent or convective lifting. As the air rises, it cools by expansion. When the lifted air cools to the dew-point temperature, condensation occurs, releasing latent heat. If atmospheric conditions are unstable, the air will continue to rise. The rising air creates an area of low-pressure at the base of the cloud, which draws in more warm, moist air, causing updrafts up to 160 kmh (100 mph). The second stage occurs when the warm air rises until it reaches the tropopause or the top is sheared off by strong winds. Here the air spreads out horizontally, creating a distinctive anvil shape. The water droplets in the cloud coalesce into larger drops, and freeze into ice once they

reach freezing altitude. If the updrafts are strong enough, these ice particles will grow until they fall as hail. During this phase of the thunderstorm, the rising air causes strong updrafts and the falling rain causes strong downdrafts. The final phase of the thunderstorm occurs when there is no longer sufficient warm, moist air to support the updrafts and the storm is dominated by downdrafts.

Thousands of thunderstorms occur on Earth at any given moment. Many are around the ITCZ. For example, the equatorial city of Kampala in Uganda, East Africa, averages a record 242 days a year of thunderstorms. In North America, most thunderstorms occur in areas dominated by mT air masses (◄Fig. 4.39).

Lightning and Thunder Lightning strikes Earth over 8 million times each day! **Lightning** refers to flashes of light caused by enormous electrical discharges—hundreds of millions of volts—that briefly heat the air to temperatures of 15,000°–30,000°C (27,000°–54,000°F). A buildup of electrical energy within a cumulonimbus cloud or between the cloud and the ground creates lightning. The violent expansion of this abruptly heated air sends shock waves through the atmosphere as the sonic boom of thunder (►Fig. 4.40).

Lightning causes nearly 200 deaths and thousands of injuries each year in the United States and Canada. Ninety percent of all strikes occur over land in response to increased convection over relatively warmer land surfaces. Strikes shift north and south with the subsolar point and the ITCZ (see http://thunder.msfc.nasa.gov/lis/).

Hail Ice pellets of **hail** form within cumulonimbus clouds. Raindrops circulate repeatedly above and below the freezing level in the cloud, adding layers of ice until updrafts in the cloud can no longer support their weight. The size of hail is related to the speed of the updrafts—faster updrafts result in larger hail. Pea-sized hail is common, although baseball-sized hail occurs around a half dozen times each year in the United States. The largest authenticated hailstone in the United States fell from a thunderstorm in Vivian, South Dakota, on

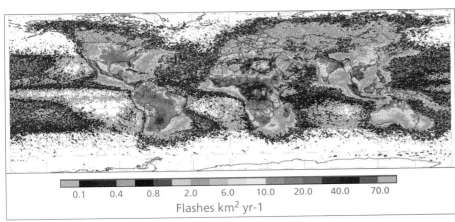

▲4.40 Global lightning strikes, 1998 to 2012, recorded by the Lightning Imaging Sensor-Optical Transient Detector (LIS-OTD).

▲4.41 Hail This 0.75 kg (1.75 lb) hailstone fell on Coffeeville, KS. Note the bands indicating trips above and below the freezing level.

July 23, 2010. It weighed 0.88 kg (1.93 lb) and measured 20.3 cm (8.0 in.) in diameter and 47.3 cm (18.6 in.) in circumference (▶Fig. 4.41). Hail causes $800 million in damage each year in the United States. The pattern of hail occurrence is similar to that of thunderstorms shown in Figure 4.39.

Atmospheric Turbulence Turbulence experienced on airplane flights is due to flying through air of different densities or through air layers moving at different speeds and directions. Thunderstorms can produce severe turbulence in the form of exceptionally strong downdrafts called *downbursts* that can bring down aircraft. Such turbulence events are short-lived and hard to detect, although NOAA's laboratories are making progress in developing forecasting methods.

geoCHECK ✔ What is hail, and how does it form?

Derechos

Straight-line winds over 93 kmh (58 mph) associated with downbursts from rapidly moving thunderstorms and bands of showers are known as **derechos**. The name derives from a Spanish word meaning "straight ahead." These derechos blast out in straight paths along the curved front of a band of severe thunderstorms.

Their highest frequency (about 70%) is from May to August in the region stretching from Iowa, across Illinois, and into the Ohio River Valley. By September and on through to April, areas of activity migrate southward to eastern Texas through Alabama. For more information and descriptions of derechos, see http://www.spc.noaa.gov/misc/AbtDerechos/derechofacts.htm.

From June 29 to June 30, 2012, a fast-moving derecho complex traveled 1000 km (600 mi) in 10 hours

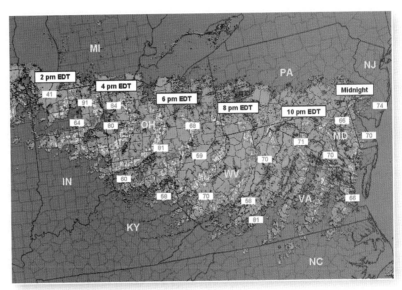

▲4.42 Doppler radar signature of a derecho Composite of hourly radar imagery shows the development of the June 29, 2012, derecho event, including selected wind gusts (mph), beginning at 2 P.M. Eastern Daylight Time (far left) and ending at midnight (far right).

(▲ Fig. 4.42). The storms raced across the Midwest, through the central Appalachians, and into the Mid-Atlantic states. Wind speeds of up to 140 kmh (87 mph) were recorded. The series of storms resulted in 22 deaths and the loss of electricity to more than 3.7 million people during a heat wave.

geoCHECK ✔ Which types of violent weather begin with thunderstorms?

geoQUIZ

1. Describe the conditions that are necessary for a thunderstorm to form.
2. Explain what causes the updrafts and downdrafts associated with thunderstorms.
3. Compare and contrast derechos and downbursts.

4.10 Tornadoes

Key Learning Concepts

▶ **Explain** how tornadoes form.
▶ **Describe** the Enhanced Fujita Scale of tornado measurement.
▶ **Describe** the distribution and incidence of tornadoes.

Under certain conditions, clouds that produce thunderstorms can also spawn tornadoes. These violent storms generate extreme winds that can destroy virtually every structure in the tornado's path. In the United States, tornadoes are common occurrences across a wide region stretching from the Great Plains and Midwest to the Southeast.

How a Tornado Forms

When updrafts within large cumulonimbus clouds combine with horizontally rotating bodies of air, tornadoes can form. Because of the effect of friction close to the ground, wind speeds aloft are higher than wind speeds close to the ground. The difference between wind speeds next to the ground and wind speeds aloft produce a horizontally rotating body of air that moves along like a rolling pin (**Fig. 4.43a**). When this horizontally rotating body of air runs into an updraft within a

cumulonimbus cloud, the horizontally rotating body of air is tipped up and rotates vertically (Fig 4.43b and c). The spinning, rising column of air forms a **mesocyclone**. Ranging up to 10 km (6 mi) in diameter, a mesocyclone rotates vertically within a supercell cloud (the parent cloud) to a height of thousands of meters. A mature mesocyclone will produce heavy rain, large hail, blustery winds, and lightning; some mature mesocyclones will generate tornado activity.

As more moist air is drawn up into the mesocyclone, more energy is liberated, and the rotation of air increases speed (▼ **Fig. 4.43**). The narrower the mesocyclone, the faster the spin of the air being sucked into the rotation. The rotation of the mesocyclone is visible, as are smaller, dark gray **funnel clouds** that pulse from the bottom side of the parent cloud. Doppler radar is used to identify tornadoes because the rotating funnel cloud creates a distinctive "hook" signature. When a funnel cloud lowers from a cumulonimbus cloud to the surface, the result is a **tornado**.

A tornado can range from a few meters to a few hundred meters in diameter and can last anywhere from a few moments to tens of minutes. Tornadoes have been increasing in average speed and duration over the last two decades. Over 13,000 houses were destroyed in Moore, Oklahoma, by a powerful tornado in May 2013. Winds exceeded 340 kmph (210 mph) in the 2.1-km-wide (1.3-mi-wide) funnel that was on the ground along a 27-km (17-mi) track (refer back to **Fig. 4.37**). A tornado's scoured path is visible on the ground in a series of semicircular "suction" prints as well as in the trail of damage. When tornado circulation occurs over water, a **waterspout** forms, and surface water is drawn some 3–5 m (10–16 ft) up into the funnel. The rest of the waterspout funnel is made visible by the rapid condensation of water vapor.

▼ 4.43 (a–c) Stages in tornado development (d) Mature tornado

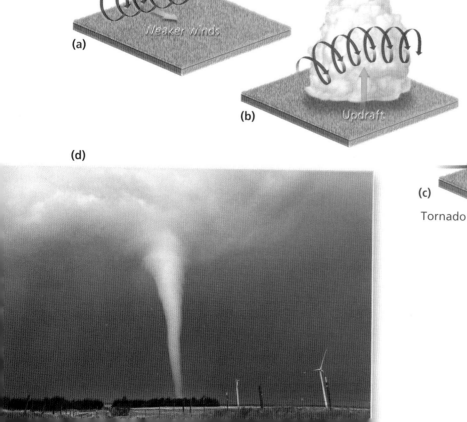

Spinning along horizontal axis

Stronger winds

Weaker winds

(a)

Thunderstorm forming

Updraft

(b)

Clouds overshoot top of thunderstorm

Anvil

Mesocyclone (3 to 10 km diameter)

Tornado

Air inflows

(c)

(d)

geo **CHECK** ✔ What is the difference between a funnel cloud and a tornado?

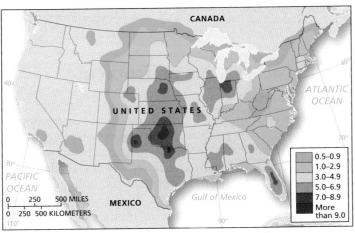

(a) Average number of tornadoes per 26,000 km² (10,000 mi²). Tornado numbers in Alaska and Hawai'i are negligible.

▲ 4.44 Tornado frequency

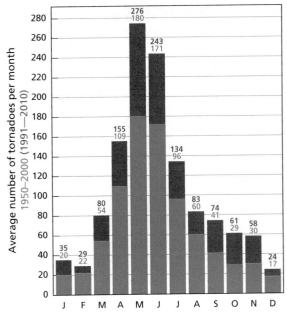

(b) Average number of U.S. tornadoes per month from 1950 to 2000, with additional updated statistics from 1991 to 2010 (in red).

Animation (MG)
Tornado Wind Patterns
http://goo.gl/tlfMNt

Tornado Measurement & Frequency

Pressures inside a tornado usually are about 10% less than those in the surrounding air. This steep pressure gradient causes high wind speeds. Theodore Fujita, a noted meteorologist from the University of Chicago, designed the Enhanced Fujita Scale, or EF Scale (▶Table 4.1), which classifies tornadoes according to wind speed.

North America experiences more tornadoes than anywhere on Earth because its latitudinal position and topography permit contrasting air masses to collide. Tornadoes have struck all 50 states (▲Fig. 4.44). Canada experiences an average of 80 observed tornadoes per year throughout its provinces and territories. Other continents experience a small number of tornadoes each year. May and June are the peak months.

The long-term annual average number of tornadoes before 1990 was 787. After 1990, this long-term average per year rose to over 1000. Since 1990, the years with the highest number of tornadoes were 2004 (1820 tornadoes), 2011 (1691 tornadoes), and 1998 (1270 tornadoes). On April 27, 2011, 319 tornadoes were sighted, the third highest number ever recorded on a single day. Potential causes of the increase in tornado frequency range from more intense thunderstorm activity to better reporting through Doppler radar and more people with video recorders.

In the United States, 56,691 tornadoes were recorded in the years from 1950 through 2012. In these 62 years, tornadoes caused over 5000 deaths, or about 85 deaths per year, as well as more than 80,000 injuries and property damage of over $28 billion. The yearly average of $500 million in damage is rising each year.

The Storm Prediction Center in Kansas City, Missouri, provides short-term forecasting for thunderstorms and tornadoes and can provide warning times of 12 to 30 minutes.

Doppler radar is useful in identifying tornadoes, but tornadoes are extremely difficult to predict. Even if all the ingredients are present, tornadoes do not necessarily form because of the complex factors involved. When they do form, they are often short-lived, with an average lifespan of under 10 minutes and an average width of just over 100 m (100 yd).

Table 4.1 The Enhanced Fujita Scale

EF-Number	3-Second-Gust Wind Speed; Damage
EF-0 Gale	105–137 kmph, 65–85 mph; *light damage*: branches broken.
EF-1 Weak	138–177 kmph, 86–110 mph; *moderate damage*: mobile homes pushed off foundations.
EF-2 Strong	178–217 kmph, 111–135 mph; *considerable damage*: roofs torn off frame houses, large trees uprooted or snapped.
EF-3 Severe	218–266 kmph, 136–165 mph; *severe damage*: trains overturned, trees uprooted, cars thrown.
EF-4 Devastating	267–322 kmph, 166–200 mph; *devastating damage*: well-built houses leveled, cars thrown.
EF-5 Incredible	More than 322 kmph, >200 mph; *incredible damage*: houses lifted and carried, car-sized missiles fly farther than 100 m.

Note: See http://www.depts.ttu.edu/weweb/Pubs/fscale/EFScale.pdf for details.

geoCHECK ✔ Which month sees the most tornadoes?

geoQUIZ

1. Explain the role of a mesocyclone in the process of tornado formation.
2. Referring to Table 4.1, compare the winds and level of damage from an EF-2 tornado with that of an EF-4 ttornado.
3. How has the incidence of tornadoes changed in recent decades? What is one possible explanation for this change?

4.11 Tropical Cyclones

Key Learning Concepts

▶ *Explain* how tropical cyclones form and move.
▶ *Describe* the physical structure of tropical cyclones and the winds, storm surge, and damage associated with these storms.
▶ *Describe* trends in the intensity of tropical cyclones and efforts to protect people and property.

B etween 1975 and 2013, over $1 trillion was lost in 151 weather events. Of these, tropical cyclones—known as hurricanes in the United States—are the most costly. Hurricane Katrina and the 2005 season alone caused $159 billion in damages, and Hurricane Sandy (2012) caused more than $75 billion in damages!

Characteristics of Tropical Cyclones

A powerful demonstration of the transfer of energy from the tropics to the poles is the **tropical cyclone**. Approximately 80 tropical cyclones occur annually worldwide, and 45 of those are powerful enough to be classified as hurricanes. A full-fledged **hurricane** has wind speeds greater than 119 kmph (74 mph, or 65 knots). They have regional names: Hurricanes occur around North America, **typhoons** in the western Pacific, and *cyclones* in Indonesia, Bangladesh, and India (▼**Fig. 4.45a**). See the National Hurricane Center at http://www.nhc.noaa.gov or the Joint Typhoon Warning Center at http://www.usno .navy.mil/JTWC for current information.

Formation Tropical cyclones form without fronts or conflicting air masses. Sea-surface temperatures over 26°C (79°F) provide the water vapor and latent heat energy to fuel these storms (◄**Fig. 4.45a**). Tropical cyclones convert heat energy from the ocean into mechanical energy in the wind—the warmer the ocean and atmosphere, the more intense and powerful the storm. Topical cyclone season for the Atlantic is from June 1 to November 30 each year.

Cyclones begin with **easterly waves**, slow-moving waves of low pressure that move from east to west in the trade-wind belt of the tropics (▶ **Fig. 4.46**). Tropical cyclones form along the eastern side of these easterly waves, a place of convergence and rainfall. Surface air flows into the low-pressure area, begins rotating, ascends, and diverges aloft. This important divergence aloft acts as a chimney, pulling more moist air into the developing system. A weak, tropical disturbance is the first step toward the development of a full-fledged hurricane (▶Table 4.2). Hurricanes are unable to form within 500 km (300 mi) of the equator because the Coriolis force is not strong enough for the storm to rotate.

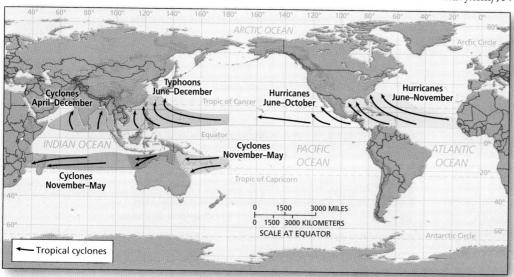

(a) Seven primary areas of tropical cyclone formation, with regional names and principal months of occurrence. Note that the general time period is the same for the Southern Hemisphere cyclones.

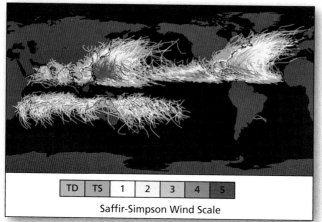

(b) Global tropical cyclone tracks from 1856 to 2013. Note the track of Hurricane Catarina in the south Atlantic.

(c) Sea-surface temperature on Aug 27, 2005. Hurricanes require sea-surface temperatures over 26°C (79°F) in order to form.

▲4.45 **Global patterns of tropical cyclones**

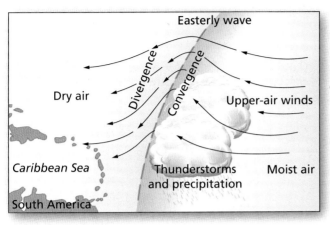

▲4.46 **Easterly waves**

Distribution & Movement Many hurricanes that strike North and Central America begin as low-pressure areas off the West African coast, become tropical depressions, and intensify into tropical storms as they cross the Atlantic. If tropical storms mature early along their track, they tend to curve northward toward the North Atlantic and miss the United States. If tropical storms mature after they reach the longitude of the Dominican Republic (70° W), then they have a higher probability of hitting the United States. After they make *landfall*, hurricanes are cut off from their supply of water vapor and begin to weaken.

Size & Physical Structure Tropical cyclones range in diameter from a compact 160 km (100 mi) to some western Pacific super typhoons that reach 1300–1600 km (800–1000 mi). Vertically, these storms occupy the full height of the troposphere.

The spiraling clouds form dense rainbands, with a central area designated the eye. Around the eye swirls a thunderstorm cloud called the eyewall, which is the area of most intense precipitation. In the midst of devastating winds and torrential rains, the eye is quiet and warm, with even a glimpse of blue sky or stars. The structure of the rainbands, central eye, and eyewall is clearly visible for Hurricane Patricia ▶ (**Fig. 4.47**). In Figure 4.48a you can see the pattern of wind movement inside a hurricane, while Figure 4.48b shows a cross section of the storm's rain bands.

Table 4.2 Tropical Cyclone Classification

Designation	Winds	Features
Tropical Disturbance	Variable	Center of low pressure
Tropical Depression	Up to 63 kmh (38 mph, 34 kts)	More organized
Tropical Storm	63-118 kmh (39-73 mph, 35-63 kts)	Circular organization
Hurricane, Typhoon, Cyclone	Greater than 119 kmh (74 mph, 65 kts)	Heavy rain, tornadoes in right-front quadrant

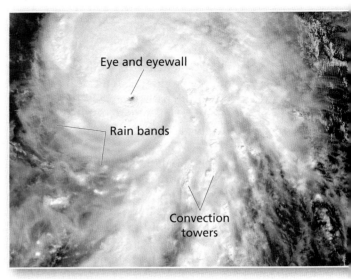

Hurricane Patricia made landfall in central Mexico on October 23, 2015 with sustained winds over 325 kmh (200 mph) and a central air pressure of 879 mb, making it the strongest Western Hemisphere hurricane ever.

▲4.47 **Tropical cyclone (top view)**

Animation MG
Hurricane Hot Towers

http://goo.gl/1GWjL0

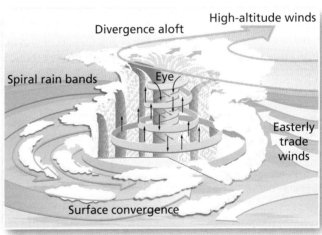

(a) Cutaway view of a mature hurricane showing the eye, rain bands, and wind patterns.

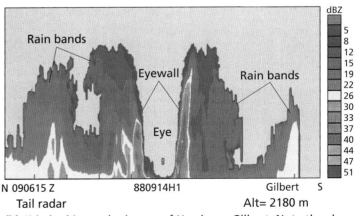

N 090615 Z   880914H1   Gilbert   S
  Tail radar   Alt= 2180 m

(b) Side-looking radar image of Hurricane Gilbert. Note the clear skies in the eye and the dense cloud bands shown in yellow and red.

▲4.48 **Inside a tropical cyclone**

4.11 (cont'd) Tropical Cyclones

Winds A tropical cyclone moves along at 16–40 kmph (10–25 mph). The strongest winds of a tropical cyclone are usually in its right-front quadrant (relative to the storm's path), where dozens of fully developed tornadoes may be embedded (▼Fig. 4.49). Hurricane Camille in 1969 had up to 100 tornadoes embedded in that quadrant.

Storm Surge The low air pressure and high winds push a bulge of water called a **storm surge** ahead of the storm. When it makes landfall, or moves ashore, the storm pushes seawater inland, causing higher water levels. Storm surges cause the majority of hurricane damage and drownings. The landfall of Hurricane Sandy in 2012 coincided with high tide to create record storm surge in New York City—as high as 4.2 m (13.9 ft) at the southern tip of Manhattan. As sea levels rise from melting ice and the expansion of sea water as it warms, storm tides will continue to increase during storm events.

Tropical Cyclone Devastation When meteorologists speak of a "category 4" hurricane, they are using the Saffir–Simpson Hurricane Wind Scale, which ranks hurricanes into five categories based on sustained wind speed (▶Table 4.3). This scale does not address other potential hurricane impacts, such as storm surge, flooding, and tornadoes. Newer building codes, such as those that have taken effect since 2000 in North Carolina, South Carolina, and Florida, will likely reduce the structural damage to newer buildings from that predicted by this scale.

Animation (MG)

Hurricane Wind Patterns

http://goo.gl/Fld02E

Table 4.3 Saffir–Simpson Hurricane Wind Scale		
Category	**Wind Speed**	**Types of Damage**
1	119–153 kmph (74–95 mph; 65–82 knots)	Some damage to homes
2	154–177 kmph (96–110 mph; 83–95 knots)	Extensive damage to homes; major roof and siding damage
3	178–208 kmph (111–129 mph; 96–112 knots)	Devastating damage; removal of roofs and gables
4	209–251 kmph (130–156 mph; 113–136 knots)	Catastrophic damage; severe damage to roofs and walls
5	> 252 kmph (> 157 mph; > 137 knots)	Catastrophic damage; total roof failure and wall collapse on high percentage of homes

Tropical cyclones are potentially the most destructive storms experienced by humans, claiming thousands of lives each year worldwide. The costs of destruction from tropical cyclones can be substantial. For example, in 2008, Hurricane Ike caused over $27 billion in damages. Ike made landfall on Galveston Island, destroying 80% to 95% of the homes on the Bolivar Peninsula (▶ Fig. 4.50).

Hurricane Sandy began in the Caribbean Sea on October 22, 2012. While over the Bahamas, it underwent complex changes, growing in diameter while weakening in intensity. Sandy moved northeastward, parallel to the eastern coast of the United States and gained intensity while passing over the warmer waters of the Gulf Stream. Sandy became an post-tropical cyclone, unofficially called "Superstorm Sandy," when it was no longer fueled by the release of latent heat of condensation close to the center of low pressure, but by conflicting air masses more similar to a midlatitude cyclone. By the time Sandy dissipated on October 31, 2012, it had become the largest Atlantic basin storm by diameter, claimed 286 lives in seven countries, affected 24 US states, and caused almost $70 billion in damage (▶Fig. 4.51).

geoCHECK ✔ What are the ingredients needed to create a hurricane?

▼4.49 **Wind patterns of a hypothetical Northern Hemisphere hurricane or typhoon**

North Carolina

South Carolina

Speed and direction of hurricane movement 50 kmph

Hurricane movement 50 kmph

Net wind 225 kmph = from the southwest

Wind on right side 175 kmph

Georgia

Hurricane movement 50 kmph

Net wind 125 kmph = from the northwest

Wind on left side 175 kmph

Florida

(a) September 9, 2008

(b) September 15, 2008

▲4.50 **The Bolivar Peninsula in Texas, before and after Hurricane Ike, September 2008** Yellow arrows point to structures that remained after the storm surge.

This connection is established for the Atlantic basin, where the number of hurricanes has increased over the past 20 years. Models suggest that the number of category 4 and 5 storms in this basin may double by the end of this century (see http://www.gfdl.noaa.gov/21st-century-projections-of-intense-hurricanes).

Another causal link between recent increases in hurricane damage and climate change is the effect of rising sea level on storm surge. Sandy's destruction was worsened by recent sea level rise—as much as 2 mm (0.08 in) per year since 1950—along the coast from North Carolina to Massachusetts.

Continued ocean warming, sea level rise, and the current trend of increasing coastal population will likely result in substantial hurricane-related property losses. Eventually, though, the more-intense storms and rising sea level could lead to shifts in population along U.S. coasts, and even the abandonment of some coastal resort communities.

Video **MG**
Making of
a Superstorm

https://goo.gl/k6HaNa

Video **MG**
Superstorm Sandy

http://goo.gl/Yj3FWj

geo**CHECK** ✔ How could climate change contribute to more powerful hurricanes?

geo**QUIZ**

1. Explain the process that forms a hurricane, beginning with an easterly wave.
2. Where would you expect to find the most intense precipitation and the strongest winds in a hurricane? Explain.
3. How are tornadoes and hurricanes similar? How are they different in terms of size and duration?

Prediction & the Future

Hurricanes are powered by the latent heat of condensation of water vapor that has evaporated from warm, tropical waters. Record sea-surface temperatures across the globe produce warmer water, and the natural mixing by storms brings up more heat. For example, Hurricane Wilma experienced a massive 100-mb central pressure drop in a little over 24 hours as it passed over record sea-surface temperatures on October 19, 2005. In the spring of 2014, there were 20 hurricane-level wind events, including 14 that met the criteria for "bombogenesis" –24 mb of pressure drop in 24 hours. Because many factors come into play in order for bombogenesis to occur, storms that form this way are often especially powerful and severe.

Research has correlated longer tropical storm lifetimes and greater intensity with rising SSTs. As oceans warm, the energy available to fuel tropical cyclones is increasing. The Tropical Prediction Center, at the National Hurricane Center in Miami, Florida, analyzed weather records for the period 1900–2006 and concluded that:

"Atlantic tropical cyclones are getting stronger on average . . . as seas warm, the ocean has more energy that can be converted to tropical cyclone wind."

▲4.51 **A roller coaster is battered by the surf in the aftermath of Hurricane Sandy**

WEATHER ⇒ HUMANS

• Frontal activity and midlatitude cyclones bring severe weather that affects transportation systems and daily life.

• Severe weather events such as ice storms, derechos, tornadoes, and tropical cyclones cause destruction and human casualties.

HUMANS ⇒ WEATHER

• Rising temperatures with climate change have caused shrinking spring snow cover in the Northern Hemisphere.

• Sea-level rise is increasing hurricane storm surge on the U.S. east coast.

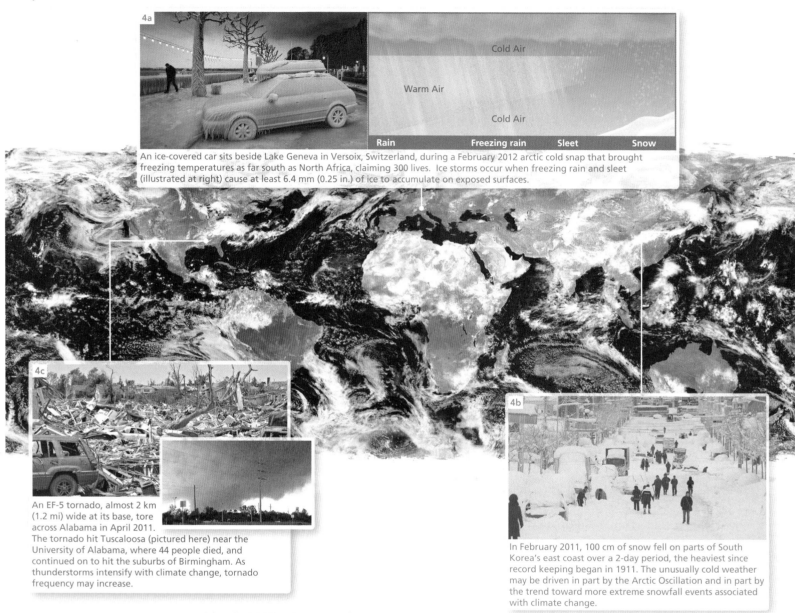

An ice-covered car sits beside Lake Geneva in Versoix, Switzerland, during a February 2012 arctic cold snap that brought freezing temperatures as far south as North Africa, claiming 300 lives. Ice storms occur when freezing rain and sleet (illustrated at right) cause at least 6.4 mm (0.25 in.) of ice to accumulate on exposed surfaces.

An EF-5 tornado, almost 2 km (1.2 mi) wide at its base, tore across Alabama in April 2011. The tornado hit Tuscaloosa (pictured here) near the University of Alabama, where 44 people died, and continued on to hit the suburbs of Birmingham. As thunderstorms intensify with climate change, tornado frequency may increase.

In February 2011, 100 cm of snow fell on parts of South Korea's east coast over a 2-day period, the heaviest since record keeping began in 1911. The unusually cold weather may be driven in part by the Arctic Oscillation and in part by the trend toward more extreme snowfall events associated with climate change.

ISSUES FOR THE 21ST CENTURY

• Global snowfall will decrease, with less snow falling during a shorter winter season; however, extreme snowfall events (blizzards) will increase in intensity. Lake-effect snowfall, transitioning to rainfall, will increase owing to increased lake temperatures.

• Increasing ocean temperatures with climate change will strengthen the intensity and frequency of tropical cyclones by the end of the century.

Looking Ahead

In the next chapter, we use the water balance model and examine inputs and outputs of the water budget. Water quality and quantity and the availability of potable water loom as major issues for the global society.

How does water behave in the atmosphere?

4.1 Water's Unique Properties

Describe the heat properties of water.

Identify the traits of water's three states: solid, liquid, and gas.

- Water exists as solid, liquid, and gas. One calorie will raise the temperature of 1 g of water (at 15°C) 1 Celsius degree, but to melt ice at 0°C requires 80 calories per gram, and to evaporate water at 100°C requires 540 cal per gram.

1. Describe the three states of matter as they apply to ice, water, and water vapor.

2. What is latent heat? How is it involved in the phase changes of water?

4.2 Humidity—Water Vapor in the Atmosphere

Define humidity, relative humidity, and dew-point temperature.

- Humidity refers to water vapor in the air. Relative humidity is the ratio (given as a percentage) of the amount of water vapor that is present in the air compared to the maximum water vapor possible in the air at a given temperature. The temperature at which a given mass of air becomes saturated and condensation begins is the dew-point temperature.

3. Define relative humidity. What does the concept represent? What is meant by the terms *saturation* and *dew-point temperature*?

4.3 Atmospheric Stability

Define atmospheric stability.

- Stability refers to the tendency of an air parcel to rise, remain in place, or fall. A parcel that is warmer than the air around it is unstable, and it will rise until it reaches an altitude where the air around it has a density and temperature similar to its own. A *stable* parcel will either stay where it is or sink. If the rate of cooling of the atmosphere is greater than the rate of cooling of a parcel of air, the atmosphere is stable.

4. Why is a lifted parcel of air stable or unstable?

5. What atmospheric temperature and moisture conditions would you expect on a day when the weather is unstable? When it is stable?

4.4 Clouds & Fog

Explain cloud formation, and identify major cloud classes and types, including fog.

Describe the two ways that raindrops form.

- Clouds are masses of water droplets that form when water vapor in saturated air condenses on condensation nuclei. Clouds are classified as high, middle, low, and vertically developed clouds. Fog is a cloud on the ground. Moisture droplets in a cloud form when saturation and the presence of condensation nuclei in air combine to cause condensation. Raindrops are formed from moisture droplets through either the collision–coalescence process or the Bergeron–Findeisen ice-crystal process.

6. What are the basic types of clouds? Describe how the basic cloud forms vary with altitude.

What are the forces that produce our weather?

4.5 Air Masses

Describe air masses that affect North America.

- Air masses are bodies of air categorized by their moisture content—**m** for maritime (wetter) and **c** for continental (drier)—and their temperature, a function of latitude—designated A (arctic), P (polar), T (tropical), E (equatorial), and AA (Antarctic). They are regional to continental in size and extend through the lower half of the troposphere.

7. Which air masses are the most significant to North America? What happens to their characteristics as they migrate to different locations?

4.6 Processes That Lift Air

Identify four types of atmospheric lifting mechanisms.

- The four main lifting mechanisms are:
  - **Convergent lifting**—air flows toward an area of low pressure.
  - **Convectional lifting**—local surface heating causes warm air to rise.
  - **Orographic lifting**—air is lifted up by a barrier such as a mountain range.
  - **Frontal lifting**—warmer air is lifted upward by cooler air.

8. How are orographic and frontal lifting similar? How are they different?

4.7 Midlatitude Cyclones

Describe the life cycle of a midlatitude cyclonic storm system.

- A **midlatitude cyclone**, or **wave cyclone**, is a vast low-pressure system that migrates across the continent, pulling air masses into conflict along fronts. It can be thought of as having a life cycle of birth, maturity, old age, and dissolution. **Cyclogenesis** is the birth of the low-pressure circulation. An **occluded front** is produced when a cold front overtakes a warm front in the maturing cyclone. These systems are guided by the jet streams of the upper troposphere along seasonally shifting storm tracks.

9. Diagram a midlatitude cyclonic storm during its open stage. Label each of the components in your illustration, and add arrows to indicate wind patterns in the system.

4.8 Weather Forecasting

List the measurable elements that contribute to weather forecasting.

- Meteorologists collect and analyze weather data, using computer models based upon past events to predict weather. Building a database of wind, pressure, temperature, moisture conditions, and other factors that influence weather is key to numerical weather prediction and the development of weather-forecasting models. Technology employed includes weather satellites; automated, ground-based sensors; advanced computer processing systems; and Doppler radar.

10. What are the main components of weather forecasting? What are the main types of weather data collected? What systems are used to collect and process these data?

What causes violent storms?

4.9 Thunderstorms, Tornadoes, & Tropical Cyclones

Identify various forms of violent weather—thunderstorms, tornadoes, and tropical cyclones.

- Thunderstorms, derechos, tornadoes, and tropical cyclones are the main types of violent weather. They are all powered by the latent heat of condensation—energy that is released when water vapor condenses.

11. What constitutes a thunderstorm? What type of cloud is involved? What type of air mass would you expect in an area of thunderstorms in North America?

12. What conditions are required to form a tropical cyclone? Where are these conditions found?

Critical Thinking

1. How does the latent heat of condensation transfer heat energy from the tropics to the poles?

2. Which cloud types show that the atmosphere is stable? Which cloud types show that the atmosphere is unstable?

3. Which lifting mechanisms would you expect to find on a tropical island? Which lifting mechanisms would you not expect to find on a tropical island?

4. How is a midlatitude cyclone similar to a tropical cyclone? How are they different?

5. How could global warming affect tropical cyclone intensity?

Visual Analysis

In this photograph of the New Jersey shore, you can see the destruction caused by Hurricane Sandy in September 2012. The New Jersey shore is barely above sea level, yet it is a popular vacation destination.

1. What are the main mechanisms mentioned in the text that destroyed houses and other buildings during this storm?

2. What physical factors associated with climate change would increase the likelihood of future hurricane damage along this coast?

▲R 4.1

Interactive Mapping
Login to the **MasteringGeography** Study Area to access **MapMaster**.

Climate Change: Global Warming & Tropical Cyclones

Global warming is increasing sea-surface temperatures. As sea-surface temperatures increase, more water vapor is available to power these storms.

- Open MapMaster in MasteringGeography.
- Select the 27°C and 30°C layers from *Global Mean Temperature for July* from the *Physical Environment* menu.
- Select the following layers from *Global Surface Warming Worst Case Projections* from the *Physical Environment* menu: 2-2.5°C, 2.5-3°C

1. By turning on and off individual *Global Surface Warming Worst Case Projections* layers, you can calculate what projected sea-surface temperatures might be. What is the temperature of the warmest projected sea-surface temperature? How much warmer than today is that?

2. Which regions of the world will be most affected by sea-surface temperature increase? What are some countries in the path of these stronger tropical cyclones, based on the projected temperatures and the tropical cyclones paths shown in Figure 4.45a?

Climate Change: Sea Level Rise

Global warming is not only causing stronger hurricanes, but also increasing sea level. This is a crucial issue given the number of people that live in low-lying coastal regions.

- In Google Earth, fly to the Bolivar *Peninsula, Texas.*
- Click *View*, and select *Historical Imagery.*

Move the slider over the years, from the earliest to the most recent imagery. Stop the slider on September 13, 2008. These images show the peninsula just after Hurricane Ike. The storm surge from Ike ranged from 4.9 to 6.5 m (16 to 22 ft).

1. What are the highest and lowest elevations you can find around Pirate Cove Street?

2. How deep would the water have been over Pirate Cove Street during maximum storm surge depth?

3. What do you notice on the ground that could tell you about the direction of wind and waves from the storm?

▲ R 4.2

Looking for additional review and test prep materials? Visit the Study Area in MasteringGeography™ to enhance your geographic literacy, spatial reasoning skills, and understanding of this chapter's content by accessing a variety of resources, including MapMaster™ interactive maps, videos, *Mobile Field Trips*, *Project Condor Quadcopter videos*, *In the News* RSS feeds, flashcards, web links, self-study quizzes, and an eText version of *Geosystems Core*.

Key Terms

adiabatic, *94*
advection fog, *98*
air mass, *100*
altocumulus, *97*
cirrus, *97*
climate, *86*
cloud, *96*
cold front, *104*
condensation nuclei, *96*
convectional lifting, *102*
convergent lifting, *102*
cumulonimbus, *97*
cumulus, *97*
derechos, *97*
dew-point
    temperature, *90*

dry adiabatic rate, *94*
easterly waves, *114*
environmental lapse
    rate, *94*
evaporation fog, *98*
fog, *14*
funnel clouds, *112*
hail, *110*
humidity, *6*
hurricane, *114*
latent heat, *88*
latent heat of
        condensation, *89*
latent heat of
        evaporation, *89*

latent heat of
    vaporization, *89*
lifting condensation
    level, *94*
lightning, *110*
mesocyclone, *112*
meteorology, *86*
midlatitude
    cyclone, *106*
moist adiabatic
    rate, *94*
moisture droplet, *96*
nimbostratus, *97*
normal lapse
    rate, *94*

occluded front, *107*
orographic
    lifting, *102*
phase change, *88*
precipitation, *96*
radiation fog, *98*
rain shadow, *104*
relative humidity, *90*
saturation, *90*
specific heat, *88*
specific humidity, *92*
squall line, *105*
stability, *94*
storm surge, *116*
stratocumulus, *97*
stratus, *97*

sublimation, *89*
thunder, *110*
tornado, *112*
tropical
    cyclone, *114*
typhoon, *114*
upslope fog, *98*
valley fog, *98*
vapor pressure, *92*
virga, *88*
warm front, *104*
waterspout, *112*
wave cyclone, *106*
weather, *86*

Stormy Weather: How do you track a midlatitude cyclone on a weather map?

Midlatitude cyclones are massive, swirling storms that bring different air masses together, causing clouds and rain. As a weather forecaster, you would identify the conditions in which these powerful storms develop, predict their paths, and warn people of their dangers.

In this lab you will use a weather map to determine current conditions at weather stations, the position and speed of movement of the warm and cold fronts, and identify the stages of midlatitude cyclone development. Before you begin this exercise, review the chapter sections on air masses, warm and cold fronts, and midlatitude cyclones.

GeoLab 4 MG
Pre-Lab Video

http://goo.gl/wrQJZ0

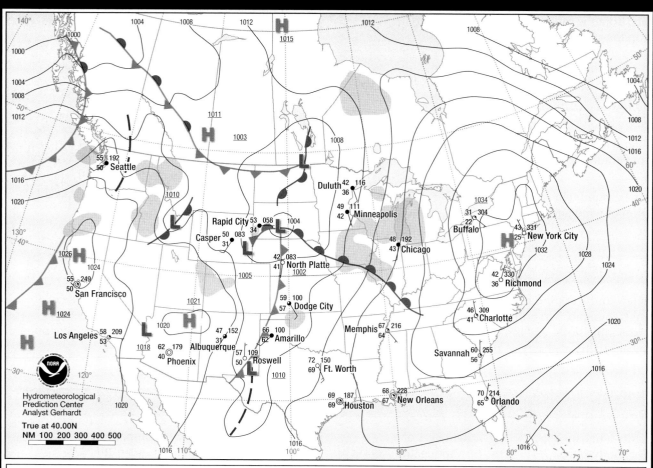

▲GL4.1 **Weather map for October 13, 2012**

Apply

You are a weather forecaster for Chicago, Illinois. You have been tasked with identifying air masses, cold and warm fronts, locating them on weather maps, and predicting the weather.

Objectives

- Interpret weather station symbols to determine current conditions
- Locate and identify warm, cold, and occluded fronts on a weather map
- Identify stages of midlatitude cyclone development

Procedure

The weather map (Fig GL 4.1) shows a midlatitude cyclone moving across the United States on October 13, 2012. Refer to the map to answer the questions below.

1. Describe the current temperature, dew-point temperature, wind speed and direction, and dominant air mass in Rapid City, SD.

2. Describe the general location of the cold

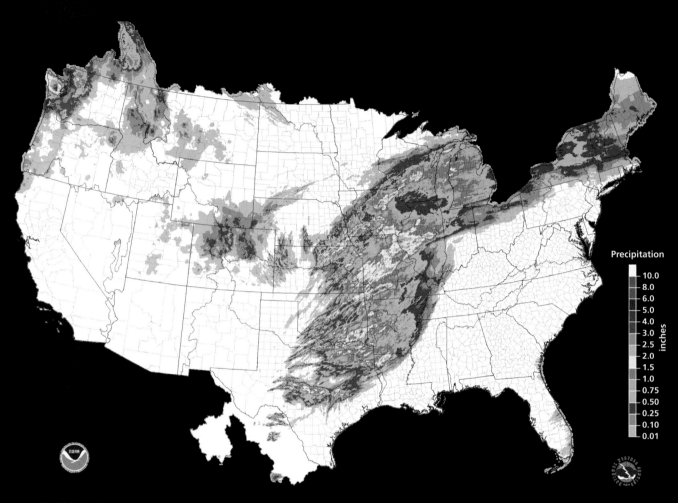

▲GL4.2 **24-hour Precipitation amounts ending 7:00 AM EST October 14, 2012**

front and warm front. Identify some states that each front is over.

3. Where is the central area of low pressure? Which cities are located by it?
4. Name a city that the occluded front is over.
5. Assuming that the cold front moves at 40 kmph, list two cities where it will be overhead in 24 hours.
6. Compare the temperature, dew-point temperature, and wind speed and direction for Pueblo, CO; Wichita, KS; and Memphis, TN. Which cities are experiencing a cT air mass? a mT air mass?

Now refer to the map (Fig GL 4.2) showing the precipitation total for the 24-hour period ending 7 A.M. October 14, 2012.

7. The precipitation in Colorado occurred

after the passage of the cold front. Given that information, which lifting mechanism or mechanisms probably produced the precipitation recorded there? Explain.
8. The midlatitude cyclone did not pass over southern Florida. Which lifting mechanism was most probably responsible for the precipitation recorded there? Explain.

Analyze & Conclude

9. Assume that the storm moves due east over the next several days. What would the temperature, dew-point temperature, wind speed and direction, and dominant air mass in Chicago, IL, be after the warm front passes through? After the cold front?
10. How does the amount of rain recorded across Texas, Oklahoma, Missouri, Iowa,

and Illinois compare with the amount of rain recorded across Wisconsin, Michigan, Ontario, and Quebec? Which type of front produced the heaviest rain? the lightest rain? Relate your answer to the properties of warm and cold fronts.

Log in MasteringGeography™ to complete the online portion of this lab, view the Pre-Lab Video, and complete the Post-Lab activity.
www.masteringgeography.com

Water Resources

5

Water is *the* essential resource for life. Plants, animals, and even humans are about 70% water. We use water for household needs, and to nourish small gardens and vast agricultural tracts. Most industrial processes require water.

Water covers 71% of Earth, the only planet with significant quantities of liquid water. Yet water is not always available where and when we want it. We rearrange surface water resources to suit our needs. We drill wells, build reservoirs, and dam and divert streams to redirect water. Fortunately, water is a renewable resource that cycles through the environment. Yet over a billion people in 80 countries lack access to safe drinking water.

Water supplies of adequate quantity and quality loom as the most important resource issue for many parts of the world in this century.

Key Concepts & Topics

Torrents of water released from China's Three Gorges Dam on the Yangtze River

5.1 Water on Earth

Key Learning Concepts

▶ *Describe* the origin of Earth's waters.
▶ *Define* the quantity and the location of Earth's freshwater supply.
▶ *Visualize* the hydrologic cycle.

In the Solar System, water occurs in significant quantities only on our planet. Much of Earth's water likely originated from asteroids and from hydrogen- and oxygen-laden debris among the materials that coalesced to form the planet. All life—and most of Earth's physical processes—are dependent on water.

▲5.1 Outgassing of water from Earth's crust in a geothermal area in the Chilean Altiplano

Origin & Distribution

Earth's hydrosphere contains about 1.36 billion cubic kilometers of water (326 million cubic miles). Where did this water come from? Early in Earth's history, **outgassing** (discharge of gases by Earth) released massive quantities of water and water vapor from deep within the crust. When the atmosphere formed, water vapor condensed and then fell to Earth in torrential rains. As Earth's surface cooled 3.8 billion years ago, the water pooled into lakes, seas, and eventually oceans. Outgassing continues to-day, adding new water to the system through volcanic eruptions, geysers, and seepage from layers 25 km (15.5 mi) below the surface. (▶Fig. 5.1). Earth also "recycles" water from subsurface deep groundwater back onto the surface.

Water is not distributed evenly across Earth (▶Fig. 5.2). The oceans contain about 97% of all water. Freshwater comprises less than 3% of all water, and most of that is locked up in ice or lies below the surface. The freshwater that terrestrial plants and animals—including humans—depend upon actually represents less than 1% of all water (▼Fig. 5.3).

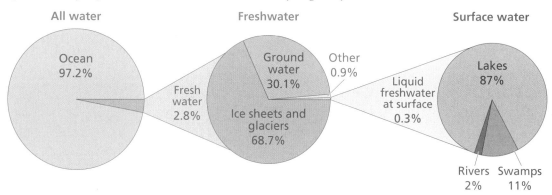

▲5.3 **Ocean and freshwater distribution on Earth** The locations and percentages of all water on Earth, with detail of the freshwater portion (surface and subsurface) and a breakdown of the surface water component.

Worldwide Equilibrium

Water is the most common compound on the surface of Earth. Although some water escapes into space or breaks down to form new compounds, the quantity remains relatively unchanged. The outgassing of pristine water replaces this "lost" water. Despite this steady-state equilibrium, change in sea level results when the amount of water stored in glaciers and ice sheets varies during periods of global warming and cooling. For example, at peak glaciation during the Pleistocene Epoch from 1.8 million to 11,500 years ago, sea level was 100 m (984 ft) lower than today. Over the past 100 years, mean sea level has risen by 20–40 cm (8–16 in.) and is still rising as higher temperatures melt more ice and also cause ocean water to thermally expand.

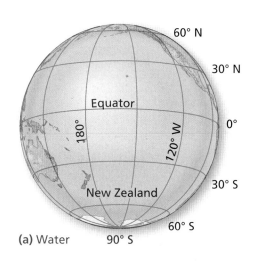

(a) Water

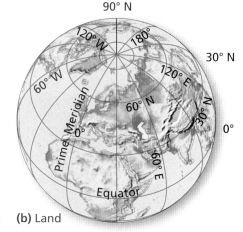

(b) Land

▲5.2 **Land and water hemispheres**

geo**CHECK** ✔ Where is most of Earth's freshwater located?

The Hydrologic Cycle

Vast currents of water, water vapor, ice, and energy continually circulate about the planet. This hydrologic cycle has operated for billions of years, from the lower atmosphere to several kilometers beneath Earth's surface. The cycle involves the circulation and transformation of water throughout Earth's atmosphere, hydrosphere, lithosphere, and biosphere.

Figure 5.4 represents a simplified model of this complex system. For the purposes of this discussion, we start in the ocean, where 97% of Earth's water resides. Here is where 86% of all evaporation occurs. The other 14% is from land, including water moving from the soil into plant roots and *transpiring* through their leaves into the atmosphere. In the figure, you can see that of the 86% total evaporation rising from the ocean, 66% combines with the 12% *advected* (moving horizontally) from the land to produce the 78% of all precipitation that falls back into the ocean. The remaining 20% of moisture evaporated from the ocean, plus 2% of land-derived moisture, produces the 22% of all precipitation that falls over land. Clearly, most continental precipitation comes from the oceanic portion of the cycle.

This model portrays the worldwide averages for the proportions of water involved in evaporation over land and water; atmospheric advection of water vapor, clouds, and precipitation; and surface runoff. In reality the different parts of the cycle vary enormously across Earth. The temperate rain forests of North America cycle over 99% more water than Africa's arid Sahara Desert, which covers a much larger area. Across the globe, the geographic imbalance in precipitation, runoff, and soil moisture influences the global distribution of plants and animals, and thus human settlement and land use.

geoCHECK ✔ What percentage of precipitation falls directly to Earth's land surface?

geoQUIZ

1. Define outgassing and explain what role it plays in the hydrologic cycle.
2. Describe how changing sea level during the 21st century will affect human settlement.
3. Identify the percent of total evaporation that rises directly from the ocean.

Video MG
Hydrological Cycle

http://goo.gl/h7pTt6

Animation MG
Earth's Water and the Hydrologic Cycle

http://goo.gl/9KPUH1

Note that all evaporation (86% + 14% = 100%) equals all precipitation (78% + 22% = 100%).

Atmospheric advection
of water vapor

20

12

Cloud
formation

20

2

Precipitation

Cloud
formation

66

22

86

Precipitation

78

14

Evaporation
Transpiration

Evaporation

Ocean

Surface runoff

Groundwater
flow

8

Continent

△ Percent (global average)

Moisture in the atmosphere carried inland by wind is balanced by surface runoff and subsurface groundwater flow when all of Earth is considered.

▲ **5.4 Hydrologic cycle model** Water travels endlessly through the hydrosphere, atmosphere, lithosphere, and biosphere. The triangles show global average values as percentages. The elongated arrows underneath the triangles indicate the direction of moisture as it progresses through this idealized hydrologic cycle.

5.2 The Soil–Water Budget Concept

Key Learning Concepts

▶ *Relate* the water-budget concept to the hydrologic cycle, water resources, and soil moisture.

▶ *Construct* the water-balance equation to account for expenditures of the water supply.

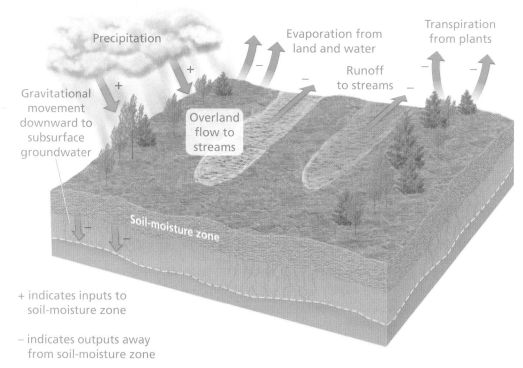

+ indicates inputs to soil-moisture zone

– indicates outputs away from soil-moisture zone

▲ **5.5 The soil-moisture environment** The principal routes for precipitation on Earth's surface include interception by plants, throughfall to the ground, overland flow to streams, transpiration from plants, evaporation from land and water, and gravitational movement downward to subsurface groundwater.

Precipitation landing on Earth's surface is distributed several ways (▶ Figure 5.5). Some returns to the atmosphere by evaporation and plant transpiration. Some exits as stream run-off and overland flow, or filters down to become subsurface groundwater. Some remains stored in the soil. A **soil-water budget** helps evaluate water resources on any scale: continental, regional, or even a single farm. We calculate budgets by determining how the incoming precipitation "supply" is distributed to satisfy the output "demand" of plants, evaporation, and soil-moisture storage.

Actual & Potential Evapotranspiration

Figure 5.6 organizes the water-balance components into an equation where the precipitation (left side) must balance with the expenditures (right side). **Precipitation** (P) is the moisture supply to Earth's surface. Some of this water leaves Earth's surface by the **evaporation** of water molecules into air, plus the **transpiration** of moisture into the atmosphere by plants. When a plant transpires, water vapor exits the plant leaf through small openings, or pores, called stomata that are usually most plentiful on the underside of its leaves. Evaporation from land and water surfaces and transpiration by plants are combined into one term—**evapotranspiration** (▶ Fig. 5.7).

It is important to distinguish **actual evapotranspiration** (ACTET) from **potential evapotranspiration** (POTET). ACTET is the amount of water that returns to the atmosphere through evapotranspiration. POTET is the amount of water that would evaporate and transpire if the moisture supply was always present, as is the case over the oceans and in rain forests. Suppose you fill a bowl with water and let the water evaporate. As long as water remains in the bowl, ACTET = POTET. However, once the water in the bowl evaporates, the atmosphere's "demand" for water remains unmet, so POTET now exceeds ACTET. Across much of Earth's land surface the atmospheric heat available for evapotranspiration exceeds the amount of moisture available (Figure 5.6). However, in moist rain forest environments, POTET and ACTET are often very similar.

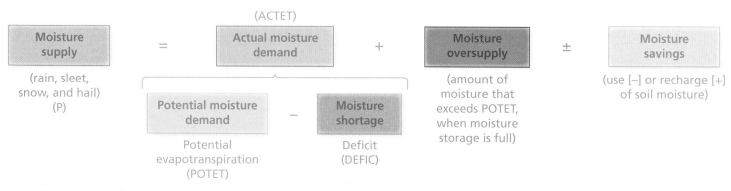

▲ **5.6 The water-balance equation explained** The outputs (components to the right of the equal sign) are an accounting of expenditures from the precipitation moisture-supply input (to the left).

▲ **5.7 Evapotranspiration** In the Arabian Desert, warm temperatures and scant rainfall create severe moisture deficits in every month. Thus POTET for these orchard and ground crops is high throughout the year. Drawing water from deep underground, farmers use drip irrigation to reduce evaporation from the soil.

Soil Moisture Precipitation and runoff add moisture to the soil (▶ Fig. 5.8). These inputs are then accounted for as **soil moisture**. This component includes *recharge* of soil moisture by precipitation and percolation into the ground, and by plant roots absorbing soil moisture as part of evapotranspiration, when sufficient solar energy is available. As the soil-moisture storage diminishes, plants eventually are unable to rely on this source of moisture.

Water Deficit (DEFIC) A natural and recurrent feature of climate, **drought** occurs when less precipitation and higher temperatures produce drier conditions over an extended period of time. The four types of droughts consider precipitation, temperature, and soil-moisture content, along with shortages resulting from human demand for water:

- *Meteorological drought* is the degree of dryness, compared to a regional average, and the duration of dry conditions.
- *Agricultural drought* occurs when shortages of precipitation and soil moisture affect crop yields.
- *Hydrological drought* relates to the effects of rain and snow shortages on water supply, such as when streamflow decreases, reservoirs drop, snowpack declines, and groundwater use increases.
- *Socioeconomic drought* results when reduced water causes resource shortages, such as declining hydropower.

geoCHECK ✔ Explain the difference between evaporation and transpiration.

▼ **5.8 Precipitation and runoff** After heavy rains in South Carolina and nearby states in October, 2015, soils in the hardest-hit areas became saturated (dark blue). Soil moisture controls how much rainfall sinks into the ground and how much becomes runoff.

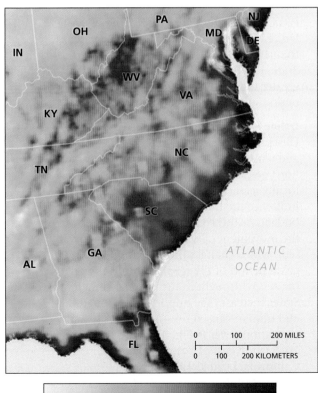

SOIL MOISTURE (m³/m³)

5.2 (cont'd) The Soil–Water Budget Concept

Applying Water Budget Concepts

Understanding the relationship between PRECIP and POTET as presented in the water balance equation in Figure 5.6 is key to understanding water budgets. PRECIP is the principal input, or supply of water. All the components on the other side of the equation are outputs, or demands, for this water:

Precipitation = Actual Evapotranspiration + Surplus ± Change In Soil-Moisture Storage

For any given location or portion of the hydrologic cycle, the water inputs are equal to the water outputs, plus or minus the change in soil-moisture storage. We will first examine the relationship between the water budgets inputs and outputs in North America. Figure 5.9 shows precipitation for North America, while Figure 5.10 presents POTET values for the same area. Comparing the POTET (demand) map with the PRECIP (supply) map, you can see how the relation between these two meteorological factors determines the remaining components of the water-balance equation. From the two maps, it's also possible to identify regions where PRECIP is greater than POTET (the eastern United States), or where POTET is greater than PRECIP (the southwestern United States). When the latter situation (POTET greater than PRECIP) persists for an extended period, the result is a drought.

Other Factors Affecting the Water Budget
Ultimately, the climatic factors of precipitation and temperature (as it affects evapotranspiration) determine the water budget. However, local vegetation, soils, and land use also influence the inputs and outputs of a water budget. As these variables change, so does the movement of water through the ecosystem, which changes the water budget for any given period of time. For example, the ongoing severe drought in the western United States increases the opportunity for natural and human-caused fires that affect the land's vegetation cover (▶Fig. 5.11). In August of 2015, a fire east of Yosemite National Park, California, burned 1488 hectares (3676 acres) of timber, brush, and grass. The

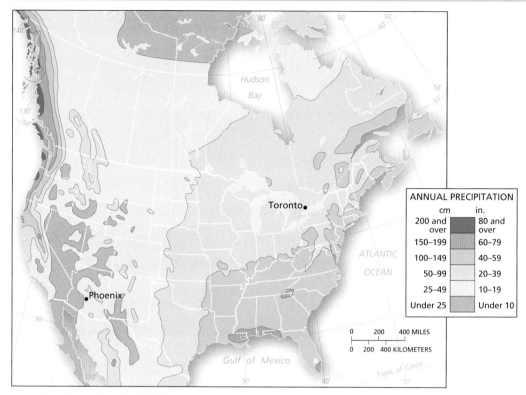

ANNUAL PRECIPITATION

cm	in.
200 and over	80 and over
150–199	60–79
100–149	40–59
50–99	20–39
25–49	10–19
Under 25	Under 10

▲**5.9 Precipitation (PRECIP) in North America–the water supply**
Note how the Pacific Northwest, the Southeast, and the Northeastern seaboard receive significantly more precipitation than the Midwest and arid Southwest.

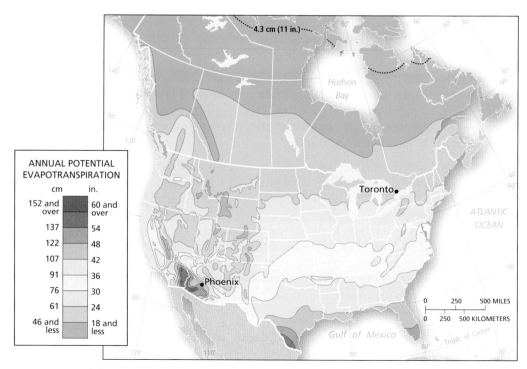

ANNUAL POTENTIAL EVAPOTRANSPIRATION

cm	in.
152 and over	60 and over
137	54
122	48
107	42
91	36
76	30
61	24
46 and less	18 and less

▲**5.10 Potential evapotranspiration (POTET) for the United States and Canada–the water demand** Note how POTET roughly correlates with latitude, and thus with insolation.

reduction of green vegetation will change the water budget for this area. With fewer plant roots to absorb moisture, precipitation will form runoff rather than be absorbed into the ground, and higher surface temperatures will promote evaporation into the atmosphere. Both factors will reduce the amount of moisture stored in the soil. Thus in this fire-altered landscape, the POTET demand will remain the same (because insolation remains the same), but any PRECIP that lands on the surface will run off and evaporate more quickly, creating a moisture deficit earlier than if the ground was covered with trees, shrubs, and grasses.

Analyzing a Water Budget: Berkeley, California Using these concepts, we graph the annual water-balance supply and demand for Berkeley, California (▼Fig. 5.12). On the graph, a comparison of PRECIP and POTET by month determines whether there is a net surplus or a net deficit for water. In Berkeley the cooler time from November to March shows a net surplus (blue area), but the warm days from April to October create a net water deficit. In this graph the soil-moisture utilization (green area) is the amount of moisture within the soil that can be seasonally utilized

▲**5.11 Vegetation and the water budget** The Walker Fire burns east of Yosemite National Park. The loss of vegetation caused by the fire will alter the water budget by increasing evaporation and runoff on the exposed land surface.

for evapotranspiration. The soil moisture storage capacity delays the transition from moisture surplus to moisture deficit.

Water budgets can be used to assess water supply and demand for any spatial scale over any period of time—from minutes to millennia. As human populations grow, balancing human water demands with that needed for Earth systems becomes more challenging. Evaluating the linkages between components of the water budget helps us understand how changes to one component affect the other components. An example is the extraction of groundwater (as when wells are drilled to provide water for irrigating crops), which also affects streamflow. This topic is discussed in the next modules.

geoCHECK ✔ Identify the two meteorological factors that contribute to drought conditions.

geoQUIZ

1. Using Figure 5.9, identify the regions where precipitation is the scarcest *and* the most abundant.
2. Explain the difference between actual evapotranspiration and potential evapotranspiration.
3. Describe how POTET in July is different between Phoenix, Arizona, and Toronto, Canada.
4. Referring to the sample water budget for Berkeley, California, in Figure 5.12, predict how water use by homeowners and the city's parks department would vary during the year. Explain.

Figure 5.12

The cool and wet season produces a moisture surplus as soil moisture is recharged and ACTET meets or exceeds POTET.

A moisture deficit occurs where POTET exceeds ACTET.

The surplus quickly fades during the warm and dry season as water in the soil is utilized.

Soil moisture (mm) — left axis: 0, 10, 20, 30, 40, 50, 60, 70, 80, 90, 100, 110, 120, 130, 140, 150, 160
Soil moisture (in.) — right axis: 1, 2, 3, 4, 5, 6

Months: J F M A M J J A S O N D

Berkeley, California

Legend:
- Surplus
- Soil-moisture utilization
- Soil-moisture recharge
- Deficit
- ●—● P
- ○—○ POTET
- ▲—▲ ACTET

▲**5.12 Sample water budget for Berkeley, California**

5.3 Surface Water & Groundwater Resources

Key Learning Concepts

▶ **Summarize** how humans have adapted surface water resources for their use.

▶ **Describe** the nature of groundwater and the structure of the groundwater environment.

Water distribution is uneven over space and time. Because humans require a steady supply, we build large-scale management projects to redistribute water or store it until needed. In this way, deficits are reduced and surpluses are held for later release. Analyzing the water balance highlights the contribution that streams and groundwater make to water resources.

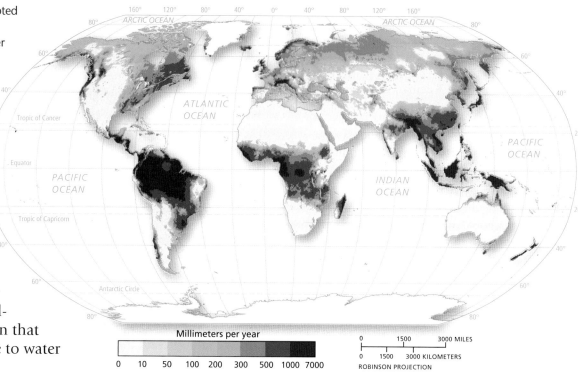

▲**5.13 Annual global runoff** Surface water runoff is an important source of the freshwater found in streams, lakes, and reservoirs.

Surface Water Resources

Streams may be *perennial* (constantly flowing) or *intermittent* (flowing occasionally). In either case, the total **streamflow** comes from surplus surface water runoff, subsurface flow, and groundwater. Figure 5.13 maps global runoff, an important component of average annual streamflow. The highest streamflow occurs along the equator within the tropics, reflecting the continual rainfall along the Intertropical Convergence Zone (ITCZ). Southeast Asia also experiences high streamflows, as do northwest coastal mountains in the Northern Hemisphere. Lower streamflows occur in Earth's subtropical deserts, rain-shadow areas, and continental interiors, particularly in Asia.

Erecting dams across rivers is the most common method humans use to contain water surplus. There are presently 40,000 dams over 15 m (49 ft)

high in the world. These dams interrupt free-flowing rivers and impound water on every continent except Antarctica. Three Gorges Dam on the Yangtze River in China is the largest dam in the world (▼**Fig. 5.14**). For all dams, society must weigh the benefits of water storage, flood control, and hydroelectric generation, against the environmental and social consequences of altering the environment of the watershed both above and below the dam.

geoCHECK ✔ Describe the difference between a perennial and intermittent stream.

▼**5.14 The Yangtze River in China's Hubei Province, before (1987) and after (2004) the Three Gorges Dam** The gray areas indicate urban development.

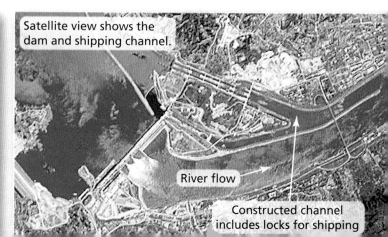

Satellite view shows the dam and shipping channel.

River flow

Constructed channel includes locks for shipping

Groundwater Resources

Within more arid regions where countries experience great seasonal fluctuations in runoff, groundwater becomes an important reserve water supply. **Groundwater** is an essential part of the hydrologic cycle, although it lies beneath the surface beyond the soil-moisture root zone. Pores in soil and rock tie groundwater to surface supplies. With the exception of glacial ice, groundwater is the largest potential freshwater source in the hydrologic cycle—larger than all surface lakes and streams combined. However, pollution can threaten groundwater quality, and over-consumption can deplete groundwater faster than it can be replaced.

Remember that *groundwater* is not an independent source of water, for it is tied to surface supplies that infiltrate below the surface in a process called **groundwater recharge**. In many regions groundwater accumulation occurred over millions of years, so people should not exceed this long-term buildup with excessive short-term demands. About half of the U.S. population derives a portion of its freshwater from groundwater sources. For example, in Nebraska, groundwater supplies 85% of water needs. Between 1950 and 2000, annual groundwater withdrawal in the United States and Canada increased more than 150%. Figure 5.15 shows potential groundwater resources in both countries, while Figure 5.16 shows global groundwater resources.

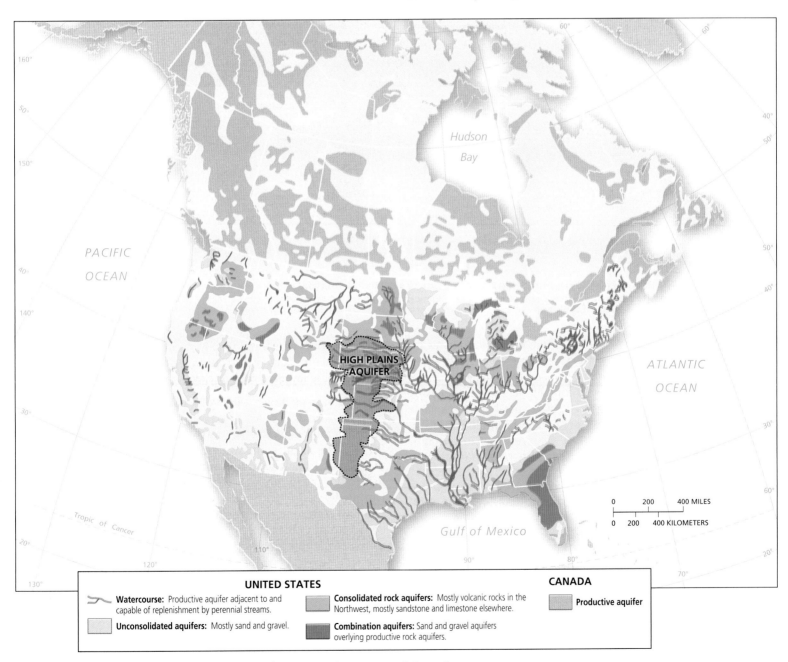

UNITED STATES

⤳ **Watercourse:** Productive aquifer adjacent to and capable of replenishment by perennial streams.

Unconsolidated aquifers: Mostly sand and gravel.

Consolidated rock aquifers: Mostly volcanic rocks in the Northwest, mostly sandstone and limestone elsewhere.

Combination aquifers: Sand and gravel aquifers overlying productive rock aquifers.

CANADA

Productive aquifer

▲ **5.15 Groundwater resource potential for the United States and Canada**
North America's groundwater is part of a vast but invisible global reservoir of freshwater.

5.3 (cont'd) Surface Water & Groundwater Resources

▲5.16 **World groundwater resources** Beneath Earth's land surface to a depth of 4 km (13,000 ft) worldwide, resides some 8,340,000 km³ (2,000,000 mi³) of water, a volume comparable to 70 times all the freshwater lakes in the world.

Groundwater begins as surplus water that percolates downward below the surface (▶Fig. 5.17). This excess surface water moves through the **zone of aeration**, where soil and rock are less than saturated, because some pore spaces contain air. The *porosity* of any rock layer depends on the arrangement, size, and shape of its particles, and the degree to which they are packed together. Subsurface rocks are *permeable* if they conduct water readily. Conversely, *impermeable* rocks, or those of low permeability, tend to obstruct the flow of water.

Eventually, water reaches the **zone of saturation** where the pores completely fill with water. Like a hard sponge made of sand, gravel, and rock, the saturation zone stores water in countless pores and voids. The upper limit of the water that collects in the zone of saturation is the **water table**, the surface where the zone of saturation and the zone of aeration are in transition (all across Figure 5.17). The slope of the water table, which generally follows the contours of the land surface, controls groundwater movement.

An **aquifer** is a subsurface rock layer that is permeable to groundwater flow (Figure 5.17). An aquifer can store and transmit large amounts of water and nourish wells and springs. The largest aquifers cover extensive regions and sustain large human populations. An *unconfined aquifer* has a permeable layer above, which allows water to pass through, and an impermeable layer below. A *confined aquifer* is bounded above and below by impermeable layers of rock or clay.

Springs, Streamflows, and Wells Groundwater tends to move toward areas of lower pressure and elevation. Where the water table intersects the surface, it creates **springs** (▶Fig. 5.18). Such an intersection also occurs in lakes and riverbeds. Ultimately, groundwater may enter stream channels to flow as surface water (stream at right in Figure 5.17). During dry periods, groundwater may sustain river flow. Humans commonly extract groundwater by drilling wells downward into the ground until they penetrate the water table and pumping the water to the surface. In some confined aquifers the water is under pressure, creating an **artesian well** where water under natural pressure may rise to the surface without pumping.

Groundwater Pollution If surface water is polluted, groundwater is also at risk. Pollution enters groundwater from many sources: industrial wells, septic tank outflows, seepage from toxic waste sites and agricultural residues (pesticides, herbicides, and fertilizers), and leakage from urban

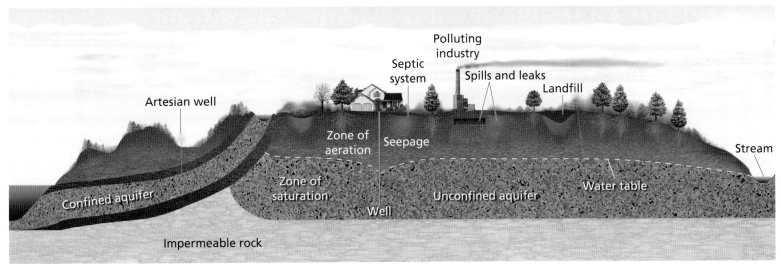

▲ **5.17 The water table and aquifers** The water table is a boundary between the zones of aeration and saturation. Beneath the zone of saturation an impermeable layer that prevents water from moving downward. An aquifer is a subsurface rock layer that is permeable to groundwater flow.

solid-waste landfills. A controversial source of groundwater contamination is *hydraulic fracturing* (fracking). This process pumps large quantities of water, sand, and chemicals under high pressure into subsurface shale rock, in order to release deposits of natural gas or oil. There is evidence that this practice will have short- and long-term consequences for the groundwater system.

geoCHECK ✔ What is an aquifer?

geoQUIZ

1. Identify three world regions in Figure 5.13 where annual streamflow is very high.
2. Explain how groundwater resources are linked to surface water supplies.
3. Suppose you are standing on a hillside that slopes down toward a stream. In what direction would groundwater move in this landscape? Explain.

An impermeable layer of rock traps the water and prevents it from draining lower into the water table.

(a) Springs are natural features that occur when water flows from an aquifer and through a permeable rock layer, to the surface.

▲ **5.18 An active spring** Springs are natural features that occur when water flows from an aquifer to the surface. This example shows a confined aquifer meeting the ground surface and forming a spring, but that this is just one condition under which a spring forms.

Animation (MG)
The Water
Table

http://goo.gl/tmd506

(b) This spring-fed creek in Nebraska flows from the High Plains aquifer.

5.4 Overuse of Groundwater – High Plains Aquifer

▶ *State* why geographers now consider groundwater a nonrenewable resource.
▶ *Identify* critical aspects of the largest aquifer in the United States.

As cities and suburbs grow and agriculture expands in dry areas where irrigation is required, society's demand for water explodes. Surface water alone is often not enough to meet this demand, and people increasingly rely on groundwater.

Groundwater Mining

As water is pumped from a well, the surrounding water table within an unconfined aquifer may become lowered. An **overdraft** occurs if the pumping rate exceeds the replenishment flow of water into the aquifer or the horizontal flow around the well. This condition, also known as **groundwater mining**, occurs today in large areas of the American Midwest, the lower Mississippi Valley, Florida, and in the Palouse region of eastern Washington State. In many places, the water table has declined more than 12 m (40 ft)—equal to the height of a four-story building (▶ **Fig. 5.20**).

Groundwater mining represents a threat to available freshwater supplies worldwide. For example, about half of India's irrigation, industrial and urban water needs is met by groundwater. In rural areas, groundwater supplies 80% of domestic water from some 3 million hand-pumped wells. And in approximately 20% of India's agricultural districts, groundwater mining through more than 17 million wells is beyond recharge rates. In the Middle East, groundwater overuse is even more severe. The groundwater beneath countries such as Israel and Saudi Arabia accumulated over tens of thousands of years, forming "fossil aquifers,"

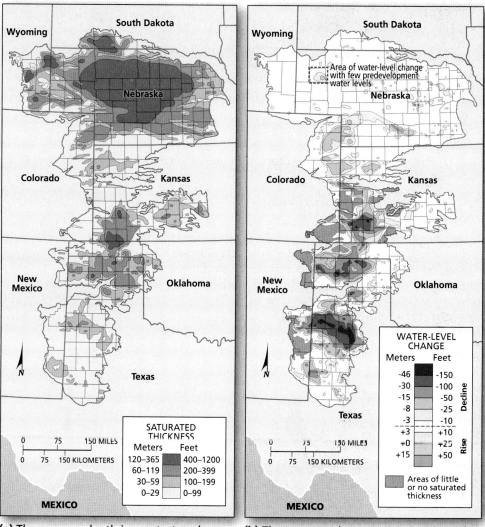

(a) The average depth is greatest under Nebraska.

(b) The greatest decrease in water level from 1950 to 2011 occurred underneath Kansas and Texas.

▲5.19 **The High Plains aquifer**

but today's increasing withdrawals are not being naturally recharged due to the desert climate—in essence, groundwater has become a *nonrenewable* resource.

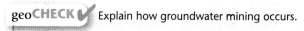 **geoCHECK** Explain how groundwater mining occurs.

High Plains Aquifer Overdraft

Groundwater mining is of special concern in the massive High Plains aquifer, where the water table has lowered over 30 m (100 ft) in some areas. The High Plains aquifer is North America's largest underground reservoir. Also known as the Ogallala aquifer, it lies beneath the American High Plains, an eight-state, 450,600-km² (174,000-mi²) area from southern South Dakota to Texas

(Fig. 5.19a). Annual precipitation over the region varies from 30 cm in the southwest to 60 cm in the northeast (12 to 24 in.).

For several hundred thousand years, meltwaters from retreating glaciers accumulated in the aquifer's sand and gravel. However, heavy mining of High Plains groundwater began 100 years ago to irrigate wheat, sorghums, cotton, corn, and about 40% of the grain fed to cattle in the United States—one-fifth of all U.S. cropland!

A USGS map of the water-level changes in the High Plains aquifer appears in Figure 5.19b. Declining

◄ **5.20 Subsidence caused by overuse of groundwater** Land in the San Joaquin Valley sank almost 30 feet between 1925 and 1977 because soils were compacted as groundwater mining lowered the water table.

(a) The route of the proposed Keystone XL Pipeline above the High Plains aquifer.

water levels are most severe in northern Texas, where the saturated thickness of the aquifer is least, through the Oklahoma panhandle and into western Kansas. The USGS estimates that recovery of the High Plains aquifer would take at least 1000 years if groundwater mining stopped today! Rising water levels have been noted in portions of south-central Nebraska and a portion of Texas due to recharge from surface irrigation, a period of above-normal precipitation years, and downward percolation from canals and reservoirs.

Obviously, billions of dollars of agricultural activity cannot be abruptly halted, but neither can profligate water mining continue. Present irrigation practices, if continued, will deplete about half of the High Plains aquifer resource (and two-thirds of the Texas portion) by 2020. Add to this the approximate 10% loss of soil moisture due to increased evapotranspiration demand caused by climatic warming for this region by 2050, as forecast by computer models, and we have a portrait of a major regional water problem and a challenge for society.

In 2005, the Keystone XL pipeline system proposed by TransCanada Corporation, emerged as a new threat to the High Plains Aquifer. The many-fingered pipeline system was designed to transport crude oil extracted from the Athabasca Tar Sands to refineries in the American Midwest, with connections to refineries in Oklahoma and Texas. Much of the proposed route passed directly above the High Plains Aquifer and would traverse the Sand Hills region of Nebraska, where the aquifer is close to the surface (▶ **Fig. 5.21**). In November 2015, the President accepted the recommendation from the U.S. State Department to deny the Keystone XL pipeline permit, citing numerous environmental concerns. However, future politicians may revive the issue, particularly if oil prices rise again.

(b) Nebraska's Sand Hills, where the water table lies close to the surface.

▲ **5.21 The proposed Keystone XL Pipeline and the landscape it would cross**

Mobile Field Trip (MG)
Oil Sands:
An Unconventional Oil

https://goo.gl/Tur5R5

geoCHECK ✔ Describe the geographic location of the High Plains aquifer.

geo**QUIZ**

1. Explain what is meant by the term "fossil aquifer."
2. Describe how climate change is influencing groundwater recharge in the High Plains aquifer.
3. Explain why in 2012, the federal government did not grant permission to route the Keystone XL pipeline through Nebraska.

5.5 Our Water Supply

▶ *Compare* regional factors that influence the global water supply.
▶ *Summarize* water supply and demand in the United States and describe pollution threats to water resources.

Quenching human thirst for adequate and high-quality water supplies will be a major issue in this century. Worldwide, per capita water use is increasing at twice the rate of population growth. Because of increased groundwater mining, water levels in major aquifers are falling (▶**Fig. 5.22**). Despite our dependency on water, accessible water supplies are not well correlated with population distribution or the regions where population growth is greatest.

▼**5.22 Groundwater depletion—a global problem** NASA determined that most of Earth's major aquifers are rapidly being depleted as populations grow and demand for water increases. NASA's satellite-based instruments measured slight variations in gravity indicating changes in the amounts of groundwater.

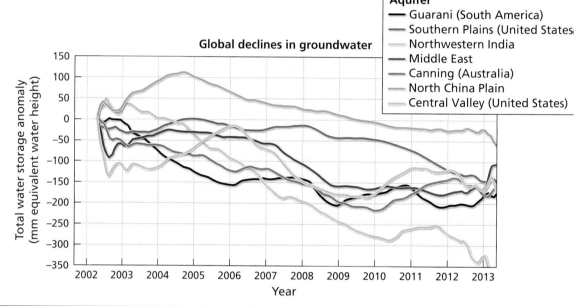

Aquifer
— Guarani (South America)
— Southern Plains (United States)
— Northwestern India
— Middle East
— Canning (Australia)
— North China Plain
— Central Valley (United States)

Global declines in groundwater

Table 5.1	Regional Comparison of Factors Influencing Global Water Supply			
Region	**2015 Population (in millions)**	**Share of Global Population**	**Share of Global Land Area**	**Share of Annual Streamflow**
Africa	1171	(% of world pop: 16)	23%	11%
Asia	4397	(60%)	33%	34%
Oceania	40	(0.5%)	6%	5%
Europe	742	(10%)	7%	8%
N. America*	484	(6.6%)	16%	15%
Central and South America	503	(6.8%)	13%	27%
Global (excluding Antarctica)	7336			

*Includes Canada, Mexico, and the United States.

Source: BGD, billion gallons per day. Population data from 2015 World Population Data Sheet (Washington, DC: Population Reference Bureau, 2015). CO_2 data from PRB 2009.

Table 5.1 portrays the unevenness of Earth's water supply in relation to the distribution of human population. These data include land area, annual streamflow, and projected population change for six world regions. Also note 2012 data on carbon dioxide emissions per person for each region. There is evidence that climate change (Chapter 7) will reduce precipitation in some areas, reducing water supplies at the same time that demand for water is rising because of economic development and higher per capita consumption.

For example, North America's mean annual stream discharge is 4310 billion gallons per day (bgd)—and Asia's is 9540 bgd. However, North America has only 6.6% of the world's population,

whereas Asia has 60%, with a population doubling time less than half that of North America. In northern China, 550 million people living in approximately 500 cities lack adequate water supplies. For comparison, note that the 1990 floods cost China $10 billion, whereas water shortages are running at more than $35 billion a year in costs to the Chinese economy. An analyst for the World Bank stated that China's water shortages pose a more serious threat than floods during this century, yet the news coverage is usually limited to large flooding events.

geoCHECK ✔ Identify how the global increase in per capita water use compares to world population growth.

Water Supply in the United States

The U.S. water budget summarized in Figure 5.23 derives from surface and groundwater sources that are fed by an average daily precipitation of 4200 bgd. That sum is based on an average annual precipitation value of 76.2 cm (30 in.) divided evenly among the 48 contiguous states. However, those 4200 billion gallons of average daily precipitation are unevenly distributed across the country and unevenly distributed throughout the year. For example, New England's water supply is so abundant that only about 1% of available water is consumed each year. The same is true in Canada, where the resource greatly exceeds that in the United States. In California, two-thirds of the population and most of the crops are located south of the where most of the precipitation falls. To compensate for this geographical predicament, California has developed the most elaborate system of aqueducts, canals, and reservoirs since Roman times.

Surface Water Pollution As we have learned, freshwater comprises just 2.78% of all water on Earth. Most of this is locked up in glacial ice. Protecting the remaining surface freshwater is a constant challenge to humans everywhere. Pollutants taint water from stationary **point sources** (PS) such as a single pipe, factory, or ship. They also enter water from **nonpoint sources** (NPS) such as runoff containing fertilizers, pesticides, animal wastes, and from oil or road salt deposited onto highways and parking lots (▶Fig. 5.24). Water moving downslope collects and transports these pollutants before depositing them into rivers, lakes, and wetlands. Groundwater is also susceptible to pollution from point and nonpoint sources, as was shown in Figure 5.17.

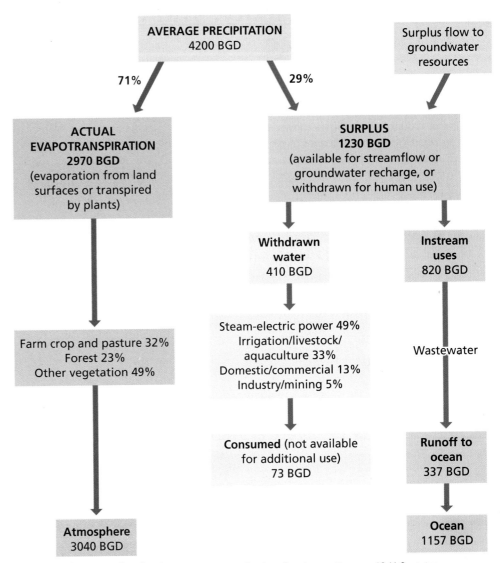

AVERAGE PRECIPITATION
4200 BGD

71% — 29%

Surplus flow to groundwater resources

ACTUAL EVAPOTRANSPIRATION
2970 BGD
(evaporation from land surfaces or transpired by plants)

SURPLUS
1230 BGD
(available for streamflow or groundwater recharge, or withdrawn for human use)

Farm crop and pasture 32%
Forest 23%
Other vegetation 49%

Withdrawn water
410 BGD

Instream uses
820 BGD

Steam-electric power 49%
Irrigation/livestock/ aquaculture 33%
Domestic/commercial 13%
Industry/mining 5%

Wastewater

Consumed (not available for additional use)
73 BGD

Runoff to ocean
337 BGD

Atmosphere
3040 BGD

Ocean
1157 BGD

▲ **5.23 U.S. water budget** The daily water budget for the contiguous 48 U.S. states exceeds 4200 billion gallons a day.

▼ **5.24 Sources of surface water pollution**

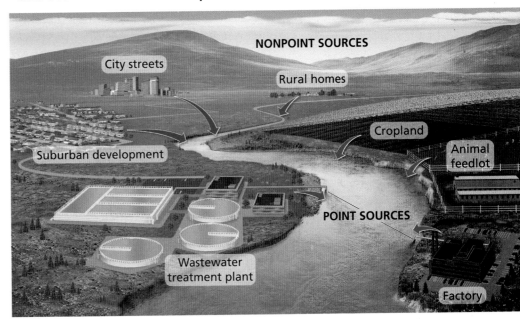

NONPOINT SOURCES
City streets
Rural homes
Suburban development
Cropland
Animal feedlot
POINT SOURCES
Wastewater treatment plant
Factory

geoCHECK ✔ What is the difference between point source and nonpoint sources of water pollution?

geoQUIZ

1. Identify the two most important factors that influence human demands for water.
2. Explain the geographical imbalance between water resources and American settlement patterns.
3. Based on Figure 5.22, roughly what proportion of surplus water is withdrawn for human use?

5.6a Water Scarcity: The Colorado River System Out of Balance

▶ *Identify* why the water budget of the Colorado River watershed is out of balance.

▶ *Evaluate* California's policies designed to combat the impact of drought on the Sierra snowpack.

▶ *Identify* critical challenges of present and future freshwater scarcity around the world.

Water resources in the American Southwest are linked to the Colorado River system. The Colorado flows almost 2317 km (1440 mi) from Colorado and Wyoming through the southwestern United States (▼Fig. 5.25). This *exotic stream* is so-called because its headwaters are in a region of water surpluses, but then the river and its tributaries flow mostly through arid lands to the sea.

The rapidly expanding urban areas of Las Vegas, Phoenix, Tucson, Denver, San Diego, and Albuquerque all depend on Colorado River water, which supplies approximately 30 million people in seven states. The Colorado River is just one example of the growing crisis of water scarcity. California's efforts to meet the water demands of its growing population and agribusiness industry during the state's ongoing drought is another. Worldwide, both developed and less developed countries are seeking solutions to similar challenges of water scarcity.

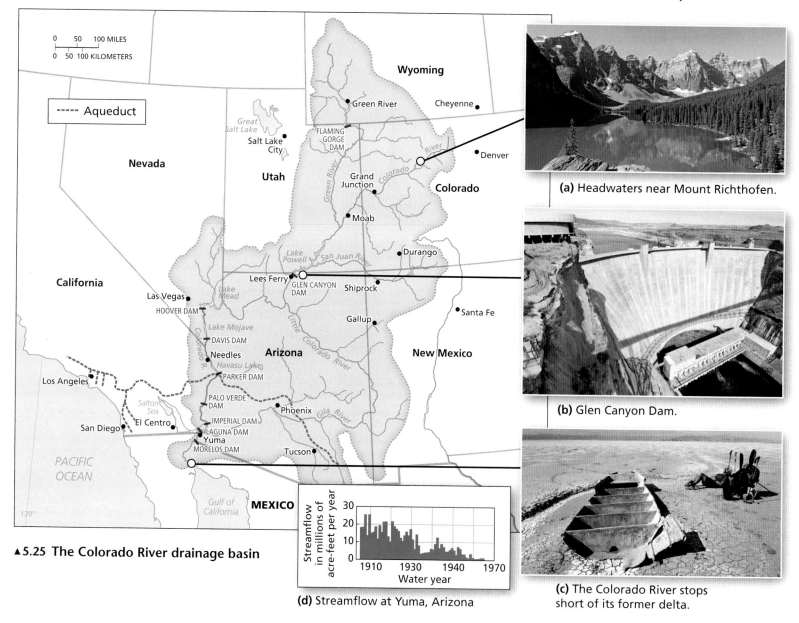

▲ **5.25 The Colorado River drainage basin**

(a) Headwaters near Mount Richthofen.

(b) Glen Canyon Dam.

(c) The Colorado River stops short of its former delta.

(d) Streamflow at Yuma, Arizona

(a) Declining Lake Mead water levels at Hoover Dam, 1931–2015.

Colorado River: Sharing a Scarce Resource

For a century, the states of the Colorado River basin have disputed how to apportion the river's water. The Colorado River Compact was signed by six of the seven basin states in 1923. The seventh state, Arizona, signed in 1944, when the United States also signed the Mexican Water Treaty. The compact was designed to equitably allocate water between the seven states. In 1928, Hoover Dam became the first major water storage project on the river. Today, six enormous dams are on the Colorado and its major tributary, the Green River.

Streamflow Variability and Drought Exotic streams have highly variable flows, and the Colorado is no exception. In addition, the river's flow reflects highly variable precipitation in the arid and semiarid Southwest. The 2003–2011 water years marked the lowest 8-year average flow in the total record. Figure 5.26a compares Lake Mead water levels at Hoover Dam in July 1983, 2007, and 2015, as evidence of increasing aridity in the Southwest causing declining Lake Mead levels (Figure 5.26b).

Presently, the seven states *ideally* want rights that total as much as 25.0 million acre feet (maf) per year. When added to the guarantee for Mexico, this comes up to 26.5 maf of yearly demand—more than the river's average annual flow. And six states assert that California's right to the water must be limited to a court-ordered 4.4 maf. No surplus exists in the river's water budget during the ongoing western drought.

Nine droughts have affected the American Southwest since A.D. 1226. However, the present drought is the first to occur with increasing human demand for water as the region's population growth continues. The drought affecting the Colorado River system began in 2000 and is ongoing. Reservoirs store water to offset variable Colorado River flows. The largest reservoirs—Lake Mead, formed by Hoover Dam, and Lake Powell, formed by Glen Canyon Dam—together account for over 80% of total system storage (Figure 5.24). However, Lake Mead and Lake Powell are in water-budget trouble, since water allocations already have exceeded the long-term average flows in the system.

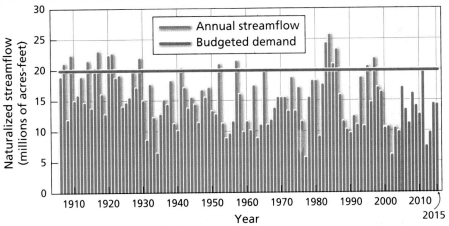

(b) Annual records of Colorado River stream flow at Lees Ferry near Page, Arizona. Note that in many years, streamflow is insufficient to meet expected demand.

▲ **5.26 Hoover Dam in Arizona, 1983, 2007, and 2015**

Effects of Climate Change on Water Resources Scientists think that climate change is an important factor in the future of the Colorado River basin's water resources. Higher temperatures and evaporation rates as well as reduced mountain snowpack and earlier spring melts have resulted in the current drought. Evidence and computer models suggest that the decade-long drought is directly linked to an expansion of subtropical dry zones with climate change. In regions such as California and the American Southwest, scientists' application of the term *drought* is changing to take into consideration the change in *permanent average climate*. When future droughts occur—for example, during a Pacific Ocean La Niña condition discussed in Chapter 4—they will be in addition to this new base state of a drier climate.

Solutions to Water Scarcity for the Colorado Basin Engineers estimate that it could take 13 normal winters of snowpack in the Rockies and basinwide precipitation to "reset" the system. Such a recovery would be much slower than previous drought recoveries because of increasing evaporation associated with increasing temperatures across the Southwest as well as swelling human water use. States in the region have not stressed conservation (using less water) and efficiency (using water more effectively) as ways to reduce the tremendous demand for water—but that may change.

geoCHECK ✔ Explain the purpose of the Colorado River Compact.

5.6b Water Scarcity: The New Normal?

Sierra Snowpack: California's Dwindling Treasure

As the western United States endured its fourth consecutive dry year, the USDA drought map in Figure 5.27 shows most of California suffering from *Exceptional Drought* (meaning these conditions exist only once in a 50 to 100 year time span). Accumulating snow in the Sierra Nevada normally accounts for 30% of California's total water supply. While dry years are not uncommon in this climate, the measured snowpack for the 2014–2015 winter was just 5% of normal, crushing the old record of 25% set in 1976–1977. The lack of snowfall resulted from a near-absence of midlatitude cyclones that normally sweep in from the Gulf of Alaska, combined with record warm winter temperatures that sped the melting of what little snow did accumulate on the ground. For the first time in the nearly century of measurements, many stations recorded no snow in the April end-of-winter surveys (▼ Fig. 5.28). This was the fourth consecutive year of significantly below-average precipitation.

Following that record-low snowpack, California mandated a 25% reduction in water use statewide, although this agreement does not apply to agricultural uses that comprise 80% of the water used in the Golden State. Throughout the state, reduced watering left lawns and playing fields brown and hardened. Over one-half million acres of farmlands went unplanted in an effort to save orchards (such as almond, walnut, orange trees) and grape vines. The farmland losses in 2015 alone are expected to total 2.7 billion dollars, and leave over 15,000 seasonal workers without jobs. Another casualty is California's 33 million acres of forest, where over 12.5 million trees have succumbed to drought stress. In 2015, scientists used aerial surveys to compile a digital map of reddish-brown trees where lack of water either killed trees outright or weakened them for infestation by beetles and disease (▶ Fig. 5.29).

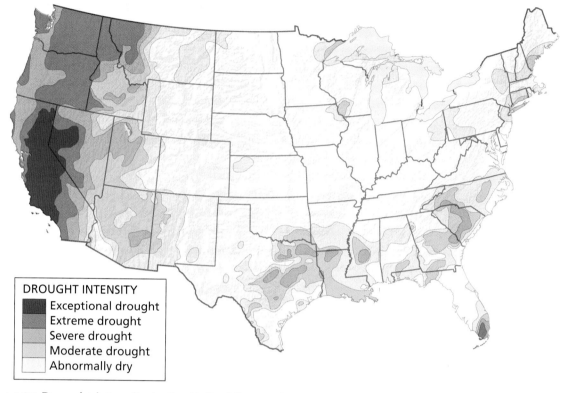

▲ 5.27 **Drought intensity in the United States, August 2015**

DROUGHT INTENSITY
- Exceptional drought
- Extreme drought
- Severe drought
- Moderate drought
- Abnormally dry

geoCHECK ✔ Describe the change in California's typical winter weather pattern that occurred in 2014–2015.

Mobile Field Trip (MG)
Moving Water Across California

https://goo.gl/7yNmJ4

▶ 5.28 **No snow**
(a) On April 1, 2015, California Governor Brown looks on as an official from the Department of Water Resources points to a mark on the measuring pole that was the lowest previous snow pack level. On this day, the surveyors recorded no snowpack following an extremely dry and warm winter.

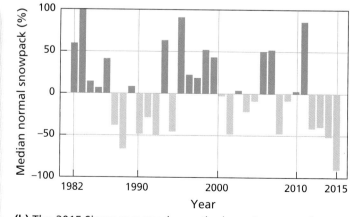

(b) The 2015 Sierra snowpack was the lowest on record.

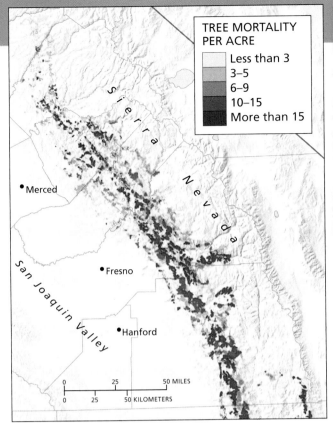

TREE MORTALITY PER ACRE

- Less than 3
- 3–5
- 6–9
- 10–15
- More than 15

(a) Forest trees mortality, 2015

▲ **5.29 Effects of the 2015 drought on California's forests**

(b) Drought kill in Sequoia National Park, California

Water Scarcity: A Global Challenge

When water supply and demand are examined in terms of water budgets, the limits of water resources become apparent. In the water-balance equation in Figure 5.6, any change in one side (such as increased demand for surplus water) must be balanced by an adjustment in the other side (such as an increase in precipitation). Water availability per person declines as population increases, and individual demand increases with economic development, affluence, and technology. World population growth since 1970 has reduced per capita water supplies by one third. Mounting water pollution in some regions also limits the water resource base.

Solutions to Water Scarcity Most of the world's major rivers and many of the largest freshwater lakes span international boundaries. One hundred and forty-five countries possess territory within an international river drainage basin. Because water flows across boundaries, it is truly a global "commons," meaning a resource that is accessible to and used by everyone. Thus international cooperation will be necessary to meet the global demand for freshwater. For example, since 1972, the United States and Canada have made a series of agreements to protect the Great Lakes, one of Earth's largest supplies of freshwater. In addition to international cooperation, two other strategies are already helping to meet this challenge.

- **Desalination** The first strategy, desalination of seawater to augment diminishing groundwater supplies, is becoming increasingly important as a freshwater source. More than 12,000 desalination plants are now in operation worldwide, and plant construction is expected to increase 140% by 2020.

Approximately 50% of all desalination plants are in the Middle East. In Saudi Arabia, 30 desalination plants currently supply 70% of the country's drinking water needs. In the United States, along the coast of southern California and in Florida, the number of desalination plants is slowly increasing. In Sydney, Australia, a reverse-osmosis desalination plant opened in 2010 and supplies about 15% of the region's water needs. However, the significant energy required to desalinate seawater, combined with undesirable waste products (mostly salt), make water conservation a more desirable option.

- **Conservation** The second strategy—water conservation—involves everyone on Earth. In parts of the United States, Australia, and the Middle East, more efficient drip irrigation techniques have reduced agricultural demands for water by 40%, without reducing crop yields (▶ **Fig. 5.30**). In affluent societies where per capita consumption is high, installing more efficient water fixtures and regularly checking for leaks, households can reduce daily indoor per capita water use by about 35%.

▲ **5.30 Drip irrigation**

geoCHECK ✔ Identify the three factors that, with each passing year, decrease the amount of water available to every person on Earth.

geoQUIZ

1. Identify the seven states and two countries that share a portion of the Colorado River's annual flow.
2. Identify three drought-related factors that are reducing annual flows in the Colorado River.
3. Explain why the 2014–2015 dry winter in California was different from prior dry periods.
4. Compare and contrast desalination and conservation as strategies for coping with water scarcity. What are the advantages and disadvantages of each?

WATER RESOURCES IMPACT HUMANS

- Freshwater, stored in lakes, rivers, and groundwater, is a critical resource for human society and life on Earth.
- Drought results in water deficits, decreasing regional water supplies and causing declines in agriculture.

Desalination is an important supplement to water supplies in regions with large variations in rainfall throughout the year and declining groundwater reserves. This plant in Barcelona, Spain, uses the process of reverse osmosis to remove salts and impurities.

HUMANS IMPACT WATER RESOURCES

- Climate change affects lake depth, thermal structure, and associated organisms.
- Water projects (dams and diversions) redistribute water over space and time.
- Groundwater overuse and pollution depletes and degrades the resource, with side effects such as collapsed aquifers and saltwater contamination.

The third largest reservoir in the world, Lake Nasser is formed by the Aswan High Dam on the Nile River in Egypt. Its water is used for agricultural, industrial, and domestic purposes, as well as for hydropower.

Blue Marble–Next Generation image shows December land surface topography and bathymetry.

The Itaipu Dam and power plant on the Paraná River bordering Brazil and Paraguay produces more electricity annually than the Three Gorges Dam in China. Itaipu Reservoir displaced over 10,000 people and submerged Guaíra Falls, formerly the world's largest waterfall by volume.

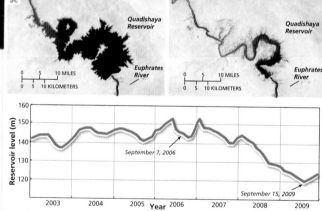

Data from *GRACE* reveal a rapid decline in reservoir levels from 2006 to 2009 along the Euphrates River in the Middle East; Quadishaya Reservoir is an example. The graph shows the surface-level decline, with dates of the images marked. About 60% of the volume loss is attributed to groundwater withdrawals in the region.

ISSUES FOR THE 21ST CENTURY

- Maintaining adequate water quantity and quality will be a major issue. Desalination will increase to augment freshwater supplies.
- Hydropower is a renewable energy resource; however, drought-related streamflow declines and drops in reservoir storage interfere with production.
- Drought in some regions will intensify, with related pressure on groundwater and surface water supplies.

Looking Ahead

In the next chapter we investigate how the world's climate systems influence the distribution of plants and animals across Earth. Climate also affects the rates of weathering and erosion. The climate also exerts a vital influence on human settlement, transportation, and our ability to grow crops. This upcoming chapter integrates all that we have learned so far.

Chapter 5 Review

Where does water exist on Earth?

5.1 Water on Earth

Describe the origin of Earth's waters.
Define the quantity and the location of Earth's freshwater supply.
Visualize the hydrologic cycle.

- Although water is the most common compound on Earth's surface, it is not distributed evenly across the planet's surface. The oceans hold 97.22% of all water, leaving just 2.78% as freshwater, and most of that is locked in ice or found below the ground. Water in all of its forms—liquid, gaseous, and solid—continually circulates about the planet as part of the hydrologic cycle, which is powered by solar energy.

1. Explain the steps necessary for ocean water to reach the interior of a continent.

5.2 The Soil-Water Budget Concept

Relate the water-budget concept to the hydrologic cycle, water resources, and soil moisture.

Construct the water-balance equation to account for expenditures of the water supply.

- Precipitation landing on Earth's surface evaporates or transpires into the atmosphere; drains into streams or filters down into groundwater; or is stored within the soil. A soil-water budget evaluates water resources at various scales. Developing a budget accounts for precipitation (PRECIP), actual (ACTET) and potential (POTET) evapotranspiration, runoff, and soil-moisture storage (STRGE).

2. Explain what the blue line signifies in Figure 5.12.
3. Describe the relationship between soil-moisture storage and plant roots.

What is the nature of groundwater?

5.3 Surface Water & Groundwater Resources

Summarize how humans have adapted surface water resources for their use.

Describe the nature of groundwater and the structure of the groundwater environment.

- Though hidden beneath Earth's surface, groundwater is the largest potential freshwater source in the hydrologic cycle. Surplus water on the surface percolates downward through the zone of aeration, and then settles into pore spaces within soil and rock. Humans often rely on existing groundwater resources, which are tied to surface supplies and annual precipitation for recharge.

4. Explain the relationship between the zone of saturation and the water table.
5. Identify the percent of the U.S. population that relies on groundwater to meet at least some of their needs?

5.4 Overuse of Groundwater–High Plains Aquifer

State why geographers now consider groundwater a nonrewable resource.

Identify critical aspects of the largest aquifer in the United States.

- Subsurface aquifers are crucial sources of freshwater on every inhabited continent. However, the extraction of groundwater often exceeds the natural rate of replenishment. Overdraft pumping is apparent on the High Plains aquifer of North America, along with other important aquifers in India, Saudi Arabia, and Israel.

6. Explain where and why some areas of the High Plains aquifer are slowly recharging.

What is the future of our water resources?

5.5 Our Water Supply

Compare regional factors that influence the global water supply.

Summarize water supply and demand in the United States and describe pollution threats to water resources.

Identify critical aspects of present and future freshwater supplies around the world.

- Supplying adequate and high-quality fresh water to Earth's growing population is the most important natural resource issue of this century. The worldwide increase in per capita water use is now twice the rate of population growth. Water resources often poorly correlate with both existing human settlement and predicted population growth. Water conservation and desalination of seawater are emerging as the most important strategies to meet the global water challenge.

7. Identify two world regions where the percent of global streamflow exceeds the percent of world population (see Figure 5.18).

5.6 Water Scarcity: The Colorado River System Out of Balance

Identify why the water budget of the Colorado River watershed is out of balance.

- The Colorado River is the most important source of freshwater for 30 million people, crops, and industry in the drought-prone American Southwest. This *exotic stream* generates water surpluses in its Rocky Mountain headwaters that then flow mostly through arid lands to the sea. A water compact between the downriver states and Mexico partitions the water to each member state, but persistent drought coupled with increasing population and agricultural demands have overtapped this limited resource.

8. Identify the climatic factor that computer models link to the decade-long drought in the Colorado River basin.

Critical Thinking Questions

1. Are groundwater resources independent of surface supplies, or are the two interrelated? Explain your answer.

2. If you lived in a hot desert, describe strategies you could take to increase ACTET around your home.

3. Explain how the point source and nonpoint sources of water pollution change when urbanization replaces agricultural land.

4. At what point does groundwater utilization become groundwater mining? Give examples of mined aquifers and identify human actions and natural events that might reverse this trend.

5. Develop a water-balance graph (using Figure 5.12 as a model), for an imaginary location with a dry and cool winter, and a wet and warm summer.

6. Compare and contrast how the population and water demands of the United States and China might change in the next half century.

7. Since wars in the 21st century may be about securing adequate water resources, what action should we take now to understand the issues and avoid the conflicts?

Visual Analysis

In this photograph of a valley deep in Pakistan's Karakoram Mountains, irrigated crops thrive in an otherwise arid landscape, as evidenced by the barren slopes above the valley.

1. Identify three ways that moisture "leaves" this valley.

2. How might irrigation water directly applied from the river alter the water table?

3. Explain how irrigating these fields will influence the ever-changing relationship between potential evapotranspiration (POTET) and actual evapotranspiration (ACTET) in the valley.

▲R5.1

Environmental Change & POTET

• Open MapMaster in MasteringGeography.

• Select *Global Environmental Issues* from the *Physical Environment* menu, and explore the various sublayers.

1. In terms of area, identify the region that is undergoing the most deforestation, and the most desertification.

2. In both of your choices, describe how the changing environment will alter potential evapotranspiration (POTET).

Explore | Use **Google Earth** to explore **Lake Chad** in West Africa.

A Disappearing Lake

- In Google Earth, fly to *Lake Chad* in *West Africa*.
- Click *Layers*, then *Primary Database*, and then select *Borders and Labels*.

Zoom in and out of this basin to visualize Lake Chad and its surrounding environment. Click *Layers*, then *Primary Database*, and then select *Borders and Labels*. Zoom out until the Atlantic Ocean (to the west) and the Red Sea (to the east) appear.

1. Describe the topography, vegetation, and likely climate of this region.

2. Zoom close to the lake, and describe the distinctive landforms found in abundance north and northeast of the lakeshore. Hypothesize what landforms these might be, and how they form and change.

3. Name the countries that either share part of Lake Chad or have streams that flow into or out of *Lake Chad*.

4. Since the early 1960s, Lake Chad has shrunk over 80% due to climate change and the growing demands of agriculture. Using what you have learned in this chapter, describe three factors that will make reversing the decline in lake volume difficult. Your response may include physical-environmental and human reasons.

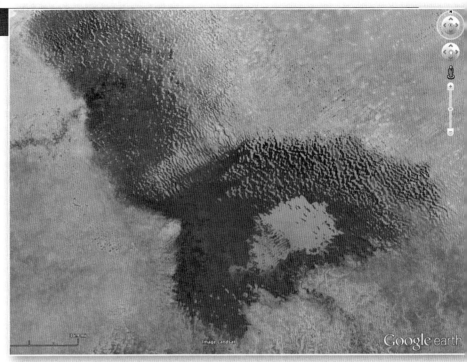

▲ R5.2

Mastering Geography™

Looking for additional review and test prep materials? Visit the Study Area in MasteringGeography™ to enhance your geographic literacy, spatial reasoning skills, and understanding of this chapter's content by accessing a variety of resources, including MapMaster interactive maps, videos, *Mobile Field Trips*, *Project Condor* Quadcopter videos, *In the News* RSS feeds, flashcards, web links, self-study quizzes, and an eText version of *Geosystems Core*.

Key Terms

actual evapotranspiration p. 128
aquifer p. 134
artesian well p. 134
desalination p. 143
drought p. 129

evaporation p. 128
groundwater p. 133
groundwater mining p. 136
groundwater recharge p. 133

nonpoint source p. 139
outgassing p. 126
overdraft p. 136
point source p. 139
potential evapotranspiration p. 128

precipitation p. 128
soil-moisture storage capacity p. 129
soil-water budget p. 128
springs p. 134
streamflow p. 132

transpiration p. 128
water table p. 134
wells p. 134
zone of aeration p. 134
zone of saturation p. 134

Good to the Last Drop: How do planners use the water-budget concept to meet community needs?

GeoLab5 MG
Pre-Lab Video

https://goo.gl/3Qb4RH

Meeting the water needs of growing communities everywhere takes considerable planning. This interdisciplinary effort usually involves people who specialize in geoscience, hydrology, climate, agriculture, waste management, computer modeling, geographic information science, and civil/environmental engineering.

Apply

You are a city manager for Kingsport, TN. You have been asked to develop a water-balance model for your city. Now that the data have been collected, you must complete the calculations necessary to develop the plan.

Objectives

- Calculate the total annual PRECIP and PO-TET for Kingsport
- Use the water-budget equation to develop a water budget
- Make recommendations for water conservation

GL5.1	
PRECIP	precipitation
ACTET	actual evapotranspiration
POTET	potential evapotranspiration
STRGE	soil-moisture storage
△STRGE	change in soil-moisture storage
DEFIC	water deficit
SURPL	water surplus

Procedure

Recall the water-budget equation variables presented in GL 5.1:

Precipitation = Actual Evapotranspiration + Surplus $\pm$ Change in Soil-Moisture Storage

Use the data presented in GL 5.2 to answer the following questions:
1. When does Kingsport experience a net supply of water? List the months.
2. With respect to water supply, what occurs during the warm summer months?
3. What is the total annual PRECIP for Kingsport? _____ mm
4. What is the total annual POTET for Kingsport? _____ mm
5. Based on these totals, how would you characterize the average soil-moisture conditions in Kingsport over the course of a year?

Take the PRECIP-POTET totals from GL 5.2 and carry them over to GL 5.3. Complete the remainder of the blanks in the water-balance table. Note that for Kingsport, the soil-moisture storage capacity is 100 mm. The examples that appear will help you visualize the remaining calculations that you must do on your own.

Calculations:

To determine PRECIP – POTET:	PRECIP – POTET
To determine STRGE:	$POTET - \dfrac{PRECIP}{POTET}$
To determine △STRGE:	$STRGE - \dfrac{PRECIP}{POTET}$
To determine ACTET	This mirrors POTET if moisture is available.
To determine DEFIC	This occurs when POTET $>$ PRECIP.
To determine SURPL	This only occurs when PRECIP – POTET is positive ($+$), or when STRGE can make up the difference.

6. Soil moisture (STRGE) remains at full field capacity through which month?
7. How much surplus is accumulated through the first five months of the year?
8. What is the net demand for water in June (PRECIP – POTET)? After this demand is satisfied in part by soil-moisture utilization, what is the remaining water in soil moisture (STRGE) to begin the month of July?
9. Calculate total annual evapotranspiration (ACTET) for Kingsport (POTET – DEFIC = ACTET).

Analyze & Conclude

10. What is the total DEFIC for the year?
11. What is the total SURPL for the year?
12. During how many months at Kingsport does POTET exceed PRECIP?
13. Your final task is to make recommendation for measures the people of Kingsport could take to conserve water during the months of water deficit. Identify at least five different measures to promote water conservation in both indoor and outdoor settings.

GL5.2

	Jan	Feb	Mar	Apr	May	Jun	Jul	Aug	Sep	Oct	Nov	Dec
PRECIP	97	99	97	84	104	97	132	112	66	66	66	99
POTET	7	8	24	57	97	132	150	133	99	55	12	7
Net supply (+)	+90	+91	+73	+27	+7					+11	+54	+92
Or demand (−)						−35	−18	−21	−33			

GL5.3

Kingsport Water Balance

	Jan	Feb	Mar	Apr	May	Jun	Jul	Aug	Sep	Oct	Nov	Dec	Total
PRECIP	97	99	97	84	104	97	132	112	66	66	66	99	1119
POTET	7	8	24	57	97	132	150	133	99	55	12	7	781
PRECIP—POTET	+90					−35							—
STRGE	100					65							—
ΔSTRGE	0					−35							—
ACTET	7					132							
DEFIC	0					0							
SURPL	90					0							

(All quantities in millimeters)

▲GL5.4 **Effects of flooding in Kingsport, TN, July, 2013.**

Log in MasteringGeography™ to complete the online portion of this lab, view the Pre-Lab Video, and complete the Post-Lab Quiz.

6 Global Climate Systems

The climate where you live may be humid with distinct seasons, or dry with consistent warmth, or moist and cool: Almost any combination is possible. Some places such as the Amazon Basin receive rainfall totaling more than 20 cm (8 in.) each month, with monthly average temperatures remaining above 27°C (80°F) year-round. Other places such as the Sahara Desert in North Africa may be rainless for a decade, with summer temperatures exceeding 37°C (100°F) almost every day. Some northern European climates have temperatures that average above freezing year-round, yet still pose severe frost problems for agriculture. Students reading this book in Singapore experience precipitation in every month, totaling 228 cm (89.7 in.), whereas university students in Karachi, Pakistan, see only 20.4 m (8 in.) of annual rainfall. Whatever the climate, humans must adapt to the specific climatic conditions where they live.

Key Concepts & Topics

1 How are climates classified?

2 What differentiates climates of tropical & midlatitude regions?

3 What differentiates climates of microthermal & cold regions?

4 What differentiates climates of dry regions?

A summer rain falls on the Wrangell Mountains, Wrangell–St. Elias National Park and Preserve, Alaska.

6.1 Components of Earth's Climate System

Key Learning Concepts

▶ **Explain** the difference between climate and weather.

▶ **Analyze** how the components of Earth's climate system interact to produce global climate patterns.

▶ **Define** the Köppen climate classification system.

Earth experiences an almost infinite variety of *weather* at any given place. But if we consider the weather over many years, including its variability and extremes, a pattern emerges that constitutes **climate**. Thus the expression "climate is what you expect, weather is what you get" is accurate. Although the weather changes all the time, we should also think of climate patterns as dynamic, but over much longer periods of time.

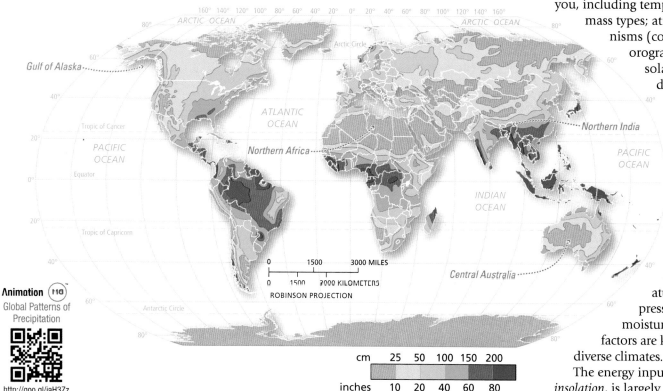

Animation 🄰🄰
Global Patterns of Precipitation

http://goo.gl/jaH3Zz

▲**6.1 Worldwide average annual precipitation** The wettest locations lie within the tropics and on the windward side of mountains (Gulf of Alaska and northern India). The driest regions are in subtropical high-pressure areas or in continental interiors far from sources of ocean moisture (northern Africa and central Australia).

The study of climate and its variability is called **climatology**. This branch of physical geography analyzes long-term weather patterns over time and space and the controls that produce Earth's diverse climatic conditions. Climates are so varied that no two places on Earth's surface experience exactly the same climate conditions. However, broad similarities among local climates permit their grouping into **climatic regions**. This chapter will focus on using a type of climatic analysis that locates areas of similar weather statistics and groups them into climatic regions. This examination will bring together the primary components of climate discussed in the first five chapters.

Components That Interact to Produce Climate

Several components of the energy–atmosphere system work together to determine Earth's climates. The two principal climatic components—temperature and precipitation— together reveal general climate types such as hot and dry tropical deserts, cold and dry polar ice sheets, or warm and wet equatorial rain forests. Figure 6.1 maps worldwide average annual precipitation. Note how these patterns reflect factors that should now be familiar to you, including temperature and pressure; air mass types; atmospheric lifting mechanisms (convergent, convectional, orographic, and frontal); and solar energy availability, which decreases from the equator to the poles.

Figure 6.5 (on pp. 154) reviews the main components of climate discussed in the previous chapters. The climate in any location results from the complex interrelationships between several factors: (1) insolation, (2) Earth's energy balance, (3) temperature, (4) air masses, (5) air pressure, and (6) atmospheric moisture and precipitation. These factors are key to understanding Earth's diverse climates.

The energy input for the climate system, *insolation*, is largely determined by latitude. The difference in insolation with latitude creates an energy imbalance between tropical and polar regions. Latitude also determines the length of day and the Sun angle, which varies throughout the year. Latitude, however, is not the only control on *temperature*, as elevation, land–water heating differences, and cloud cover also play a role.

Regions of higher or lower *air pressure* result from differences in insolation and temperature. These pressure differences help to define air masses that also have distinct temperature and moisture characteristics. Air pressure patterns and air masses both play a crucial role in creating climates. The air pressure patterns presented in Figure 6.5 and Figure 3.10 impact

▲6.2 **The Pasto River roars over Pailón del Diablo waterfall, Ecuador** The cascading torrent and the dense tree cover both reflect the tropical rain forest climate.

atmospheric circulation and the movement of air masses by creating atmospheric "super highways" that girdle the Earth. Along these highways stream air masses of distinctive temperature and moisture content (see Fig. 6.5). Interacting air masses also produce fronts and cyclonic storms. Lastly, flowing air masses, responding to pressure differences, transport and release variable amounts of precipitation in all its forms across Earth's surface. These variations contribute to unique climates. For example, equatorial low-pressure zones create consistently wet climates (◄**Fig. 6.2**), while subtropical high-pressure zones with their descending and warming air masses, create areas of permanent water deficit (►**Fig. 6.3**).

Köppen Climate Classification System

German climatologist Wladimir Köppen (1846–1940, and pronounced KUR-pen) devised the most widespread system used to classify world climate. Geosystems Core uses a compromise classification system, combining the causative physical factors to explore "why" a certain mix of climatic ingredients occurs in certain locations. This is added to the Köppen system based on statistics or other data that measure the observed effects.

For the *Geosystems Core* classification, we focus on temperature and precipitation measurements to classify world climates. The following basic climate categories provide the structure for our discussion in this chapter:

- **Tropical** (tropical latitudes, warm and rainy year-round or with rainy and dry seasons)
- **Mesothermal** (midlatitudes, mild winters; *meso-* means "middle," so *mesothermal* refers to climates between the extreme temperatures of tropical and polar climates)
- **Microthermal** (mid- and high latitudes, cold winters; *micro-* means "small," so *microthermal* refers to climates with lower temperatures)
- **Polar** (high latitudes and polar regions; cold year-round, no true summer)

▲6.3 **The barren hills above Death Valley, California, reflect an arid desert climate**

- **Highland** (high elevations at all latitudes)
- **Dry** (permanent moisture deficits at all latitudes; the only primary climate category determined mainly by moisture criteria)

For each climate category, information on a specific location can be summarized in a **climograph**, a graph that shows monthly temperature and precipitation for a weather station (▼**Fig. 6.4**). Notice that the climographs also include the atmospheric circulation systems that influence precipitation. For example, in Figure 6.4, you can see that the subtropical high and cyclonic storm tracks influence Madrid's climate.

geoCHECK ✔ How do air pressure patterns and air masses influence climate?

Climograph for Madrid, Spain
Lat/long: 40° 28′ N, 3° 33′ E

▲6.4 **Sample Climograph** A climograph plots monthly temperature and precipitation values for a selected station.

6.1 (cont'd) Components of Earth's Climate System

Earth's climate system is the result of interactions among several components (▶**Fig. 6.5**). These include the input and transfer of energy from the Sun (1 and 2); the resulting changes in atmospheric temperature and pressure (3 and 4); the movements and interactions of air masses (5); and the transfer of water—as vapor, liquid, or solid—throughout the system (6).

geoCHECK ✔ Describe the information presented on a climograph.

geoQUIZ

1. Identify the primary data used to develop the Köppen climate classification system.
2. Describe the geographic range of the mesothermal and microthermal climates.
3. Identify the only primary climate type that is based on moisture criteria in the form of water deficits as well as temperature.
4. What is the basis for classifying climates into regions and climate types?

Video MG
Supercomputing the Climate

http://goo.gl/r0763E

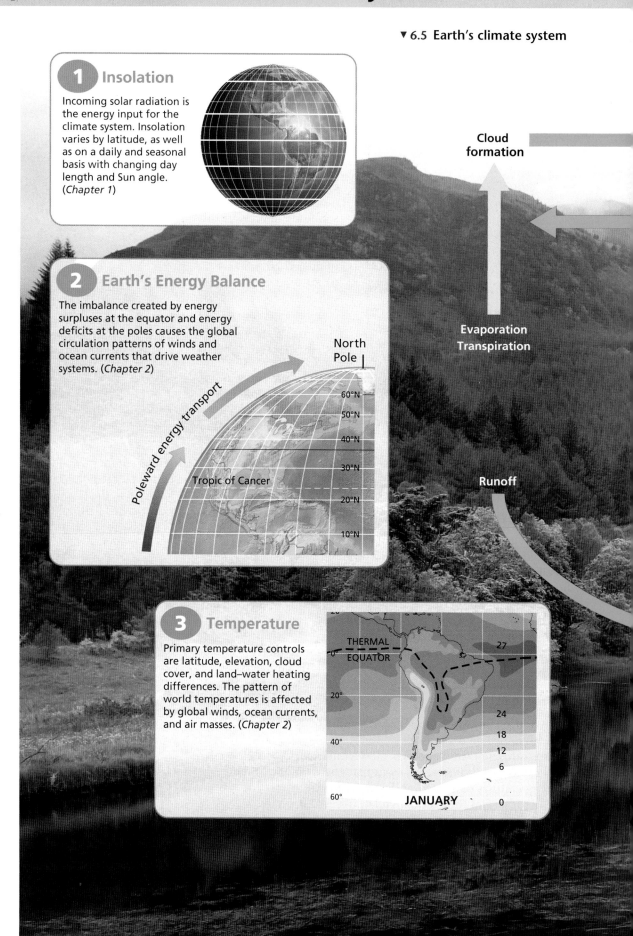

▼ **6.5 Earth's climate system**

1 Insolation

Incoming solar radiation is the energy input for the climate system. Insolation varies by latitude, as well as on a daily and seasonal basis with changing day length and Sun angle. (*Chapter 1*)

2 Earth's Energy Balance

The imbalance created by energy surpluses at the equator and energy deficits at the poles causes the global circulation patterns of winds and ocean currents that drive weather systems. (*Chapter 2*)

Poleward energy transport

North Pole

60°N
50°N
40°N
30°N
20°N
10°N

Tropic of Cancer

3 Temperature

Primary temperature controls are latitude, elevation, cloud cover, and land–water heating differences. The pattern of world temperatures is affected by global winds, ocean currents, and air masses. (*Chapter 2*)

THERMAL EQUATOR

0°
20°
40°
60°

27
24
18
12
6
0

JANUARY

Cloud formation

Evaporation Transpiration

Runoff

Cloud formation

Atmospheric movement of water vapor

Precipitation

Evaporation

4 Air Pressure

Winds flow from areas of high pressure to areas of low pressure. The equatorial low creates a belt of wet climates. Subtropical highs create areas of dry climates. Pressure patterns influence atmospheric circulation and movement of air masses. Oceanic circulation and multiyear oscillations in pressure and temperature patterns over the oceans also affect weather and climate. (*Chapter 3*)

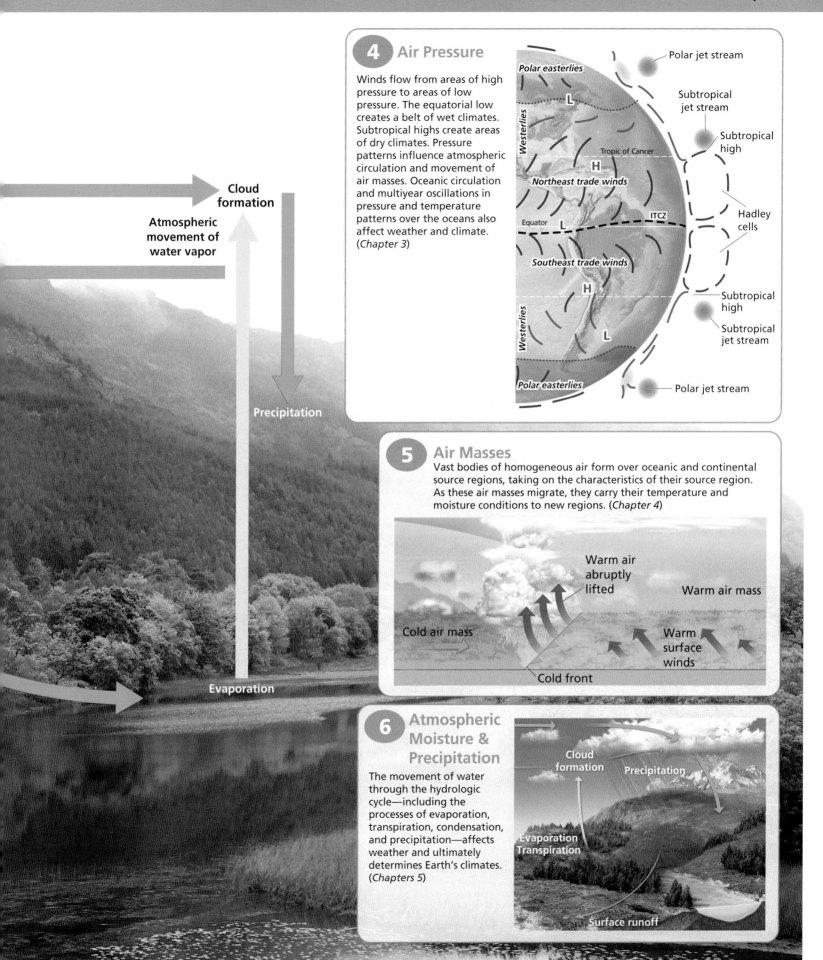

Polar easterlies

L

Westerlies

Tropic of Cancer

H

Northeast trade winds

Equator   L   ITCZ

Southeast trade winds

H

Westerlies

L

Polar easterlies

Polar jet stream

Subtropical jet stream

Subtropical high

Hadley cells

Subtropical high

Subtropical jet stream

Polar jet stream

5 Air Masses

Vast bodies of homogeneous air form over oceanic and continental source regions, taking on the characteristics of their source region. As these air masses migrate, they carry their temperature and moisture conditions to new regions. (*Chapter 4*)

Warm air abruptly lifted

Warm air mass

Cold air mass

Warm surface winds

Cold front

6 Atmospheric Moisture & Precipitation

The movement of water through the hydrologic cycle—including the processes of evaporation, transpiration, condensation, and precipitation—affects weather and ultimately determines Earth's climates. (*Chapters 5*)

Cloud formation

Precipitation

Evaporation Transpiration

Surface runoff

6.2 Climate Classification System

Key Learning Concepts

▶ **Describe** the *principal* climate classification categories.
▶ **Locate** the Köppen climate classification regions on a world map.

The world climate classification map in Figure 6.6a portrays the geographic range of the various climates: tropical, dry, mesothermal (midlatitude), microthermal (high latitude), and polar/highland climates. Temperature and precipitation

characteristics are the main factors that determine these climates. Also shown on this map are selected atmospheric and ocean circulation patterns that modify the various climates. These patterns include air masses, nearshore ocean currents, air pressure systems, and the January and July locations of the intertropical convergence zone (ITCZ).

▼**6.6 World climate classification** The climate classification presented here is a modified form of the Köppen system.

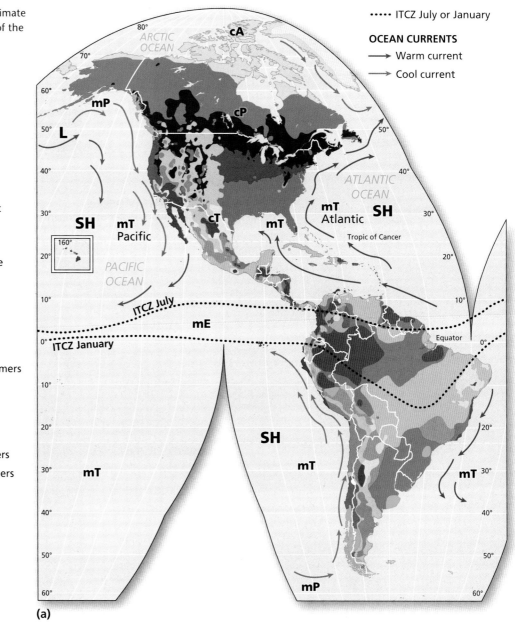

(a)

As you learn the spatial categories and boundaries of the modified Köppen system presented here, remember that the boundaries of climate regions really are transition zones of gradual change. The overall patterns of boundary lines are more important than their precise placement. For example, in Figure 6.6b you can see the transition from the subtropical hot desert to the tropical savanna.

geoQUIZ

1. Using Figure 6.6a, identify the important climate differences between the Amazon Basin and the southern tip of South America.
2. Describe the primary difference between the ocean current that flows into the coast of Texas and the ocean current that flows along the coastline of Washington and Oregon.
3. Identify the air mass that circulates just north of Africa's Sahel Region (use Fig. 6.6a and b).

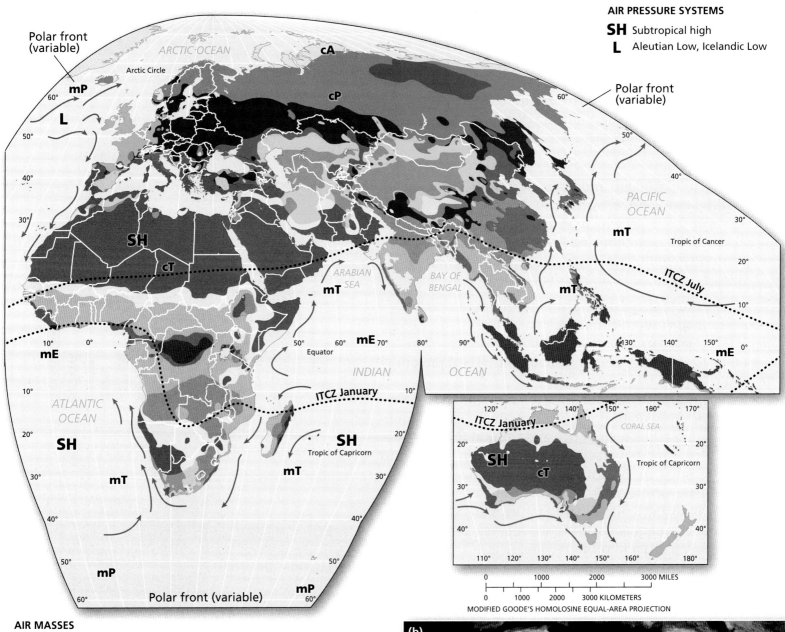

AIR PRESSURE SYSTEMS
SH Subtropical high
L Aleutian Low, Icelandic Low

MODIFIED GOODE'S HOMOLOSINE EQUAL-AREA PROJECTION

AIR MASSES

mP Maritime polar (cool, humid)	**mT** Maritime tropical (warm, humid)	**mE** Maritime equatorial (warm, wet)
cP Continental polar (cool, cold, dry)	**cT** Continental tropical (hot, dry summer only)	**cA** Continental arctic (very cold, dry)

(b)

Arid Desert (SH)

Semiarid Steppe

Tropical Savanna

6.3 Tropical Climates (Tropical Latitudes)

Key Learning Concepts

▶ **Explain** the precipitation and temperature criteria used to determine tropical climates.

▶ **Locate** tropical climate regions on a world map.

Tropical climates are the most extensive on Earth, covering about 36% of ocean and land areas (▶ **Fig. 6.7**). The tropical climates straddle the Equator from 20° N to 20° S, roughly between the Tropics of Cancer and Capricorn. Tropical climates stretch northward to the tip of Florida and south-central Mexico, central India, and Southeast Asia and southward to northern Australia, Madagascar, central Africa, and southern Brazil. These climates truly are winterless. A number of factors combine to produce tropical climates, including

- consistent day length and insolation, which produce steadily warm temperatures;
- effects of the ITCZ, which brings rain as it shifts seasonally with the high Sun; and
- warm ocean temperatures and unstable maritime air masses.

Tropical climates have three distinct regimes (a *climate regime* is a distinct set of climatic characteristics): *tropical rain forest* (ITCZ present all year), *tropical monsoon* (ITCZ present 6 to 12 months), and *tropical savanna* (ITCZ present less than 6 months).

The tropical climates are home to approximately 37% of the world's population. Within the rain forest and monsoon climates, crops such as rice and tropical fruits such as mango, cacao, and bananas thrive in the warm and humid growing seasons. In the drier but still warm tropical savanna climate, cereal crops such as wheat are possible. In the tropical savanna climate of East Africa, free-roaming wildlife such as elephants, lions, wildebeest, and giraffe thrive within parks and protected areas.

Tropical Rain Forest Climates

All *tropical rain forest* climates are constantly moist and warm. Convectional thunderstorms, triggered by local heating and trade-wind convergence, peak each day from mid-afternoon to late evening. Precipitation follows the migrating ITCZ northward and southward with the summer Sun. Water surpluses are enormous, creating the world's greatest stream discharges in the Amazon and Congo River basins.

High rainfall sustains the lush, broadleaf evergreen tree growth of the Earth's equatorial and tropical rain forests. Uaupés, Brazil (▶**Fig. 6.8**), is characteristic of a tropical rain forest. On the climograph, you can see that the lowest-precipitation month receives nearly 15 cm (6 in.), and

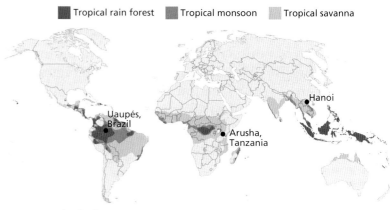

■ Tropical rain forest **■ Tropical monsoon** **□ Tropical savanna**

▲**6.7 Tropical climates**

geoCHECK ✔ Analyze what triggers the northward-to-southward shift in tropical rain forest precipitation.

(a)

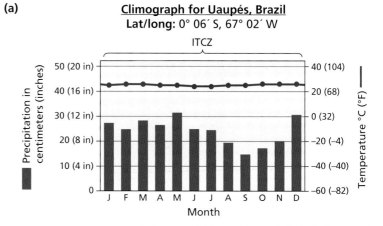

Climograph for Uaupés, Brazil
Lat/long: 0° 06′ S, 67° 02′ W

▲**6.8 Tropical rain forest climate**

(b) The rain forest along the Rio Negro River, Amazonas state, Brazil. The ITCZ is responsible for the precipitation pattern throughout the year.

the annual temperature range is barely 2°C (3.6°F). In the tropics, the diurnal (day-to-night) temperature range exceeds the annual monthly temperature range. The only interruption in the distribution of tropical rain forest climates across the equatorial region is in the highlands of the South American Andes and in East Africa, where cooler air at higher elevations produces lower temperatures (see Fig. 6.6).

Tropical Monsoon Climates

The *tropical monsoon* climates feature a dry season that lasts 1 or more months when the ITCZ is not overhead. You will recall from Chapter 3, that monsoons are seasonal rains produced by shifting winds associated with changing atmospheric pressure patterns. Rain brought by the ITCZ and the monsoon falls in these areas from 6 to 11 months of the year. These climates lies principally along coastal areas and experience seasonal variation of wind and precipitation. Hanoi, Vietnam, is an example of this climate type (▼ Fig. 6.9).

geoCHECK ✔ Why do tropical monsoon climates have a short dry season?

▼6.9 Tropical monsoon climate

(a)

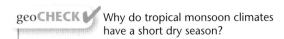

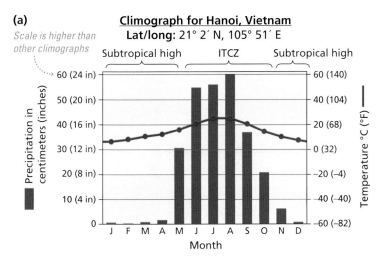

Climograph for Hanoi, Vietnam
Lat/long: 21° 2′ N, 105° 51′ E

Scale is higher than other climographs

(b) Torrents of monsoon rain near Hanoi, Vietnam.

Tropical Savanna Climates

The *tropical savanna* climates lie poleward of the tropical rain forest climates. The ITCZ reaches these climate regions for about 6 months or less of the year as it migrates with the summer Sun. Summers are wetter than winters, because convectional rains accompany the shifting ITCZ when it is overhead. This produces dry conditions when the ITCZ is farthest away and high pressure dominates. Thus the natural moisture demand exceeds the natural moisture supply in winter, causing water-budget deficits.

Temperatures vary more in tropical savanna climates than in tropical rain forest regions. The tropical savanna can have two temperature maximums during the year, because the Sun's direct rays are overhead twice—before and after the summer solstice in each hemisphere as the Sun moves between the equator and the tropics. Grasslands with scattered trees, drought resistant to cope with the highly variable precipitation, characterize the tropical savanna regions. Arusha, Tanzania, is a tropical savanna city (▼ Fig. 6.10). Temperatures for this region are consistent with tropical climates. Note the marked dryness from June to October, which results from changing dominant pressure systems rather than annual changes in temperature. This region is near the transition to the drier desert hot steppe climates to the northeast.

geoCHECK ✔ When does moisture demand exceed precipitation in the tropical savanna climate?

geoQUIZ

1. Describe the location of the Earth's three tropical climates.
2. Why does the tropical savanna regime experience two annual temperature maximums?
3. Explain why the vegetation of the tropical rain forest is different from the vegetation of the tropical savanna regions.

▼6.10 Tropical savanna climate

(a)

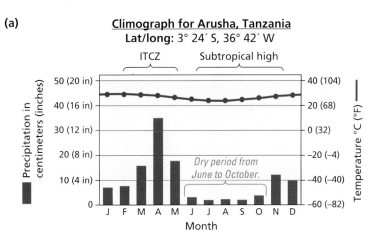

Climograph for Arusha, Tanzania
Lat/long: 3° 24′ S, 36° 42′ W

Dry period from June to October.

(b) Characteristic landscape in the Serengeti National Park, Tanzania, near Arusha, with plants adapted to seasonally dry water budgets. The shifting ITCZ and the subtropical high are responsible for the precipitation pattern throughout the year.

6.4 Mesothermal Climates (Midlatitudes, Mild Winters)

Key Learning Concepts

▶ *Explain* the precipitation and temperature criteria used to differentiate mesothermal climates.

▶ *Locate* mesothermal climate regions on a world map.

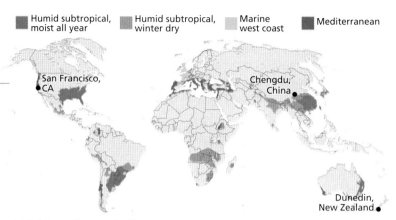

▲**6.11 Mesothermal climates**

The term *mesothermal* ("middle temperature") describes the warm and temperate climates where true seasonality begins. These are also referred to as the *temperate* climate regions (▶Fig. 6.11). More than 40% of the world's population resides in these climates, which occupy 27% of Earth's land and sea surface. The mesothermal climates (warm to moderate winters) are regions of great weather variability, for these are the latitudes of greatest air mass interaction. The factors that help define these climates include the following:

- Increasing variability of day length and insolation move poleward from the tropics. This "latitudinal effect" causes summers to transition from hot to warm to cool.
- Shifting maritime and continental air masses are guided by upper-air westerly winds.
- Variable precipitation results from the migration of cyclonic (low-pressure) and anticyclonic (high-pressure) systems, bringing changeable weather conditions and air mass conflicts.
- Variable sea-surface temperatures impact air mass strength: Cooler temperatures along west coasts weaken air masses, and warmer temperatures along east coasts strengthen air masses.

Mesothermal climates are humid, except where subtropical high pressure produces dry summer conditions. Their four distinct regimes based on precipitation variability are *humid subtropical hot-summer* (moist all year), *humid subtropical winter-dry* (hot to warm summers; Asia), *marine west coast* (warm to cool summers, moist all year), and *Mediterranean dry-summer* (warm to hot summers).

Roughly 40% of Earth's human population lives in the mesothermal climate zones. In these highly variable climates, humans cultivate most of the world's grain (such as corn, wheat, barley) and vegetable (tomatoes, onions, squash) crops. The Mediterranean climate is ideal for nuts, fruits, and many vegetable crops, if water for irrigation is available.

geoCHECK ✔ Explain how the term "mesothermal" applies to climate classification.

Humid Subtropical Climates

The **humid subtropical hot-summer** climates either are moist all year or have a distinct winter-dry period. In summer, warm and unstable maritime air produces convectional showers over land. In fall, winter, and spring, maritime tropical and continental polar air masses interact, generating frontal activity and frequent midlatitude cyclonic storms. Precipitation averages 100 to 200 cm (40 to 80 in.) a year.

Humid subtropical winter-dry climates are related to the winter-dry, seasonal pulse of the monsoons. They extend poleward from

▼**6.12 Humid subtropical winter-dry climate**

(a)

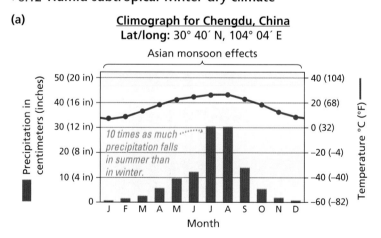

(b) Chengdu, China. As with tropical savanna climates, the shifting ITCZ and the subtropical high determine the precipitation pattern.

tropical savanna climates and have a summer month that receives 10 times more precipitation than their driest winter month. The abundant moisture and long agricultural growing seasons of the *humid subtropical hot-summer* and *humid subtropical winter-dry* climates sustain the largest human populations on Earth. These concentrations occur in north-central India, eastern China, and the southeastern United States (◄**Fig. 6.12**).

geoCHECK ✔ Describe the two ways moisture reaches the humid subtropical climate regions.

Marine West Coast Climates

The *marine west coast* climates of mild winters and cool summers are characteristic of Europe and other middle- to high-latitude west coasts (Fig. 6.10). Cool, moist, and unstable maritime polar air masses dominate these climates. Weather systems forming along the polar front and maritime polar air masses move into these regions throughout the year. Coastal fog, annually totaling 30 to 60 days, is a part of the moderating marine influence. The climograph for Dunedin, New Zealand, demonstrates the moderate temperature patterns and the annual temperature range for a *marine west coast* city in the Southern Hemisphere (▼**Fig. 6.13**).

geoCHECK ✔ List the temperature and moisture characteristics of a marine west coast climate.

Mediterranean Dry-Summer Climates

The criteria for a **Mediterranean dry-summer** climate specify that at least 70% of annual precipitation occurs during the winter months. This is in contrast to the precipitation pattern in most of the world, where summer-maximum precipitation is normal. This climate occurs in just 2% of Earth's land surface, along the western margins of North America, central Chile, and the southwestern tip of Africa, as well as across southern Australia and the namesake Mediterranean Basin.

Across the planet during summer, shifting cells of subtropical high pressure block moisture-bearing winds from adjacent regions. This shifting of stable, warm air to hot, dry air over an area in summer and away from that area in winter creates a distinct dry-summer and wet-winter pattern. The dry summer brings water-balance deficits. Along some coastlines, summer fog moderates the temperatures (▼**Fig. 6.14**).

geoCHECK ✔ Explain the distinctive precipitation characteristic of the *Mediterranean dry-summer*.

geoQUIZ

1. Identify the climate types that occupy the most land and ocean area on Earth.
2. Explain the risks of living in the climates that support the largest human populations on Earth.
3. Describe the role that offshore ocean currents play in the distribution of the marine west coast climates.

▼**6.13 Southern Hemisphere marine west coast climate**

(a)

Climograph for Dunedin, New Zealand
Lat/long: 45° 54′ S, 170° 31′ E

Cyclonic storm tracks

(b) Meadow, forest, and mountains on South Island, New Zealand.

▼**6.14 Mediterranean climate, California**

(a)

Climograph for San Francisco, California
Lat/long: 37° 37′ N, 122° 23′ W

Cyclonic storm tracks | Subtropical high | Cyclonic storm tracks

Most precipitation occurs in winter.

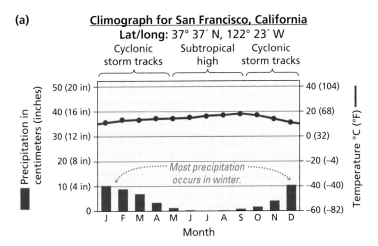

(b) Central California Mediterranean landscape of oak savanna.

6.5 Microthermal Climates (Midlatitudes & High Latitudes, Cold Winters)

Key Learning Concepts

▶ *State* the precipitation and temperature criteria used to distinguish microthermal climates.

▶ *Locate* microthermal climate regions on a world map.

The *microthermal* climates are cool temperate to cold. These climates have a winter season with some summer warmth. Approximately 21% of Earth's land surface is influenced by these climates, almost all in the Northern Hemisphere (▶**Fig. 6.15**). These climates occur poleward of the mesothermal climates and experience great temperature ranges related to continentality (Chapter 2) and air mass conflicts (Chapter 4). Temperatures decrease with increasing latitude and toward the interior of continental landmasses. The result of this pattern is intensely cold winters. The very cold and dry winter, in part due to an association with the Asian monsoon, is what distinguishes microthermal climates from the moist-all-year mesothermal climates to the south. The factors that help to determine microthermal climates include the following:

- Increasing differences in day length and seasonality due to increasing latitude produce greater diurnal and annual temperature ranges.
- Increasing latitude results in decreasing Sun angles and highly variable day length. Thus the summers become progressivly cooler moving northward, while the winters turn cold.
- Precipitation results from cyclonic air mass conflicts in winter and convectional thunderstorms in summer.
- Continental interiors serve as source regions for intense continental polar air masses that dominate winter, blocking cyclonic storms.
- An Asian winter-dry pattern is associated with continental high pressure and related air masses, increasing from the Ural Mountains eastward to the Pacific Ocean.

Microthermal climates have four distinct regimes based on increasing cold with latitude and precipitation variability: *humid continental hot-summer* (Chicago, New York); *humid continental mild-summer* (Duluth, Toronto, Moscow); *subarctic cool-summer* (Churchill); and the formidable extremes of frigid *subarctic with very cold winters* (Verkhoyansk and northern Siberia).

Approximately 11% of the of the world's population, almost all in the Northern Hemisphere, lives in microthermal

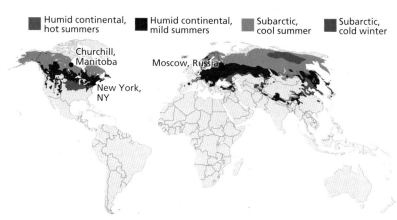

▲**6.15 Microthermal climates**

Legend: Humid continental, hot summers · Humid continental, mild summers · Subarctic, cool summer · Subarctic, cold winter

climates. Some of the most productive fisheries are found along the cooler waters of these higher-latitude coastlines, while in inland areas farmers grow hardy grain crops such as wheat and barley.

geoCHECK ✔ Why do microthermal climates occur almost entirely in the Northern Hemisphere?

▼**6.16 Humid continental hot-summer climate, New York**

(a)
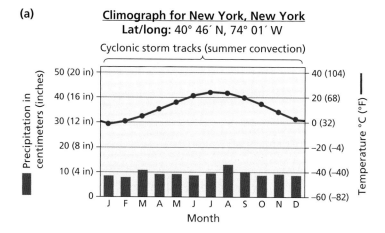

Climograph for New York, New York
Lat/long: 40° 46′ N, 74° 01′ W

(b) New York City's Central Park in winter

Humid Continental Hot-Summer Climates

The distribution of annual precipitation is what differentiates *humid continental hot-summer climates* from other microthermal climates. In the summer, maritime tropical air masses influence both *humid continental moist-all-year* and *humid continental winter-dry* climates. In North America, frequent weather activity results from the interaction of conflicting air masses—maritime tropical and continental polar—especially in winter (◄**Fig. 6.16**). In the United States, the *humid continental hot-summer* region extends from the East Coast westward to about the 100th meridian of longitude that runs north–south down the middle of the states of North and South Dakota, Nebraska, Kansas, and Texas. West of this meridian, the summers are much drier.

geoCHECK ✔ Why does the annual moisture falling on the American Great Plains decline west of the 100th meridian?

Humid Continental Mild-Summer Climates

Precipitation is less than in the hot-summer regions to the south. However, heavier snowfall is important to soil-moisture recharge when it melts. Characteristic cities include Duluth, Minnesota, and Moscow, Russia (▼**Fig. 6.17**).

geoCHECK ✔ Identify the geographic location of the humid continental mild-summer climates.

▼**6.17 Humid continental mild-summer climate**

(a)

Climograph for Moscow, Russia
Lat/long: 55° 45′ N, 37° 34′ E

Continental air mass (summer convection)

(b) Landscape between Moscow and St. Petersburg along the Volga River.

Subarctic Climates

Farther poleward, seasonal change becomes greater. The short growing season is more intense during long summer days. The *subarctic climates* include vast stretches of Alaska, Canada, and northern Scandinavia with their cool summers, and Siberian Russia with its very cold winters. Areas receive 25 cm (10 in.) or more of precipitation a year on the northern continental margins and are blanketed by forests. Precipitation and potential evapotranspiration are low, so soils are generally moist and either partially or totally frozen beneath the surface.

Churchill, Canada, has average monthly temperatures below freezing for 7 months of the year (▼**Fig. 6.18**). Churchill is representative of the *subarctic cool-summer* climate, with an annual temperature range of 40°C (72°F) and low precipitation of 44.3 cm (17.4 in.). The *subarctic climate with very cold winters* that feature a dry and very cold winter occur only within Russia. These areas experience an average temperature below freezing for 7 months and minimum temperatures of below –68°C (–90°F). Yet, summer-maximum temperatures can exceed 37°C (98°F). An example of this extreme is Verkhoyansk, Siberia, which has the world's greatest annual temperature range from winter to summer: a remarkable 63C° (113.4F°).

geoCHECK ✔ Identify the country with the coldest winters on Earth.

geoQUIZ

1. Explain why microthermal climates experience such great temperature ranges.
2. Identify the geographic range of the *humid continental hot-summer* region in the United States.
3. Why are soils in subarctic climates usually moist or frozen?

▼**6.18 Extreme subarctic cold-winter climate**

(a)

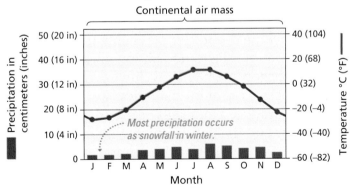

Climograph for Churchill, Manitoba, Canada
Lat/long: 58° 45′ N, 94° 04′ W

Continental air mass

Most precipitation occurs as snowfall in winter.

(b) Churchill and other port facilities on Hudson Bay may expand with renewed interest in mineral and petroleum reserves in subarctic regions.

6.6 Polar & Highland Climates

Key Learning Concepts

▶ *Explain* the precipitation and temperature criteria used to determine polar and highland climates.

▶ *Locate* the polar and highland climate regions on a world map.

A polar ice cap climate occurs on the upper slopes of Denali, illustrating the highland climate effect.

The polar climates have no true summer. The South Pole lies at the center of the Antarctic continent, which is surrounded by the Southern Ocean. The North Pole region comprises the Arctic Ocean, surrounded by the continents of North America and Eurasia. Although poleward of the Arctic and Antarctic Circles, summer day length increases to 24 hours, and the average monthly temperatures never rise above 10°C (50°F). These temperatures are too cold for tree growth. Principal factors that cause polar climates include the following:

- Low Sun altitude even during the long summer days, which is the principal climatic factor
- Extremes of day length between winter and summer, which determine insolation
- Extremely low humidity, producing low precipitation—the Earth's frozen deserts
- Surface albedo impacts, as light-colored surfaces of ice and snow reflect substantial energy away from the ground, thus reducing net radiation

Polar climates have three regimes: **tundra** (high latitude or high elevation); **ice caps** and **ice sheets** (continuously frozen until recent melting due to climate change); and *polar marine* (oceanic association, slight moderation of extreme cold). Also in this climate category are highland climates that occur on Earth's mountain ranges. Even at low latitudes, the cooling effects of elevation can produce tundra, glaciers, and polar conditions.

These polar and highland climates are home to less than 2% of the world's population. The most important economic activities are fishing, hunting, herding animals (in Eurasia), and participating in booming oil and gas extraction.

Tundra Climates

The tundra climates are almost entirely in the Northern Hemisphere, except for elevated mountain locations in the Southern Hemisphere and a portion of the Antarctic Peninsula. Here the land is under snow cover for 8 to 10 months, with the warmest month above 0°C (32°F), yet never above 10°C (50°F). The summit of Mount Washington in New Hampshire (1914 m, or 6280 ft) offers a highland tundra climate on a small scale. On a larger scale, an ice-free area of tundra and rock that is the size of California exists along the coasts of Greenland. In spring, when the snow melts, numerous stunted sedges, mosses, flowering plants, and lichens appear and persist through the short summer. By September, the tundra everywhere turns golden as the long winter approaches.

▲**6.19 Tundra and highland climates, Alaska** Late September with fall colors on the tundra below Mt. Denali (6,194 m/20,322 feet) in Denali National Park, Alaska. A polar ice cap climate occurs on the upper slopes of Mt. Denali, illustrating the highland climate effect.

In high-latitude mountains such as Denali (6194 m, or 20,322 feet), an **ice cap climate** occurs above the tundra, illustrating the highland climate effect (▲**Fig. 6.19**). The highland effect can even produce an ice cap climate near the equator, but one must climb (5181 m, or 17,000 ft.) to reach it (▼**Fig. 6.20**). Global warming is bringing dramatic changes to the tundra and highland climates. In parts of Canada and Alaska, record temperatures as much as 5° to 10°C (9° to 18°F) above average are now common. Warming Arctic temperatures are thawing permafrost in many locations, and there is also firm evidence that the treeline is beginning to move northward.

geoCHECK ✔ What is the summer temperature range of the tundra climate?

▼**6.20 Ice on the equator** Mt. Cotopaxi (5897 m, or 19,347 ft.), Ecuador

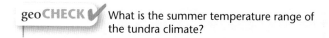

Ice Cap & Ice Sheet Climates

Earth's two ice sheets cover the Antarctic continent and most of Greenland. Here and the North Pole experience an **ice sheet climate** dominated by dry, frigid air masses where all months average below freezing. In fact, winter minimum temperatures in central Antarctica (July) frequently drop below the temperature of solid carbon dioxide or "dry ice" (–78°C, or –109°F).

The area of the North Pole is a sea covered by ice, whereas Antarctica is a substantial continental landmass. Antarctica is constantly snow-covered but receives less than 8 cm (3 in.) of precipitation each year. However, Antarctic ice has accumulated to several kilometers deep and is the largest repository of freshwater on Earth (▼Fig. 6.21).

geoCHECK ✔ Identify the defining temperature trait of the ice sheet climate.

▲**6.21 Ice cap and ice sheet climates, Antarctica** Antarctica's ice sheet, along with Greenland's, make up Earth's largest reservoir of freshwater.

▲**6.22 Polar and highland climates**

Polar Marine Climates

Polar marine climates (▼Fig. 6.22) are more moderate than other polar climates in winter, with no month below –7°C (20°F), but they are still colder than *tundra* climates. Because of marine influences, annual temperature range is usually 8.5C° (15.3F°). This climate exists along the Bering Sea, on the southern tip of Greenland, and in northern Iceland, Norway, and the Southern Hemisphere, generally over oceans between 50° S and 60° S (▼Fig. 6.23).

▼**6.23 South Georgia Island, a polar marine climate**

(a)

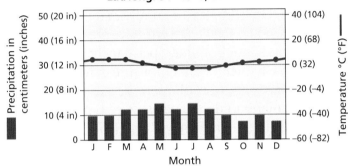

(b) Polar marine climate in Grytviken, an isolated settlement in South Georgia, in the South Atlantic Ocean.

geoCHECK ✔ Identify the geographic regions that experience an ice sheet climate.

geoQUIZ

1. Explain why, despite long summer days, polar climates have no true summer.
2. Summarize why some midlatitude mountains have temperature characteristics similar to those of a polar climate.
3. How does the surface albedo influence ice cap and ice sheet climates?

6.7 Dry Climates (Permanent Moisture Deficits)

Key Learning Concepts

▶ **Explain** the precipitation and temperature criteria used to distinguish dry climates.

▶ **Locate** dry climate regions on a world map.

The Earth's dry climates cover broad regions between 15° and 30° N and S (Fig. 6.6a). These regions occupy more than 35% of Earth's land area and are the most extensive climate over land. The dry climates are the only climate type in which low precipitation causes permanent moisture deficits. Dry climates are subdivided into arid deserts and steppes (▶Fig. 6.24). Deserts have greater moisture deficits than steppes, but both have permanent water shortages. Recall from Chapter 5 that a moisture deficit develops when the potential evapotranspiration exceeds moisture supply. Water demand in dry climates exceeds precipitation water supply, creating permanent water deficits.

The term **steppe** refers to the vast semiarid grasslands of Eastern Europe and Asia. The North American short-grass prairie and African savanna are also classified as steppes. A **steppe climate** is considered too dry to support forest, but too moist to be a desert. Several factors produce dry land climates:

- The dominant presence of dry, subsiding air in subtropical high-pressure systems
- Location in mountain rain shadows, where dry air subsides after dropping moisture on the windward slopes

- Location in continental interiors far from moisture-bearing air masses, particularly Central Asia
- Shifting subtropical high-pressure systems, creating semiarid steppes around the periphery of arid deserts

Latitude and the amount of moisture deficits determine the distribution of four distinct dry climate regimes. Arid climates, where the demand for moisture exceeds the precipitation water supply throughout the year, include *tropical, subtropical hot desert*, and *midlatitude cold desert*. Semiarid climates, where the precipitation is roughly one-half of the moisture demand, include *tropical, subtropical hot steppe*, and *midlatitude cold steppe*.

Desert and steppe regions are biologically rich environments home to varied plants, animals, and 10% of humanity. Oasis farming (fruits, vegetables, and nuts), animal herding, and mining predominate. But these climates are prone to long periods of drought, requiring special adaptations for plants, animals, and humans.

geoCHECK ✔ Why do dry climates occur in much of Central Asia?

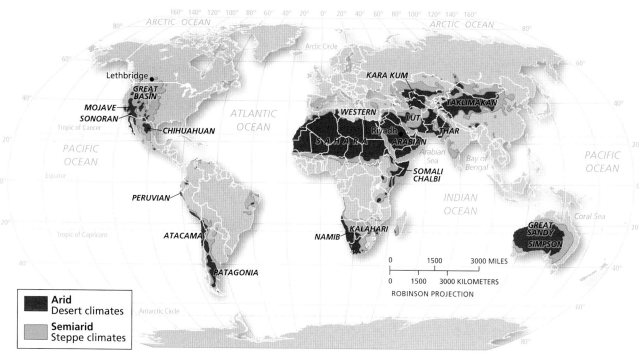

▲**6.24 Arid desert and semiarid steppe climates** Earth's dry climates experience permanent moisture deficits.

Tropical, Subtropical Hot Desert Climates

Tropical and *subtropical hot desert climates* feature annual average temperatures above 18°C (64.4°F). They occur on the western sides of continents and in Egypt, Somalia, and Saudi Arabia as you can see in Figure 6.24. Rainfall arrives from local summer convectional showers. Annual rainfall varies from almost nothing to 35 cm (14 in.). A representative *subtropical hot desert* city is Riyadh, Saudi Arabia (▼Fig. 6.25). South of the Sahara Desert is the Sahel, a semiarid steppe region that forms a transition zone between the intensely hot and dry Sahara to the north, and the cooler and wetter tropical savanna to south. The southward expansion of desert conditions following persistent drought is leaving many African peoples on land that no longer experiences the rainfall of just three decades ago. Other factors contributing to desertification in the Sahel are population increases, deforestation and overgrazing, poverty, and the lack of a regional environmental policy.

geoCHECK ✔ Describe the climate north and south of Africa's Sahel region.

Midlatitude Cold Desert Climates

Midlatitude cold desert climates cover a small area. Included are the southern border of Russia, along with Asia's Taklimakan Desert and Mongolia. Other areas include central Nevada, high-elevation areas of the American Southwest, and Patagonia in Argentina. These areas feature lower temperatures and moisture demand, with annual average rainfall only about 15 cm (6 in.).

geoCHECK ✔ Identify three continents where a midlatitude cold desert climate is found.

Tropical, Subtropical Hot Steppe Climates

Tropical, subtropical hot steppe climates generally exist around the periphery of hot deserts, where shifting subtropical high-pressure cells create a distinct summer-dry and winter-wet pattern. Average annual precipitation in *subtropical hot steppe* areas is usually below 60 cm (23.6 in.). This climate occurs around the Sahara's periphery and in the Iran, Afghanistan, Turkmenistan, and Kazakhstan region. In the Southern Hemisphere, interior Australia experiences this climate.

geoCHECK ✔ Why do tropical and subtropical hot steppe climates usually encircle hot deserts?

Midlatitude Cold Steppe Climates

The **midlatitude cold steppe climates** occur poleward of about 30° latitude and the *midlatitude cold desert* climates (▼Fig. 6.26). Such midlatitude steppes are not generally found in the Southern Hemisphere. As with other dry climate regions, rainfall in the steppes is widely variable and undependable, ranging from 20 to 40 cm (7.9 to 15.7 in.). Both convectional thunderstorms and cyclonic storm tracks can bring rainfall.

geoCHECK ✔ Identify where the *midlatitude cold steppe climate* occurs.

geoQUIZ

1. Compare and contrast a desert and a steppe.
2. How is a climate-related problem affecting residents of the Sahel region of Africa?
3. Explain how subtropical high-pressure cells influence tropical and subtropical hot steppe climates.

▼6.25 Tropical, subtropical hot desert climate

(a)
Climograph for Riyadh, Saudi Arabia
Lat/long: 24° 42′ N, 46° 43′ E

Average annual temperature exceeds 18°C.

(b) The Arabian Desert.

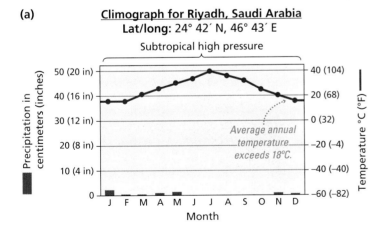

▼6.26 Midlatitude cold steppe climate, Canada

(a)
Climograph for Lethbridge, Alberta, Canada
Lat/long: 49° 42′ N, 110° 50′ W

Continental air mass (summer convection) (winter cP air mass)

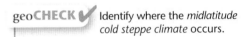

Rainfall is highly variable.

(b) Grain elevators highlight an Alberta, Canada, landscape.

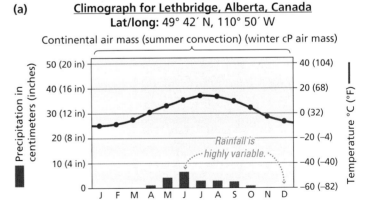

CLIMATES IMPACT HUMANS

• Climates affect many facets of human society, including agriculture, water availability, and natural hazards such as floods, droughts and heat waves.

HUMANS IMPACT CLIMATES

• Anthropogenic climate change is altering Earth systems that affect temperature and moisture, and therefore climate.

6c

A recent survey revealed that the area covered by polar climate types is shrinking as temperatures rise. In parts of Greenland, ice breakup has increased summer tourism, stimulating the economy, but has made hunting difficult for locals as animals shift their seasonal distributions.

6a

In the Eurasian Arctic tundra, 30 years of warming temperatures have allowed willow and alder shrubs to grow into small trees. This trend may lead to changes in regional albedo as trees darken the landscape and cause more sunlight to be absorbed.

6d

ISSUES FOR THE 21ST CENTURY

• Human-caused global warming is driving a poleward shift in the boundaries of climate regions.

As the tropics expand poleward, storm systems shift toward midlatitudes, and the subtropics become drier. Drought in Texas has devastated crops, affected beef production, and lowered reservoir levels. In this photo, a farmer surveys his cotton field. Under normal conditions, plants would be at knee height.

6b

Dengue fever, carried by the *Aedes aegypti* mosquito, is one of several diseases spreading into new areas as climatic conditions change. Dengue is now in previously unaffected parts of India, and in Nepal and Bhutan. In the United States, dengue is still uncommon, but reported cases are rising.

Looking Ahead

In the next chapter we explain the causes and potential consequences of climate change. We will investigate the evidence for climate change and present current forecasts for the rate of change. Our changing climate will influence almost every process discussed in this book, from the rates of weathering and erosion to the distribution of life on Earth.

How are climates classified?

6.1 Components of Earth's Climate System

Explain the difference between climate and weather.
Define the Köppen climate classification system.

- *Climate* is a synthesis of weather phenomena at many scales, from planetary to local, in contrast to weather, which is the condition of the atmosphere at any given time and place. Climatic conditions can be grouped by similarities of temperature and precipitation into climatic regions, which are the foundation of the Köppen climate classification system.

1. Describe the five principal components that determine the Earth's major climate groups.

6.2 Climate Classification System

Describe the principal Köppen climate classification categories.
Locate these distinctive regions on a world map.

- The Köppen climate classification system is based on temperature and precipitation measurements that help explain the geographic distribution of climates. Five basic climate categories and their regional types are recognized. The boundaries between the climate zones are not fixed, but are transition zones of gradual change.

2. List and discuss each of the principal climate categories, and identify the one in which you live.

3. Explain the characteristic that makes the highland climate unique among world climates.

What differentiates climates of tropical & mid-latitude regions?

6.3 Tropical Climates (Tropical Latitudes)

Explain the precipitation and temperature criteria used to determine tropical climates.
Locate these regions on a world map.

- Tropical climates are are found between the Tropics of Cancer and Capricorn. Consistent day length and insolation produce steady warm temperatures. There is no winter season. The shifting ITCZ creates subtle differences in annual rainfall, resulting in three distinctive tropical climates: *rain forest, monsoon,* and *savanna.*

4. Explain why the tropical climates have no winter season.

5. Describe the annual migration of the ITCZ and how this movement impacts the tropical climates.

6.4 Mesothermal Climates (Midlatitudes, Mild Winters)

Explain the precipitation and temperature criteria used to differentiate mesothermal climates.
Locate these regions on a world map.

- Mesothermal climates are found in the warm and temperate middle latitudes, where the annual changes in insolation create distinct seasons. The warm summers transition into warm to moderate winters, with the colder regimes occurring near the transition zone into the microthermal climates. These are regions of great weather variability due to consistent air mass interaction from the adjacent tropical and microthermal regions.

6. Mesothermal climates occupy the second-largest portion of Earth's entire surface. Describe their temperature, moisture, and precipitation characteristics.

7. Identify the precipitation trait that distinguishes the Mediterranean climate from the other mesothermal climates.

What differentiates climates of microthermal & cold regions?

6.5 Microthermal Climates (Midlatitudes & High Latitudes, Cold Winters)

Explain the precipitation and temperature criteria used to distinguish microthermal climates.
Locate these regions on a world map.

- These cool temperate to cold climates have a winter season with moderate summer warmth. They occur poleward of the mesothermal climates and experience great temperature ranges due to continentality in the large landmasses of North America and Eurasia. Precipitation arrives from cyclonic air mass conflicts in winter and convectional thunderstorms in summer.

8. Describe the primary climatic features that differentiate microthermal climates from the mesothermal climates to the south.

6.6 Polar & Highland Climates

Explain the precipitation and temperature criteria used to determine *polar* and *highland* climates.
Locate these regions on a world map.

- Because the Sun angles are so low, these high-latitude climates have no true summer. The winter is dark for 3 full months. The cold temperatures and minimal precipitation prevent tree growth. Because of the colder temperatures at high elevation, mountain environments at any latitude become polar in nature, with short growing seasons and year-round cool to cold temperatures.

9. Describe the primary difference between *tundra* climates and *ice cap* climates.

10. Explain how proximity to an ocean influences climate in the highest latitudes.

What differentiates climates of dry regions?

6.7 Dry Climates (Permanent Moisture Deficits)

Explain the precipitation and temperature criteria used to distinguish dry climates.
Locate these regions on a world map.

- Dry climates occupy 35% of Earth's land surface, more than any other climate. Dry climates are divided into *desert* and *steppe* (grasslands). Both have severe moisture deficits, but *steppe* climates have higher precipitation than arid deserts. All dry climates result from either subsiding and warming air systems, extreme subtropical high pressure, or a location in continental interiors.

11. Describe at least three locations where desert and steppe climates occur on Earth, and provide the reasons for their existence in these locations.

12. Describe why *midlatitude cold steppe* and *desert* climates are uncommon in the Southern Hemisphere.

Critical Thinking

1. If you did not know the location of a climograph station, how could you use the information presented to deduce the climate region to which it belonged? Give an example.

2. Using Africa's tropical climates as an example, characterize the climates produced by the seasonal shifting of the ITCZ.

3. Explain the distribution of the humid subtropical hot-summer and Mediterranean dry-summer climates in North America, and the difference in precipitation patterns between the two types.

4. Describe at least three locations where desert and steppe climates occur on Earth, and explain how both the amount and the timing of precipitation play a role in determining these classifications.

5. Identify where the coldest region on Earth is located, and explain two reasons for this extreme temperature regime.

Visual Analysis

This photograph was taken in early June on the southwest coastline of Myanmar, at the edge of the Andaman Sea (part of the Indian Ocean). Based on this information, use the full information presented on the Köppen climate map in Figure 6.5 to answer the following questions.

1. Why do you think the villagers moved their fishing boats onto the beach?

2. What event does this photograph capture? Explain your answer. (Hints: Identify the type of air mass, climate name, and relative location—north or south—of the ITCZ.

3. What is the likely temperature (warm or cool) and the cardinal direction of the ocean current?

4. Use your knowledge about the climate to identify the vegetation type pictured in the background (Hint: Use the World Vegetation Map in Chapter 14).

▲ **R6.1 Fishing village, southwest coast of Myanmar**

(MG) Interactive Mapping | Login to the **MasteringGeography** Study Area to access **MapMaster**

Climate & Human Conflict

The ongoing civil strife in Iraq and Syria has troops from many nations constantly on the move. The many-sided civil conflicts have also forced over 3 million people from their homes. Many refugees must flee on foot into neighboring countries.

- Open MapMaster™ in MasteringGeography™.
- Select Southwest Asia and North Africa, and select *Climate and Environmental Issues* from the *Physical Environment* menu.

1. Identify the climates, along with the specific seasonal temperature and precipitation characteristics, of this conflicted region.

2. Describe the climate-related challenges that are faced daily by thousands of people who suddenly find themselves homeless and traveling across the landscape.

3. If you were suddenly forced to flee from your village in the middle of Syria, in which direction would you travel? Explain your choice.

Explore | Use **Google Earth** to explore Earth's Climate Zones.

Visualizing Climate Transition Zones

Seeing Earth from space allows us to visualize the transition zones between Earth's different climate regions. Use Google Earth and the world climate classification map (Fig. 6.5) to answer the questions.

- In Google Earth, fly to *Australia*.
- Click *View*, and select *Borders and Labels*.

Zoom the slider in and out to navigate around this large country.

1. With the arrow pointer, trace the ring of steppe that surrounds the arid desert climate. Zoom in to identify at least five place names associated with each climate region.

2. Identify the types of climates that occur along the line of the January ITCZ. For each climate, describe the temperature and precipitation characteristics.

3. Use the Search box to locate *Cairns, Queensland, Australia*, in the northwest corner of the country. Note the different climate transitions on Figure 6.5, and then "tour" this part of Australia by zooming in and out in all four cardinal directions. How and why does the vegetation change in the climate transition zones, especially as you "fly" south along the coast to *Townsville*, and east into the interior?

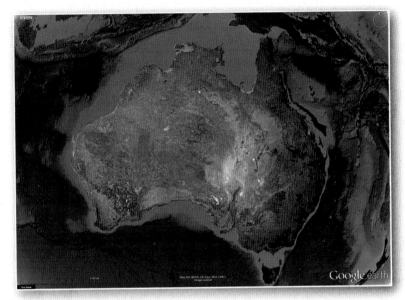

▲ **R6.2 Australia**

MasteringGeography™

Looking for additional review and test prep materials? Visit the Study Area in MasteringGeography™ to enhance your geographic literacy, spatial reasoning skills, and understanding of this chapter's content by accessing a variety of resources, including

MapMaster™ interactive maps, videos, *Mobile Field Trips, Project Condor* Quadcopter videos, *In the News* RSS feeds, flashcards, web links, self-study quizzes, and an eText version of *Geosystems Core*.

Key Terms

climate, p. 152
climatic regions, p. 152
climatology, p. 152
climograph, p. 153
desert climates (arid and semiarid), p. 167

dry climates, p. 153
highland climates, p. 153
humid subtropical hot-summer climate, p. 160

ice cap and ice sheet climates, p. 165
Köppen climate classification, p. 169
Mediterranean dry-summer climates, p. 161

mesothermal climates, p. 153
microthermal climates, p. 153
midlatitude cold steppe climates, p. 167

polar climates, p. 153
steppe, p. 166
tundra, p. 164

Industrial Location: Which Location Is Optimum?

Selecting a location to manufacture consumer goods involves many factors. These include proximity to markets, access to raw materials, labor supply, transportation corridors, and climate. The climate is important, because it can influence some of the other factors. Moreover, if any production occurs outside, then temperature and precipitation may influence the production of goods in a timely and efficient manner.

GeoLab6 (MG)
Pre-Lab Video

https://goo.gl/gN7R3L

Apply

As a consulting climatologist, you must prepare an analysis of three potential locations for producing large prefabricated homes. While some manufacturing takes place indoors, two key steps occur outside and with raw materials that are sensitive to high temperatures (because metal expands) and precipitation (because wood may swell, and uncoated metal rusts). The company is considering three sites that meet their needs (Figs. GL 6.1 and GL 6.2). The last consideration is climate, and they have asked you to evaluate the data.

Objectives

Graph data on temperature and precipitation to produce climographs.

Analyze climate data to identify the benefits and limitations of each site and make a final recommendation on site selection.

Procedure

1. Using the average monthly temperature and precipitation data in Figure GL 6.4. develop climographs for Sacramento, Tucson, and Buffalo. Draw a line connecting the upper data points of the temperature data to each month. With a different color, repeat this process along the upper points of the precipitation data.
2. For the three cities:
   a. compute the average annual rainfall, and average temperature.
   b. identify the highest and lowest monthly rainfall.
   c. identify the highest and lowest monthly temperature.
3. For each city, write a short summary that describes the climate characteristics for each of the four seasons.

Analyze & Conclude

4. Given the production constraints listed above, prepare a short synopsis of the climatic benefits and drawbacks of each site.

5. From a climate standpoint—with no other considerations—rank one site over the other two, and explain your choice.
6. Recommend two other cities in North America that from a climate perspective, might be more suitable than these three locations (Fig. 6.5 might be helpful here). Explain your choices.
7. While climate is an important element to consider when manufacturing prefabricated homes, other factors will also influence the final decision. Three of these factors are listed below. Carefully consider each one, and then decide if this new aspect changes the recommendation you made when the decision was based on climate alone. Explain your reasoning.
   a. Distance from large population centers that may purchase your product.
   b. Proximity to railroad networks that deliver heavy raw materials used to manufacture your product.
   c. Access to shipping lanes that will enable the overseas export of this bulky product.

Table GL6.1

	Jan.	Feb.	Mar.	Apr.	May	Jun.	Jul.	Aug.	Sep.	Oct.	Nov.	Dec.
Sacramento, CA Temperature °F	47	51	55	59	66	72	75	75	72	64	54	46
Precipitation (in inches)	3.62	3.46	2.76	1.14	0.67	0.2	0.04	0.04	0.28	0/94	2.09	3.27
Tucson, AZ Temperature °F	52	58	62	68	77	86	89	87	82	73	61	54
Precipitation (in inches)	1.02	0.94	0.87	0.31	0.2	0.28	1.93	2.24	1.22	1.22	0.67	1.02
Buffalo, NY Temperature °F	25	26	34	46	57	66	71	70	62	51	41	30
Precipitation (in inches)	3.19	2.48	2.87	2.99	3.46	3.66	3.23	3.27	3.9	3.5	4.02	3.9

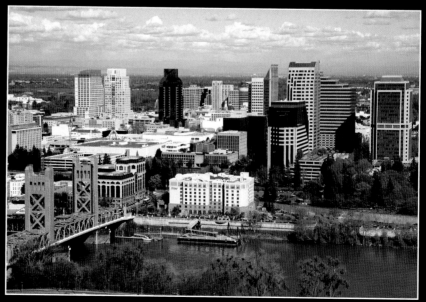

▲ GL6.1 **The Sacramento River flows west through Sacramento, California**

Climograph for Sacramento, California

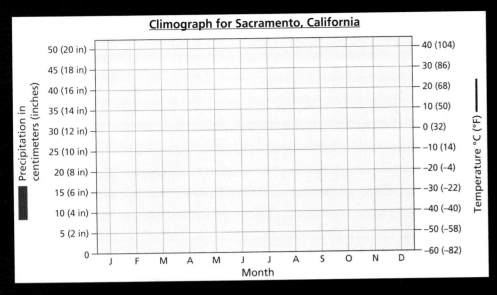

Climograph for Tucson, Arizona

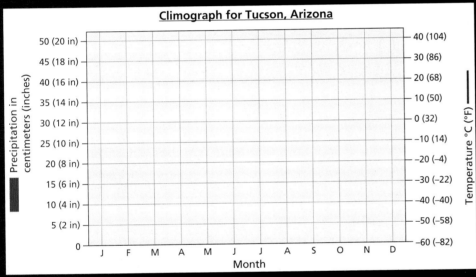

Climograph for Buffalo, New York

▲GL6.4 **Climographs for Sacramento, Tucson, and Buffalo**

▲GL6.2 **Saguaro cactus dot the desert landscape on the outskirts of Tucson, Arizona**

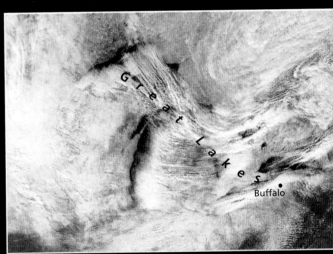

▲GL6.5 **A satellite image of a "Lake Effect" snowstorm over the Great Lakes and Buffalo, NY**

Log in MasteringGeography™ to complete the online portion of this lab, view the Pre-Lab Video, and complete the Post-Lab Quiz.
www.masteringgeography.com

7

Climate Change

Increasingly, the dramatic images of drought, floods, and wildfires that you see in the news media are visible expressions of climate change. Everything we have learned in *Geosystems Core* up to this point sets the stage for our exploration of climate change, one of the most critical issues facing humankind in the 21st century. Today, the study of climate change is an integral part of physical geography and Earth systems science. Three key elements of climate change science are the study of past climates, the measurement of current climatic changes, and the modeling and projection of future climate scenarios. We revisit some of the principles of the scientific method as we explore global climate change and address some of the confusion that complicates the public discussion of this topic.

Key Concepts & Topics

Birthday Canyon is 46m (150 ft) deep and has been carved by meltwater on the Greenland Ice Sheet.

7.1 Deciphering Past Climates

Key Learning Concepts

▶ **Describe** scientific tools used to study paleoclimatology.

To understand present climate changes, we need to understand past climates and how scientists reconstruct them. Scientists find clues to past climates in a variety of natural materials: fossil plankton, gas bubbles in glacial ice, ocean-bottom sediments, fossil pollen from ancient plants, growth rings in trees, mineral formations in caves, and corals (▼Fig. 7.1). Scientists analyze these materials to establish a chronology of climates from thousands to millions of years in the past.

The study of Earth's past climates is the science of **paleoclimatology**, which tells us that Earth's climate has changed over hundreds of millions of years. **Climate change science** is the interdisciplinary study of the causes and consequences of changing climate, affecting all Earth systems and the sustainability of human societies. To learn about climates that existed before human record keeping began, scientists use **proxy methods** instead of direct measurements. A *climate proxy* is a record of something that can tell us about the climate. For example, by studying the pattern of widths of tree rings, we can know what precipitation patterns were like thousands of years ago.

Rocks and fossils help geologists understand and reconstruct climates over time spans of millions of years. For example, tropical plant fossils preserved in rock indicate warmer climate conditions, ripple marks indicate a river or ocean environment, the presence of rocks called evaporites indicate that conditions changed from wetter to much drier, and ocean-dwelling fossils indicate ancient marine environments. Climate reconstructions show that Earth's climate has cycled between periods that were colder and warmer than today during the past several million years.

▼7.1 Sources of climate data

(a) Fossil pollen, as seen under a microscope.

(b) Coral reef cores

(c) Ocean-bottom sediment cores

(d) Ice cores. Project Leader Dorthe Dahl-Jensen, from the University of Copenhagen, holds up the last section of ice core, showing the bands of dust and debris.

▼7.2 The *U.S. JOIDES Resolution* drilling ship The JOIDES Resolution, and ships like her, have travelled the globe to obtain valuable ocean-sediment core samples.

Methods for Long-Term Climate Reconstruction

Materials deposited year after year in layers provide scientists with a major source of evidence for Earth's past climate conditions. Scientists obtain *core samples* of these materials for analysis (◄Fig. 7.2). A *core* is a long cylinder of material extracted with a hollow drill, for example from ocean-bottom sediments, ice sheets, or even trees. These cores may contain fossils, air bubbles, and other materials. Scientists use **isotope analysis**, which looks at the relative amounts of isotopes in a substance. *Isotopes* are different forms of the same chemical element. Scientists can reconstruct past temperature conditions by studying the ratios of certain isotopes found in glacial ice, ocean sediments, and other materials.

◄7.3 Ice core analysis

(a) Sunlight reveals layers of snow and ice as it shines through the wall of a snow pit in Greenland.

(b) Light shines through a thin section from the ice core, revealing air bubbles trapped within the ice that indicate the composition of past atmospheres.

Ice Cores In the regions of the world cold enough to have snow year-round, snow accumulates in layers that eventually become glacial ice (▲Fig. 7.3a). The thickest and oldest ice occurs in the vast ice sheets of Greenland and Antarctica. Ice cores also contain gas bubbles that tell us about the concentration of gases when the bubbles were sealed into the ice in the past (▲Fig. 7.3b).

Scientists study the oxygen isotope ratios in the ice to determine past climates. For example, oxygen isotopes are a valuable proxy for studying air temperatures. Water mainly contains two different oxygen isotopes, oxygen-16 or "light" oxygen, and oxygen-18 or "heavy" oxygen, and very small amounts of other oxygen isotopes. The light oxygen, ^{16}O, makes up 99.76% of all oxygen atoms, and heavy oxygen, ^{18}O makes up 0.20% of oxygen atoms, with other oxygen isotopes making up the remainder. Water containing the light ^{16}O evaporates more easily than the heavy ^{18}O. It takes more energy to evaporate the heavier ^{18}O, so ice with higher levels of ^{18}O is formed during warmer conditions (▼Fig. 7.4).

Ocean Sediment Cores Oxygen isotopes are found not only in water molecules, but also in the calcium carbonate ($CaCO_3$) in the shells of marine microorganisms called *foraminifera*. These are some of the world's most abundant shelled marine organisms and live from the equator to the poles. After their death, their shells fall to the bottom of the ocean and build up in layers of sediment. By extracting a core of these ocean-floor sediments, scientists can determine the isotope ratio of seawater at the time the shells were formed (Fig. 7.4b). Foraminifera shells with a high $^{18}O/^{16}O$ ratio were formed during cold periods; those with low ratios were formed during warm periods.

Over the past 50 years, scientists have obtained more than 35,000 core samples (see www.oceandrilling.org). Recent improvements in both isotope analysis techniques and the quality of ocean core samples have led to improved resolution of climate records for the past 70 million years.

geoCHECK ✔ A higher ratio of which oxygen isotope in ice cores is associated with warmer conditions? Explain.

geoQUIZ

1. Give three examples of climate proxies.
2. How can scientists determine past climate conditions by examining layers of rocks?
3. Describe how paleoclimatologists use ice cores to reconstruct past climate conditions.

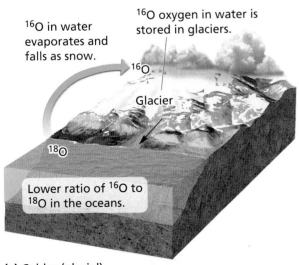

^{16}O in water evaporates and falls as snow.

^{16}O oxygen in water is stored in glaciers.

^{16}O

Glacier

^{18}O

Lower ratio of ^{16}O to ^{18}O in the oceans.

(a) Colder (glacial)

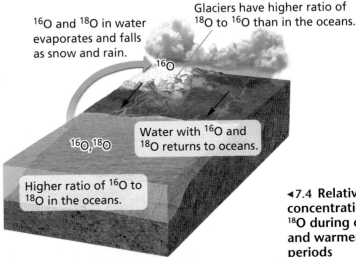

^{16}O and ^{18}O in water evaporates and falls as snow and rain.

Glaciers have higher ratio of ^{18}O to ^{16}O than in the oceans.

^{16}O

Water with ^{16}O and ^{18}O returns to oceans.

$^{16}O, ^{18}O$

Higher ratio of ^{16}O to ^{18}O in the oceans.

(b) Warmer (interglacial)

◄7.4 Relative oceanic concentrations of ^{16}O and ^{18}O during colder (glacial) and warmer (interglacial) periods

7.2 Climates of Earth's Past

Key Learning Concepts

▶ *Summarize* Earth's long-term climatic history since the Paleocene–Eocene Thermal Maximum.

Animation (MG)
Global Warming, Greenhouse Gases

http://goo.gl/cTHCHK

Animation (MG)
End of the Last Ice Age

http://goo.gl/XB2LMA

Climate reconstructions using fossils, deep-ocean sediment cores, and other evidence reveal long-term changes in Earth's climate. Over the past 4.56 billion years, climate has varied greatly, with periods warmer than present by 14°C (25.2°F) and periods when most of Earth was covered with ice (▼Fig. 7.5). One such colder episode, 2.4 billion years ago, is thought to have occurred when oxygen produced by microorganisms combined with methane, removing the methane from the atmosphere. The loss of methane, an important greenhouse gas, would have dramatically cooled Earth. Later episodes of global ice coverage from 800 to 600 million years ago (mya) could have occurred because of positive feedback from growing ice sheets in the tropics, which had formed because of a different continental configuration and lower carbon dioxide (CO_2) levels in the atmosphere. (Refer to Introduction to Physical Geography, p. I-1, to review positive feedback.)

Here, we focus on the last 70 million years because geologists have reconstructed the climates of this time span in some detail, the relative position of the continents has been stable, and because these conditions set the stage for more recent climates. In addition, the rapid change in climate about 70 million years ago (often abbreviated as m.y.a.) gives us insight into current climate changes.

The recent climates of the past 5 million years have seen the evolution of modern humans and the rise of agriculture and civilization. This period also saw the onset of an ice age, an extended period of cold (not a single brief cold spell), in some cases lasting several million years. An ice age is a time of generally cold climate that includes one or more *glacials* (glacial periods, characterized by glacial advance) interrupted by brief warm periods known as *interglacials*. The most recent ice age, known as the Pleistocene epoch, lasted from about 2.5 mya to about 11,700 years ago.

Ordovician period

▼**7.5 Earth's temperature over the past 500 million years, compared to the 1960–1990 average** Notice the general cooling trend from the Eocene to the Pleistocene.

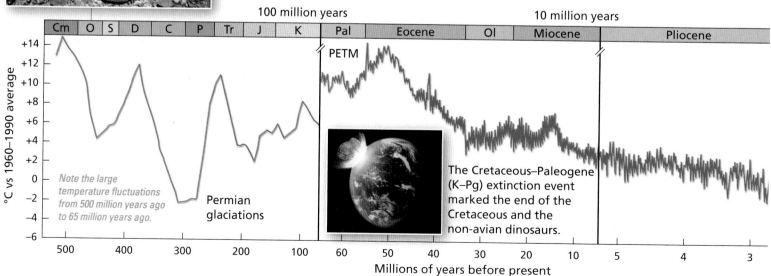

PETM

Note the large temperature fluctuations from 500 million years ago to 65 million years ago.

Permian glaciations

The Cretaceous–Paleogene (K–Pg) extinction event marked the end of the Cretaceous and the non-avian dinosaurs.

70 mya: Rapid Warming, Followed by Gradual Cooling

Over the span of 70 million years, we see that Earth's climate was much warmer in the distant past, and tropical conditions extended to higher latitudes than today. Since the warmer times of about 50 mya, climate has generally cooled. The exact reasons are presently unclear, but the cooling was probably the result of a combination of factors including the absorption of CO_2 by the newly uplifted rocks of the Himalayas.

About 56 mya, a distinct, short period of rapid warming occurred known as the Paleocene–Eocene thermal maximum, or PETM. The event is named after the geologic time intervals during which it occurred (see Chapter 8 for the geologic time scale). Scientists think this temperature maximum was caused by a sudden increase in atmospheric carbon. While the cause of the increase in atmospheric carbon is unknown, one leading hypothesis is that the carbon came from the melting of methane hydrates in the oceans. Methane hydrates are ice-like chemical compounds that are stable when frozen and under high pressure. However, if there was an event that warmed the ocean, such as an abrupt change in ocean currents, the hydrates would melt and release large quantities of methane, a short-lived, yet potent greenhouse gas.

The PETM rise in atmospheric carbon probably happened over a period of about 20,000 years or less—a "sudden" increase in terms of the vast scale of geologic time. The current increase in atmospheric CO_2 (discussed in Chapter 1, Module 1.6) is occurring at a more rapid pace. Scientists estimate that the amount of carbon that entered the atmosphere during the PETM is similar to the amount of carbon that human activity would release to the atmosphere with the burning of all Earth's fossil-fuel reserves. The post-PETM climate was quite different from our present climate, with crocodiles swimming off Greenland and tropical palm forests in Wyoming.

geoCHECK ✔ Why is it important to understand why the PETM occurred?

5 mya: Fluctuating Temperatures

Analysis of fossil microorganisms in ocean sediment cores reveals that over the last 5 million years, there have been a series of cooler and warmer periods. These ocean core records show nearly identical trends as ice core records.

As discussed earlier, ice cores provide data on the concentration of CO_2 and methane in the atmosphere. Figure 7.5 shows the changing concentrations of those two greenhouse gases and changing temperatures over the last 650,000 years. There is a close correlation between the two gas concentrations and between the gases and temperature on the graphs. The changes in greenhouse gas concentrations lag behind the temperature changes, generally by about 1000 years, which shows the presence and importance of climate feedbacks.

The last time temperatures were similar to the present-day interglacial period was about 125,000 years ago, during which time temperatures were warmer than at present. Notably, atmospheric carbon dioxide was below 300 ppm, a lower level than expected. Scientists interpret the lower-than-expected CO_2 level as resulting from the ocean's absorption of excessive atmospheric CO_2. We discuss the movement of CO_2 within Earth's carbon cycle later in this chapter.

geoCHECK ✔ How do the temperatures of today's climate compare with the temperatures from 70 to 5 mya?

geoQUIZ

1. What makes the PETM different from the average climate from 70 to 50 mya?
2. Is there good or poor agreement between ice core and ocean core data for the past 5 million years? Explain.
3. What were temperatures like 125,000 years ago? Why do scientists study that time period?

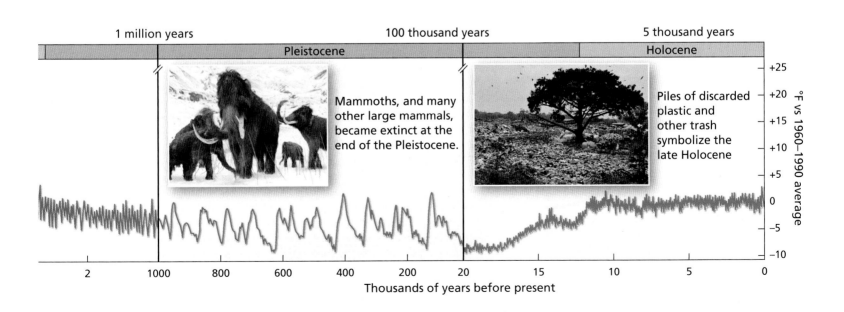

1 million years   100 thousand years   5 thousand years

Pleistocene   Holocene

Mammoths, and many other large mammals, became extinct at the end of the Pleistocene.

Piles of discarded plastic and other trash symbolize the late Holocene

°F vs 1960–1990 average

+25 +20 +15 +10 +5 0 −5 −10

2   1000   800   600   400   200   20   15   10   5   0

Thousands of years before present

7.3 Climates of the Last 20,000 Years

Key Learning Concepts

▶ **List** methods used to reconstruct climates during the relatively recent past.

▶ **Summarize** major climate trends and events over the past 20,000 years.

Based on the paleoclimatic evidence just discussed, scientists know that Earth has undergone long-term climate cycles that included conditions warmer and colder than today. Using other indicators, they have also determined climatic trends on shorter timescales, on the order of hundreds or thousands of years.

Methods for Short-Term Climate Reconstruction

The tools for short-term climate analysis consist mainly of radiocarbon dating and the analysis of growth rings of trees. Data from lake cores, limestone in caves, and corals are also used.

Carbon Isotope Analysis Carbon, like oxygen, has several isotopes. By studying the ratio of carbon-12 (^{12}C) to carbon-13 (^{13}C), scientists can determine if conditions were warmer and drier or cooler and wetter. The radioactive isotope carbon-14 (^{14}C) can also be used to date organic material up to 50,000 years old. Half of the ^{14}C present decays to nitrogen-14 (^{14}N) in 5730 years. This interval is the *half-life*, or the time that it takes for half of the sample to undergo radioactive decay (▼Fig. 7.6).

Tree Rings Most trees outside of the tropics add a growth ring of new wood each year. These rings are easily observed in a cross section of a tree trunk or in a core sample analyzed in a laboratory (▼Fig. 7.7). Wider rings suggest good growth conditions, while narrower rings suggest harsher, drier, or warmer conditions. The dating of tree rings by these methods is *dendrochronology*. By correlating multiple tree records and local climate records, scientists can create a continuous record of climatic conditions over 10,000 years long for an area. Long-lived tree species are most useful for tree-ring studies. For example, bristlecone pines in the western United States live up to 5000 years. Evidence from tree rings in the U.S. Southwest is important in comparing the magnitude of recent droughts with past droughts.

Other Sources of Data Cores taken from lake sediment, limestone formations in caves, and corals all provide scientists with additional data about Earth's climate (▶Fig. 7.8). Lake sediments contain pollen, fossils, and charcoal, all of which can be radiocarbon dated. By studying pollen in lake and ice sediments, scientists can identify the species of plants from the pollen, as well as the age of the pollen. By studying plant communities, scientists can determine what the climate was like at a given time. For example, the recent pollen found in Clear Lake in California reflects the present-day oak grassland plant community, while older pollen shows there were pine trees, indicating a cooler, wetter period. Scientists use carbon isotope analysis to date the calcium carbonate that makes up limestone formations in caves and coral reefs.

geoCHECK ✔ What can tree rings tell us about past climates?

Video (MG)

18,000 Years of Pine Pollen

http://goo.gl/iCAe1

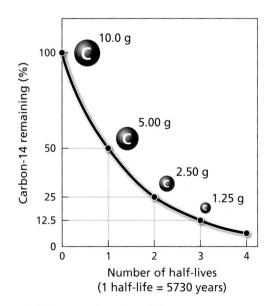

▲**7.6 Decay of carbon-14** Carbon-14 decays to nitrogen-14 with a half-life of 5730 years.

Wider rings indicate good growing conditions

Narrower rings indicate harsh growing conditions

▲**7.7 Tree rings in trunk cross section** Trees generally add one growth ring each year. The size and character of annual growth rings indicate growing conditions.

(a) Limestone formations in Royal Cave, Buchanan, Australia

Limestone formations

▲7.8 **Limestone cavern** Like tree rings, the layers that make up these limestone formations provide evidence of changing climate conditions.

(b) Growth bands in a cross section of cave limestone

Growth bands

that occurred from the last glacial maximum to about 15,000 years ago (▼Fig. 7.9).

Cooling Dip: The Younger Dryas About 14,000 years ago, average temperatures abruptly increased for several thousand years and then dropped again during the colder period known as the *Younger Dryas*, named for an arctic flower (Dryas, *Dryas octopetala*) that flourished during this period. The abrupt warming about 11,700 years ago marked the end of the most recent period of glaciation. Note in Figure 7.9 that less snow accumulates during colder glacial periods. Remember that cold air can have less water vapor than warm air, so less snow falls during glacial periods, even though a greater volume of ice may be present over Earth's surface.

Warm Spell: The Medieval Climate Anomaly From A.D. 800 to 1200, there was a mild climatic episode that affected the North Atlantic region, the *Medieval Climate Anomaly*. An anomaly occurs when there is a departure from normal conditions. During this time, there were warmer temperatures—as warm as or warmer than today—in some regions, and cooler temperatures in other regions. The warmth over the North Atlantic region allowed a variety of crops to grow at higher latitudes in Europe, shifting settlement patterns northward. For example, the Vikings settled Iceland and coastal areas of Greenland during this period.

Chilling Out: The Little Ice Age From approximately A.D. 1250 through about 1850, temperatures cooled globally during a period known as the *Little Ice Age*. Winter ice was more extensive in the North Atlantic Ocean, and expanding glaciers in western Europe blocked mountain passes. During the coldest years, snowlines in Europe were about 200 m (650 ft) in elevation. This period included many short-term climate fluctuations that lasted only decades and are probably related to volcanic activity and multiyear changes in global circulation patterns. After the Little Ice Age, temperatures steadily warmed. With growing human population and the onset of the Industrial Revolution, warming has continued—a trend that is accelerating today.

Short-Term Climate Trends

The Pleistocene epoch, Earth's most recent period of repeated glaciations, began 2.5 mya. The last glacial period lasted from about 110,000 years ago to about 11,700 years ago, with the *last glacial maximum*, the time when ice extent in the last glacial period was greatest, occurring about 20,000 years ago (see Figure 7.12 in the next module). The climate record for the past 20,000 years reveals a period of cold temperatures and little snow accumulation

geoCHECK ✔ When was the last glacial maximum?

geoQUIZ
1. How can scientists determine past climate conditions by examining pollen? What are two types of information they can learn from pollen?
2. When conditions are colder, would you expect snowfall to increase or decrease? Explain why this occurs.
3. Briefly describe Earth's climate history over the past 14,000 years.

Abrupt temperature rises occurred about 14,000 years ago and again about 12,000 years ago at the end of the Younger Dryas.

Temperature

Medieval Climate Anomaly

Snow accumulation

Periods of colder temperatures occurred during the last glacial maximum and the Younger Dryas.

Temperature in central Greenland in °C (°F): −25 (−13), −35 (−31), −45 (−49), −55 (−67), −60 (−76)

Snow accumulation in m/yr (ft/yr): 0.35 (1.15), 0.25 (0.82), 0.15 (0.49), 0.05 (0.16)

Time (thousands of years before present): 20 — Last glacial maximum, 15 — Younger Dryas, 10, 5, 0 — Little Ice Age

◄7.9 **The past 20,000 years of temperature and snow accumulation** Evidence from Greenland ice cores shows periods of colder temperatures occurring during the last glacial maximum and the Younger Dryas. The ice cores also show an abrupt temperature rise occurring about 14,000 years ago and again about 12,000 years ago at the end of the Younger Dryas. Although this graph uses ice core data, these temperature trends correlate with other climate proxy records.

7.4 Mechanisms of Natural Climate Change

Key Learning Concepts

▶ **Analyze** several natural factors that influence Earth's climate.

Using the methods just discussed, paleoclimatologists have reconstructed Earth's climate history for the last 650,000 years, as you saw in Figure 7.5. Over this time span, patterns of 20,000 to 100,000 years are apparent. Scientists have evaluated a number of natural mechanisms that affect Earth's climate and probably cause these long-term cyclical climate variations.

Animation MG
Earth-Sun
Relations

http://goo.gl/XVJd3y

Animation MG
Orbital Variations &
Climate Change

http://goo.gl/p08UPy

Solar Variability

Energy from the Sun is the most important driver of the Earth–atmosphere climate system. The Sun's energy varies over several timescales, and these variations affect climate. For example, solar output has increased by one third since the formation of our solar system. Other variations on the scale of thousands of years are linked to changes in the Sun's magnetic field. On a time frame of decades, slight variations in solar output are linked to sunspot activity (▼ **Fig. 7.10**). Note that in Figure 7.10a the number of sunspots ranges from 1 to over 200, the total range of insolation has varied by approximately 0.25 W/m².

High sunspot activity matches high solar output, which scientists have linked with slightly higher temperatures in the past. A period of low sunspot activity from 1645 to 1715 called the Maunder minimum was one of the coldest periods of the Little Ice Age (▼ **Fig. 7.10b**), although the locally colder conditions were probably more a result of volcanic activity. However, temperatures increased during the period of low sunspot activity from 2005 to 2010. This lack of correlation between sunspot activity and temperature indicates that the causality is not definite. Scientists agree that insolation is not the primary driver of recent global warming trends (as an example, see www.giss.nasa.gov/research/news/20120130b/), shown in Figure 7.10c.

geoCHECK ✔ How have changes in solar output corresponded with changes to Earth's temperatures in the past? How have they related to Earth's temperatures from 2005 to 2010?

▼**7.10 400 years of Sunspot Observations and Northern Hemispheric Temperature reconstruction**

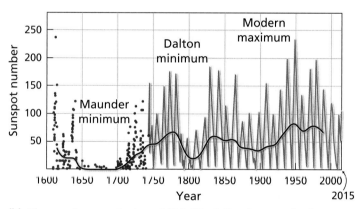

(b) Observed sunspots since 1600. Black line is smoothed average number of sunspots. During the Maunder minimum (shown as red dots), sunspots were rare and infrequent.

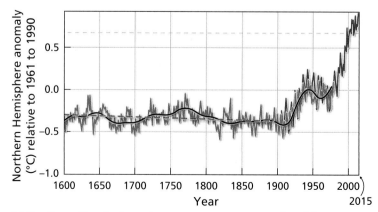

(c) Northern Hemisphere temperature reconstruction (blue line) and instrument data (red line) from 1600. Black line is smoothed average temperature line and dashed purple line is the trend line from AD 1000 to 1850.

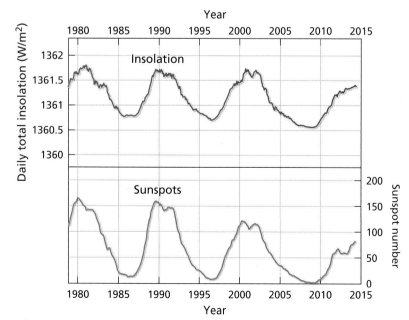

(a) Insolation and sunspot activity.

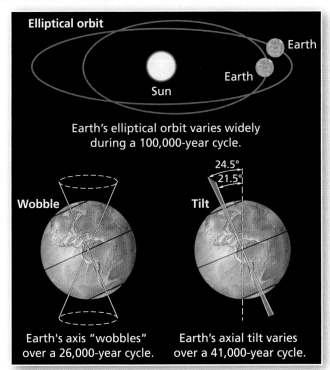

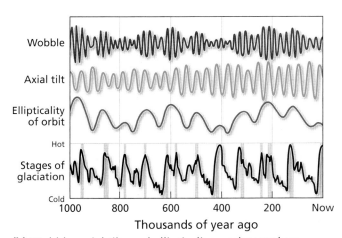

(a) The three Earth–Sun relationship factors of the Milankovitch cycles.

(b) Wobble, axial tilt, and ellipticity work together to create the patterns of glaciation.

▲ **7.11 Astronomic factors that may affect broad climatic cycles**

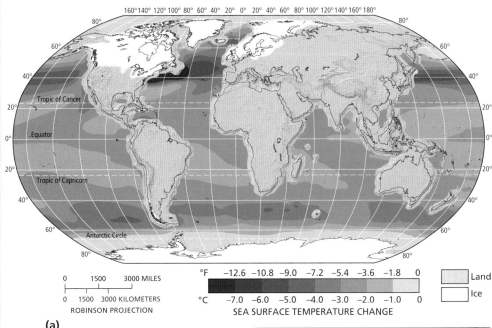

(a)

▲ **Figure 7.12 Sea-surface temperatures and lowered sea level** During colder glacial conditions, (a) global sea-surface temperatures and sea levels were lower than today, and (b) dry land connected the British Isles to Europe.

- Earth's axis "wobbles" through a 26,000-year cycle, in a movement much like that of a spinning top winding down (◄Fig. 7.11a). Earth's wobble, known as *precession*, changes the orientation of hemispheres and landmasses to the Sun.
- Earth's axial tilt is now about 23.5°, but the amount of tilt varies from 21.5° to 24.5° over a 41,000-year period (Fig. 7.11a).

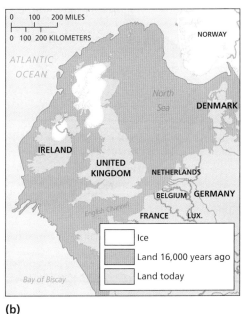

(b)

Earth's Orbital Cycles

Earth–Sun relationships affect energy and seasons on Earth and are another possible factor in climate change. These relationships include Earth's changing distance from the Sun, Earth's orientation to the Sun, and Earth's varying axial tilt (▲Fig. 7.11). Milutin Milankovitch (1879–1958), a Serbian astronomer, studied the changes in Earth's orbit around the Sun and identified regular cycles that relate to climatic patterns:

- Earth's elliptical orbit about the Sun changes over a 100,000-year cycle in which the shape of the ellipse varies by more than 17.7 million kilometers (11 million miles), from a shape that is nearly circular to one that is more elliptical (▲Fig. 7.11a).

Although the scientific community at first rejected many of Milankovitch's ideas, scientists today accept these orbital variations, now called **Milankovitch cycles**, as a factor in glacial–interglacial cycles. Ice core evidence from Greenland and Antarctica and from sediments of Lake Baikal in Russia has confirmed a roughly 100,000-year climatic cycle (◄Fig. 7.11b). Other evidence supports the effect of shorter-term cycles of roughly 40,000 and 20,000 years on climate. Figure 7.12a shows how much colder ocean temperatures were during the last glacial maximum. During this glacial maximum sea levels were much lower, allowing humans to live on what is now the ocean floor (▲Fig. 7.12b).

geoCHECK ✔ What are the three components of the Milankovitch cycles and what are their timescales?

7.4 (cont'd) Mechanisms of Natural Climate Change

Continental Position & Topography

Earth's continents have moved together and apart, creating different arrangements throughout geologic history (▼Fig. 7.13). The movement of the continents affects climate, since landmasses have strong effects on atmospheric and oceanic circulation. For example, the uplift of the Himalayas altered the path of the polar jet stream, as well as created wetter conditions on the windward slope and a massive rain shadow to their north. As discussed in previous chapters, the relative proportions of land and ocean area affect surface albedo, as does the position of landmasses relative to the poles or equator. The position of the continents also impacts

ocean currents, which redistribute heat throughout the world's oceans. The closing of the Panamanian land bridge radically changed ocean currents (▼Fig. 7.14). Warm currents in the Atlantic were redirected to the north rather than continuing on to Asia, forming the Gulf Stream that now warms Europe in winter by as much as 10°C (18°F). The new Gulf Stream would also have increased available water vapor that could fall as rain. Finally, the movement of continental plates causes mountain building, and seafloor spreading involves volcanic activity and changes the size and shape of ocean basins. These processes affect Earth's climate system. For example, high-elevation mountain ranges accumulate snow and ice during glacial periods, increasing albedo. The volcanic activity that accompanies seafloor spreading releases large quantities of CO_2.

geoCHECK ✔ How could the formation of new high-elevation regions affect Earth's albedo?

(a) Pangea and climate zones, Triassic period, 220 mya.

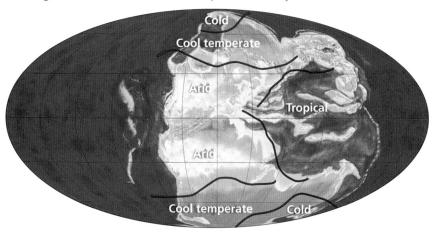

(b) Continents and climate zones today.

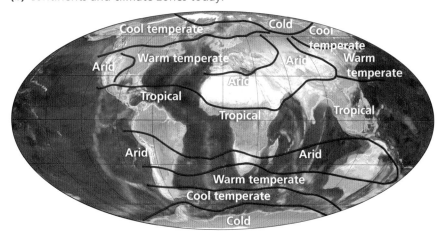

▲**7.13 Continental movements and global climate patterns** The positions of the continents influence ocean temperatures and climate patterns. When the continents moved together to form the supercontinent Pangea, huge belts of arid climates dominated that landmass. See 8.13 in Chapter 8 for other configurations of Earth's continents.

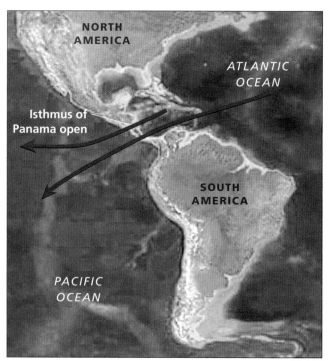

(a) The Miocene Epoch 20 mya. Warm currents flow from the Atlantic to the Pacific; the North Atlantic is cool.

▲**7.14 Effect of the Panamanian land bridge on climate** The ocean currents that influence climate changed dramatically when continental movements and volcanic activity connected North and South America.

Atmospheric Gases & Aerosols

Natural processes release gases and small particles called **aerosols** into Earth's atmosphere with varying impacts on climate. Natural outgassing from Earth's interior through volcanoes and vents in the ocean floor is the primary natural source of CO_2 emissions. Water vapor is a natural greenhouse gas. Over long periods of time, higher levels of greenhouse gases generally correlate with warmer interglacials, and lower levels correlate with colder glacials. Earth's surface heats or cools in response to changes in greenhouse gas concentrations.

Volcanic eruptions also produce aerosols linked to climatic cooling. Aerosols ejected into the stratosphere increase albedo so that more insolation is reflected and less solar energy reaches Earth's surface. Sulfur aerosol accumulations affect temperatures on timescales of months to years. For example, the 1991 Mount Pinatubo eruption lowered temperatures for 2 years (▶**Fig. 7.15**). Scientific evidence also suggests that large volcanic eruptions may have triggered the Little Ice Age.

geoCHECK ✔ How do volcanoes act to both increase and decrease temperatures on Earth?

geoQUIZ

1. What are the minimum and maximum numbers of sunspots from 1980 to 2010? What are the maximum and minumum values for insolation over this time period?
2. How did the formation of the Panamanian land bridge affect climate in Europe?
3. Describe two ways that plate tectonics affects climate.

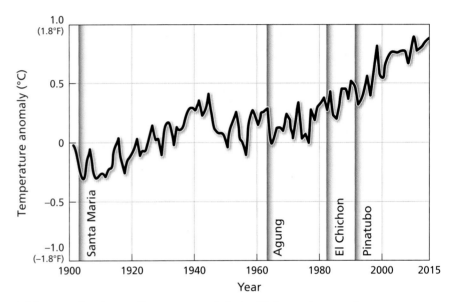

(a) Large volcanic eruptions are associated with lower temperatures for several years.

(b) The Panamanian land bridge today. The land bridge caused warm currents to flow toward the north Atlantic, warming eastern North America and Western Europe.

(b) Mount Pinatubo erupting in 1991.

▲**7.15 Global temperatures and volcanic eruptions**

7.5 Climate Feedbacks & the Carbon Budget

Key Learning Concepts

▶ **Explain** two main climate feedbacks—ice–albedo feedback and water vapor feedback.

▶ **Analyze** the carbon cycle in terms of feedback mechanisms that affect Earth's climate.

In addition to responding to the natural factors discussed in Module 7.4, Earth's climate system changes in response to a number of feedback mechanisms. As discussed in the introductory chapter, systems can produce outputs that sometimes influence their own operations via positive or negative feedback loops. Positive feedback amplifies system changes and tends to destabilize the system. Negative feedback reduces system changes and tends to stabilize the system. **Climate feedbacks** are processes that either amplify or reduce climatic trends toward either warming or cooling. Scientists now know that greenhouse gases can cause feedback loops that amplify climatic trends and can potentially drive climatic change.

The Ice–Albedo Feedback

A good example of a positive climate feedback is *ice–albedo* feedback (▼ Fig. 7.16). As snow and ice melt, bare rock is exposed, which absorbs more energy from the Sun, which causes warmer temperatures, which melts more snow and ice. This feedback is accelerating the current trend of global warming as ice melts in the Arctic. However, ice–albedo feedback can also amplify global cooling because lower temperatures lead to more snow and ice cover, which increases albedo, or reflectivity, and causes less sunlight to be absorbed by Earth's surface. Scientists think that the ice–albedo feedback may have amplified global cooling following the volcanic eruptions at the start of the Little Ice Age. As atmospheric aerosols increased and temperatures decreased, more ice formed, further increasing albedo, leading to further cooling and more ice formation. These conditions persisted until the recent increase in anthropogenic greenhouse gases that began with the Industrial Revolution in the 1800s.

Video (MG)
Taking Earth's
Temperature

http://goo.gl/qCy4ev

geoCHECK ✔ What are two possible positive feedback loops associated with snow and ice?

▼ **7.16 Ice-albedo melt ponds**

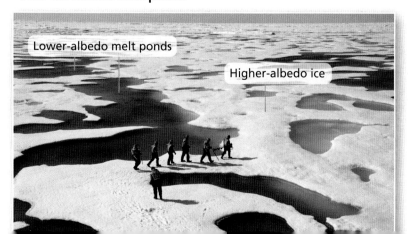

Lower-albedo melt ponds

Higher-albedo ice

White and blue areas have high concentrations of water vapor.

The brightest white areas are thunderclouds.

Dark regions are relatively dry.

▲ **7.17 Water vapor in the atmosphere** As the atmosphere warms, the water vapor content can increase, which warms the atmosphere.

Water Vapor Feedback

Water vapor is the most abundant natural greenhouse gas in the Earth–atmosphere system (▲ Fig. 7.17). Water vapor feedback is a result of the effect of air temperature on the amount of water vapor that air can absorb (discussed in Chapter 5). As air temperature rises, evaporation increases, because the capacity to absorb water vapor is greater for warm air than for cooler air. Thus, more water enters the atmosphere from land and ocean surfaces, humidity increases, and greenhouse warming accelerates. As temperatures increase further, more water vapor enters the atmosphere, greenhouse warming further increases, and the positive feedback continues.

A factor that complicates the water vapor feedback is the role of clouds in Earth's energy budget. As atmospheric water vapor increases, higher rates of condensation will lead to more cloud formation. Low, thick cloud cover increases the albedo of the atmosphere and has a cooling effect on Earth. In contrast, high, thin clouds can cause warming because they allow energy in from the Sun, but absorb and re-radiate heat from Earth's surface.

geoCHECK ✔ Is water vapor feedback positive or negative feedback? Explain.

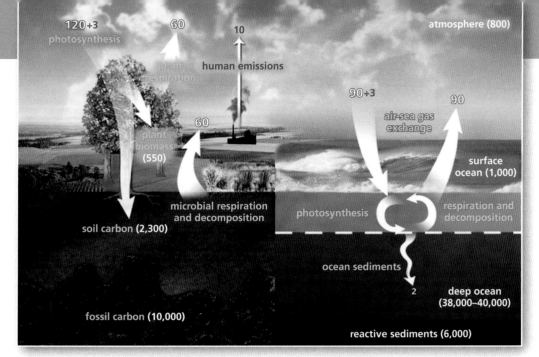

120+3
photosynthesis
60
10
human emissions
plant
respiration
60
plant
biomass
(550)
atmosphere (800)
90+3
air-sea gas
exchange
90
surface
ocean (1,000)
microbial respiration
and decomposition
photosynthesis
respiration and
decomposition
soil carbon (2,300)
ocean sediments
2
deep ocean
(38,000–40,000)
fossil carbon (10,000)
reactive sediments (6,000)

▲7.18 **The global carbon cycle** Biological processes remove (photosynthesis) and add (respiration, decomposition) carbon from the atmosphere on a time scale of days to thousands of years, while physical processes such as formation and weathering of sedimentary rocks add and remove carbon over time scales of millions of years. The numbers are in gigatons (one gigaton is one billion tons).

Earth's Carbon Budget

Many climate feedbacks involve the movement of carbon through Earth systems in the *carbon cycle*. **Carbon sinks**, or carbon *reservoirs*, store carbon released from various sources. The exchange of carbon between Earth systems is the **global carbon budget**, which should naturally remain balanced as carbon moves between sources and sinks. Figure 7.18 illustrates the components, both natural and anthropogenic, of Earth's carbon budget and carbon sinks.

Human Impacts on Carbon Cycle Humans began clearing forests for agriculture thousands of years ago, which reduced one of Earth's natural carbon sinks (forests) and transferred carbon to the atmosphere. With the Industrial Revolution, around 1850, the burning of fossil fuels became a large source of atmospheric CO_2, transferring carbon stored in rock into gaseous carbon. This process is continuing at an accelerating rate (▶Fig. 7.19). The Keeling curve, named after Dr. Charles Keeling, shows the steady increase in CO_2 levels from 1959 to the present.

Oceans and the Carbon Cycle Given the large concentrations of CO_2 released by human activities, scientists have wondered why the amount of CO_2 in Earth's atmosphere is not higher. Studies suggest that uptake of carbon by the oceans is offsetting some of the atmospheric increase. Seawater absorbs atmospheric CO_2, forming carbonic acid (H_2CO_3), in the process of *ocean acidification*. The increased acidity harms marine organisms, such as corals and some types of plankton that have calcium carbonate shells. Scientists estimate that the oceans have absorbed some 50% of the rising concentrations of atmospheric carbon, slowing the warming of the atmosphere. However, as the oceans increase in temperature, their ability to dissolve CO_2 decreases. Thus, as global air and ocean temperatures warm, more CO_2 will likely remain in the atmosphere.

Ecosystems and the Carbon Cycle Uptake of excess carbon is also occurring as increased CO_2 levels in the atmosphere enhance photosynthesis in plants, allowing them to grow more quickly and store more carbon. However, as temperatures increase, rates of photosynthesis will decrease, slowing plants' uptake of carbon. Also, many human practices—for example, overgrazing and massive deforestation in the tropics—reduce the capacity of land ecosystems as carbon sinks. The uneven line on the graph shows the importance of photosynthesis in affecting global CO_2 levels. The highest values during the year are usually in May and the lowest values are usually in October. The annual variation between spring and fall due to seasonal changes in vegetation cover in the higher latitudes in the Northern hemisphere. Vegetation is dormant during the Northern Hemisphere winter so CO_2 levels climb, and in spring, when plants take up more CO_2, levels fall slightly.

geoCHECK ✔ How is the increase in atmospheric CO_2 related to the rate of plant growth?

geoQUIZ

1. How could ice–albedo feedback have contributed to the cold conditions of the Little Ice Age?
2. How do different types of clouds affect surface temperatures differently?
3. How is the increase in the amount of atmospheric CO_2 affecting marine organisms? Explain your answer.

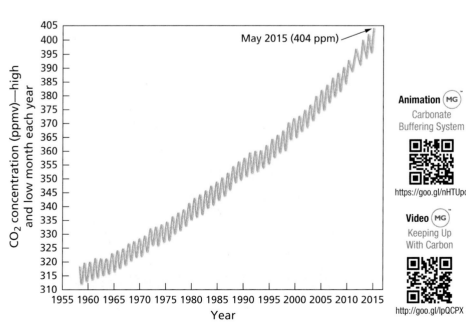

May 2015 (404 ppm)

CO_2 concentration (ppmv)—high and low month each year

Year

Animation MG
Carbonate
Buffering System

https://goo.gl/nHTUpq

Video MG
Keeping Up
With Carbon

http://goo.gl/lpQCPX

▲7.19 **The Keeling curve** The graph represents 58 years of CO_2 data measured on Mauna Loa, a Hawaiian volcano far from sources of pollution. During May 2015, CO_2 jumped to a new record of more than 403.9 ppm.

7.6 Evidence for Present Climate Change

Key Learning Concepts

▶ **List** the key lines of evidence for present global climate change.

The evidence for climate change comes from a variety of measurements showing global trends over the past century, and especially over the last two or three decades. Data from weather stations, orbiting satellites, weather balloons, ships, buoys, and aircraft confirm climatic warming. Figure 7.20 illustrates the four main, measurable indicators of warming: rising temperatures, melting ice, rising sea levels, and increased atmospheric water vapor.

Rising Temperatures

Long-term climate reconstructions of temperature show that the present is the warmest time in the last 120,000 years (see Figure 7.7) and suggest that the increase in temperature during the 20th century is the largest to occur in any century over the past 1000 years. The temperature data unmistakably show a warming trend (▼Fig. 7.21). Since 1880, in the Northern Hemisphere, the period from 2000 to 2010 was the warmest decade, and 2015 was the warmest year ever recorded. Figure 7.22 shows global temperatures from 1880 to 2015.

▼ **7.20 Key indicators of climatic warming** The climate is changing as demonstrated by rising air and ocean temperatures, shrinking glacial and sea ice, rising sea level, and increasing atmospheric water vapor.

Increasing atmospheric water vapor

Higher air temperature in the lower troposphere

Melting of glacial ice

Higher Air temperature over land

Decreasing sea-ice extent

Higher air temperature over oceans

Reduced snow cover

Higher sea-surface temperature

Increasing temperatures over land and oceans

Rising sea levels

Increasing ocean heat content

Each of the past three decades has been the warmest decade ever recorded. Scientists have measured record-setting summer daytime temperatures in many countries. In the 1950s, the ratio of record-high temperatures to record-low temperatures was 1.09:1; however, during 2000 to 2012, the ratio had increased to over 9:1! In 2012, 90% of the new records set in the United States were new high-temperature records. (See http://www3.epa.gov /climatechange/science/indicators/weather-climate /high-low-temps.html.)

Average ocean temperatures have also increased by 0.13°C (0.23°F) per decade from 1901 to 2012 as oceans absorbed heat from the atmosphere. This rise has warmed the upper 700 m (2296 ft) of the oceans. (www.ncdc.noaa.gov/indicators/, click "warming climate").

geoCHECK ✔ Where does the decade you were born in rank among the warmest decades?

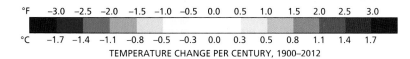

No data

°F −3.0 −2.5 −2.0 −1.5 −1.0 −0.5 0.0 0.5 1.0 1.5 2.0 2.5 3.0

°C −1.7 −1.4 −1.1 −0.8 −0.5 −0.3 0.0 0.3 0.5 0.8 1.1 1.4 1.7

TEMPERATURE CHANGE PER CENTURY, 1900–2012

◄**7.21 Change in global surface temperature, 1901–2012**
Note the greater changes in the mid and high latitudes.

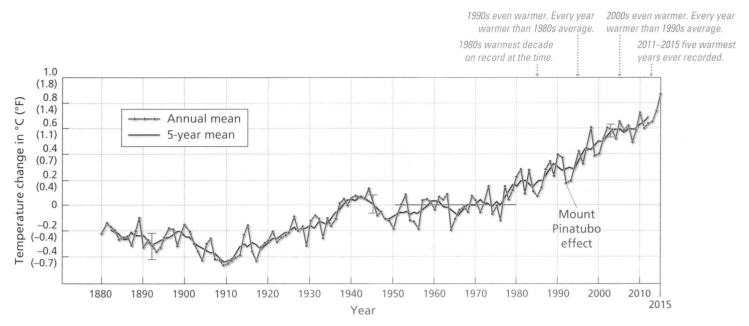

1980s warmest decade on record at the time.

1990s even warmer. Every year warmer than 1980s average.

2000s even warmer. Every year warmer than 1990s average.

2011–2015 five warmest years ever recorded.

▲ 7.22 **Global land–ocean temperature trends, 1880–2010** The graph shows change in global surface temperatures relative to the 1951–1980 global average. The gray bars represent uncertainty in the measurements. Note the inclusion of both annual average temperature anomalies and 5-year mean temperature anomalies; together, they give a sense of overall trends.

Melting Ice

The heating of Earth's atmosphere and oceans is causing sea ice and land ice to melt. Chapter 12 discusses the characteristics of Earth's cryosphere.

Sea Ice Sea ice is composed of frozen seawater, unlike ice shelves or icebergs, which are made of freshwater. When sea ice melts, it doesn't increase sea level, because it is already in the ocean. However, as high-albedo ice melts, it exposes lower-albedo ocean water, discussed previously as the warming version of ice–albedo feedback.

The extent of Arctic sea ice varies over the course of a year as well as from year to year (▼ Fig. 7.23). Both the summer minimum area and the winter maximum have decreased since 1979. The summer minimum has decreased by 11% per decade, compared to the 1979 to 2000 average. The accelerating decline and recent record losses suggest that the Arctic Ocean could experience an ice-free summer within a few decades, which would accelerate global warming further.

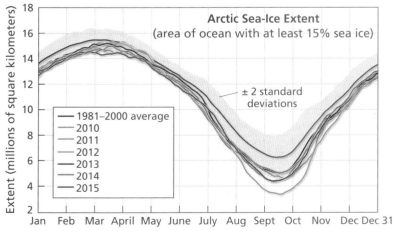

(a) Arctic sea reached its lowest extent in the satellite record in 2007 and 2012.

(b) Image shows the 2012 record low compared to the average low since 2000.

Animation MG
Arctic Sea Ice Decline

▲ 7.23 **Arctic sea-ice extent** Summer melting reduces Arctic sea ice to a minimum each year in September. Average minimum summer ice extent has decreased significantly since 1979.

https://goo.gl/kqfOzu

7.6 (cont'd) Evidence for Present Climate Change

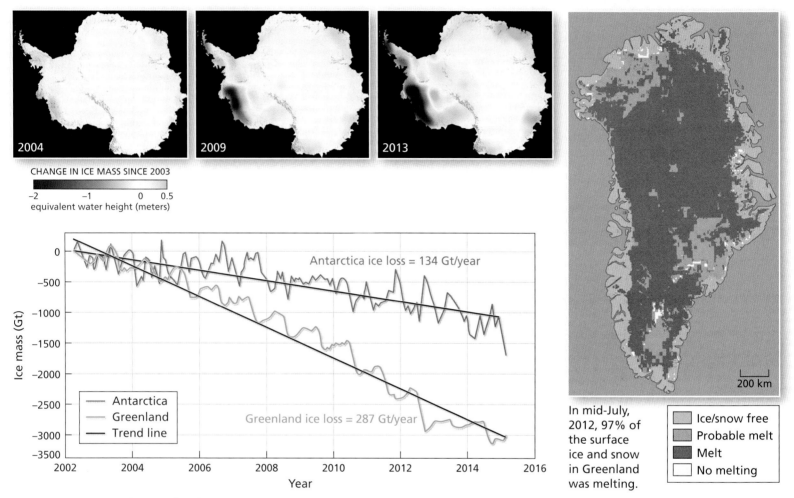

▲**7.24 Antarctic and Greenland surface temperatures and ice loss** Glaciers grow when annual winter snowfall exceeds annual summer melt; when snowfall and melting are equal, glacial mass balance is zero (see the discussion in Chapter 12).

Glacial Ice and Permafrost Land ice occurs in the form of glaciers, ice sheets, ice caps, ice fields, and frozen ground. These freshwater ice masses are found at high latitudes and high elevations. As temperatures rise in Earth's atmosphere, glaciers are losing mass and shrinking in size (▲Fig. 7.24).

Earth's two largest ice sheets, in Greenland and Antarctica, are also losing mass. Summer melt on the Greenland Ice Sheet increased 30% from 1979 to 2006. On average about half of the surface of the ice sheet now experiences some melting during the summer months. In July 2012, satellite data showed that 97% of the ice sheet's surface was covered by meltwater, the greatest extent in the 30-year satellite record. Recent satellite measurements of the West Antarctic Ice Sheet show that portions are in what scientists studying the ice sheet have called "irreversible decline." Part of these glaciers float on the ocean, forming ice shelves that are attached to the continental ice on land. As the ice shelves melt, the ice behind the shelves flows more quickly into the sea. If glacial retreat then

destabilizes other areas of the ice sheet, the overall sea-level rise could be much greater.

Permafrost (ground that remains frozen throughout the year) covers roughly 24% of the Northern Hemisphere. Permafrost is thawing in the Arctic at accelerating rates, releasing massive amounts of methane into the atmosphere (►Fig. 7.25). There are 800 gigatons of carbon in our atmosphere, but there are 1400 gigatons of carbon stored in the permafrost. Recall that methane is a greenhouse gas that contributes to global warming. Scientists now estimate that between one and two thirds of Arctic permafrost, which took tens of thousands of years to form, will thaw over the next 200 years, if not sooner. Warming land is also causing the thaw of methane hydrates stored in permafrost, amplifying the effect.

geoCHECK ✔ How do sea ice loss and glacial ice loss affect sea-level rise differently?

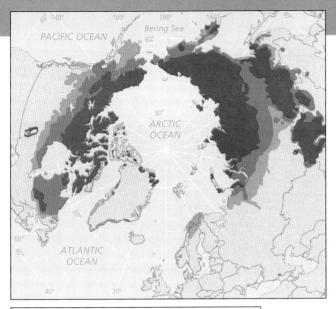

PROJECTED PERMAFROST THAWING

	Thawed by 2050		Still frozen in 2100
	Thawed by 2100		

(a) Projected regions of thawing permafrost.

▲**7.25 Thawing permafrost and climate change** As permafrost melts, the greenhouse gas methane is released from frozen Arctic soils.

(b) Trees fall over when their supporting permafrost thaws, creating "drunken forests."

Mobile Field Trip (MG)
Climate Change
in the Arctic

https://goo.gl/X0dlg8

(c) Methane lies under Arctic lakebeds and is highly flammable.

Rising Sea Levels

Scientists use mean sea level as the reference for elevations on Earth. **Mean sea level** is a value based on average tidal levels recorded hourly at a given site over many years. Although MSL varies globally, long-term sea level changes represent changes in the volume of water in the oceans, which are related primarily to temperature.

Sea level is now rising more quickly than scientists had predicted, and the rate appears to be accelerating (◄Fig. 7.26). During the last century, sea level rose 17–21 cm (6.7–8.3 in.). From 1993 to 2013, satellite data show that sea level rose 3.16 mm (0.12 in.) per year, twice as fast as the average rate from 1901 to 2010. Sea level is also rising unevenly. For instance, it is rising more quickly along the U.S. East Coast than along the U.S. West Coast (see Fig 7.42). Sea-level rise has already begun to affect low-lying Pacific islands, such as Tuvalu (▼Fig. 7.27).

Two thirds of current sea-level rise is due to the melting of ice on the land. The other one third comes from the expansion of seawater as it absorbs heat from the atmosphere.

geoCHECK ✔ What are the main causes of sea level rise?

▼**7.27 Sea-level rise in Tuvalu** This flooding was caused by the normal daily high tide, not storm surge.

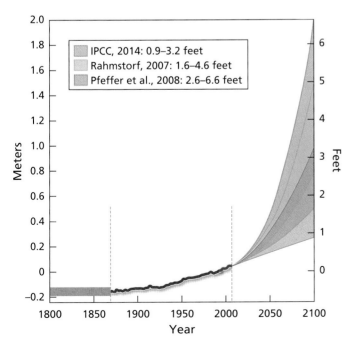

IPCC, 2014: 0.9–3.2 feet
Rahmstorf, 2007: 1.6–4.6 feet
Pfeffer et al., 2008: 2.6–6.6 feet

▲**7.26 Rate of global mean sea-level change, 1992–2013**
Sea-level trends as measured by *TOPEX/Poseidon*, *Jason-1*, and *Jason-2* satellites vary with geographic location.

7.6 (cont'd) Evidence for Present Climate Change

Increased Atmospheric Water Vapor

Since 1973, global average specific humidity has increased by about 0.1 g of water vapor per kilogram of air per decade. This change is consistent with rising air temperatures, since warm air has a greater capacity to absorb water vapor. More water vapor in the atmosphere can lead to "extreme" events involving temperature, precipitation, and storm intensity, since extreme weather such as hurricanes (▶Fig. 7.28), thunderstorms, and tornadoes (▼Fig. 7.29), are all powered by the latent heat of condensation (refer to Chapter 3). The Annual Climate Extremes Index (CEI) for the United States, which tracks extreme events since 1900, shows such an increase during the past four decades

▲7.29 **F4 tornado in South Dakota**

▲7.28 **Typhoon Maysak, May 31, 2015** This storm was the first of many record breaking tropical cyclones in 2015. Cyclones and tornadoes are powered by the latent heat of condensation so higher amounts of atmospheric water vapor results in more powerful storms.

(see the data at www.ncdc.noaa.gov/extremes/cei/graph/cei/01-12). Since 1959, precipitation falling during the heaviest rainfall events has increased, especially since 1991 (▼Fig 7.30a). Similarly, one index of hurricane strength has increased along with sea-surface temperatures in the North Atlantic (▼Fig. 7.30b).

▼7.30 **Increases in precipitation events and hurricane strength** Evidence suggests that (a) heavy precipitation events are becoming more frequent and (b) North Atlantic hurricanes are becoming stronger.

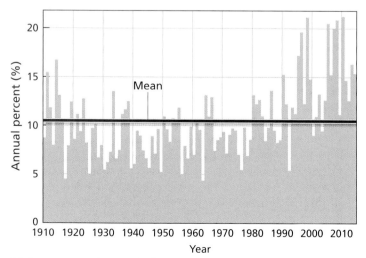

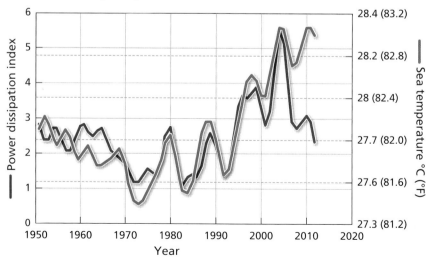

(a) Percentage changes in heavy precipitation events in the United States, 1910-2014. Heavy precipitation events are the heaviest 1% of all daily events.

(b) North Atlantic hurricane strength and sea surface temperature. This figure compares the Power Dissipation Index (PDI) with North Atlantic sea surface temperatures. The PDI includes cyclone strength, duration, and frequency. The lines are a five-year average, plotted at the middle year.

Extreme Events

According to the World Meteorological Organization, the decade from 2001 to 2010 showed evidence of a worldwide increase in extreme events—notably, heat waves, droughts increased precipitation, and floods (▼Fig. 7.31). Studies published in 2014 of extreme weather and climate events found that anthropogenic climate change was a contributing factor in Hurricane Sandy in 2012; super typhoon Haiyan in 2013; the European, Asian, and Australian heat waves of 2013; and the ongoing drought in California that began in 2013 (▼Fig. 7.32). However, to assess extreme weather trends and definitively link these events to climate change requires data for a longer time frame than is now available.

▲7.31 **2014 flooding in the United Kingdom**

 Why can't we say for certain that climate change is linked to extreme weather events?

geoQUIZ

1. How do present temperatures compare with temperatures during the past 120,000 years?
2. What would the consequences on global temperatures be if large amounts of permafrost melted? Explain.
3. What climate change factors would cause increased precipitation and flooding? What climate change factors would make droughts worse?

▼7.32 **Global pattern of drought, as measured by the drought index**

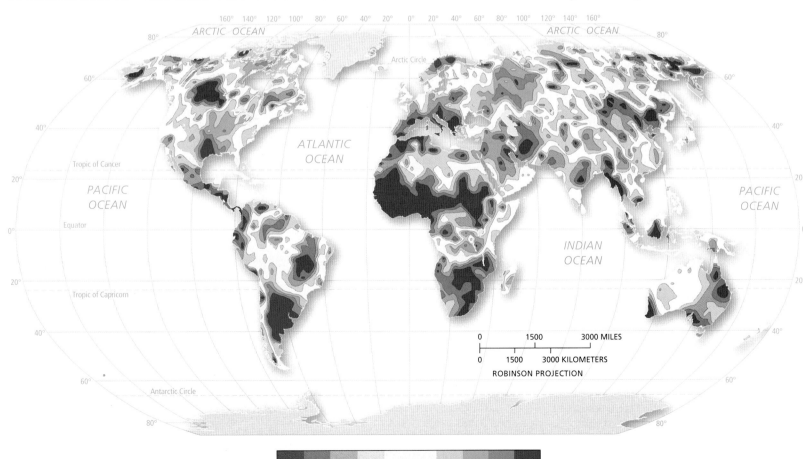

-4	-3	-2	-1	0	1	2	3	4	

DROUGHT SEVERITY INDEX
(PRECIPITATION TREND SINCE 1900)

7.7 Causes of Present Climate Change

Key Learning Concepts

▶ **Compare** the different greenhouse gases in terms of residence time and contribution to warming.

▶ **Evaluate** the warming or cooling effects of factors that affect Earth's energy balance.

RESIDENCE TIME AND GLOBAL WARMING POTENTIAL OF ANTHROPOGENIC GREENHOUSE GASES

Gas	CO_2	CH_4	N_2O	CFC-11	HFC-134a	CF_4
Residence time, years	Multiple	12	121	45	13	50,000
Global warming potential of a pulse of this greenhouse gas compared to CO_2						
After 20 years	1	86	268	7,020	3,790	4,950
After 100 years	1	34	298	5,350	1,550	7,350
After 500 years	1	8	153	1,620	435	11,200

▲7.33 **Residence times and global warming potential for greenhouse gases** The amount of warming for each gas is a function of how long each gas stays in the atmosphere, as well as how much heat each gas absorbs. The residence time in the atmosphere and global warming potential (GWP) relative to CO_2 are shown for each gas.

Scientists agree that rising concentrations of greenhouse gases in the atmosphere are the primary cause of recent worldwide temperature increases. CO_2 emissions from the burning of fossil fuels—primarily coal, oil, and natural gas—have increased with growing population and rising living standards. Changes in the amount of CO_2 in the atmosphere match changes in Earth's average surface temperature. After comparing the natural and anthropogenic factors, scientists have determined that the current warming is due to human actions.

Contributions of Greenhouse Gases

Scientists have known since the 1800s that CO_2 in the atmosphere absorbs heat and that higher levels of CO_2 result in warmer temperatures. Increasing concentrations of greenhouse gases absorb more longwave radiation, warming the atmosphere. Today's CO_2 and methane levels far exceed the natural range over the past hundreds of thousands of years.

The contribution of each greenhouse gas toward warming the atmosphere depends on how much energy the gas absorbs and how long the gas resides in the atmosphere. The primary greenhouse gases in Earth's atmosphere are water vapor (H_2O), carbon dioxide (CO_2), methane (CH_4), nitrous oxide (N_2O), and certain gases containing fluorine, chlorine, and bromine. Of these, water vapor is the most abundant. However, water vapor has a short residence time (about 90 days). CO_2, in contrast, has a longer residence time—50 to 200 years (▶Fig. 7.33).

Carbon Dioxide CO_2 emissions from human sources include burning fossil fuels, biomass burning (such as the burning of solid waste for fuel), forest removal, industrial agriculture, and cement production. Fossil-fuel burning accounts for over 70% of the total. CO_2 emissions rose an average of 1.1% per year from 1990 to 1999. From 2000 to 2010, the rate increased to 2.7% per year and is growing (▶Fig. 7.34a).

Recently, scientists have used carbon isotope analysis to determine that most of the CO_2 increase comes from humans' burning of fossil fuels. One scientist aware of the effects of CO_2 on temperature in the 1800s was Dr. Svante Arrhenius. In 1896 he calculated that the effect of doubled atmospheric CO_2 to be an increase of 4°C, remarkably close to the figure obtained by today's sophisticated computer models.

Methane After carbon dioxide, methane is the second most prevalent greenhouse gas produced by human activities. Today, methane levels are increasing more quickly than CO_2 levels. Methane levels over the past 800,000 years haven't exceeded 750 parts per billion (ppb) until modern times. The present levels are 1890 ppb (Fig. 7.34b).

Methane has a residence time of about 12 years in the atmosphere, much shorter than that of CO_2. However, over a 100-year timescale, methane is 25 times more effective at trapping heat than CO_2. About two thirds of

▼7.34 **Concentrations of carbon dioxide, methane, nitrous oxide, and fluorinated gases since 1978 and their main sources** Gas concentrations are in parts per billion (ppb) or parts per trillion (ppt), indicating the number of molecules of each gas per billion or trillion molecules of air

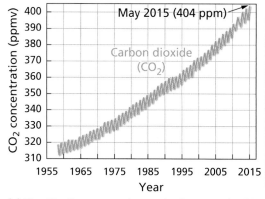

(a) The Keeling curve shows the increase in CO_2 due to burning fossil fuels and industries such as cement making, shown here.

atmospheric methane is from anthropogenic sources. Of the anthropogenic methane released, about 20% is from livestock; about 20% is from the mining of coal, oil, and natural gas, including shale gas extraction; about 12% is from rice farming; and about 8% is from the burning of vegetation in fires. Methane is released naturally from wetlands and termites.

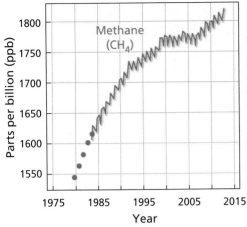

(b) Natural gas and petroleum systems are the main source of methane, with agriculture coming a close second.

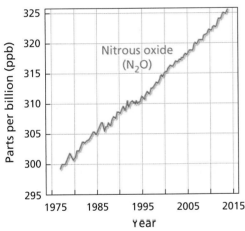

(c) Synthetic fertilizers are the main source of nitrous oxide emissions in the United States.

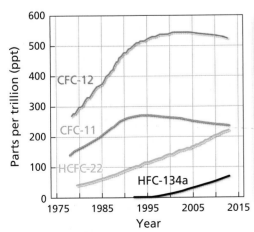

(d) Fluorinated gases are used as aerosol propellants and solvents and are also used in air conditioning and heating systems and refrigerators, like these shown at a disposal site in Germany.

Nitrous Oxide The third most important greenhouse gas produced by human activity is nitrous oxide (N_2O), which increased 19% in atmospheric concentration since 1750 and is now higher than at any time in the past 10,000 years (Fig. 7.34c). Nitrous oxide has a lifetime in the atmosphere of about 120 years—giving it a high global warming potential.

The recent rise in concentrations of nitrous oxide is mainly due to agricultural activities. Earth's nitrogen cycle, as well as wastewater management and fossil-fuel burning, release smaller amounts.

Chlorofluorocarbons and Related Gases Human activities produce certain gases containing fluorine, chlorine, or bromine. These gases are the most potent greenhouse gases with the longest atmospheric residence times. The most important of these are chlorofluorocarbons (CFCs). Atmospheric concentrations of CFC-12 and CFC-11 have decreased in recent years due to the Montreal Protocol (Fig. 7.34d; also see the discussion of stratospheric ozone in Chapter 1). However, hydrofluorocarbons, used as substitutes for CFCs, have been increasing since the early 1990s.

geoCHECK ✔ What are the two factors that determine the amount of warming generated by a greenhouse gas?

Video MG
Temperature & Agriculture

http://goo.gl/nlp7A2

7.7 (cont'd) Causes of Present Climate Change

Factors that Affect Earth's Energy Balance

We learned in Chapter 1 that Earth's energy balance is theoretically zero, meaning that the amount of energy arriving at Earth's surface is equal to the amount of energy eventually radiated back to space. However, Earth's climate has cycled through periods where this balance is not achieved, and Earth systems are either gaining or losing heat. Factors called **radiative forcings** (also called *climate forcings*) can contribute to a positive energy balance, leading to warming, or to a negative energy balance, leading to cooling.

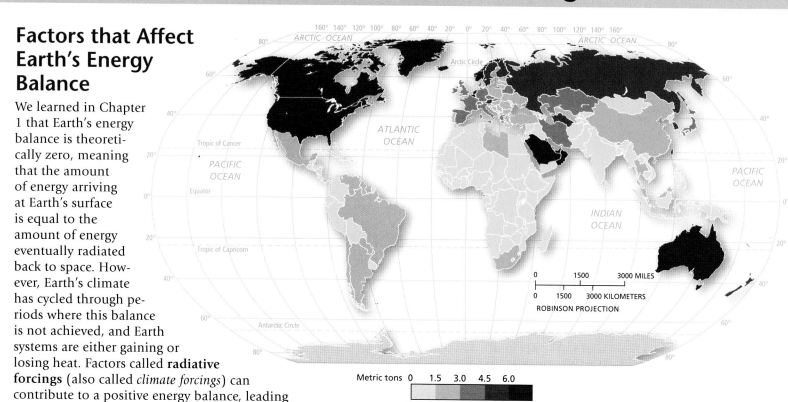

Metric tons 0   1.5   3.0   4.5   6.0

U.S. tons 0   1.6   3.3   5.0   6.6

TONS OF OIL EQUIVALENT (TOE) PER CAPITA, 2011

▲7.35 World energy consumption by region

Anthropogenic Greenhouse Gases An increase in greenhouse gases, largely from the burning of fossil fuels to power our global economy, has changed Earth's energy balance (▲ **Fig. 7.35**). Scientists have measured the radiative forcing of greenhouse gases on Earth's energy budget, using watts of energy per square meter of Earth's surface (W/m$_2$), since 1979. Figure 7.36, which compares the radiative forcing exerted by greenhouse gases, shows that CO_2 is the dominant gas affecting Earth's energy budget. On the right side of the figure is the Annual Greenhouse Gas Index (AGGI), as measured by the National Oceanic and Atmospheric Administration (NOAA), which reached 1.32 in 2012. This indicator converts the total radiative forcing for each gas into an index by using the ratio of the radiative forcing for a particular year compared to the radiative forcing in 1990 (the baseline year). The graph shows that radiative forcing has increased steadily for all gases, with the proportion attributed to CO_2 increasing the most.

Comparison of Radiative Forcing Factors The Intergovernmental Panel on Climate Change (IPCC; (see Module 7.9) has estimated the amount of radiative forcing of climate between the years 1750 and 2011 for a number of natural and anthropogenic factors (▶ **Fig. 7.37**). This analysis revealed that greenhouse gases are responsible for about 73% of the total atmospheric warming. The second most important positive forcing factor is tropospheric ozone (in contrast to stratospheric ozone, which has a negative forcing, or cooling, effect on climate). Other factors causing positive forcing are tropospheric ozone, stratospheric water vapor, black carbon on snow, and contrail cirrus clouds. The highest negative forcing of climate is from anthropogenic

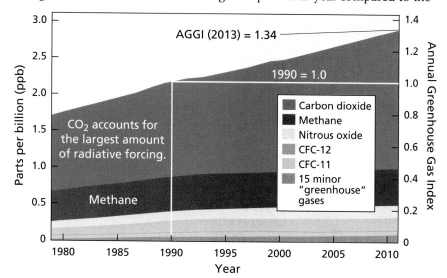

◀**7.36 Greenhouse gases: relative percentages of radiative forcing** The colored areas indicate the amount of radiative forcing accounted for by each gas, based on the concentrations present in Earth's atmosphere. Note that CO_2 accounts for the largest amount of radiative forcing. The right side of the graph shows radiative forcing converted to the Annual Greenhouse Gas Index (AGGI), set at a value of 1.0 in 1990. In 2013, the AGGI was 1.34, an increase of over 30% in 22 years.

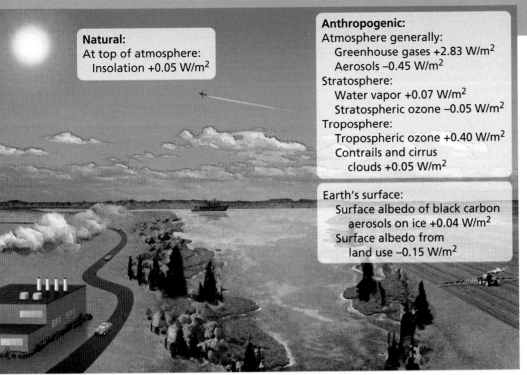

Natural:
At top of atmosphere:
Insolation +0.05 W/m²

Anthropogenic:
Atmosphere generally:
Greenhouse gases +2.83 W/m²
Aerosols −0.45 W/m²
Stratosphere:
Water vapor +0.07 W/m²
Stratospheric ozone −0.05 W/m²
Troposphere:
Tropospheric ozone +0.40 W/m²
Contrails and cirrus
clouds +0.05 W/m²

Earth's surface:
Surface albedo of black carbon
aerosols on ice +0.04 W/m²
Surface albedo from
land use −0.15 W/m²

▲7.37 **Forcing values** Positive and negative forcing values for natural and anthropogenic factors. Positive forcing leads to climatic warming; negative forcing leads t cooling.

aerosols such as sulfates, with a smaller amount coming from natural aerosols such as sea spray or dust particles.

The IPCC analysis included one important natural forcing factor for climate—insolation. The overall effect of solar output was a positive forcing of 0.05 W/m², a small amount compared to the overall 2.3 W/m² of forcing caused by the combined anthropogenic factors in the analysis.

Adding It All Up Figure 7.38 compares global temperature anomalies, insolation, volcanic activity, natural climate variability, and anthropogenic contributions. Global surface temperatures have increased by 0.8° C from 1880 to the present (▶Fig. 7.38a). Insolation values vary on an 11-year cycle and are currently in the low-output period of the cycle (▶Fig. 7.38b). Volcanic activity has accounted for several periods of lower temperatures, especially from the eruption of Krakatau in 1883 and Mount Pinatubo in 1990 (▶Fig. 7.38c). Internal variability in climate is largely due to El Niño events. This variability has produced temperature anomalies of ±0.2° C. The effects of El Nino on higher temperatures can be seen here, especially during the 1997–1998 El Niño (▶Fig. 7.38d). The largest and most important factor is the anthropogenic component, which includes warming from greenhouse gases and cooling from atmospheric aerosols (▶Fig. 7.38e). From this analysis it is clear that the dominant factor in our current climate change is human contributions of greenhouse gases to positive radiative forcings.

geoCHECK What climate forcing factor has the highest positive radiative forcing? The lowest radiative forcing?

geoQUIZ

1. How do scientists know that most of the CO₂ currently emitted is from the burning of fossil fuels?
2. Compare the forcing effects of CO₂ and methane. How do they compare in terms of forcing and residence time?
3. Compare the amount of anthropogenic forcing with the amount of natural forcing. Which is larger, and by how much?

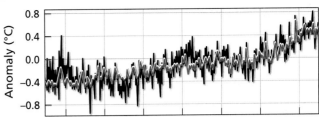

(a) Comparison of actual temperature anomalies (black line) to modeled temperature anomalies (red line) for the time period 1870-2010. The zero line represents the average global surface temperature for 1961 to 1990. The red line represents the sum of the natural (b, c, d) and anthropogenic factors (e).

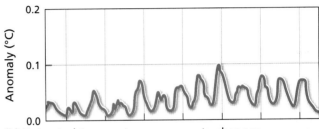

(b) Estimated temperature response to changes in insolation.

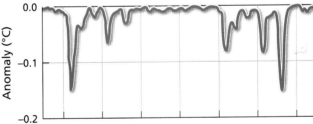

(c) Estimated temperature response to volcanic activity.

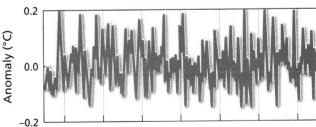

(d) Estimated temperature variability due to the El Niño-Southern Oscillation.

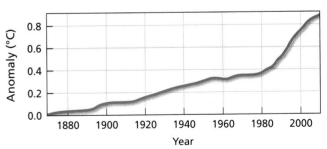

(e) Estimated temperature response to anthropogenic forcing, consisting of warming from greenhouse gases and cooling from aerosols.

▲7.38 **Global surface temperature anomalies, 1870–2010** Natural (solar, volcanic, and internal) and anthropogenic factors influenced the anomalies on the graphs. Note that the graphs have different scales on the vertical axis. Positive anomalies indicate warming and negative anomalies indicate cooling.

7.8 Climate Models & Forecasts

Key Learning Concepts

▶ **Discuss** how climate models have been used to analyze and predict climate change.

▶ **Summarize** some climate projections based on general circulation models.

Scientists use records of the past, current measurements, and computer models of climate to study past climates and forecast future changes. A climate model is a mathematical representation of the interacting factors that make up Earth's climate systems, including the atmosphere, oceans, and ice. Some of the most complex computer models are **general circulation models (GCM)**, which were originally based on weather-forecasting models.

Climate models divide the planet into a series of three-dimensional boxes (▼**Fig. 7.39**). These boxes extend upward into the atmosphere and downward into the oceans. Each box begins with a defined set of initial conditions. The GCM then calculates the interactions between these boxes over time in order to model the climate.

Many GCMs have submodels that represent the atmosphere, ocean, land cover, ice and snow, and the biosphere. The most sophisticated models, *Atmosphere–Ocean General Circulation Models (AOGCMs)*, join atmosphere and ocean submodels with other types of submodels. At least a dozen GCMs are now in operation around the world. As computing power has grown, so has the sophistication and resolution of GCMs (▼**Fig. 7.40**). Early GCMs had a spatial resolution of 1000 km (600 mi) and between 2 and 10 vertical levels, but the latest models have resolutions as small as 100 km (60 mi) with up to 20 vertical levels.

Radiative Forcing Scenarios

Scientists can use GCMs to answer the question of whether the current warming is due to anthropogenic or natural causes. When scientists run climate simulations comparing the effects of anthropogenic and natural factors, the most accurate simulations are those that include both sets of factors. Models that include only the natural factors of

Horizontal grid
(Latitude-longitude)

Vertical grid
(Height or pressure)

The initial conditions for each box represent the temperature, air pressure, humidity, wind speed and direction, cloud type, and other attributes, such as water temperature and currents, for that box.

Shortwave solar radiation; includes visible light

Longwave radiation; heat

Atmosphere

Clouds

Exchange of heat & CO_2

Vegetation

Exchange of heat & CO_2

Soil

Sea Ice

Ocean

Lithosphere

Ice Sheets

▲7.39 **General circulation model (GCM)**

▼7.40 **NASA's Discover supercomputer is made of almost 80,000 cores**

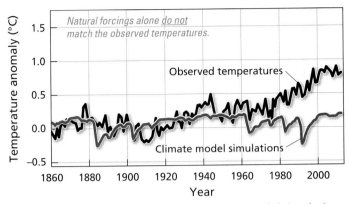

Natural forcings alone *do not* match the observed temperatures.

Observed temperatures

Climate model simulations

(a) Natural forcing. The red line shows climate model simulations of only natural forcings. The black line is observed temperatures. The natural forcing factors include solar activity and volcanic activity, which alone do not explain the temperature increases.

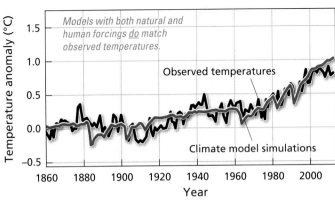

Models with both natural and human forcings *do* match observed temperatures.

Observed temperatures

Climate model simulations

(b) Natural and human forcing. The red line shows climate model simulations with both natural and human forcings. The black line is observed temperatures.

▲7.41 **Natural and human forcing** Scientists believe the recent warming is because of both natural and human factors.

solar variability and volcanoes do not accurately model the current warming trend (▲Fig. 7.41). The IPCC Fifth Assessment Report concluded that it is 99% probable that after 1950, humans are the strongest changers of climate.

geoCHECK ✔ How accurate are models that look only at natural forcing factors, compared with models that look at natural and anthropogenic forcing factors?

Future Temperature Scenarios

GCMs do not predict specific temperatures, but they do offer different scenarios of future global warming based on choices that society makes, such as how quickly we limit greenhouse gas emissions. GCM projections match warming patterns observed since 1990, and various forecast scenarios predict temperature change during this century. Figure 7.44 lists future conditions, based upon the IPCC's confidence levels of 99% confident and 90% confident.

Figure 7.42 shows two temperature scenarios presented in the IPCC Fifth Assessment Report. Each Representative Concentration Pathway, or RCP, is identified by the approximate radiative forcing for the year 2100 as compared to 1750. For example, RCP2.6 is based on 2.6 W/m² of forcing, which is higher than the current net forcing of 2.3 W/m². Each RCP is based on societal choices that lead to certain levels of greenhouse gas emissions, land use, and air pollutants that combine to produce the forcing value. For RCP2.6—the lowest level—forcing peaks before 2020 and declines before 2100. RCP4.5 (not shown) represents forcing peaking in 2040 and then declining. For RCP6.0 (not shown), radiative forcing peaks in 2080, and RCP8.5 assumes that emissions continue to rise through the 21st century.

geoCHECK ✔ What do the numbers in the RCPs (such as RCP4.5) indicate?

Animation MG
Supercomputing the climate

http://goo.gl/r0763E

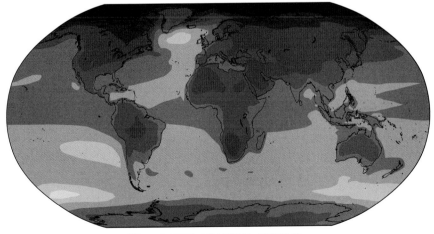

(a) STRONG MITIGATION (RCP2.6)

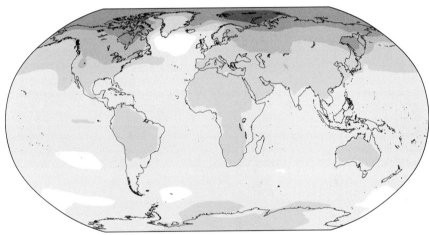

(b) BUSINESS-AS-USUAL (RCP8.5)

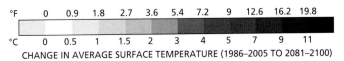

°F 0   0.9   1.8   2.7   3.6   5.4   7.2   9   12.6   16.2   19.8

°C 0   0.5   1   1.5   2   3   4   5   7   9   11

CHANGE IN AVERAGE SURFACE TEMPERATURE (1986–2005 TO 2081–2100)

▲7.42 **Model-projected warming by the end of this century for (a) strong mitigation and (b) business as usual scenarios**

7.8 (cont'd) Climate Models & Forecasts

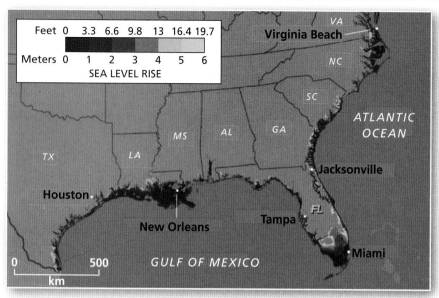

Video MG
Superstorm Sandy
https://goo.gl/k6HaNa

▲7.43 U.S. coastal areas and sea-level rise The map shows areas that would be inundated with a 1–6 m (3.2–19.6 ft) rise in sea level. Labeled cities in vulnerable areas have populations greater than 300,000 (as of the 2000 Census). Even a small sea-level rise would bring higher tides and higher storm surges to many regions, particularly lowland coastal farming valleys, and densely populated cities such as New York, Miami, and New Orleans in the United States, Mumbai and Kolkata in India, and Shanghai in China.

Sea-Level Projections

In 2012, NOAA scientists developed scenarios for sea-level rise based on present ice sheet, mountain glacier, and ice cap losses worldwide (Table 7.1). In the United States, sea level is rising more quickly on the East Coast than

▼7.44 Hurricane Sandy floods New York City Higher sea level enhanced coastal flooding during Hurricane Sandy in 2014.

Table 7.1 Global sea-level rise scenarios

Scenario	Sea-level rise by 2100, in meters (ft)	Potential effects
High	2.0 (6.6)	Over 500 million people displaced
Intermediate to high	1.2 (3.9)	
	1.0 (3.2)	130 million people displaced; some Pacific island nations uninhabitable
Intermediate to low	0.5 (1.6)	Property valued at $6 trillion exposed to coastal flooding in Baltimore, Boston, New York, Philadelphia, and Providence, RI.
	0.3 (1.0)	Shoreline retreat of 30 m (98 ft.)
Low	0.2 (0.7)	

on the West Coast. The largest areas of inundation at a rise of only 1 m are around the low-elevation cities of Miami and New Orleans (◄Fig. 7.43). Many states, including California, are using a 1.4-m sea-level rise this century as a standard for planning purposes.

Even a small sea-level rise would bring higher tides and higher storm surges to many regions, particularly lowland coastal farming valleys and densely populated cities such as New York (◄Fig. 7.44), Miami, and New Orleans in the United States; Mumbai and Kolkata in India; and Shanghai in China. The social and economic consequences will especially affect small, low-elevation island states. A flood of environmental refugees driven by climate change could continue for decades. Sea-level increases will continue beyond 2100, even if greenhouse gas concentrations were stabilized today. To see an interactive map of the effects of rising sea level on different areas of the world, go to http://geology.com/sea-level-rise/.

Possible Climate Futures

Overall, how will climate changes throughout the 21st century affect Earth systems? Synthesizing vast amounts of data and related analyses, the IPCC has projected a wide range of effects attributable to climate change on weather events, permafrost and glacial ice, ocean chemistry and circulation–and more. The IPCC rated these projections according to the likelihood that they will occur (►Fig. 7.45). Impacts on our water supply, for example, will undoubtedly have serious implications for Earth's growing human population later in the century (▼Fig. 7.46).

geoCHECK ✔ What would happen to future sea-level rise if greenhouse gas emissions were stabilized right now?

geoQUIZ

1. Briefly describe how a GCM works.
2. Which RCP is closest to matching our current radiative forcing? Which RCP is closest to matching our projected radiative forcing, based upon our current efforts to reduce greenhouse gas emissions?
3. Compare the effects on human populations of 0.5 m, 1.0 m, and 2.0 m of sea-level rise.

IPCC PROJECTIONS FOR THE LATE 21ST CENTURY

This table outlines the IPCC's projections for the late 21st century, ranked in decreasing order of certainty.

	VIRTUALLY CERTAIN
• Cold days and nights will be warmer and less frequent over most land areas • Hot days and nights will be warmer and more frequent over most land areas • The extent of permafrost will decline • Ocean acidification will increase as the atmosphere accumulates CO_2 • Northern hemisphere glaciation will not initiate before the year 3000 • Global mean sea level will rise and continue to do so for many centuries	**99% certain**
• Arctic sea ice cover will continue to shrink and thin, and northern hemisphere spring snow cover will decrease • The dissolved oxygen content of the ocean will decrease by a few percent • The rate of increase in atmospheric CO_2, methane, and nitrous oxide will reach levels unprecedented in the last 10,000 years • The frequency of warm spells and heat waves will increase • The frequency of heavy precipitation events will increase • Precipitation amounts will increase in high latitudes • The ocean's conveyor-belt circulation will weaken • The rate of sea level rise will exceed that of the late 20th century • Extreme high sea-level events will increase, as will ocean wave heights of mid-latitude storms	**VERY LIKELY** **90% certain**

▲7.45 **Scenarios for surface warming during this century** Temperature change is relative to the 1986–2005 average temperature.

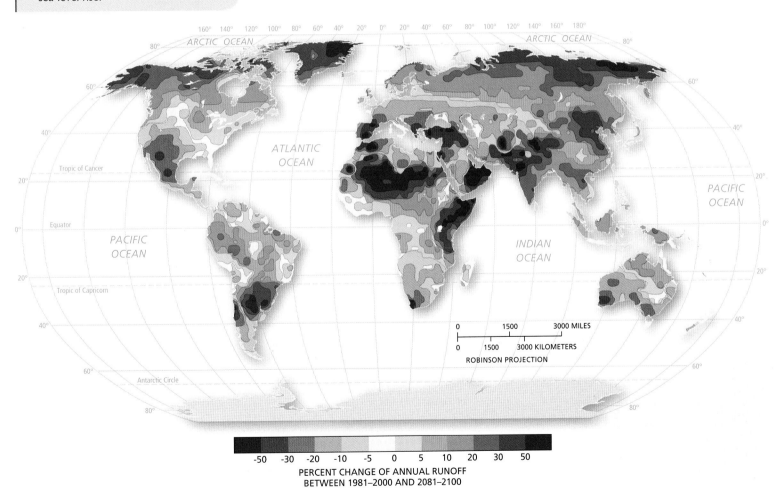

PERCENT CHANGE OF ANNUAL RUNOFF
BETWEEN 1981–2000 AND 2081–2100

▲7.46 **Future climate change impacts on water** As the distribution of precipitation changes, patterns of runoff will shift as well, leading to water shortages in some areas while increasing the danger of flooding and landslides in other areas.

7.9 Debating Climate Change

Key Learning Concepts

▶ **Describe** the scientific consensus on climate change and the reasons why some remain skeptical about it.

▶ **Discuss** some of the key questions about climate change that scientists answered using scientific methods.

Over several decades, thousands of climate scientists worldwide made observations, collected and analyzed data, developed computer models, and tested hypotheses in an effort to understand climate change. As a result of these efforts, climate scientists have now reached an overwhelming consensus that human activities are causing climate change (▶Fig. 7.47).

Weighing the Evidence

Since its establishment in the late 1980s, the IPCC is the world's foremost scientific group reporting on climate change. The IPCC has presented the emerging consensus on climate change in a series of reports drawing on international scientific expertise from many scientific disciplines. The 2013–2014 IPCC Fifth Assessment Report announced that scientists are now 95%–100% certain that human activities are the primary cause of present climate change. Numerous policy statements and position papers from professional organizations (for example the American Association for the Advancement of Science, the American Chemical Society, the American Geophysical Union, the American Meteorological Society, the Association of American Geographers, and the Geological Society of America) also support this consensus (▼Fig. 7.48). Not one national or major scientific organization disputes the consensus view on anthropogenic climate change.

The IPCC, sponsored by the United Nations and the World Meteorological Organization (WMO), is a global collaboration

▲**7.47 International cooperation to address climate change** UN Secretary General Ban Ki Moon (center), France's President François Hollande (third from right), and UN climate leaders celebrate after the adoption of a historic global warming pact at the COP21 Climate Conference in Paris, on December 12, 2015.

of scientists and policy experts that coordinate global climate change research. Its reports represent peer-reviewed, consensus opinions among experts in the scientific community. In 2007, the IPCC shared the Nobel Peace Prize for its two decades of work on global climate change science. Table 7.2 summarizes some important findings from the 2013–2014 IPCC Fifth Assessment Report.

geoCHECK ✔ How certain is the IPCC that humans are changing the climate?

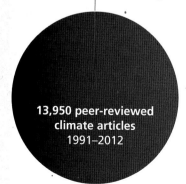

24 reject global warming

13,950 peer-reviewed climate articles 1991–2012

▲**7.48 Scientific consensus on global warming** The overwhelming majority of scientific papers and their authors agree that humans are responsible for warming the planet.

Table 7.2 Summary points for the working groups of the IPCC Fifth Assessment Report*

Working Group I: *The Physical Science Basis*	Working Group II: *Impacts, Adaptation, Vulnerability*	Working Group III: *Mitigation of Climate Change*
Warming of the climate system is unequivocal and it is extremely likely that humans are the dominant cause (95%–100% certainty). Many observed changes since the 1950s (warming atmosphere and oceans, melting snow and ice, rising sea level, rising greenhouse gas concentrations) are unprecedented over decades to millennia.	Human interference with the climate system is occurring, and climate change poses risks for human and natural systems. Increasing magnitudes of warming increase the likelihood of severe, pervasive, and irreversible impacts on species, ecosystems, crop yields, human health, natural hazards, and food security.	Effective mitigation of climate change requires collective action at the global scale because most greenhouse gases, (GHGs) accumulate over time and mix globally. International cooperation is needed to effectively mitigate GHG emissions and address other climate change issues.

*The free *Summary for Policy Makers* for each AR5 Working Group is available at www.ipcc.ch/report/ar5/.

Skepticism

Despite the consensus among scientists, the general public often thinks that there is still considerable controversy around climate change. Some disagree that climate change is occurring, while others doubt that its cause is anthropogenic. The fuel for the continuing "debate" on this topic appears to come, at least partly, from media coverage that at times is biased, alarmist, or factually incorrect. The bias often reflects the influence of special-interest groups, such as corporate interests whose financial gains are at stake if climate change solutions are imposed. In other cases, errors come from misinterpretation of the facts, sometimes as interpreted and sensationalized by blogs and other social media.

For example, the winter of 2014 brought unusually cold weather to the northeastern United States, leading some to conclude that the cold spell meant global warming must not be real. In response, climate scientists have pointed out that worldwide, 2014 was on average the warmest year since temperature record keeping began, a fact that outweighs the colder than normal temperatures in a particular region (▲Fig. 7.49).

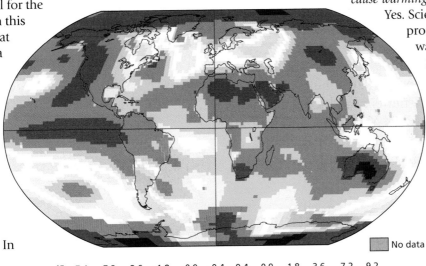

°F -7.4 -7.2 -3.6 -1.8 -0.9 -0.4 0.4 0.9 1.8 3.6 7.2 9.2

°C -4.1 -4 -2 -1 -0.50 -0.2 0.2 0.5 1.0 2.0 4.0 5.1

OCT 2015 SURFACE TEMPERATURE ANOMALY VS 1951–1980

No data

▲**7.49 Large-scale temperature pattern in January 2014** Interpreting the global distribution of temperatures in 2014 might depend on your location.

 geoCHECK ✔ What are some of the reasons that individuals are skeptical about the reality of climate change?

Taking a Position on Climate Change

Having an informed position on climate change requires an understanding of Earth's physical laws and system operations and an awareness of the scientific evidence and ongoing research. In the Introduction, we discussed the scientific process, which encourages peer evaluation, criticism, cautious skepticism, and a willingness to reevaluate evidence and formulate new hypotheses. Many would argue that skepticism concerning climate change is simply part of this process. However, climate scientists are overwhelmingly in agreement: The case for anthropogenic climate change

has become more convincing as scientists gather new data and complete more research and as we witness actual events in the environment. When considering the facts behind climate change, several key questions can help guide you to an informed position based on a scientific approach:

- *Does increasing atmospheric carbon dioxide in the atmosphere cause warming temperatures?*
  Yes. Scientists have understood the physical processes related to atmospheric CO_2 warming the lower atmosphere for almost 100 years, long before the effects of global warming became apparent.
- *Does the rise of global temperatures cause global climate change?*
  Yes. Global warming is an unusually rapid increase in Earth's average surface temperature. Scientists know, based on physical laws and empirical evidence, that global warming changes precipitation patterns, causes ice melt, lengthens growing seasons, and appears to increase extreme weather events.
- *Have human activities increased the amount of greenhouse gases in the atmosphere?*
  Yes. As discussed earlier, scientists have established that the recent increase in atmospheric CO_2 originates from fossil-fuel burning and other human activities. They now know that human sources account for a large and growing percentage of these CO_2 concentrations.
- *If climate change on Earth has occurred in the past, then why are the present conditions problematic?*
  CO_2 concentrations are today rising more quickly than is seen throughout most of the long-term climate record. This is the first time in human history that 7.3 billion people are depending upon the climate to not change and radically disrupt the agricultural systems that we all depend upon for food.
- *Can scientists definitively attribute the changes we are seeing in climate (including extreme events and weather anomalies) to anthropogenic causes alone?*
  Not with 100% certainty, although an increasing amount of research indicates this to be the case.

geoCHECK ✔ Why are scientists concerned by the present climate change, given that climate has changed in the past?

geoQUIZ

1. Why do you think there is ongoing debate around the science of climate change?
2. Why does the IPCC believe that action on climate change requires international cooperation?

7.10 Addressing Climate Change

Key Learning Concepts

▶ **Describe** several mitigation measures to slow rates of climate change.

▶ **List** actions individuals can take to help reduce greenhouse gas emissions.

Given the scientific consensus on climate change, taking action must focus on lowering atmospheric CO_2. Opportunities to reduce CO_2 are readily available and have additional *cobenefits* beyond slowing climate change, such as cost savings. For example, the cobenefits of reducing greenhouse gas emissions include improved air quality and human health; reduced oil import costs and related oil-tanker spills; and increased renewable and sustainable energy development.

Climate Change Action: Global, National, & Local

In 1997, 84 countries signed the *Kyoto Protocol*, a legally binding agreement to reduce emissions of greenhouse gases. The United States signed the treaty, but never ratified it. (See http://unfccc.int/kyoto_protocol/items/2830.php.) Since the Kyoto conference, there has been only limited international progress. However, there has been great progress at the national and local levels.

For example, Germany and Denmark have plans to generate 60% and 100% of their electrical power from renewable sources by 2050. Many large urban areas are also working to reduce greenhouse gas emissions. According to the nonprofit Carbon Disclosure Project (CDP), 110 cities reported in 2013 that actions to mitigate climate change have saved money, attracted new businesses, and improved the health of residents. (See the CDP report at www.cdproject.net/CDPResults/CDP-Cities-2013-Global-Report.pdf.)

Delaying action on climate change may cost much more than acting now. A U.S. Department of Energy study found that the United States could meet the targets of the Kyoto Protocol with cash savings ranging from $7 billion to $34 billion a year.

On a collective level, human society must strive to mitigate and adapt to climate change. Farmers can use methods that retain more carbon in the soil and plant crop varieties bred to withstand heat, drought, or flooding (▶Fig. 7.50). Coastal developers' plans must allow for rising sea levels. All societies can promote efficient water use, especially in areas prone to drought. These are only a few examples among many for mitigating and adapting to climate change. (For more information, see www.unep.org/climatechange/mitigation/ and http://climate.nasa.gov/solutions.)

Governments and businesses are now planning for climate change impacts. For example, New York City is spending over $1 billion, in response to the damage from Hurricane Sandy in 2012, to raise flood walls, bury equipment, and make other changes to prevent future damage from extreme weather events. The potential for water scarcity

and food shortages in hotter climates with more frequent droughts is emerging as a critical issue for the global community. However, unless greenhouse gas emissions are curbed substantially, the effects of climate change on Earth systems could outpace efforts to adapt. If efforts to achieve broad, global reductions in greenhouse gas emissions fail, then large-scale geoengineering projects might hold some promise on mitigating climate change (▼Fig. 7.51).

geoCHECK ✔ How much money would the United States have spent or saved if we had implemented the Kyoto Protocol?

▲**7.50 Sustainable agriculture to reduce greenhouse emissions** No-till farming leaves crop residue on the fields, rather than plowing it under. No-till farming increases carbon content in the soil, reduces nitrous oxide emissions, provides cover for beneficial insects, reduces water use, and reduces soil erosion.

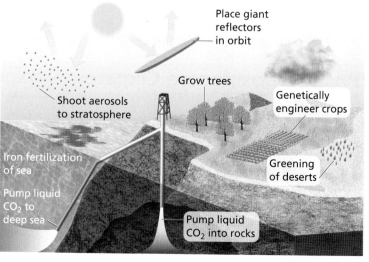

Place giant reflectors in orbit

Grow trees

Genetically engineer crops

Shoot aerosols to stratosphere

Iron fertilization of sea

Greening of deserts

Pump liquid CO_2 to deep sea

Pump liquid CO_2 into rocks

▲**7.51 Geoengineering solutions to climate change**

Climate Change Action: What Can You Do?

On an individual level, what can you do to address climate change? The principal way to slow the pace of climate change—from the individual to the international level—is to reduce carbon emissions from burning fossil fuels. One way to begin is to examine the sustainability of your daily practices and reduce your carbon footprint (▼Fig 7.52). Visit MasteringGeography™ to find links to online footprint calculators that will allow you to calculate your carbon emissions (carbon footprint) or your spatial requirements (ecological footprint), based on your lifestyle. One important goal is to reduce atmospheric concentrations of CO_2 to 350 ppm.

Approaching a "Climatic Threshold" Present greenhouse gas concentrations will remain in the atmosphere for many decades to come, but the time for action is now. Scientists describe 450 ppm as a possible *climatic threshold* at which Earth systems would transition into a chaotic mode, and 350 ppm as a long-term goal. With accelerating CO_2 emissions, this 450 ppm threshold could occur as soon as the 2020s. The goal of avoiding this threshold is slowing the rate of change and delaying the worst consequences of our current path. The information presented in this chapter is offered in the hope of providing motivation and empowerment—personally, locally, regionally, nationally, and globally.

geoCHECK ✔ Which of the changes above could you make most easily? Which change would be most difficult?

Use energy wisely
Turn down heat and turn up air conditioning settings when you leave the house; insulate your home

Drive less; walk and bike more
Savings: One lb of carbon for every mile

Reduce; reuse; recycle
Savings: Every 2 glass bottles recycled saves 2 lbs of carbon

geoQUIZ

1. What are cobenefits? Briefly discuss several cobenefits that would come from reducing greenhouse gas emissions.
2. What is the current level of CO_2? What level is a major potential tipping point? Explain.
3. What actions have your city or state taken to reduce climate change? What do you think the next step for your city should be?

Use Energy Star products
Look for the Energy Star label on appliances, electronics, light bulbs, and heating and cooling equipment. One compact fluorescent bulb saves 100 lbs of carbon.

Goal:
350 ppm CO_2
in atmosphere

Landscape wisely; plant trees
Plant trees to shade your house. Savings: 700 to 7000 lbs of carbon over the tree's lifetime

Website MG
Carbon Footprint Calculator

https://goo.gl/uXo7V0

Use renewable energy
Purchase renewable energy for your house, or generate your own power using solar or wind

Think globally; act locally
Buy local produce and other food; reduce meat consumption

◄7.52 Individual actions to reduce atmospheric CO_2 Which of these suggested actions are already part of your lifestyle? Can you expand on these possibilities for reducing your personal contribution to greenhouse gas emissions? All of your actions and decisions have positive and negative consequences for Earth's environment and for our changing climate. Remember that actions taken on an individual level by millions of people can be effective in slowing climate change for present and future generations.

CLIMATE CHANGE IMPACTS HUMANS

- Climate change affects all Earth systems.
- Climate change drives weather and triggers extreme events, such as drought, heat waves, storm surge, and sea-level encroachment, which cause human hardship and fatalities.

HUMANS IMPACT CLIMATE CHANGE

- Anthropogenic activities produce greenhouse gases that alter Earth's radiation balance and induce climate change.
- Cost-effective greenhouse gas reduction strategies are ready and could slow the rate of climate change.

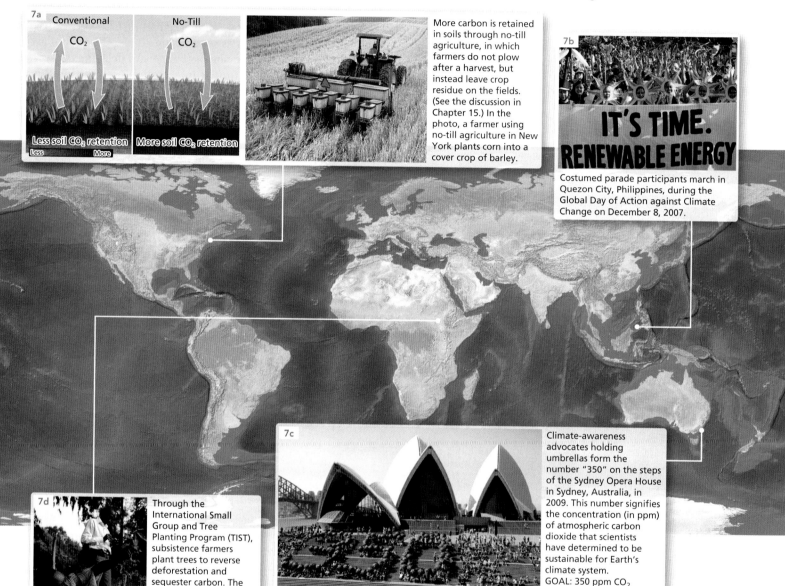

7a

Conventional | No-Till

CO_2

Less soil CO_2 retention Less   More

More soil CO_2 retention

More carbon is retained in soils through no-till agriculture, in which farmers do not plow after a harvest, but instead leave crop residue on the fields. (See the discussion in Chapter 15.) In the photo, a farmer using no-till agriculture in New York plants corn into a cover crop of barley.

7b

IT'S TIME. RENEWABLE ENERGY

Costumed parade participants march in Quezon City, Philippines, during the Global Day of Action against Climate Change on December 8, 2007.

7c

Climate-awareness advocates holding umbrellas form the number "350" on the steps of the Sydney Opera House in Sydney, Australia, in 2009. This number signifies the concentration (in ppm) of atmospheric carbon dioxide that scientists have determined to be sustainable for Earth's climate system.
GOAL: 350 ppm CO_2

7d

Through the International Small Group and Tree Planting Program (TIST), subsistence farmers plant trees to reverse deforestation and sequester carbon. The farmers are earning greenhouse gas credits that translate into small cash stipends. Planting trees is a simple and effective way to help combat climate change.

ISSUES FOR THE 21ST CENTURY

How will human society curb greenhouse gas emissions and mitigate change effects? Some examples:

- Using agricultural practices that help soils retain carbon.
- Planting trees to help sequester carbon in terrestrial ecosystems.
- Reducing use of fossil fuels, supporting renewable energy, and changing lifestyles to use fewer resources.
- Protecting and restoring natural ecosystems that are carbon reservoirs.

Looking Ahead

In the next chapter, we investigate Earth's geologic history, internal structure, and plate tectonics. The theory of plate tectonics provides a unified explanation for the patterns of earthquakes and volcanoes, as well as mountain building processes.

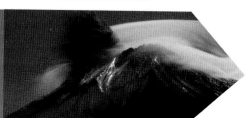

Chapter 7 Review

How do we know about past climates?

7.1 Deciphering Past Climates

Describe scientific tools used to study paleoclimatology.

- To understand present climate changes, we need to understand past climates and how scientists study past climates. The study of past climates is *paleoclimatology*. Since scientists do not have direct measurements for past climates, they use proxy methods, information about past environments, such as geologic evidence and ice or ocean sediment cores. Earth's climate has cycled between periods both colder and warmer than today over the past several million years, including extended cold spells called ice ages.

1. Why are proxy methods important for studying past climates?

2. Give two examples of climate proxies and how they are used.

7.2 Climates of Earth's Past

Summarize Earth's climatic history since the Paleocene-Eocene thermal maximum.

- Over the span of 70 mya to 50 mya. Earth's climate was much warmer than today. The climate has generally cooled since then. A short period of rapid warming occurred 56 mya, called the PETM, might have been caused by the release of methane hydrates. Over the past 5 mya there have been a series of warmer and cooler periods.

3. Explain why understanding the PETM is important in understanding current climate change. Compare the time scales of both events.

7.3 Climates of the Last 20,000 Years

List methods used to reconstruct climates during the relatively recent past.

Summarize major climate trends and events over the past 20,000 years.

- The tools for short-term climate analysis consist mainly of isotope analysis, radiocarbon dating, and the growth rings of trees. Data from lake cores, limestone in caves, and corals are also used.

- The Pleistocene, the most recent period of glaciations, began 2.5 mya The last glacial period lasted from 110,000 to about 11,700 years ago, with the last glacial maximum occurring about 20,000 years ago.

4. How can trees that live for only a few thousand years be used to recreate the past 10,000 years of climate?

5. Summarize the general temperature trends and the possible causes for those trends, for the past 1000 years.

Why has climate changed in the past?

7.4 Mechanisms of Natural Climate Change

Analyze several natural factors that influence Earth's climate.

- Several natural mechanisms can cause climatic fluctuations. The Sun's output varies over time, but this variation has not been definitely linked to climate change. Earth's orbital cycles and Earth–Sun relationships, called Milankovitch cycles, appear to affect Earth's glacial and interglacial cycles. Continental position and atmospheric aerosols, such as those produced by volcanic eruptions, are other natural factors that affect climate.

6. List and explain four possible mechanisms of natural climate change.

7.5 Climate Feedbacks & the Carbon Budget

Explain two main climate feedbacks—ice-albedo feedback and water vapor feedback.

Analyze the carbon cycle in terms of feedback mechanisms that affect Earth's climate.

- Climate feedbacks either amplify or reduce climatic trends toward warming or cooling. The ice-albedo feedback loop is a positive feedback that can either lead to warmer or cooler temperatures depending on initial conditions. The water vapor feedback is a positive feedback loop that can lead to higher temperatures. Many climate feedbacks involve the movement of carbon through Earth systems, between carbon sources where carbon is released, and carbon sinks where carbon is stored.

7. Explain how ice-albedo feedback can cause warming or cooling.

8. How will increasing temperatures decrease rates of carbon storage in two carbon sinks?

How do we know that climate is changing now?

7.6 Evidence for Present Climate Change

List the key lines of evidence for present global climate change.

- Several indicators provide evidence of climate warming: increasing air temperatures over land and oceans, increasing sea-surface temperatures and ocean heat content, melting glacial ice and sea ice, rising global sea level, and increasing specific humidity. One third of sea-level rise is due to thermal expansion of seawater.

9. List and discuss the key lines of evidence of present climate change.

7.7 Causes of Present Climate Change

Compare the different greenhouse gases in terms of residence time and contribution to warming.

Evaluate the warming or cooling effects of factors that affect Earth's energy balance.

- The scientific consensus is that present climate change is caused primarily by increased concentrations of atmospheric greenhouse gases resulting from human activities, including carbon dioxide, methane, and nitrous oxide. The increase in these gases is causing a positive radiative forcing (or climate forcing), which in turn disturbs the Earth–atmosphere energy balance. Studies show that CO_2 has a stronger radiative forcing than other greenhouse gases and other natural and anthropogenic factors that force climate.

10. Compare and contrast the relative forcing and residence times of CO_2 and methane.

11. Which is the most powerful positive climate forcing agent? Which is the most powerful negative radiative forcing factor?

How do scientists model climate change?

7.8   Climate Models & Forecasts

Discuss how climate models have been used to analyze and predict climate change.

Summarize some climate projections based on general circulation models.

- A general circulation model (GCM) is a complex computerized climate model used to assess past climatic trends and their causes and to project future changes in climate. The most sophisticated atmosphere and ocean submodels are known as Atmosphere–Ocean General Circulation Models (AOGCMs). Climate models show that positive radiative forcing is caused by anthropogenic greenhouse gases rather than natural factors such as solar variability and volcanic aerosols.

12. Explain how a GCM works and why an AOCGM is more sophisticated than a GCM.

13. What is the current net amount of climate forcing? What is the projected temperature change by 2100, if we follow the RCP closest to the current amount of climate forcing?

How are governments and people responding to climate change?

7.9   Debating Climate Change

Describe several mitigation measures to slow rates of climate change.

- The 2013–2014 IPCC Fifth Assessment Report is 95%–100% certain that human activities are the primary cause of present climate change. While there is consensus among scientists, the general public often thinks that there is still considerable controversy. The fuel for the continuing "debate" on this topic appears to come, at least partly, from media coverage that at times is biased, alarmist, or factually incorrect.

14. Exactly how certain is the IPCC that humans are altering the climate?

7.10   Addressing Climate Change

Describe several mitigation measures to slow rates of climate change.

List actions individuals can take to help reduce greenhouse gas emissions.

- Actions taken on an individual level by millions of people can slow the pace of climate change for us and for future generations. The principal way we can do this—as individuals, as a country, and as an international community—is to reduce carbon emissions, especially in our burning of fossil fuels.

15. How can reducing greenhouse gas emissions produce cobenefits? Give several examples of cobenefits.

16. If you were a city manager, what programs would you promote to combat climate change? Would you concentrate on reducing sources of warming or on increasing carbon sinks or would you do something else entirely?

Key Terms

aerosols p. 185
carbon sink p. 187
climate change science
   p. 176

climate feedback
   p. 186
general circulation
   model p. 198

global carbon budget
   p. 187
isotope analysis
   p. 176

Milankovitch cycles
   p. 183
paleoclimatology
   p. 176

proxy method p. 176
radiative forcing p. 196

Critical Thinking

1. After reviewing Table 7.1 showing positive and negative forcing factors, which positive factors would you suggest reducing first? Explain your answer.

2. Explain how the water vapor feedback loop would change as temperatures increase. Is this a positive or negative feedback loop?

3. Discuss what you feel is the most compelling evidence for anthropogenic climate change. What makes this more compelling than other types of evidence?

4. What do you think is the greatest barrier to broader public understanding of the causes and probable effects of climate change? How would you overcome this barrier to communicate the actual state of our climate?

5. Entrepreneurs are advised to work up an "elevator pitch," a 30-second summary of their idea for a new product or company. What's your elevator pitch for the causes and probable effects of climate change?

Visual Analysis

Glaciers in Alaska have been retreating dramatically due to warming temperatures. The Muir Glacier is a good example of this.

1. Examine the two photographs and describe the changes observed.

2. What are two examples visible in the photographs that show how much conditions have changed from 1941 to 2004?

(a)   (b)

▲R7.1 **Muir Glacier** (a) 1941 and (b) 2004.

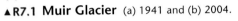

(MG) Interactive Mapping | Login to the **MasteringGeography** Study Area to access **MapMaster**.

Climate Change

Earth's climate is changing, but not all locations will change equally. Some locations will change much more than others.

- Open MapMaster in MasteringGeography™.
- Select *Global Surface Warming Worst Case Projections* from the *Physical Environment* menu. Explore the sublayers of different temperature change projections.

1. What is the largest projected change for the land in the Northern Hemisphere? What is the largest projected change

for the land in the Southern Hemisphere? What is the projected change for the Hawaiian Islands? For your home town?

2. Describe the pattern of projected change, as a function of latitude and continentality. What are the characteristics of the locations with the highest amount of projected change? Locations with the lowest amount of projected change?

Explore | Use **Google Earth** to explore the **Glaciers of Alaska**.

Over 95% of glaciers are in retreat worldwide. Glaciers in Alaska are no exception. Search for the *Columbia Glacier, Alaska*. Zoom in until you can see where the end of the glacier meets the sea. Use the *Add Path* tool to trace the outline of the end of the glacier. Turn on *Historical Imagery* (the clock button), and go back to 11/27/2007. Use the *Add Path* tool again to draw the outline of the end of the glacier.

1. Use the *Show Ruler* tool to measure the retreat from 2007 to 2013 at several places. What is the maximum and minimum retreat?

2. How many miles or kilometers per year has the glacier been retreating?

3. If the glacier continues to retreat at this rate, how long until the retreat equals your daily commute to school?

► R7.2

Mastering Geography™

Looking for additional review and test prep materials? Visit the Study Area in MasteringGeography™ to enhance your geographic literacy, spatial reasoning skills, and understanding of this chapter's content by accessing a variety of resources, including

MapMaster™ interactive maps, videos, *Mobile Field Trips*, *Project Condor* Quadcopter Videos, *In the News* RSS feeds, flashcards, web links, self-study quizzes, and an eText version of *Geosystems Core*.

Getting to the Core: Reconstructing Past Climates

To understand present climate changes, we need to understand the climates of the past, especially climates that occurred before record keeping with standardized instruments, which began almost 140 years ago. Ice core projects in Greenland have produced data spanning more than 120,000 years. One such effort, completed in 2004, was part of the European Project for Ice Coring in Antarctica (EPICA). The Dome C ice core reached a depth of 3270 m (10,729 ft) and produced the longest ice core record yet: 800,000 years of Earth's past climate history. This record was correlated with a core of 400,000 years from the nearby Vostok Station and matched with ocean sediment core records to provide scientists with a sound reconstruction of climate changes throughout this time period.

GeoLab7 (MG)
Pre-Lab Video

https://goo.gl/dVmDOs

Apply

Scientists look at ice core data to determine past temperatures and atmospheric compositions. You have been tasked with evaluating ice core records from Antarctica to compare each of them with the other and to current conditions.

Objectives

- **Identify** patterns of CO_2 and temperature in ice core data.
- **Evaluate** whether past climate changes have been global or hemispheric.

Procedure

Examine Figure GL7.1. This EPICA ice core is 3270 m (10,729 ft) long. So far, scientists have analyzed the top 3189.45 m (10,462 ft), revealing the past 801,588 years of climate. The figure shows CO_2 in parts per million (ppm) and temperature anomalies—the change from normal—in Celsius degrees relative to the average for the past 1000 years.

1. What are the maximum and minimum temperature anomalies recorded? What is the range of temperature anomalies over this record?

2. How many interglacials (times where the temperature deviation reached 0°C or greater) occurred in the past 450,000 years? What is the average time between interglacials?

3. How much has CO_2 varied in the EPICA record? What were the highest and lowest values? What was the range of CO_2 values?

4. How many times has CO_2 exceeded 270 ppm in the past 450,000 years?

5. What is the average period from peak to peak with regard to CO_2 and for temperature for the past 450,000 years?

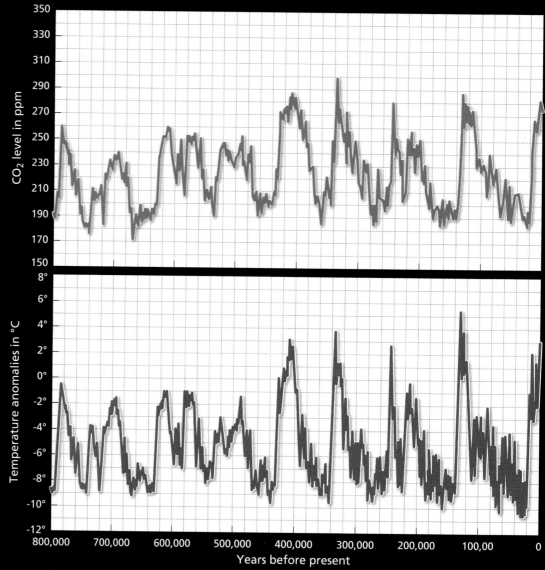

▲ Figure GL7.1 **EPICA Ice core record of CO_2 levels and temperature anomalies**

Examine Figure GL7.2. The Vostok ice core is 3310 m (10,860 ft) long. The full record extends back to 422,000 years before the present. The figure shows CO_2 in parts per million (ppm) and temperature anomalies in Celsius degrees.

6. How many years of data are displayed from the Vostok ice core?
7. How many times has the temperature anomaly exceeded 0°C in the Vostok record?
8. What are the maximum and minimum temperature anomalies recorded? What is the range of temperature anomalies in the Vostok record?
9. How many interglacials (times where the temperature deviation reached 0°C

or greater) do you detect over the past 450,000 years? What is the average time between interglacials?
10. How much has CO_2 varied in the Vostok record? What were the highest and lowest values? What was the range of CO_2 values?

Analyze & Conclude

11. What is the time period from peak to peak with regard to CO_2 and for temperature for the Vostok record? How does this compare with the period (time from peak to peak) of the EPICA record?
12. Why is it important to compare records from different locations? What other locations would you examine to study climate?

13. Interglacial–glacial climate cycles often appear saw-toothed, with an abrupt change followed by a more gradual change. When do we see an abrupt change: when going from interglacial to glacial times or when going from glacial to interglacial times? When do we see a gradual change? How do these temperature changes compare to the changes of the last 100 years?
14. What was the warmest anomaly in either the EPICA or Vostok record? What was the level of CO_2 at that time? How does that level of CO_2 compare with the current level of CO_2 in our atmosphere? Do the highest temperatures always correlate with the highest CO_2 levels?

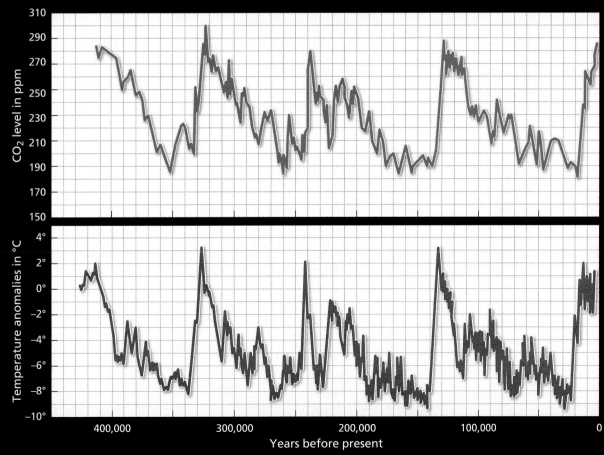

▲Figure GL7.2 **Vostok ice core record of CO_2 levels and temperature anomalies**

8

Tectonics, Earthquakes, & Volcanism

During the twentieth century, new research revolutionized our understanding of Earth science. These discoveries explained how volcanoes such as Ecuador's Tungurahua were formed and how the continents and oceans came to their present arrangement. The theory of plate tectonics—the main subject of this chapter—provides a unified explanation for the patterns of earthquakes, volcanoes, and mountain building that dominate our planet. Events such as the March 2011 earthquake and tsunami that devastated Japan turn world attention to the Earth sciences, including physical geography, for explanations.

Key Concepts & Topics

Mount Tungurahua in Ecuador is one of many active volcanoes in the Andes mountain chain that extends along the western coast of South America.

8.1 The Vast Span of Geologic Time

Key Learning Concepts

▶ **Describe** the geologic time scale.
▶ **Distinguish** between relative and absolute time.
▶ **Explain** how the principle of uniformitarianism helps geologists interpret Earth's history.

Earth is about 4.6 billion years old. The Moon, about 30 million years younger, formed when a Mars-sized object struck the early Earth and threw debris into orbit that then coalesced into the Moon. Geologists have used data from rocks to reconstruct these and other key events in Earth's history.

Eons, Eras, Periods, & Epochs

The **geologic time scale** is a summary timeline of all Earth history (▼Fig. 8.1). It names the time intervals for each segment of Earth's history, from vast *eons* through briefer *eras*, *periods*, and *epochs*. Breaks between some of the major time intervals also mark the five major mass extinctions, when the total number of living species dropped dramatically. These range from a mass extinction 440 million years ago (m.y.a.) to the present-day sixth episode of extinctions caused by modern civilization.

geoCHECK ✔ Referring to Figure 8.1, list the geologic periods that make up the Cenozoic era.

Absolute Time & Relative Time

The geologic time scale depicts two kinds of time: **absolute time** is the actual number of years before the present, and **relative time** is the sequence of events—what happened in what order. Scientific methods such as radiometric dating (Chapter 7) determine absolute age, the actual "millions of years

ago" shown on the time scale. In contrast, relative dating is based on the relative positions of rock strata (layers) above or below each other. The study of the relative positions of these strata is called *stratigraphy*. To establish relative age based on stratigraphy, geologists apply the **principle of superposition**, which states that *rock and sediment always are arranged with the youngest beds "superposed" toward the top of a rock formation and the oldest at the base, if they have not been disturbed.* Thus if you look at the stratigraphy of the Grand Canyon, the youngest layers are found at the canyon rim, and the oldest layers are found at the bottom (▶Fig. 8.2).

In the Grand Canyon, the principle of superposition places the extremely ancient strata of the Precambrian at the bottom of the canyon and the more recent, but still ancient, strata of the Permian period at the top. Rocks younger than the Permian, however, are missing from the canyon because they have worn away over time. Important time clues—namely, *fossils*, the remains of ancient plants and animals—lie embedded within the canyon walls. Since approximately 4 billion years ago, life has left its evolving imprint in Earth's rocks.

We now live in the Holocene epoch, which began about 11,500 years ago. Geologists are debating whether to add another epoch to the geologic time scale—the Anthropocene—beginning around the year 1800, which marks the start of the Industrial Revolution in Europe. As the impacts of humans on Earth systems increases, namely deforestation, land clearing for agriculture, and fossil-fuel burning, numerous scientists now agree that we are in a new epoch called the Anthropocene.

geoCHECK ✔ What is the difference between absolute and relative time?

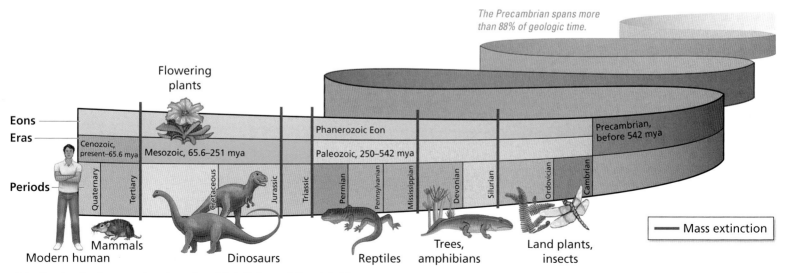

The Precambrian spans more than 88% of geologic time.

▲**8.1 Geologic time scale, showing highlights of Earth's history**

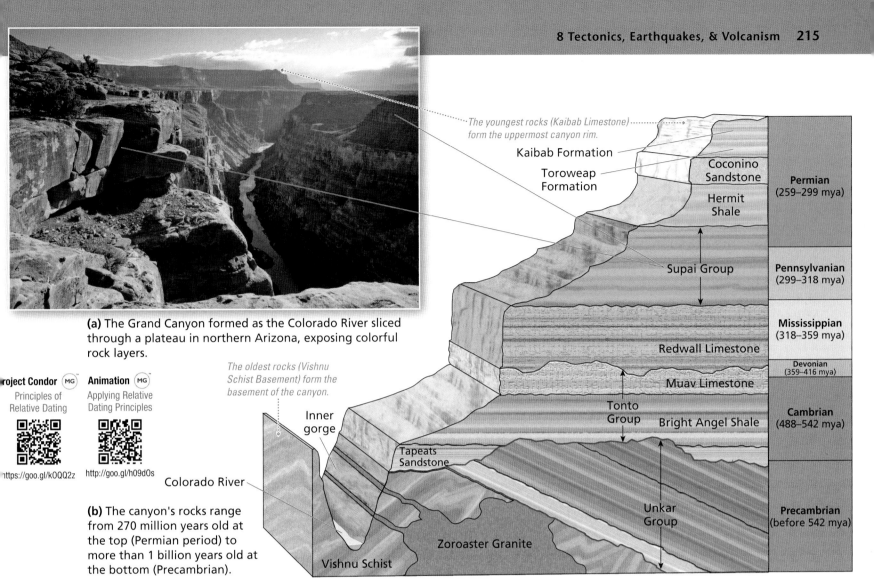

(a) The Grand Canyon formed as the Colorado River sliced through a plateau in northern Arizona, exposing colorful rock layers.

The youngest rocks (Kaibab Limestone) form the uppermost canyon rim.

Kaibab Formation
Toroweap Formation
Coconino Sandstone
Hermit Shale
Supai Group

Permian (259–299 mya)
Pennsylvanian (299–318 mya)

Redwall Limestone

Mississippian (318–359 mya)
Devonian (359–416 mya)

Muav Limestone

Tonto Group
Bright Angel Shale

Cambrian (488–542 mya)

Tapeats Sandstone

The oldest rocks (Vishnu Schist Basement) form the basement of the canyon.

Inner gorge

Colorado River

Zoroaster Granite

Unkar Group

Precambrian (before 542 mya)

Vishnu Schist

Project Condor (MG)
Principles of Relative Dating
https://goo.gl/k0QQ2z

Animation (MG)
Applying Relative Dating Principles
http://goo.gl/h09d0s

(b) The canyon's rocks range from 270 million years old at the top (Permian period) to more than 1 billion years old at the bottom (Precambrian).

▲8.2 **Superposition of rock layers in the Grand Canyon** The different types of rock that line the Grand Canyon formed during different geologic periods.

The Principle of Uniformitarianism

A guiding principle of Earth science is **uniformitarianism**— "the present is the key to the past"—which assumes that *the same physical processes active in the environment today have been operating throughout geologic time* (▼Fig. 8.3). For example, if streams carve valleys now, they must have done so 500 m.y.a.

▼8.3 **Uniformitarianism in a Landscape** Although the ages and types of rocks vary by location, the same physical processes of erosion, transport, and deposition of sediment active in the environment today have been operating throughout geologic time. This stream-carved valley is in the Pamir Mountains of Central Asia.

Evidence from the landscape record of volcanic eruptions, earthquakes, and processes that shape Earth's surface supports uniformitarianism. Although the ages and types of rock differ from place to place, the processes that form, erode, and deposit these rocks and sediment are *uniformly* similar. Uniformitarianism refers to the actual physical processes that create our different landforms, while geologic time provides the millions of years during which these processes operate. James Hutton first proposed this concept in his *Theory of the Earth* (1795). Charles Lyell further explored the implications of uniformitarianism in *Principles of Geology* (1830).

Mobile Field Trip (MG)
Desert Geomorphology

https://goo.gl/XngZ60

 geoCHECK ✔ How do Earth scientists define uniformitarianism?

geoQUIZ

1. Imagine looking at a canyon wall of layered rock strata. According to the principle of superposition, how does the relative age of strata at the top of the canyon compare with the age of strata at the bottom of the canyon?
2. Explain why scientists may add a new geologic epoch—the Anthropocene—to the geologic time scale.
3. What does the principle of uniformitarianism suggest about how river deltas form in different parts of the world, such as the Nile Delta in Egypt and the Mississippi Delta in North America?

8.2 Earth History & Interior

Key Learning Concepts

▶ **Summarize** how and when Earth formed.

▶ **Describe** Earth's layered structure.

Our knowledge of Earth's interior comes from analyzing surface features such as road cuts, landslide scars, volcanic flows, and the stratigraphy of deep canyons. Scientists also gain information from seismic waves that travel at different rates through the interior layers of Earth. When scientists analyze these different sources of information, a clear picture of Earth's interior emerges. The structure of the interior also reflects the processes that formed our planet.

Earth's Formation

Along with the other planets and the Sun, Earth is thought to have condensed from a nebula of dust, gas, and icy comets about 4.6 billion years ago. Rocks in Western Australia date to between 4.2 and 4.4 billion years old and are possibly the oldest materials in Earth's crust.

As Earth solidified, gravity sorted materials by density. Denser substances such as iron gravitated slowly to its center, and less dense elements, such as oxygen, silicon, and aluminum, slowly welled upward to the surface and concentrated in the crust. Consequently, Earth's interior is sorted

into roughly concentric layers, each one distinct in either chemical composition or temperature (▼ **Fig. 8.4a**). Thus Earth's core is nearly twice as dense as the mantle because it is composed of denser metallic iron-nickel alloy.

Earth's Internal Energy Source Heat left over from the planet's formation and heat produced by the decay of radioactive isotopes make Earth's interior extremely hot. This heat energy migrates outward from the center by conduction and by convection (review heat transfer in Chapter 2). This escaping heat—from Earth's origin and from radioactive decay—is the energy that drives plate tectonics. As explained below, this heat energy also affects processes on Earth's surface.

geoCHECK ✔ Why does Earth's interior have a layered structure?

Earth's Core & Magnetism

At Earth's center lies the dense, metallic core, divided into an inner core and outer core (▼ **Fig. 8.4b**). The core makes up one third of Earth's entire mass, but only one sixth of its volume. The inner core of solid impure iron is well above the melting temperature of iron at surface conditions, but remains solid because of tremendous pressure caused by the weight of the overlying materials. The outer core is molten iron and has a lower density than the inner core. The rotation of Earth affects the fluid outer core, generating Earth's magnetic field. Recall from Chapter 1 that the magnetosphere surrounds and protects Earth from solar wind and cosmic radiation. A compass needle points to magnetic north because of Earth's magnetic field (▶ **Fig. 8.5**).

geoCHECK ✔ What do scientists think causes Earth's magnetic field?

Earth's Mantle

Above the core lies the mantle, a layer of hot, but mostly solid material that represents about 80% of Earth's total volume (Fig. 8.4b). The mantle is rich in oxides of iron and magnesium and in *silicates*—silicon dioxide, a component of many rocks and minerals. The extremely high temperature and pressure in the lower mantle softens and deforms solid rock material over millions of years. Just as boiling water in a pot rises to the surface, rocks in the warmer lower mantle rise upward where they slowly cool as their heat is transferred to the upper mantle (▶ **Fig. 8.6**). As the rock material cools, its density increases, and the rocks begin to sink and move downward, also aided by gravity. The up-and-down movement of rock material in Earth's interior is the result of these **convection currents**, which power the process of plate tectonics discussed ahead.

▼ **8.4 Earth in cross-section**

The cutaway (b) shows how Earth's interior from the inner core to the crust is layered with solid and liquid material the outer layers (c) of ocean and continental crust and the upper mantle sit atop the asthenosphere.

Oceanic crust

Continental crust

Lithosphere

Mohorovičić discontinuity

Uppermost mantle (rigid)

Asthenosphere (plastic)

(c)

Lithosphere Crust
Athenosphere

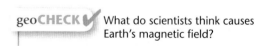

Upper mantle

70 km
250 km

Mantle

Lower mantle

2900 km

Outer core

Core

5150 km

Inner core (solid)

6370 km (Earth's center)

(a)

(b)

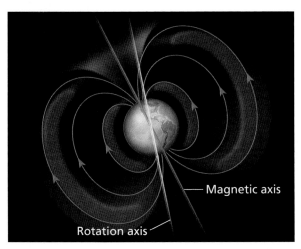

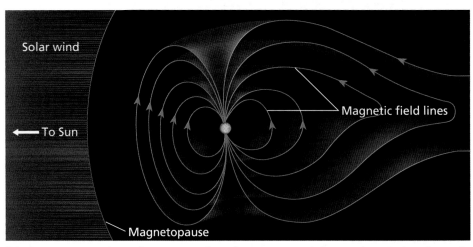

(a) Earth has geographic poles and magnetic poles. The magnetic field draws a compass needle to point to magnetic north.

(b) The magnetosphere that surrounds and protects Earth from solar wind and cosmic radiation. The magnetosphere encircling Earth is the strongest of all the solid planets.

▲ **8.5 Earth's magnetic field** Scientists think motions within Earth's molten outer core produce the planet's magnetic field.

In the upper mantle, closer to Earth's crust, the temperature and pressure decline, so the rocks are more rigid. The asthenosphere (from Greek *asthenos*, meaning "weak") lies just below the lithosphere and extends from about 70 km beneath Earth's surface to a depth of 250 km (43 mi to 155 mi). The asthenosphere is the least rigid part of the mantle. Geologists describe the rock of the asthenosphere as being "plastic." This means that its weak material can bend or flow slowly over long periods of time.

The asthenosphere contains pockets of increased heat from radioactive decay. This partly explains why about 10% of this layer is molten in uneven patterns and hot spots and why slow convective currents occur in its hot, plastic material. The resulting slow movement in the asthenosphere disturbs the overlying crust and creates tectonic folding, faulting, and deformation of surface rocks.

geoCHECK ✔ Describe the characteristics of the asthenosphere.

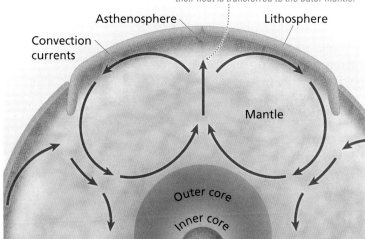

▲ **8.6 Convection currents in Earth's interior** Differences in temperature and density within the mantle produce convection currents, in which hot, but solid material slowly rises through the mantle as cooler material near the top sinks downward.

Earth's Lithosphere & Crust

Above the asthenosphere, the uppermost 70 km (43 mi) of mantle and Earth's crust are collectively called the lithosphere (Fig. 8.4b). This layer "floats" on the asthenosphere and is broken into "plates" (discussed in Module 8.4) that are moved during tectonic processes. The **Mohorovičić discontinuity**, or **Moho**, separates the crust from the mantle. The Moho is named for the Croatian scientist who determined that seismic waves generated by earthquakes change at this boundary because of the change in Earth materials and densities.

Above the Moho lies the crust—Earth's rocky outer shell made up of the continents and the rocks that form the bottom of ocean basins. The crust represents only a

fraction of Earth's overall diameter and varies in thickness. Crustal areas below mountain masses are 50–60 km (31–37 mi) in thickness. Crustal areas beneath continental interiors are about 30 km (l19 mi) in thickness. Oceanic crust averages only 5 km (3 mi). Figure 8.4c illustrates the relation of the crust to the rest of the lithosphere and to the asthenosphere below.

geoCHECK ✔ Identify the layers immediately above and below the Moho.

geoQUIZ

1. Describe the main layers that comprise Earth's interior.
2. How does Earth's inner core differ from the outer core?
3. What is the relationship between the asthenosphere, the Moho, and the crust?

8.3 The Rock Cycle

Key Learning Concepts

▶ **Analyze** how Earth's geologic cycle relates the tectonic, rock, and hydrologic cycles.

▶ **Explain** how rocks cycle through geologic time.

▶ **Differentiate** the processes that create igneous, sedimentary, and metamorphic rocks.

▼**8.7 The geologic cycle** A systems model comprising the hydrologic, rock, and tectonic cycles, the geologic cycle shows how the interaction of exogenic (external) and endogenic (internal) forces creates distinctive landscapes.

While the internal forces such as plate motions and volcanism build landforms, surface processes such as weathering and erosion work to reshape them. Although the processes occur slowly over geologic time, the result is a restless planet where these two opposing forces interact in a process called the **geologic cycle**. Earth's internal heat and external solar energy fuel the geologic cycle, which is also influenced by the force of gravity.

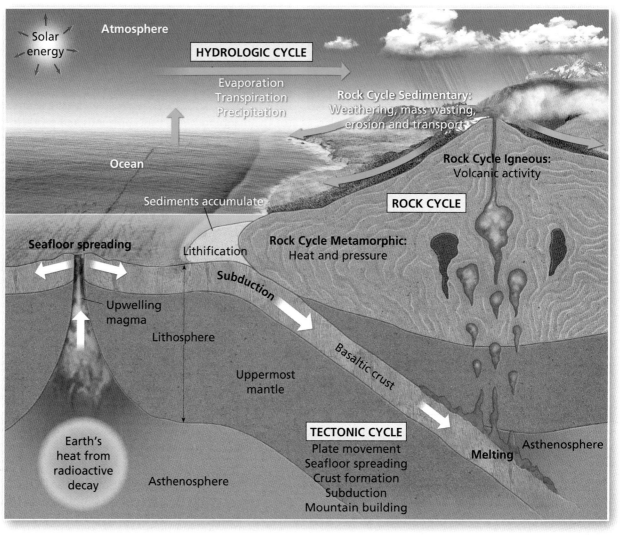

Subcycles within the Geologic Cycle

Three subcycles compose the geologic cycle (▲ Fig. 8.7). The *rock cycle* explained in the next section produces the igneous, metamorphic, and sedimentary rocks found in the crust (▶ Fig. 8.9). The *tectonic cycle* brings heat energy and new material to the surface and recycles material, creating movement and deformation of the crust as shown in Figure 8.8. The *hydrologic cycle* processes Earth materials with the chemical and physical action of water, ice, and wind. This chapter explores the rock cycle. (You learned about the hydrologic cycle in Chapter 5).

geoCHECK ✔ Explain the roles of endogenic and exogenic processes in the geologic cycle.

▶ **8.8 From seafloor to summit** Jebel Hafeet soars 1240 m (4068 ft) above the Arabian Desert. The colliding Arabian and Eurasian tectonic plates pushed up sedimentary rocks, containing fossils—easily spotted in the rocks—that formed in an ancient sea.

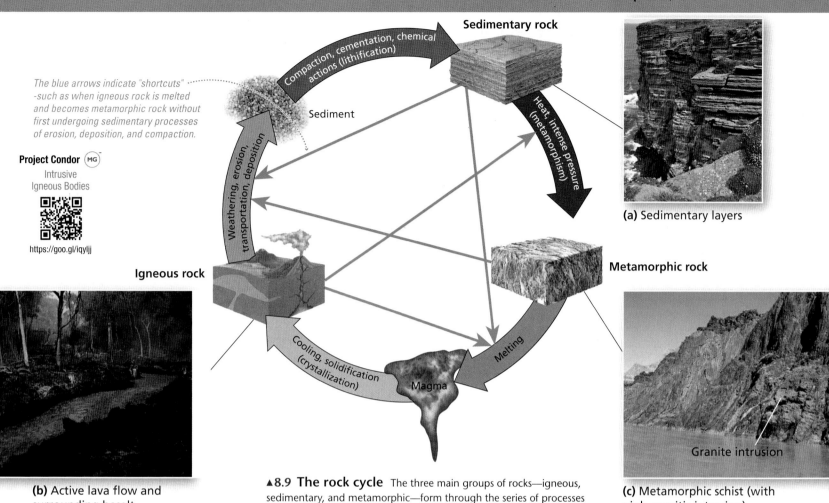

The blue arrows indicate "shortcuts" -such as when igneous rock is melted and becomes metamorphic rock without first undergoing sedimentary processes of erosion, deposition, and compaction.

Project Condor (MG)
Intrusive
Igneous Bodies

https://goo.gl/iqyljj

Sedimentary rock

Compaction, cementation, chemical actions (lithification)

Sediment

Weathering, erosion, transportation, deposition

Heat, intense pressure (metamorphism)

Igneous rock

Metamorphic rock

Cooling, solidification (crystallization)

Melting

Magma

(a) Sedimentary layers

(b) Active lava flow and surrounding basalt

▲8.9 **The rock cycle**   The three main groups of rocks—igneous, sedimentary, and metamorphic—form through the series of processes that makes up the rock cycle.

Granite intrusion

(c) Metamorphic schist (with pink granitic intrusion)

Minerals, Rocks, & the Rock Cycle

The **rock cycle** is a natural recycling process by which rocks form and then, over time, change and re-form into another type of rock (▲ **Fig. 8.9**). To understand the rock cycle, you need to become familiar with basic Earth materials—elements, minerals, and rocks. Eight natural elements (oxygen, silicon, aluminum, iron, calcium, sodium, potassium, and magnesium) compose 98.5% of Earth's crust by weight. These elements combine to form many minerals.

What Is a Mineral? A **mineral** is an inorganic (nonliving) natural substance with a specific chemical formula and usually a crystalline structure. This combination gives each mineral distinct physical properties such as hardness and density. For example, silicon dioxide, with a distinctive six-sided crystal, is the common mineral you know as quartz.

What Is a Rock? A **rock** is an assemblage of materials, including mineral and organic matter, bound together as part of the lithosphere. Examples of rocks range from granite (containing three minerals), to rock salt (made of one mineral), to coal (made of solid organic material). There are thousands of kinds of rocks, but all fall into one of three broad classes, depending how they formed.

Pathways of the Rock Cycle A series of processes called the rock cycle forms these three groups of rocks:

- **Igneous rocks** form from material that melts under the high temperature and pressure found inside the mantle and crust. Igneous rocks are further subdivided into two types. **Intrusive igneous rocks** form from mantle material injected into the crust, where it cools and hardens beneath the surface. These rocks then reach the surface after subsequent uplift and erosion. **Extrusive igneous rocks** form through volcanic activity when molten material from the mantle melts through the crust and onto Earth's surface, where the molten material hardens into solid rock.

- **Sedimentary rocks** often form from fragments of eroded rocks (sediments) that are deposited in low depressions such as valleys, lakes, or oceans, and then are compacted to form new rocks. Other types of sedimentary rocks form through chemical and biological processes. Chemical sedimentary rocks form when materials dissolve in water and precipitate (harden) into rock. Organic sedimentary rocks form when material from plant and animal remains compact into sediment, or when organisms still living secrete material that hardens into rock.

- **Metamorphic rocks** are any rocks that chemically re-form under tremendous pressure and accompanying heat that occurs as a result of deep burial inside the crust, or—on a smaller scale—through "contact metamorphism," the intense heat resulting from direct contact with magma as it moves into the mantle or flows onto Earth's surface.

geoCHECK ✔   Identify the processes that form igneous, sedimentary, and metamorphic rocks.

8.3 (cont'd) The Rock Cycle

Igneous Processes

You can think of the rock cycle as beginning when an igneous rock solidifies and crystallizes from molten magma that forms beneath the surface (refer to Fig. 8.9). **Magma** is molten rock that contains dissolved gases under tremendous pressure. **Lava** is magma that reaches the surface. Either magma *intrudes* into crustal rocks, cooling and hardening below the surface to form intrusive igneous rock such as granite, or it *extrudes* onto the surface as lava and cools to form extrusive igneous rock, such as basalt. The texture of igneous rocks is based on the size of their crystals, which is directly related to how the rocks formed. Slow cooling, intrusive igneous rocks such as granite have larger, coarse-grained crystals (▶ **Fig. 8.10a**). Faster cooling rocks result in smaller, fine-grained crystals, as in extrusive igneous rocks such as basalt and rhyolite (Fig. 8.10b, c). Darker igneous rocks contain more iron, while lighter igneous rocks contain more silicates.

A mass of slow cooling intrusive igneous rock forms a **pluton**. Multiple plutons larger than 100 km² (40 mi²) in area compose a **batholith**. Huge batholiths, exposed at the surface by uplift and erosion, often form the core of a mountain range, such as the Sierra Nevada.

Igneous rocks make up 90% of Earth's crust, although sedimentary rocks, soil, or oceans frequently cover them. Continental crust is made up largely of light-colored, lower-density rocks such as granite, an intrusive igneous rock. Oceanic crust is made up of dark-colored, higher-density rocks such as basalt, an extrusive igneous rock.

 Compare and contrast the formation and characteristics of granite and basalt. Which rock is more likely to be found in a batholith?

(a) Granite   **(b)** Basalt flows   **(c)** Rhyolite

 ▲8.10 **Igneous rocks**

Clastic & Organic Sedimentary Rocks Clastic means rock fragments, and once deposited in a low basin or water body, this sediment compacts under the weight of overlying layers, cements, and hardens into *clastic* sedimentary rock, such as conglomerate, sandstone or shale. A similar process forms sedimentary rocks made of organic materials such as the shells and skeletons of marine organisms (coquina limestone) and the decayed remains of plants (coal).

Chemical Sedimentary Rocks In marine and saline inland sea environments, some minerals separate out from the water to form solid sedimentary deposits, such as calcium carbonate, which forms the sedimentary rock limestone. Chemical sediments from inorganic sources are also deposited when water evaporates and leaves behind a residue of salts. These *evaporites* may exist as common salts, such as potash or sodium chloride (table salt). Chemical deposition also occurs in the water of natural hot springs from chemical reactions between minerals and oxygen.

Sedimentary Rock Strata Layered sedimentary rocks record the past. Because sedimentary rocks are formed in a sequence of layers from the oldest rocks on the bottom to the youngest on the top, the sequence of layers yields clues about how past environments and living things changed over time. Figure 8.11 shows sandstone formed in a marine environment. Note the multiple layers in the formation and how differently they resist weathering processes.

Sedimentary Processes

The rock cycle continues as sedimentary rocks form from pieces of other rocks (refer to Fig. 8.9). Most sedimentary rocks form when eroding fragments of existing rock or organic materials are transported downslope by water, ice, wind, and gravity.

 List the steps in the process that forms a clastic sedimentary rock.

1 Eroded sediments end up in the water and begin to settle (**sedimentation**)

Land   Water

2 With time, more layers pile up and press down the lower layers (**compaction**)

3 More layers (strata) and further compaction forces out water from the layers

4 Salt crystals glue the layers together (**cementation**) forming sedimentary rock

(a)

▼ 8.11 **Sedimentary rocks** (a) Sediment deposited by water, wind, or glacial ice builds up in layers, which are compacted and cemented to form solid rock. (b) Layers of sedimentary rock strata are clearly visible along this shoreline of Khor ash-Sham, in Oman's Musandam Peninsula.

(b)

▼ 8.12 Metamorphic rocks

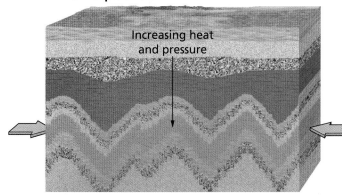

Increasing heat and pressure

(a) Regional metamorphism Compression forces in the crust, the weight of overlying rock and the heat of Earth's interior combine to cause metamorphism of deeply buried rocks.

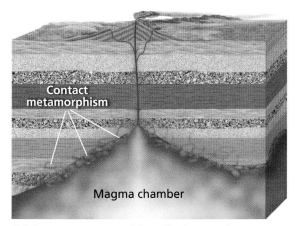

Contact metamorphism

Magma chamber

(b) Contact metamorphism The intense heat of contact metamorphism quickly alters neighboring rocks.

(c) Foliated rock Gneiss forms when granite or sedimentary rocks undergo extreme heat and pressure.

(e) Nonfoliated rock Marble forms when limestone is exposed to high temperature and pressure. The strength and resistance to erosion makes this non-foliated metamorphic rock an excellent building material.

Metamorphic Processes

Through the rock cycle, any rock can become a metamorphic rock (refer to Fig. 8.9). Metamorphic rocks have undergone intense heat and pressure that alter their physical and chemical properties. This process typically makes rocks harder and more resistant to weathering and erosion.

Regional metamorphism occurs when rocks buried inside the crust are subjected to high temperatures and pressures over millions of years (▼ Fig. 8.12a). This can take place when rocks compress during tectonic collisions between slabs of Earth's lithosphere. The shearing and stressing along earthquake fault zones is another cause of metamorphism.

Regional metamorphism can also occur when the weight of sediments collecting in broad depressions in Earth's crust creates enough pressure in the bottommost layers to cause metamorphism. However, on a smaller scale, magma rising into the crust may "cook" adjacent rock, a process known as *contact metamorphism* (Fig. 8.12b).

Metamorphic rocks have textures that are *foliated* or *nonfoliated*, depending on the arrangement of minerals after metamorphism. *Foliated* rocks (Fig. 8.12c) have thin wavy lines, as seen in slate (formed from shale) and gneiss (formed from granite). *Nonfoliated* rocks (Figs. 8.12d and 8.12e) have a crystalline texture, as seen in marble (formed from limestone) and quartzite (formed from sandstone).

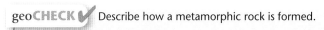

geoCHECK ✔ Describe how a metamorphic rock is formed.

(d) Taj Mahal The marble walls of India's Taj Mahal have withstood the elements for over 360 years.

Animation (MG)
Foliation of Metamorphic Rock

http://goo.gl/zDmhVN

geoQUIZ

1. What role does each of the three subcycles play within the geologic cycle?
2. What is the relationship between elements, minerals, and rocks?
3. Use examples to describe the main differences between igneous, sedimentary, and metamorphic rocks.

8.4 Plate Tectonics

Key Learning Concepts

▶ *Summarize* Wegener's hypothesis of continental drift, the formation and breakup of Pangaea, and why scientists at the time rejected the hypothesis.

▶ *Describe* how the processes of plate tectonics transform Earth's surface over time.

Looking at a world map, you may have noticed that some continents have matching shapes like pieces of a jigsaw puzzle—particularly South America and Africa. Scientists had wondered about this "fit" of the continents since the first accurate world maps were produced hundreds of years ago.

Wegener's Hypothesis of Continental Drift

In 1912, German geophysicist Alfred Wegener proposed a hypothesis to explain the continental puzzle: that the continents had moved together by the end of the Paleozoic Era, forming the supercontinent of **Pangaea,** which then started to break apart near the beginning of the Mesozoic Era (▶ **Fig. 8.13**). According to Wegener's hypothesis, the moving continents slowly plowed across the seafloor in the process of *continental drift*. As proof that the now widely separated landmasses had once been joined together, Wegener cited several types of evidence, including matching rock formations on opposite sides of the Atlantic Ocean and matching fossils, from Africa and South America, of organisms that could not have migrated across oceans (▼ **Fig. 8.14**). Wegener could not, however, provide a plausible mechanism to explain why continental drift occurred. Most scientists of Wegener's time rejected the hypothesis, because it lacked a mechanism for driving continental movement.

geoCHECK ✔ Explain how fossil evidence supports the existence and subsequent breakup of Pangaea.

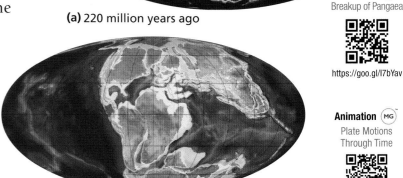

(a) 220 million years ago

Animation (MG)
Breakup of Pangaea

https://goo.gl/I7bYav

Animation (MG)
Plate Motions Through Time

http://goo.gl/quab8Y

(b) 135 million years ago

(c) Earth today

▲ **8.13 Continents adrift, from Pangaea to the present** The formation and breakup of the supercontinent Pangaea were part of a repeated cycle in which pieces of the lithosphere move together, split apart, and eventually reform again. Over the 4 billion years of Earth history, this cycle may have repeated itself a dozen times.

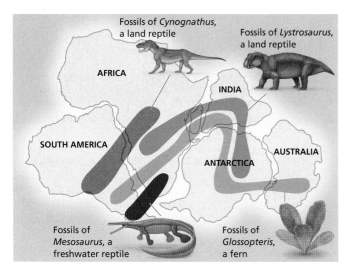

Fossils of *Cynognathus,* a land reptile
Fossils of *Lystrosaurus,* a land reptile
AFRICA
INDIA
SOUTH AMERICA
AUSTRALIA
ANTARCTICA
Fossils of *Mesosaurus,* a freshwater reptile
Fossils of *Glossopteris,* a fern

▲ **8.14 Fossil evidence for plate tectonics** As Pangaea broke apart, the drifting continents marooned species on separate landmasses. These species left fossil evidence on every continent of their earlier distribution.

The Theory of Plate Tectonics

Today, we know that most of Wegener's hypothesis was correct: Continental pieces once did fit together, and they not only migrated to their present locations, but also continue moving at an average rate of about 6 cm (2.4 in.) per year. Since the 1950s and 1960s, modern science built the theory of **plate tectonics,** the now universally accepted scientific theory that the lithosphere is divided into several moving plates that float on the asthenosphere (above the mantle) and along whose boundaries occur the formation of new crust, mountain building, and the seismic activity that causes earthquakes.

Lithospheric Plates The continents move as part of pieces of lithosphere called **lithospheric plates,** also called tectonic plates. These enormous and unevenly shaped slabs of the outer crust and upper mantle are usually composed of both continental and oceanic lithosphere (as shown in ▶ **Fig. 8.15**). Plates can vary greatly in size from 300 km (186 mi) across, to those that cover entire continents. Plates vary in thickness from less than 15 km (9 mi) in oceanic lithosphere, to 200 km (120 mi) for interior continental lithosphere. Oceanic lithosphere is made up mostly of basalt, whereas continental lithosphere has a foundation of mostly granitic-type rocks.

Earth's present lithosphere is divided into at least 14 plates, of which about half are major and half are minor in terms of area (▶ **Fig. 8.15**). Hundreds of smaller pieces and perhaps dozens of microplates migrating together make up these broad, moving plates. Arrows in the figure indicate the direction in which each plate is presently moving, and the length of the arrows suggests the relative rate of movement during the past 20 million years.

Processes of Plate Tectonics and Their Effects

The word *tectonic*, from the Greek *tektonikùs*, meaning "building," refers to changes in the configuration of Earth's crust as a result of internal forces. Plate tectonics includes several processes: upwelling of magma, lithospheric plate movements, and seafloor spreading and subduction (processes that create and destroy the seafloor). The effects of plate motions include earthquakes, volcanic activity, and deformations of the lithosphere, such as warping, folding, and faulting, that result in mountain building. Figure 8.16 shows how the processes of plate tectonics form and cause changes in Earth's continental and oceanic lithosphere. You will learn more about these processes and their effects throughout the rest of this chapter.

geoCHECK ✔ In your own words, define "plate tectonics."

▶ **8.16 Overview of plate tectonics** As a result of plate tectonics, processes of Earth's interior, such as upwelling magma, change the surface, producing features such as the rock of Earth's ocean floor and chains of volcanic mountains.

Legend:
— Plate boundary
→ Plate motion
▲▲ Subduction zone
---- Collision zone
⊥ Spreading ridge offset by transform faults
······ Deep-ocean trench

▲ **8.15 Earth's major lithospheric plates** As the plates move, their interactions slowly change Earth's surface. The arrows in the figure indicate the direction in which each plate is presently moving, and the length of the arrows suggests the relative rate of movement during the past 20 million years.

Animation (MG)
Transform Faults
http://goo.gl/KXG42e

Animation (MG)
Forming a Divergent Boundary
http://goo.gl/8XVrzZ

Animation (MG)
Motion at Plate Boundaries
http://goo.gl/LNnG80

(b) Like the seam of a baseball, the boundary between the North American and Eurasian plates winds through the North Atlantic Ocean.

(c) This rockwalled valley in Iceland is part of a zone where two plates meet.

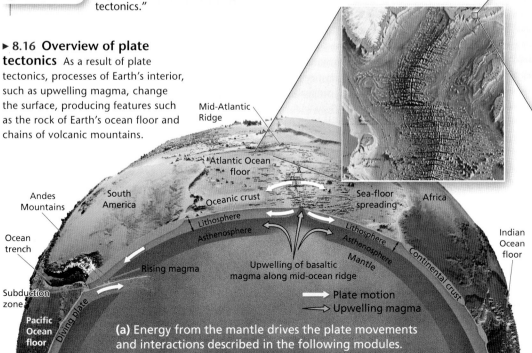

(a) Energy from the mantle drives the plate movements and interactions described in the following modules.

geoQUIZ

1. Why did scientists of Wegener's time reject his hypothesis of continental drift?
2. Compare and contrast continental and oceanic lithosphere.
3. How are earthquakes, volcanic eruptions, and faulting linked to plate tectonics?

8.5 Seafloor Spreading & Subduction Zones

Key Learning Concepts

▶ **Describe** the roles of seafloor spreading and subduction in forming and destroying the ocean floor.
▶ **Explain** how geomagnetism supports the concept of seafloor spreading.
▶ **State** the relationship between subduction, earthquakes, and volcanoes.
▶ **Identify** two mechanisms of plate motion.

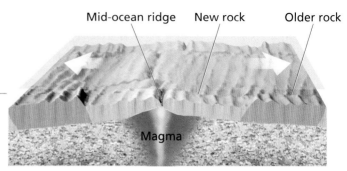

Mid-ocean ridge   New rock   Older rock

Magma

(a) Upwelling magma erupts along the spreading center, hardening to form new oceanic lithosphere.

For hundreds of years, people sailed Earth's oceans but knew almost nothing about the topography of the seafloor. Only in the mid-20th century did mapping the seafloor on a wide scale using SONAR (SOund NAvigation Ranging) become feasible. This technique involves bouncing sound waves off the ocean floor, then measuring the time it takes for them to return. The first maps of the seafloor revealed an interconnected worldwide mountain chain, forming a ridge some 64,000 km (40,000 mi) long and averaging more than 1000 km (600 mi) in width. This feature, called the *mid-ocean ridge system*, provided important evidence for the theory of plate tectonics.

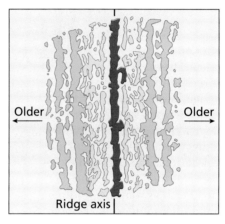

(b) Strips of magnetized rock on either side of the mid-ocean ridge form a matching pattern based on the rock's age.

Older   Older

Ridge axis

▲ **8.17 The process of sea floor spreading** Volcanic activity along Earth's mid-ocean ridge system produces the rock that makes up the ocean floor and drives the process of seafloor spreading. Map (c) on the facing page shows the result in terms of the age of the ocean floor.

Seafloor Spreading

The key to establishing the theory of continental drift was a better understanding of the seafloor crust. As scientists acquired information about the depth variations of the ocean floor, they discovered an interconnected worldwide mountain chain, termed **mid-ocean ridges** (also called spreading centers). These submerged ridges occur at divergent plate boundaries, cracks in the lithosphere where plates move apart and pockets of magma form beneath the surface, as shown in Figure 8.17a. Some magma rises and erupts through fractures and small volcanoes along the ridge, forming new seafloor. As the plates continue to move apart, more magma rises from below to fill the gaps. The upwelling of magma in the ocean floor creates new areas of ocean crust. This process, called **seafloor spreading**, is the mechanism that builds mid-ocean ridges and drives continental moment. Over millions of years, seafloor spreading has produced the oceanic lithosphere covering more than 70% of Earth's surface.

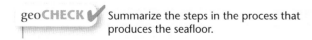 **geoCHECK** ✔ Summarize the steps in the process that produces the seafloor.

Evidence for Seafloor Spreading: Geomagnetism

geomagnetism is the study of phenomena associated with Earth's magnetic field, discussed earlier. In igneous rocks, geomagnetism affects certain minerals, thus providing evidence

for seafloor spreading. As lava erupts along a mid-ocean ridge, magnetic particles in the cooling lava align toward magnetic north. As the lava hardens, these particles lock in this alignment, creating an ongoing magnetic record in the lithosphere of Earth's polarity (▲ **Fig. 8.17b**). Earth's magnetic field has a property called *polarity*, in which the magnetic poles have either positive or negative polarity. Over geologic time, the magnetic field's polarity reverses (switches from positive to negative, or vice versa) at irregular intervals. This chronology of reversals in Earth's magnetic field produces a clear record of ages for rocks on the ocean floor. Notice in Figure 8.17b that the magnetic record pattern of strips of rock is the same on both sides of the ridge. This symmetrical pattern reflects the fact that the corresponding strips on each side of the ridge were formed at the same time. Thus the youngest lithosphere anywhere on Earth is the new lava at the spreading centers of the mid-ocean ridges. With increasing distance from these centers, the lithosphere grows steadily older (Fig. 8.17c).

Overall, the seafloor is relatively young. Nowhere is it more than 200 million years old—remarkable when you remember that Earth's age is 4.6 billion years.

geoCHECK ✔ How does seafloor rock become magnetized?

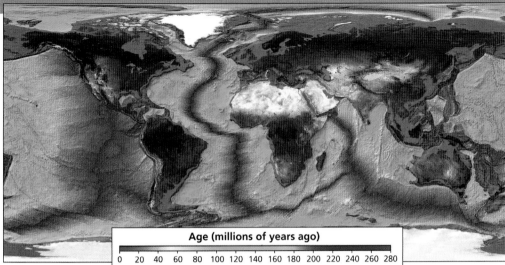

Age (millions of years ago)

| 0 | 20 | 40 | 60 | 80 | 100 | 120 | 140 | 160 | 180 | 200 | 220 | 240 | 260 | 280 |

(c) The youngest rocks (red) are along the mid-ocean ridges. Moving away from the ridges, the rocks become progressively older (blue).

Subduction Zones

What explains the absence of old oceanic lithosphere? In contrast to the older continental lithosphere, the oceanic lithosphere is short-lived because the lithosphere formed along the mid-ocean ridges sinks into **deep-ocean trenches** elsewhere. On the left side of Figure 8.18a, note how one plate is diving beneath another into the mantle. Recall that the basaltic ocean crust has a greater density than the lighter continental crust. As a result, when they slowly collide, the denser ocean floor will slide beneath the less-dense continental crust, thus forming a **subduction zone,** or deep-ocean trench.

Subduction occurs where plates are colliding. The denser subducting slab of lithosphere exerts a gravitational pull on the rest of the plate—pulling the plate into the trench. The subducted portion slides down into the asthenosphere, where it remelts and eventually recycles as magma, rising again toward the surface through deep fissures and cracks in crustal rock. This process effectively destroys older oceanic lithosphere.

Major subduction zones and their trenches occur around the edges of the Pacific plate. Volcanic mountains such as the Andes in South America, the Cascade Range from northern California to the Canadian border, and the Aleutian Islands in Alaska form inland of these subduction zones as a result of rising plumes of magma (Figure 8.18b, c). As subduction occurs, plate motions trigger powerful earthquakes. The strongest earthquakes ever recorded have occurred in subduction zones.

geoCHECK ✔ Explain what occurs in a subduction zone.

▼ 8.18 Subduction in Alaska

(a) For 2,500 km (1552 m) the Gulf of Alaska and the Aleutian Islands form the boundary where the more dense Pacific Plate is subducting underneath the less dense (and lighter) North American Plate.

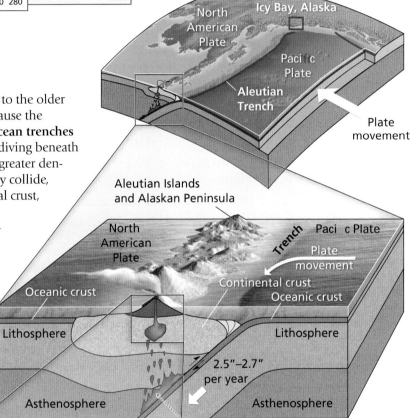

North American Plate

Icy Bay, Alaska

Pacific Plate

Aleutian Trench

Plate movement

Aleutian Islands and Alaskan Peninsula

North American Plate

Trench   Pacific Plate

Plate movement

Oceanic crust

Continental crust

Oceanic crust

Lithosphere

Lithosphere

2.5"–2.7" per year

Asthenosphere

Asthenosphere

As the Pacific Plate descends into the Aleutian Trench, heat from the Earth's interior melts the former ocean floor.

(b) Liquid magma rises to the surface in the form a string of 80 volcanoes, including Mt. Augustine shown here.

(c) Continental and ocean sediments from both plates are deposited onto the North American Plate, where folding and faulting uplifts it above sea level in Icy Bay, Alaska.

geoQUIZ

1. How does geomagnetism provide evidence for plate tectonics?
2. Within seafloor-spreading zones, where are the youngest, and the oldest, rocks located? Explain.
3. What is the geologic relationship between volcanic chains, such as the Andes and Cascade Range, and subduction zones?

8.6 Plate Boundaries

Key Learning Concepts

▶ ***Describe*** the types of plate boundaries and the movement associated with each.

▶ ***Explain*** how earthquake and volcanic activity relate to plate boundaries.

▶ ***Identify*** the cause and effect of hot spots.

Plate boundaries are dynamic places where lithospheric plates interact, producing profound changes in Earth's surface. For example, new lithosphere forms where plates pull apart, but where plates meet, lithosphere is altered by either sliding down into the asthenosphere or colliding upward to form mountains. Where plates slide past each other, lithosphere is neither created nor destroyed. Frequent earthquakes occur at all three types of plate boundaries. What causes plate movements? Driven by the heat energy of Earth's interior, plate motions result from differences in temperature and density within the plates themselves.

Mechanisms of Plate Motion

Mid-ocean ridges and subduction zones are critical to two processes that scientists think drive plate motion (▼ Fig. 8.19), along with the mantle convection currents resulting from differences in rock temperatures and density and the role of gravity discussed earlier in Module 8.2. The addition of new rock along a mid-ocean ridge pushes the plate away from the ridge in the process of **ridge push.** Because the mid-ocean ridges are elevated, ridge push causes the oceanic lithosphere to slide "downhill" toward a trench. Along the way, the plate cools and becomes more dense. Reaching the trench, gravity pulls the plate downward in the process of **slab pull.**

Animation (MG) Convection within the Mantle

geoCHECK ✔ What force is responsible for ridge push and slab pull?

http://goo.gl/gUxbH9

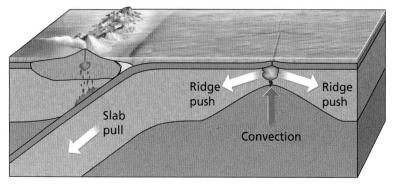

▲ **8.19 Ridge push and slab pull** Gravity drives ridge push and slab pull, in which oceanic lithosphere cools and becomes denser as it moves away from spreading centers and is eventually subducted into the mantle.

Interactions at Plate Boundaries

The block diagrams in Figure 8.20 show the three general types of plate motion and interaction that occur along the boundaries.

Divergent Boundaries Plates pull apart along *divergent boundaries*. These boundaries occur at seafloor spreading centers, where upwelling material from the mantle forms new seafloor and lithospheric plates are under tension and spread (or "pull") apart. Whereas most divergent boundaries occur at mid-ocean ridges, a few occur within continents themselves. An example is the Great Rift Valley of East Africa, where crust is rifting apart.

Convergent Boundaries Continental and oceanic lithosphere collide along *convergent boundaries*. These are zones where subduction can occur and where compression forces squeeze the crust until it fractures, causing earthquakes, volcanic activity, and mountain formation (▼ Fig. 8.20). Examples include the ocean off the west coast of South and Central America, along the Aleutian trenches, and along the east coast of Japan, where the magnitude 9.0 earthquake struck in 2011. Areas where two plates of continental crust collide include the *collision zone* between India and Asia. Areas where oceanic plates collide are found along the deep trenches in the western Pacific Ocean.

Transform Boundaries Where plates slide laterally past one another at right angles, *transform boundaries* occur. Along transform boundaries, plates neither diverge nor converge, and there are usually no volcanic eruptions. These are the right-angle fractures stretching across the mid-ocean ridge system worldwide.

Mid-ocean ridges intersect with transform faults (also called strike-slip faults) ranging from 300 km (186 mi) to over 1000 km (over 600 mi) long. These faults are parallel to the direction of plate movement. The resulting fracture zone is active only along the fault section between ridges. The motion produces horizontal displacement; no new crust is formed or old crust subducted.

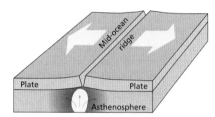

(a) Divergent plate boundary— plates diverge in areas of seafloor spreading at mid-ocean ridges and along rift valleys in continental crust.

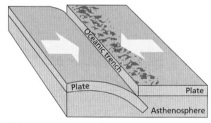

(b) Convergent plate boundary— plates converge, producing a subduction zone. Coastal area features mountains, volcanoes, and earthquakes.

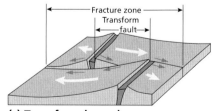

(c) Transform boundary— plates slide past each other, forming a fracture zone including a transform fault in which plates move past each other in opposite directions; along the fracture zone outside of the active fault, plates move in the same direction.

▲ **8.20 Movements of lithospheric plates** The three types of plate interactions are (a) divergent, (b) convergent, and (c) transform.

geoCHECK ✔ List and describe the three types of plate boundaries.

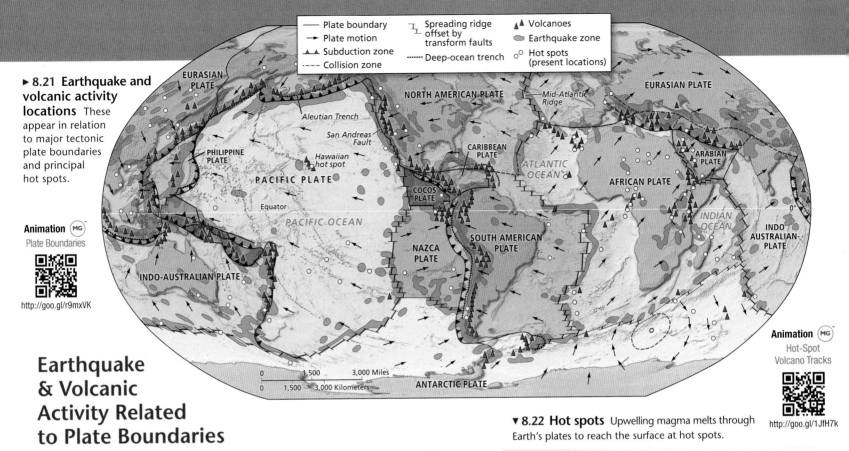

► **8.21 Earthquake and volcanic activity locations** These appear in relation to major tectonic plate boundaries and principal hot spots.

Legend:
— Plate boundary
→ Plate motion
▲▲ Subduction zone
----- Collision zone
⊥ Spreading ridge offset by transform faults
······· Deep-ocean trench
▲▲ Volcanoes
⬭ Earthquake zone
∘∘ Hot spots (present locations)

Animation (MG)
Plate Boundaries

http://goo.gl/r9mxVK

Animation (MG)
Hot-Spot Volcano Tracks

http://goo.gl/1JfH7k

Earthquake & Volcanic Activity Related to Plate Boundaries

Plate boundaries are the primary location of earthquakes and volcanoes. The correlation of these phenomena is important to the theory of plate tectonics for two reasons. First, the correlation proves that moving plates "shake" the earth, causing earthquakes. Secondly, rock material subducting into the asthenosphere becomes molten and then injects upward back into the crust as new intrusive igneous rock or extrusive volcanic rock. For example, in Figure 8.21 notice the "Ring of Fire" surrounding the Pacific Basin, named for its many volcanoes. In several subduction zones, the subducting edge of the Pacific Plate thrusts deep into the crust and mantle and produces upwelling magma that forms active volcanoes along the Pacific Rim. Such processes occur at other plate boundaries throughout the world.

geoCHECK ✔ Summarize the process by which plate tectonics "recycles" rock along convergent boundaries.

Hot Spots

Sites, not associated with plate boundaries, that send upwelling magma to the surface are called **hot spots**. Scientists estimate there are 50 to 100 hot spots across Earth's surface. They occur beneath both oceanic and continental crust, and some are found far from plate boundaries. Some hot spots anchor deep in the stiff lower mantle, remaining fixed relative to migrating plates; others appear to be above plumes that move by themselves or shift with plate motion. Thus, the area of a plate that is above a hot spot is heated for the brief geologic time—a few hundred thousand or million years—when that part of the plate is there.

The Hawaiian Islands illustrates volcanic activity at a hot spot. The Pacific Plate has been moving across a hot, upward-erupting plume over the last 80 million years, creating an island chain stretching northwestward away from the fixed hot spot (► **Fig. 8.22**). The oldest island of the chain is Kauai (5 million years), now eroded into deep canyons and valleys. The big island of Hawaii, the newest, currently sits above the hot spot, and took less than 1 million years to form.

▼ **8.22 Hot spots** Upwelling magma melts through Earth's plates to reach the surface at hot spots.

Kauai 5,100,000 yrs. old
Oahu 3,000,000 yrs. old
Molokai 1,800,000 yrs. old
Maui 1,320,000 yrs. old
Hawai'i 0-400,000 yrs. old

(a) Age of Hawai'ian islands.

(b) Eroded coastal cliffs of windward Oahu.

(c) Fresh lava on Hawai'i reaches the sea.

geoCHECK ✔ What is a hot spot?

geoQUIZ

1. Describe the three types of plate boundaries, and identify at least one place in the world where they are found.
2. Identify the specific characteristics of plate boundaries that produce earthquakes and volcanoes.
3. Explain why some volcanic chains, such as the Hawaiian Islands, are found far from a plate boundary.

8.7 Deformation, Folding, & Faulting

Key Learning Concepts

▶ *Identify* the forces that deform the crust and the effects of deformation.
▶ *Describe* folding and warping and their resulting landforms.
▶ *Compare and contrast* the different types of faults.

Plate motions, ultimately driven by mantle convection, create tremendous forces that change Earth's crust. Over time, these forces wrinkle the crust like a rug, bend it like a soft candy bar, or break it like a cracker, gradually changing the appearance of Earth's surface.

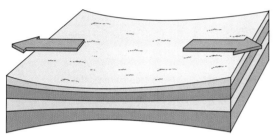

(a) Tension

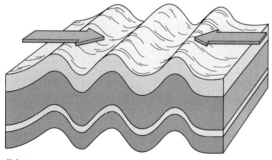

(b) Compression

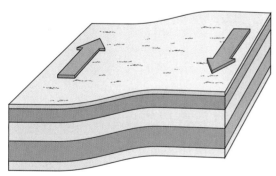

(c) Shear

▲ **8.23 Three kinds of stress and strain and the resulting surface expressions**

Project Condor (MG)
Monoclines of the
Colorado Plateau

https://goo.gl/cpvmoj

Animation (MG)
Folds

http://goo.gl/40nQS4

Project Condor (MG)
Anticlines and
Synclines

https://goo.gl/yvMydB

How Stress Affects Rock: Deformation

Rocks in Earth's crust undergo powerful stress caused by tectonic forces, gravity, and the weight of overlying rocks. Figure 8.23 shows three types of differential stress: *tension* stretches rock, *compression* shortens it, and *shear* twists or tears it. Rocks respond to this *strain* by *folding* (bending) or *faulting* (breaking). In other words, stress is a force acting on the rock, and the resulting strain is the deformation in the rock. Whether a rock bends or breaks depends on rock composition, the amount of pressure, and whether the rock is *brittle* or *ductile*. Brittle rocks fracture easily, while ductile rocks bend slowly. The effects of stress on rocks are clearly visible in the landforms we see today, especially in mountains.

geoCHECK ✔ Construct a table listing the three types of stress and how each deforms rocks.

Folding & Broad Warping

Folding occurs when rocks are deformed as a result of compressional stress and shortening. We can visualize this process by stacking sections of thick fabric on a table and slowly pushing on opposite ends of the stack. The cloth layers will bend and rumple into folds similar to those in the landscape (▼ Fig. 8.24). An arch-shaped upward fold is an **anticline**, in which the rock layers slope downward away from an imaginary center axis. A trough-shaped downward fold is a **syncline**, in which the layers slope upward away from the center axis.

In addition to folding, broad *warping* can cause continental crust to arch upward as if pushed from below. Features formed by warping can be small, fold-like structures called *basins* and *domes*, or regional features the size of the Ozark Mountains in Arkansas and Missouri, the Colorado Plateau, and the Richat Dome in Mauritania.

geoCHECK ✔ How do you distinguish between an anticline and syncline?

▼ **8.24 Folded landscapes** Compression forces can fold the crust, producing anticlines and synclines.

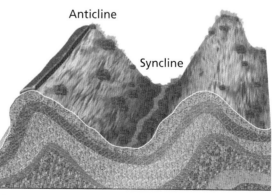

Anticline

Syncline

(a) Compression forces cause the folding that produces anticlines and synclines.

(b) The folded Appalachian Mountains of the Cumberland Gap on the tri-state borders of Kentucky, Tennessee, and Virginia.

Faulting

When rocks are stressed beyond a certain point the strain creates a break, or fracture. Rocks on either side of the fracture move relative to the other side in a process known as **faulting**. Thus, *fault zones* are areas where fractures in the rock demonstrate crustal movement. At the moment of fracture, a sharp release of energy occurs, producing an **earthquake**.

The two sides of a fault move along a *fault plane*. Three types of faults are distinguished by direction of movement along the fault as well as the tilt and orientation of the fault plane.

Normal Faults A **normal fault** forms when rocks are pulled apart by *tensional* stress (▶ Fig. 8.25a). When the break occurs, rock on one side moves vertically along an inclined fault plane. The downward-shifting side is the *hanging wall*; it drops relative to the *footwall block*. A cliff formed by faulting is called a *fault scarp*, or *escarpment*.

Reverse Faults *Compressional* forces associated with converging motion that forces rocks to move *upward* along the fault plane produce a **thrust** or **reverse fault**, in which the hanging wall shifts up and over the footwall (Fig. 8.25b). On the surface, thrust faults appear similar to a normal fault, although more collapse and landslides may occur from the hanging wall. In the Alps, thrust faults result from compressional forces of the colliding African and Eurasian Plates. Beneath the Los Angeles Basin, thrust faults caused many earthquakes, including the $30 billion 1994 Northridge earthquake.

Strike-Slip Faults A **strike-slip** fault (also called *transform fault*) forms when rocks are torn by *lateral-shearing* stress (Fig. 8.25c). The horizontal movement along a strike-slip fault produces movement that is *right-lateral* or *left-lateral,* depending on the motion perceived on one side of the fault relative to the other side. Although strike-slip faults do not produce scarps, they can create linear valleys such as those along the San Andreas fault system of California.

geoCHECK ✔ Compare and contrast the structure and movement of a normal fault with that of a reverse (or thrust) fault.

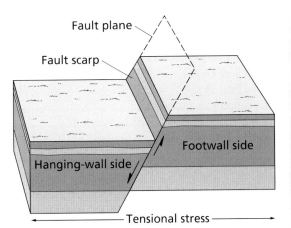

Fault plane
Fault scarp
Footwall side
Hanging-wall side
— Tensional stress —

(a) Normal fault (tension, hanging wall slides down relative to footwall)

(d) Sierra Nevada, California

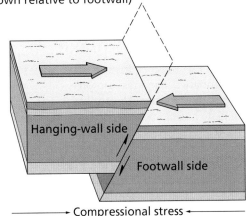

Hanging-wall side
Footwall side
— Compressional stress ◄

(b) Thrust or reverse fault (compression, hanging wall slides up relative to footwall)

(e) Mount Galatea, Alberta, Canada

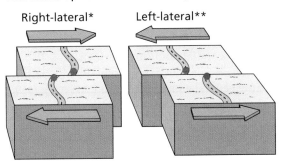

Right-lateral*   Left-lateral**

(c) Strike-slip fault (shearing, side-to-side motion)

   * Viewed from either dot on each road, movement of opposite side is *to the right*.
   ** Viewed from either dot on each road, movement of opposite side is *to the left*.

(f) San Andreas fault, California

▲ 8.25 **Types of faults** (d) Normal fault (e) Thrust fault (f) Strike-slip fault.

geoQUIZ

1. Describe the process that creates a landscape of anticlines and synclines.
2. Explain why a normal fault develops a fault scarp, but a strike-slip fault does not.
3. Compare and contrast the effects of warping, folding, and faulting on a landscape.

Mobile Field Trip (MG)
San Andreas Fault

https://goo.gl/Fi67ZP

Project Condor (MG)
Faults versus Joints

https://goo.gl/JTM0Pi

Animation (MG)
Transform Faults

http://goo.gl/KXG42e

8.8 Earthquakes

Key Learning Concepts

▶ **Explain** what happens during an earthquake.

▶ **Distinguish** between earthquake intensity and magnitude.

▶ **Discuss** why scientists cannot yet predict earthquakes.

▶ **Give examples** of earthquake hazards in different parts of the world.

Tremendous stored energy exists along plate boundaries and faults. Recall that *stress* is the force of plates moving in opposite directions that builds *strain*, which causes deformation in rocks. But if the strain exceeds the strength of the rock, the rock eventually will break—"Earthquake!"

Before, During, & After an Earthquake

Crustal plates do not glide smoothly past one another. Instead, tremendous friction exists along plate boundaries. The stress, or force, of plate motion builds strain, or deformation, in the rocks until friction is overcome and the sides along plate boundaries or fault lines suddenly break loose. The sharp release of energy that occurs at the moment of fracture, producing **seismic waves**, is an **earthquake**, or *quake*. Earthquakes are most often caused by fault ruptures, but can also be caused by volcanic events. The two sides of the fault plane then lurch into new positions, moving distances ranging from centimeters to several meters, and release enormous amounts of seismic energy into the surrounding crust. Seismic waves transmit this energy throughout the planet, diminishing with distance.

The subsurface area along a fault plane, where an earthquake begins, is the *focus*, or hypocenter, of an earthquake (▶ Fig. 8.26). The area at the surface directly above the focus is the *epicenter*. Shock waves produced by an earthquake radiate outward through the crust from the focus and epicenter. An *aftershock* may occur after the main shock, sharing the same general area of the epicenter. Some aftershocks rival the main earthquake in magnitude.

 geoCHECK Explain the role of seismic waves in an earthquake.

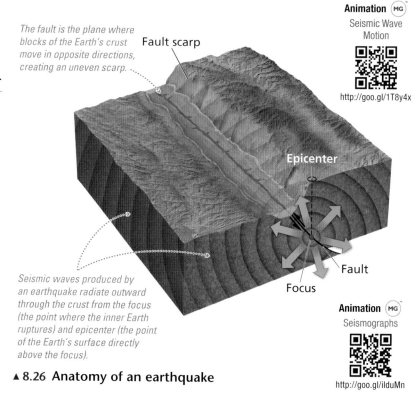

Animation (MG) Seismic Wave Motion http://goo.gl/1T8y4x

The fault is the plane where blocks of the Earth's crust move in opposite directions, creating an uneven scarp.

Fault scarp

Epicenter

Seismic waves produced by an earthquake radiate outward through the crust from the focus (the point where the inner Earth ruptures) and epicenter (the point of the Earth's surface directly above the focus).

Fault

Focus

Animation (MG) Seismographs http://goo.gl/ilduMn

▲ **8.26 Anatomy of an earthquake**

Earthquake Intensity & Magnitude

A worldwide network of more than 4000 **seismograph** instruments records vibrations transmitted as waves of energy throughout Earth. Using seismographs and actual observations, scientists rate earthquakes on either a *qualitative* damage intensity scale, or a *quantitative* scale that measures the magnitude of energy released.

Modified Mercalli Scale Developed in 1903, the *Mercalli intensity scale* is a *qualitative* scale that uses Roman numerals from I to XII to represent intensities from "barely felt" to "catastrophic total destruction." Table 8.1 shows this scale and the number of earthquakes in each category that scientists expect each year.

Table 8.1 Magnitude, Intensity, & Frequency of Earthquakes				
Description	**Effects on Populated Areas**	**Moment Magnitude Scale**	**Modified Mercalli Scale**	**Number per Year***
Great	Damage nearly total	8.0 and higher	XII	1
Major	Great damage	7–7.9	X–XI	17
Strong	Considerable to serious damage to buildings; railroad tracks bent	6–6.9	VIII–IX	134
Moderate	Felt by all, with slight building damage	5–5.9	V–VII	1,319
Light	Felt by some to felt by many	4–4.9	III–IV	13,000 (estimated)
Minor	Slight, some feel it	3–3.9	I–II	130,000 (estimated)
Very minor	Not felt, but recorded	2–2.9	None–I	1,300,000 (estimated)

*Based on observations since 1990.
Source: USGS, Earthquake Information Center.

Richter Scale In 1935, Charles Richter designed a *quantitative* system to estimate earthquake magnitude. In this method, a seismograph located at least 100 km (62 mi) from the epicenter of the quake records the amplitude of seismic waves. The measurement is then charted on the **Richter scale**. The scale is logarithmic: Each whole number on it represents a 10-fold increase in the measured wave amplitude (equal to a 31.5-fold increase in energy released). Thus, a magnitude of 3.0 on the Richter scale represents 31.5 times more energy than a 2.0 and 992 times more energy than a 1.0.

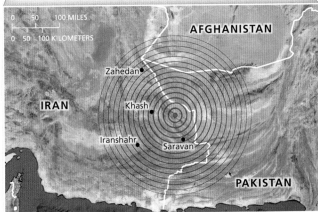

(a) Seismic hazard risk zones.

Moment Magnitude Scale The **moment magnitude scale**, in use since 1993, is more accurate for determining the magnitude of large earthquakes than is the Richter scale. Moment magnitude considers the amount of fault slippage produced by the earthquake, the size of the surface (or subsurface) area that ruptured, and the nature of the materials that faulted, such as whether they exhibit brittle or ductile characteristics. The moment magnitude scale takes into account extreme ground acceleration (movement upward), which the Richter scale method underestimated. A reassessment of past earthquakes has changed the magnitude rating of some. As an example, the 1964 earthquake at Prince William Sound in Alaska had a Richter magnitude of 8.6, but on the moment magnitude scale, it increased to M 9.2.

(b) Seismic waves from a 2013 earthquake in Iran.

Earthquake Prediction

Figure 8.27 plots global earthquake hazards and shows the waves from a 2013 quake in Iran. Note that almost all earthquakes occur along plate boundaries where the plates diverge, converge, or slide past each other. The challenge is to discover how to predict the *specific time and place* for an earthquake in the short term in regions of prior earthquake experience. One approach, the study of *paleoseismology*, examines the history of each plate boundary in order to determine the frequency of past earthquakes. A second approach observes and measures phenomena that might precede an earthquake. For example, the affected region may tilt and swell in response to strain—changes that can be measured by tiltmeters. Another indicator is an increase in radon (a naturally occurring, slightly radioactive gas) dissolved in groundwater. Even with these advances, earthquakes are complex natural events that remain extremely difficult to predict.

(c) A woman sits in front of her devastated home after a 2013 earthquake in Iran.

▲ **8.27 Global seismic hazards**

geoCHECK ✔ Describe two approaches taken in efforts to predict earthquakes.

8.8 (cont'd) Earthquakes

Earthquake Hazards & Safety

Earthquakes are often terrifying events. Unlike other natural disasters such as floods, windstorms, and volcanic eruptions, the Earth shakes with no warning over large geographic areas. Although ground shaking is the initial hazard, surface displacement through faulting, liquefaction of soil (when water-saturated soils suddenly turn into liquid mud when an earthquake strikes), tsunamis, and flooding often accompany earthquakes.

Since 1900, over 1.5 million people have perished from earthquakes around the world (▶ Fig. 8.28). However, earthquake magnitude is but one factor in determining the overall number of casualties. Other dynamics include population density, building codes, the time of day, and other events triggered by the earthquake, such as tsunamis, fires, and debris flows. These factors are discussed with the examples below.

Nepal, 2015 In the spring of 2015, a 7.8 magnitude earthquake and numerous aftershocks leveled parts of Kathmandu, the capital city of Nepal (▼ Fig. 8.29). Over 8000 people died within the city limits and in villages throughout this Himalayan country. Substandard building codes led to the collapse of both new and ancient structures, and Nepal's remote location complicated rescue and relief efforts. After the quake, heavy rains drenched thousands of suddenly homeless people, contributing to the high death toll.

Megathrust Earthquakes Powerful *megathrust* earthquakes occur in convergent subduction zones where one tectonic plate is thrust under another. These "Great Quakes" can generate a magnitude 9.0 or greater event. Recent examples include Japan 2012, Indonesia 2010, Chile 1960, and Alaska, 1964. In the United States, only the Alaska-Aleutian and Cascadia (southern British Columbia to northern California)

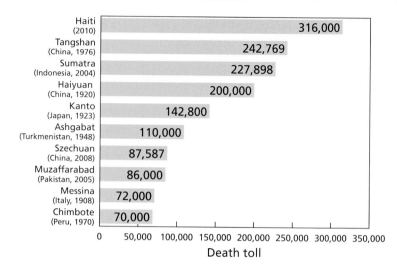

Location	Death toll
Haiti (2010)	316,000
Tangshan (China, 1976)	242,769
Sumatra (Indonesia, 2004)	227,898
Haiyuan (China, 1920)	200,000
Kanto (Japan, 1923)	142,800
Ashgabat (Turkmenistan, 1948)	110,000
Szechuan (China, 2008)	87,587
Muzaffarabad (Pakistan, 2005)	86,000
Messina (Italy, 1908)	72,000
Chimbote (Peru, 1970)	70,000

Death toll

▲ **8.28 Major earthquakes' death toll** Collapsing buildings and tsunamis cause many earthquake-related casualties.

subduction zones are capable of generating megathrust quakes. The immense energy released during these seismic events often triggers accompanying disasters such as earth and debris flows, flooding, and tsunamis (▼ Fig. 8.30).

Virginia, 2011 Some earthquakes occur distant from plate boundaries. In 2011 a 5.8 earthquake centered 64 km (40 mi) northwest of Richmond, Virginia, shook the eastern United States. It was the region's most powerful temblor since 1897 and sent people scurrying from homes and offices from Maine to South Carolina (▶ Fig. 8.31). The earthquake forced the evacuation of many government buildings in Washington, D.C., cracked foundation stones in the Washington Monument, and temporarily closed nuclear power reactors in Virginia while officials checked for damage.

▼ **8.29 Nepal, 2015** Damage was severe because most structures in Nepal are not designed to withstand earthquakes.

(a) The tsunami that followed the 2011 earthquake devastated settlements along Japan's northern Pacific coast.

▲ **8.30 Japan, 2011**

San Andreas Fault, California Earthquakes are frequent in a zone along the San Andreas fault, the transform boundary where the North American and Pacific Plates meet. In 1989, a magnitude 6.9 quake rocked California's San Francisco Bay Area, causing 6 billion dollars in damage and claiming 67 lives—47 of them when a section of the Bay Bridge collapsed (▶ Fig. 8.32). The 1989 earthquake is often compared to the 1906 "Great Quake" and fires that killed over 3000 people and destroyed 80% of San Francisco. Although the 7.9 magnitude of the 1906 temblor was much greater, the lower death toll in 1989 is attributed to progress in developing and enforcing better building codes and earthquake safety procedures.

geoCHECK ✔ Where do megathrust earthquakes occur?

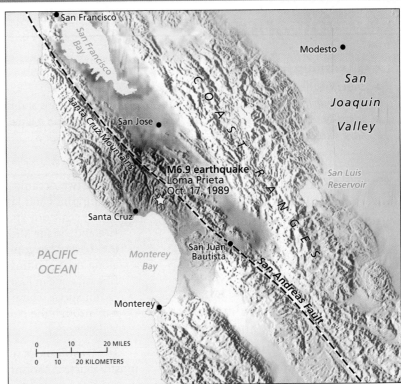

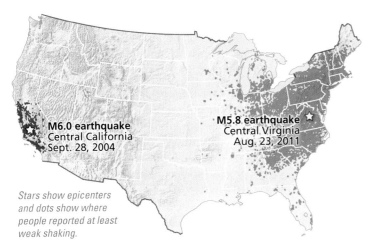

Stars show epicenters and dots show where people reported at least weak shaking.

▲ **8.31 Mapping Earthquake Shaking** The United States Geologic Survey maps reports of earthquake shaking. Notice how the 2011 Virginia quake was felt across a much wider area than the 2004 California quake of similar magnitude.

PERCEIVED SHAKING	Not tell	Weak	Light	Moderate	Strong	Very strong	Severe	Violent	Extreme
POTENTIAL DAMAGE	None	None	None	Very light	Light	Moderate	Moderate/Heavy	Heavy	Very Heavy
INSTRUMENTAL INTENSITY	I	II–III	IV	V	VI	VII	VIII	IX	X+

(a)

(b)

▲ **8.32 San Francisco Bay Area, 1989** The magnitude 6.9 Loma Prieta earthquake caused significant damage in scattered parts of the San Francisco Bay area. (a) After the 1989 earthquake, geologists used the Modified Mercalli scale to rate the amount of shaking across the area affected by the quake. (b) Damage from the 1989 earthquake.

(b) Almost nothing was left of Japan after the tsunami struck.

geoQUIZ

1. What physical factors trigger an earthquake?
2. What is the relationship between an earthquake's focus and epicenter?
3. Describe the difference between earthquake intensity and magnitude.

8.9 Volcanoes

Key Learning Concepts

▶ **Describe** the distribution of volcanic activity.
▶ **Distinguish** between effusive eruptions and explosive eruptions and the volcanic landforms each produces.
▶ **Identify** the structural features of a volcano, types of lava, and related volcanic landforms.
▶ **Give examples** of volcano hazards in different parts of the world

Over 1300 volcanic cones and mountains exist on Earth, of which 600 are classified as **active**, meaning that they have erupted in the last 10,000 years. Globally, about 50 volcanoes erupt on Earth's land surface each year. Volcanoes are spectacular to witness from a safe distance, but worldwide, more than 500 million people live within a volcanic hazard zone.

Distribution of Volcanic Activity

Volcanic eruptions reveal Earth's internal energy and match plate tectonic activity (▼ Fig. 8.33). Most volcanic activity occurs in three settings:

- Along subduction zones where continental and oceanic plates converge (the Cascade Range in the Pacific Northwest and the Andes of South America) or

where two oceanic plates converge (Philippines and Japan).
- Along sea floor spreading centers (Iceland, on the Mid-Atlantic Ridge; off the coast of Oregon and Washington) and along areas of rifting on continental plates (the rift zone in East Africa).
- At hot spots, where plumes of magma rise through the crust (Hawaii and Yellowstone).

The 70 volcanoes in North America are mostly inactive. Mount St. Helens in Washington is the most active.

geoCHECK ✔ Compare and contrast volcanic activity at divergent and convergent boundaries.

Types of Volcanic Activity

The eruption type and resulting lava is determined by the magma's mineral composition, which is related to its source, and by the magma's viscosity. Viscosity is the magma's "thickness," or resistance to flow, which ranges from low viscosity (very fluid) to high viscosity (thick and slow flowing).

Effusive Eruptions Originating from the asthenosphere and upper mantle, **effusive eruptions** are relatively gentle, such as lava flows on the seafloor, or occur in hot spots such as Hawaii and Iceland. Gases readily escape from the fluid magma, producing little volcanic debris, although trapped gases can generate lava fountains. The low-viscosity lava cools into a basaltic rock, low in silica and rich in iron and magnesium. Repeated eruptions create a gently sloping **shield volcano** (▶ Fig. 8.34c).

Explosive Eruptions Magma produced by subducting oceanic plates is more viscous and forms **explosive eruptions.** The high silica and aluminum content produces magma so viscous that it prevents the gases from escaping by blocking the magma conduit. Pressure of the trapped gases increases until it causes an explosive eruption. Repeated eruptions of alternating ash, rock, and lava form a steep-sided and more conical **composite volcano**, also called a *stratovolcano* (Fig. 8.34b). Explosive eruptions produce less lava than effusive eruptions, but more **pyroclastics**—ash, dust, cinders, scoria, pumice, and aerial bombs.

▼ **8.33 Tectonic settings of volcanic activity** Wherever plates pull apart or where oceanic plates are subducted, volcanic activity is likely to occur.

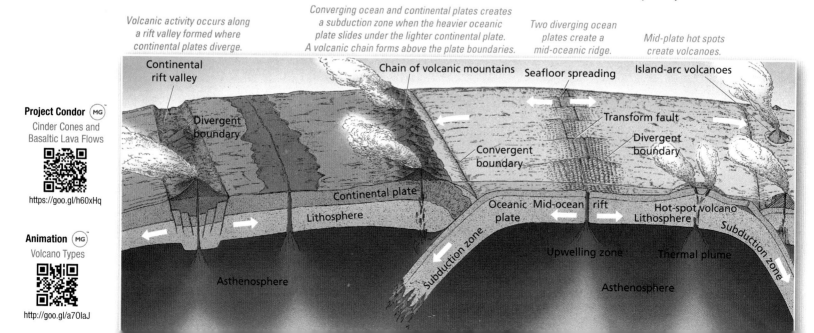

Volcanic activity occurs along a rift valley formed where continental plates diverge.

Converging ocean and continental plates creates a subduction zone when the heavier oceanic plate slides under the lighter continental plate. A volcanic chain forms above the plate boundaries.

Two diverging ocean plates create a mid-oceanic ridge.

Mid-plate hot spots create volcanoes.

Continental rift valley
Chain of volcanic mountains
Seafloor spreading
Island-arc volcanoes
Divergent boundary
Transform fault
Divergent boundary
Convergent boundary
Continental plate
Oceanic plate
Mid-ocean rift
Hot-spot volcano
Lithosphere
Lithosphere
Subduction zone
Upwelling zone
Thermal plume
Subduction zone
Asthenosphere
Asthenosphere

Project Condor (MG)
Cinder Cones and Basaltic Lava Flows

https://goo.gl/h60xHq

Animation (MG)
Volcano Types

http://goo.gl/a7OlaJ

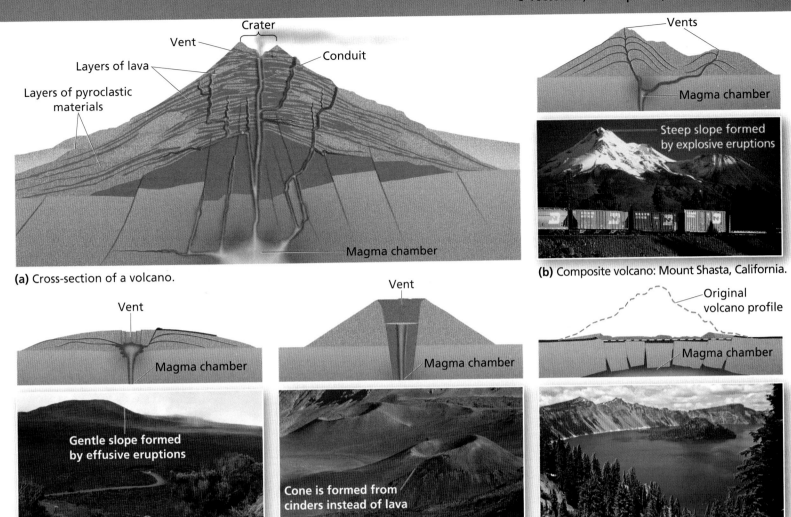

(a) Cross-section of a volcano.

(b) Composite volcano: Mount Shasta, California.

(c) Shield volcano. Road to the Piton de la Fournaise volcano, located on the island of Reunion in the Indian Ocean.

(d) Kama'oli'i and Pu'u o Pele cinder cones.

(e) Crater Lake in Oregon is a caldera.

▲ **8.34 Volcano structure and types**

In addition to lava and pyroclastic materials, a volcanic eruption can spread particulates, gases, and aerosol clouds and can alter surface albedo or atmospheric albedo. Ash plumes reflecting sunlight are what typically cause post-eruption cooling.

geoCHECK ✔ List and describe the different types of pyroclastics.

Volcanic Features

A **volcano** forms at the end of a vent that rises from the asthenosphere and upper mantle through the crust to create a volcanic mountain, as shown in Figure 8.34a. A **crater** usually forms on or near the summit. Magma rises and collects in a chamber deep below the volcano until the tremendous heat and pressure is sufficient to cause an eruption.

Recall that magma that reaches the surface is called lava. Lava, gases, and pyroclastics, eject through the vent to build volcanic landforms. Flowing basaltic lava takes two principal forms that differ in texture—both have Hawaiian names (▶ **Fig. 8.35**). Rough and jagged basalt with sharp edges is **aa**. The rough texture occurs when lava loses trapped gases, flows slowly, and develops a thick skin that cracks into jagged surfaces. In contrast,

pahoehoe forms a thin crust of "ropy" folds. Both forms can erupt together, and pahoehoe can turn into aa downslope.

A **cinder cone** is a cone-shaped hill usually less than 450 m (1500 ft) high, with a truncated top formed from cinders (a type of pyroclastic rock full of air bubbles) that accumulate during moderately explosive eruptions (Fig. 8.34d). A **caldera** (Spanish for "kettle") is a large depression that forms when summit material on a volcano collapses inward after an eruption (Fig. 8.34e).

geoCHECK ✔ Explain the difference between a cinder cone and a caldera.

▼ **8.35 Two types of basaltic lava from Hawaii** Aa (left) has a rough texture, while pahoehoe (right) has smoothly folded texture.

8.9 (cont'd) Volcanoes

Volcano Hazards & Monitoring

The danger to humans from active volcanoes includes flowing lava, volcanic bombs, and explosive ash and pyroclastic materials, among others. Volcanism also triggers earth and debris flows, flooding from melting ice and snow, and earthquakes. In the 20th century alone, almost 100,000 people died from volcanic activity, and another 5.6 million people suffered some degree of injury, evacuations, and property damage. In the past 30 years alone, volcanism has killed 29,000 people, forced 800,000 evacuations, and caused $3 billion in damage, spurring researchers' efforts to improve hazard mapping and enhance early warning systems.

Monitoring Volcanoes Unlike earthquakes, which cannot be accurately predicted, reliable indications usually precede volcanic eruptions. When a volcano begins to show signs of activity, scientists monitor data on gas emissions, ground deformation, and seismic activity (▶ Fig. 8.36). Governments use this data to determine the timing and nature of warnings intended to mitigate the loss of life and property. The following examples portray how ongoing volcanic activity threatens the well-being of human life and property (▼ Fig. 8.37). (Note that Figure 8.40 shows the locations of these recent eruptions.)

Andes Mountains Tungurahua is an active stratovolcano in the Andes Mountains of South America. In February 2014, an eruption sent clouds of ash and other pyroclastic material 8 km (5 mi) into the atmosphere, wreaking havoc over the Ecuadorian cities of Baños and Ambato (▼ Fig. 8.37). Dormant until 1999, Tungurahua, meaning "throat of fire" in the indigenous Quechua language, is one of eight active volcanoes in Ecuador. In 2015, it continued to spew ash and sulfur dioxide

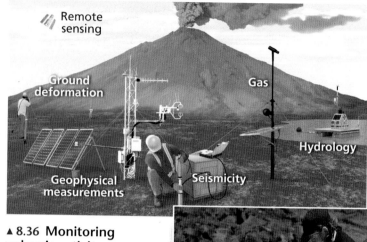

▲ **8.36 Monitoring volcanic activity** Geologists use a variety of instruments to monitor the physical changes that precede a volcanic eruption.

Mobile Field Trip (MG)

Kilauea

https://goo.gl/Z0UzsU

and trigger small earthquakes as molten lava flowed from the mantle into volcanic vents near the earth's surface.

Farther south in the Andean chain, Calbuco, a dormant volcano in Chile, suddenly erupted in April 2015. The plume of smoke and ash reached 10 km (6 mi) into the atmosphere (▼ Fig. 8.38). Within days, the drifting cloud deposited up to 1 m (3.3 ft) of ash on roadways and homes, forcing the evacuation of people and farm animals.

▼ **8.37 Andean outburst, Ecuador, 2014** Tungurahua is an active composite volcano in the Andes Mountains of South America. This February 2014 eruption sent clouds of ash and other pyroclastic material to 8 km (5 mi) in altitude, causing havoc over the Ecuadorian cities of Baños and Ambato.

▼ **8.38 Ash deposits, Chile, 2014**

(a)

Indonesia In 2014, eruptions of volcanic gas and ash from erupting Mount Sinabung rose up to 2000 m (6561 ft) above dozens of small villages, sending lava, ash, and other pyroclastic material hurtling toward people, livestock, and fields (▼ **Fig. 8.39**). Located on the Pacific Ring of Fire, the many islands that compose Indonesia are home to 147 active volcanoes.

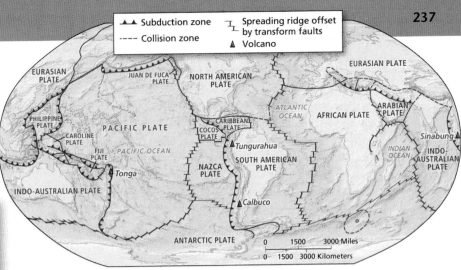

▲ **8.40 Recent volcanic eruptions**

▲ **8.39 Fleeing Mount Sinabung, Indonesia, 2014** Eruptions of volcanic gas and ash rose up to 2000 m (6561ft) above dozens of small villages. On this day, lava and pyroclastic material flowed downslope just 4.5 km (2.8 mi) away.

▶ **8.41 New island rises in the South Pacific Ocean, 2014** (a) An undersea volcano erupted in the Polynesian island Kingdom of Tonga. (b) The new landscape measures 3 km² (1.1 mi²) in size. The highest peak is approximately 250 m (820 ft) above the ocean. Note how rainfall is actively carving channels into the soft volcanic rock.

(a)

Tonga In December 2014, an undersea volcano erupted in the Kingdom of Tonga, a group of islands in the South Pacific Ocean. For over a month, periodic eruptions spewed lava and ash up to 4500 m (14,765 ft) into the sky. When the eruption ended in January 2015, a new island measuring 1 km (.6 mi) wide and 250 m (820 ft) high rose above the ocean surface (▶ **Fig. 8.41**). Although already home to shorebirds, persistent rainfall combined with a weak underlying geologic structure may level the island within a year.

(b)

(b)   (c)

geo**CHECK** ✔ What data make predicting volcanic eruptions more feasible than predicting earthquakes?

geo**QUIZ**

1. What is the relationship between the location of volcanic activity and plate tectonics?
2. Define viscosity and explain its relevance to the two main types of volcanic eruption.
3. Describe the difference between a shield volcano and a composite volcano, and provide an example of each.

8.10 Mountain Building

Key Learning Concepts

▶ **Define** orogenesis.

▶ **Explain** how mountains of different types support the plate tectonics model.

The geologic term for mountain building is **orogenesis**, meaning the birth of mountains (*oros* comes from the Greek for "mountain"). An *orogeny* is a mountain-building episode, occurring over millions of years as a large-scale deformation and uplift of the crust. For example, the Sierra Nevada of California formed when a granite batholith was exposed by erosion following uplift. No orogeny is a simple event: Movements of the crust along faults, the convergence of plates, and volcanic activity can be involved. Many orogenies occurred in Earth's past, and the processes continue today.

Types of Orogenesis

Earth's mountain belts correlate well with the plate tectonics model. Three types of tectonic activity cause orogenesis along convergent plate boundaries. These processes are illustrated in Figure 8.43.

Mountains From Oceanic–Continental Plate Collisions

Figure 8.43 illustrates three types of tectonic activity that cause orogenesis along convergent plate margins. The first type involves oceanic plate–continental plate collisions (▶ **Fig. 8.43a**) that occur along the Pacific coast of the Americas and has formed the Andes, Sierra of Central America, Rockies, and other western mountains. Folded sedimentary formations, with intrusions of magma forming granitic plutons compose these mountains. Also note the volcanic activity inland from the subduction zone and composite volcanoes.

Mountains From Oceanic–Oceanic Plate Collisions

Another type of orogenesis, oceanic plate–oceanic plate collisions (▶ **Fig. 8.43b**), produces volcanic island arcs, such as Indonesia and Japan. The same process formed the island arcs that continue from the southwestern Pacific to the western Pacific, the Philippines, the Kurils, and portions of the Aleutians. Both of these collision types are active around the Pacific Rim's "Ring of Fire." These processes are thermal in nature, because the subducting plate melts and the magma then rises back toward the surface as molten rock.

Mountains From Continental–Continental Plate Collisions

A third type of orogenesis involves continental plate–continental plate collisions. In these instances, large masses of continental crust are mechanically subjected to intense folding, overthrusting, faulting, and uplifting (▶ **Fig. 8.43c**). The converging plates crush and deform both marine sediments and basaltic oceanic crust, pushing them up as the mountains grow. The European Alps and American Appalachian Mountains result from such compression, with associated crustal shortening and overturned folds.

The collision of India with Eurasia is uplifting the Himalayan-Karakoram-Hindu Kush-Pamir, and Tien Shan chains (Fig. 8.43c). The Indian Plate is moving northwestward into the Eurasian Plate up to 6 cm (2.3 in) a year. Over 40 million years, this collision has shortened the continental crust by 1000 km (600 mi), produced thrust faults at depths of 40 km (25 mi), and caused severe earthquakes. This region includes all 10 of the highest above-sea-level peaks on Earth.

Faulted Landscapes

Across some landscapes, pairs of normal faults act together to form a distinctive terrain made up of parallel mountains, often called *fault-block mountains*, separated by valleys. The term **horst** applies to upward-faulted blocks that form the mountains; **graben** refers to downward-faulted blocks that form the valleys (▼ **Fig. 8.42**). Examples include the Great Rift Valley of East Africa, associated with crustal spreading, and the Basin and Range landscapes of the American West.

geoCHECK ✔ List and describe the three types of orogenesis.

▼ **8.42 Faulted landscapes** (a) Paired faults produce fault-block mountains in a horst-and-graben landscape. (b) Death Valley features parallel horsts, separated by the down-dropped graben.

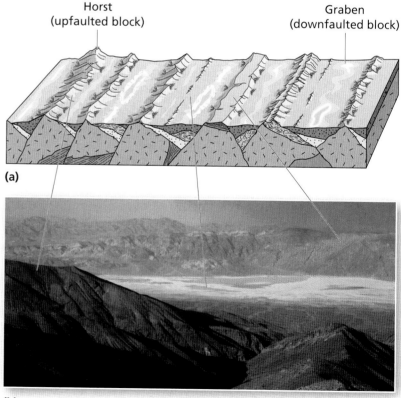

Horst (upfaulted block)   Graben (downfaulted block)

(a)

(b)

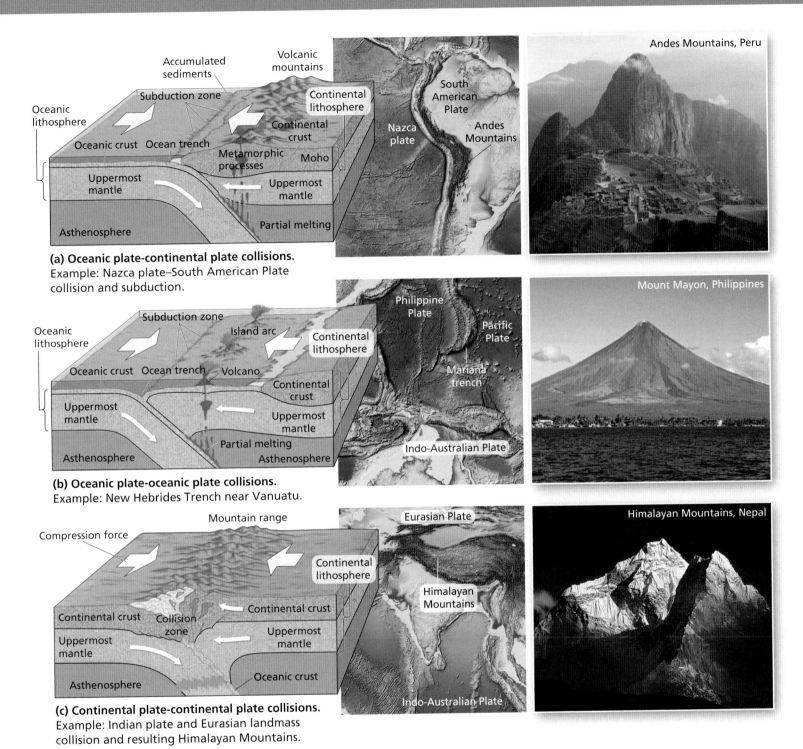

(a) Oceanic plate-continental plate collisions.
Example: Nazca plate–South American Plate collision and subduction.

(b) Oceanic plate-oceanic plate collisions.
Example: New Hebrides Trench near Vanuatu.

(c) Continental plate-continental plate collisions.
Example: Indian plate and Eurasian landmass collision and resulting Himalayan Mountains.

▲ 8.43 **Three types of orogenesis caused by plate convergence** (a) oceanic plate–continental plate collisions (example: Nazca plate–South American Plate collision and subduction); (b) oceanic plate–oceanic plate collisions (example: New Hebrides Trench near Vanuatu; and (c) continental plate–continental plate collision (example: Indian Plate and Eurasian landmass collision and resulting Himalayan Mountains).

Project Condor (MG)
Continental Rifting

https://goo.gl/fGliRV

geoQUIZ

1. What process created the "Ring of Fire" orogenic region?
2. What type of orogenesis produces intensely folded mountains?
3. Referring to Figure 8.42, summarize the process that forms fault-block mountains and their related valleys.

EARTH'S INTERIOR PROCESSES IMPACT HUMANS

• Earth's interior processes cause natural hazards such as earthquakes and volcanic events that affect humans and ecosystems.
• Rocks provide materials for human use; geothermal power is a renewable resource.

HUMANS IMPACT ENDOGENIC PROCESSES

• Wells drilled into Earth's crust in association with oil and gas drilling or for enhanced geothermal systems may cause earthquakes.

8a

Hydrothermal activity produces hot springs and associated travertine deposits in Yellowstone National Park, Wyoming, which sits above a stationary hot spot in Earth's crust. Grand Prismatic Spring, pictured here, is the largest hot spring in the United States. The geysers and thermal features of this area draw over 3 million visitors each year.

8b

The Mid-Atlantic Ridge system surfaces at Thingvellir, Iceland, now a tourist destination. The rifts mark the divergent boundary separating the North American and Eurasian plates.

This National Geophysical Data Center image combines land topography and ocean bathymetry to show Earth's relief.

8d

In April 2013, the Nevada Desert Peak EGS became the first U.S. enhanced geothermal project to supply electricity to the power grid.

ISSUES FOR THE 21ST CENTURY

• Geothermal capacity will continue to be explored as an alternative energy source to fossil fuels.
• Mapping of tectonically active regions will continue to inform policy actions with regard to seismic hazards.

8c

Uluru, also known as Ayers Rock, is probably Australia's best-known landmark. This steep-sided isolated sandstone feature, about 3.5 km (2.2 mi) long and 1.9 km (1.2 mi) wide, was formed from Earth's interior and surface processes and has cultural significance for the Aboriginal people.

Looking Ahead

In the next chapter, we investigate geomorphology—the science of surface landforms. These processes include the weathering and erosion of the uplifted landscapes that formed the subject of this chapter.

Chapter 8 Review

What are Earth's History, Interior Structure, & Materials?

8.1 The Vast Span of Geologic Time

Describe the geologic time scale.

Distinguish between relative and absolute time.

Explain how the principle of uniformitarianism helps geologists interpret Earth's history.

- The *geologic time scale* is a summary timeline of all Earth history. Earth is about 4.6 million years old. The time scale is based on the *absolute* age or *relative* age of rock strata. Of these, relative dating supports the principle of *superposition*. A guiding principle of Earth science is uniformitarianism, the idea that "the present is the key to the past."

1. How are fossils and the principle of superposition used to help determine the difference between absolute time and relative time?

8.2 Earth's History & Interior

Summarize how and when Earth formed.

Describe Earth's layered structure.

- Earth's interior is made up of layers, each one distinct in either chemical composition or temperature. Heat energy moves outward from the center by conduction and convection. Earth's layers include the inner core, outer core, mantle, and crust. Near the top of the mantle is the asthenosphere, a plastic-like layer that underlies the rigid lithosphere and outer crust.

2. What is the relationship between the asthenosphere, the Moho, and the crust?

8.3 Cycles in Earth Systems

Analyze how Earth's geologic cycle relates the tectonic, rock, and hydrologic cycles.

Explain how rocks cycle through geologic time. **differentiate** the processes that create igneous, sedimentary and metamorphic rocks.

- The geologic cycle is the endless tug-of-war between the endogenic (interior Earth) forces that build landforms and the exogenic ones that erode them. This cycle can be further subdivided into the rock, tectonic, and hydrologic cycles. This chapter emphasizes the rock cycle, which forms the three main classes of rocks: igneous, sedimentary, and metamorphic.

3. Identify the two broad subdivisions of igneous rocks, then describe how one of these rocks could transition into a sedimentary rock, and then again, how the sedimentary rock could transition into a metamorphic rock. Include the environments where these transitions occur, such as an ocean trench or river delta.

How Does Plate Tectonics Explain Changes in Earth's Surface?

8.4 Plate Tectonics

Summarize Wegener's hypothesis of continental drift, the formation and breakup of Pangaea, and why scientists at the time rejected the hypothesis.

Describe how the processes of plate tectonics transform Earth's surface over time.

- Modern science has established that there is a cycle in which continents collide, forming supercontinents that then move apart and eventually re-form again. This process, called plate tectonics, includes the upwelling of magma; lithospheric plate movements; seafloor spreading and lithospheric subduction; earthquakes; volcanic activity; and lithospheric deformation such as warping, folding, and faulting.

4. How are earthquakes, volcanic eruptions, and faulting linked to plate tectonics?

8.5 SeaFloor Spreading & Subduction Zones

Describe the roles of seafloor spreading and subduction in forming and destroying the ocean floor.

Explain how geomagnetism supports the concept of seafloor spreading.

State the relationship between subduction, earthquakes, and volcanoes.

Identify two mechanisms of plate motion.

- On the seafloor, interconnected mid-ocean ridges occur at divergent plate boundaries, where plates move apart and pockets of magma form beneath the surface. Magma erupts through fractures and small volcanoes along the ridge, forming new seafloor. As the magma cools, bits of iron align with the magnetic North Pole, providing a record of seafloor age. Oceanic lithosphere is subducted beneath deep ocean trenches in subduction zones.

5. Why do some tectonic plates "subduct" underneath other plates?

8.6 Plate Boundaries

Describe the types of plate boundaries and the movement associated with each.

Explain how earthquake and volcanic activity relate to plate boundaries.

Identify the cause and effect of hot spots.

- At plate boundaries, plates meet and then either slide down into the asthenosphere or collide upward to form mountains. Plate motions—divergent, convergent, or transform—largely determine the landforms on Earth's surface. Plate boundaries are also the primary location of earthquake and volcanic activity. In the middle of some oceanic and continental plates, hot spots send upwelling magma to the surface.

6. How would the regional landforms of the Himalaya–Karakoram Mountains and the African Rift Zone be different if their respective plate motions suddenly reversed their directions?

How Do Plate Motions Affect Earth's Crust?

8.7 Deformation, Folding, & Faulting

Identify the forces that deform the crust and the effects of deformation.

Describe folding and warping and their resulting landforms.

Compare and contrast the different types of faults.

- All rocks undergo powerful stress by tectonic forces, gravity, and the weight of overlying rocks. Rocks respond to this strain by warping, folding, or faulting. These processes closely correlate with plate boundaries, and they are strong evidence for plate tectonics.

7. What is the difference between warping, folding, and faulting in a landscape?

8.8 Earthquakes

Explain what happens during an earthquake.

Distinguish between earthquake intensity and magnitude.

Discuss why scientists cannot yet predict earthquakes.

Give examples of earthquake hazards in different parts of the world.

- Earthquakes occur mostly along plate boundaries. Earthquakes occur when friction is overcome and the sides along plate boundaries or fault lines suddenly shift. Seismic waves then radiate outward from the focus and the epicenter, carrying energy throughout the planet. A worldwide network of seismographs record the transmitted waves.

8. Where and why are humans more likely to experience earthquakes and earthquake damage?

8.9 Volcanoes

Describe the distribution of volcanic activity.

Distinguish between effusive eruptions and explosive eruptions and the volcanic landforms each produces.

Identify the structural features of a volcano, types of lava, and related volcanic landforms.

Give examples of volcano hazards in different parts of the world.

- Volcanic activity occurs along subduction zones where continental and oceanic plates converge, in seafloor spreading zones, and at hot spots. The eruption type and the properties of the resulting lava are determined by the magma's chemistry. Volcanic activity appears in two main forms: effusive eruptions that produce shield volcanoes and explosive eruptions that produce composite cones and pyroclastic material.

9. Describe the three geologic settings where most volcanic activity occurs.

8.10 Mountain Building

Define orogenesis.

Explain how mountains of different types support the plate tectonics model.

- An *orogeny* is a mountain-building episode, occurring over millions of years as a large-scale deformation and uplift of the crust. The resulting volcanic or folded and faulted mountain chains are found at the plate margins. Parallel mountain ranges formed by pairs of normal faults occur in some regions, forming distinctive landscapes of horsts and grabens.

10. Describe how two colliding continental plates produce a landscape different from the one produced by an oceanic hot spot.

Critical Thinking

1. Draw a simple sketch of the Earth's interior, label each layer, and list the physical characteristics, temperature, composition, and depth of each layer on your drawing.

2. Describe how an intrusive igneous rock forms. How would this same rock transition into a sedimentary rock? How could your new sedimentary rock then become a metamorphic rock?

3. Explain how different types of plate boundaries produce different types of landforms and landscapes.

4. Diagram a simple folded landscape in cross section, and identify two features created by the folded strata.

5. Where do you expect to find areas of volcanic activity in the world? Explain your answer.

Visual Analysis

Figure R8.1 looks into the inner gorge of the Grand Canyon.

1. What does the principle of superposition tell us about where the oldest rocks, and the youngest rocks, are found in this photograph?

2. What likely tectonic forces mentioned in the text caused the rock layers in the middle of the photograph to tilt from left (higher) to right (lower)?

3. What suggests that the lowermost rocks might be a different type of rock (metamorphic, igneous, or sedimentary), from the rocks that lie on top of them?

▲ R8.1

Explore | Use **Google Earth** to explore the **Himalayas**.

Viewing Earth from space allows us to visualize tectonic plates on a continental/oceanic scale. In Google Earth, make sure that the *"Borders and Labels"* are checked, but leave other categories unchecked. Fly to *Nepal* and locate the capital, *Kathmandu*, from which you can view the Ganges River Plain to the south, the Himalaya in the middle, and the Tibetan Plateau to the north. Locate the city of *Musahri* on the Ganges Plain at the bottom of the image. While watching the elevation change (displayed at the bottom of the Google Earth screen), slide the cursor northward to Kathmandu, then east-northeast to Mt. Everest, marked with a green mountain symbol. From Mt. Everest, slide the cursor north across the Himalaya to the Tibetan Plateau.

1. How do topography and elevation change between India and Tibet?

2. How does plate tectonics explain these changes?

▲ R8.2

(MG) Interactive Mapping | Login to the **MasteringGeography** Study Area to access **MapMaster**.

Comparing Earthquakes & Population Density in North America

- Open: MapMaster in MasteringGeography
- In the *Physical Environment* categories, Select: *Earthquake Hazard Areas* (U.S.). Next, turn on the *Population* categories, and select *Population Density*.

1. Compare the relationship between earthquake frequency and population density in the western and eastern United States. In which parts of the country are large numbers of people *most* at risk from earthquakes?

2. In which parts of the country are people *least* at risk from earthquakes?

MasteringGeography™

Looking for additional review and test prep materials? Visit the Study Area in MasteringGeography™ to enhance your geographic literacy, spatial reasoning skills, and understanding of this chapter's content by accessing a variety of resources, including MapMaster interactive maps, videos, *In the News* RSS feeds, flashcards, web links, self-study quizzes, and an eText version of *Geosystems Core*.

Key Terms

aa, p. 25
absolute time, p. 214
anticline, p. 18
batholith, p. 220
caldera, p. 235
cinder cone, p. 235
composite volcano
    p. 234
convection current,
    p. 216
crater p. 235
deep ocean trench, p. 225
earthquake, p. 229
effusive eruption, p. 234

explosive eruption, p. 234
extrusive igneous, p. 219
faulting, p. 229
folding, p. 18
geologic cycle p. 218
geologic time scale,
    p. 214
geomagnetism, p. 224
graben, p. 238
horst, p. 238
hot spot p. 227
igneous rock, p. 219
intrusive igneous, p. 219
lava, p. 220

lithospheric plates
    p. 222
magma p. 220
metamorphic rock
    p. 219
mid-ocean ridges,
    p. 224
mineral, p. 219
Mohorovičić
    discontinuity, p. 217
Moho, p. 217
moment magnitude scale
    p. 231
normal fault p. 229

orogenesis p. 238
Pangaea, p. 222
pahoehoe p. 222
pluton, p. 220
principle of
    superposition, p. 214
pyroclastic, p. 234
relative time, p. 214
reverse fault p. 229
Richter scale p. 231
ridge push p. 231
rock cycle p. 219
rock, p. 9
seafloor spreading, p. 224

sedimentary rock, p. 219
seismic wave, p. 230
seismograph, p. 230
slab pulls p. 226
strike-slip p. 229
subduction zone,
    p. 225
syncline, p. 18
theory of plate tectonics,
    p. 222
thrust p. 229
uniformitarianism,
    p. 215
volcano p. 234

Life on the Edge: How do plate boundaries put humans at risk?

Figure GL 8.1 shows tectonic features in the recent geologic time of the late Cenozoic Era. Note that in Figure GL 8.1, each arrow represents 20 million years of movement. Recall the specific types of plate boundaries: convergent (subduction); divergent (seafloor spreading); and transform (lateral motion between crests of spreading ridges producing transform, or strike-slip, faults).

GeoLab8 (MG)
Pre-Lab Video

https://goo.gl/DOYwiS

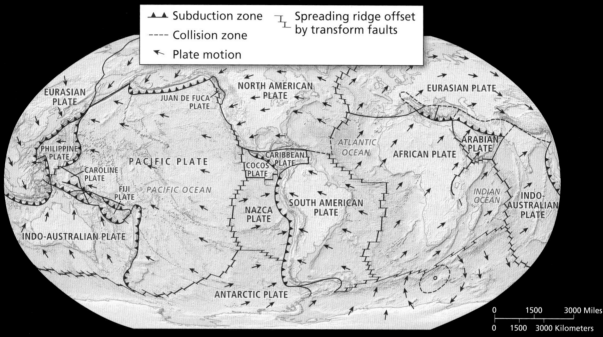

▲GL8.1 **The continents today: plate boundary interactions**

Apply

In this lab you will first use the map of late Cenozoic Era tectonic movements to learn about the principal motions of plates and plate boundaries. The second section involves using a hazard map, photograph, and tectonic diagram to assess the varied risks of living close to a plate boundary.

Objectives

- Identify the major plates.
- Locate and analyze the three types of plate movement and their boundaries.
- Explain why humans living on a plate boundary are subjected to multiple hazards.

Procedure

Part I

1. Describe the following plate boundaries in terms of the three types of plate boundary interactions shown in Figure GL8.1.
   a. The Nazca and South American plates
   b. The Caribbean and Atlantic Ocean plates
   c. The Philippine and Eurasian plates
   d. The Indo-Australian and Eurasian plates
2. Which plates converge near Japan? What explains the existence along this boundary of the islands that make up Japan?
3. Which type of landform are the colliding Juan De Fuca and North American plates likely producing? Explain your answer.
4. Looking at Figure GL 8.2a and b, identify where sedimentary, igneous, and meta-

Part II

Plate boundaries are active geologic zones where endogenic and exogenic processes occur. Taken together, they pose substantial risks for humans who live on the edge of plate boundaries.

Use Figure GL 8.2 and the Chapter 8 modules to answer the following questions:

5. How does the type of plate boundary influence the number and type of natural hazards that threaten the Seattle area?

6. Among the many natural hazards found in the Pacific Northwest, which one, usually triggered by earthquakes on the plate boundary, does not occur in inland locations?

7. If during the next million years, the North American, Pacific, and Juan de Fuca plates slowly shifted their direction relative to one another, would the Seattle region become a safer place for humans to live?

Analyze & Conclude

8. What are the advantages and disadvantages of living in the natural environment that results from this plate boundary?

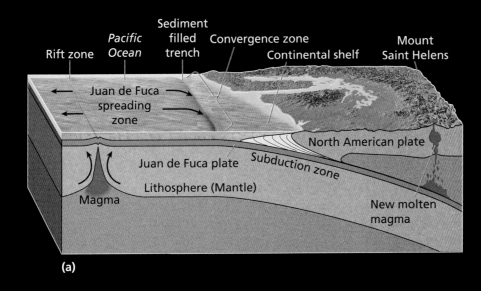

(a)

(b)

▲GL8.2 **(a) Juan de Fuca plate and subduction zone (b) Downtown Seattle, in the shadow of Mt. Rainier (c) Pacific Northwest earthquake hazard.**

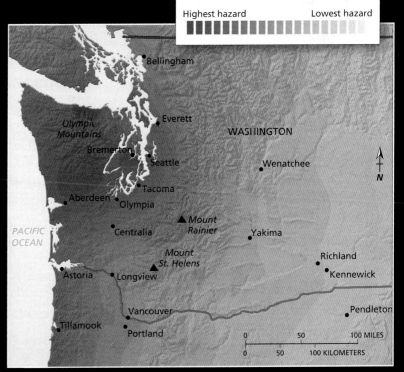

(c)

Log in MasteringGeography™ to complete the online portion of this lab, view the Pre-Lab Video, and complete the Post-Lab activity.
www.masteringgeography.com

9 Weathering & Mass Movement

The rugged landscape of Arches National Park is one of contrasting colors and spectacular landforms unlike any other in the world. The unique features are the result of weathering—the physical and chemical processes that break rock into fragments and particles. The iconic 15-story-tall Delicate Arch is dramatic evidence of how weathering can proceed at different rates, often on rock masses of similar composition. In this case, more resistant rock on top of the arch protects the weaker strata below. Weathering also releases minerals from bedrock for soil formation, which is vital for plant growth. In limestone regions, chemical weathering produces an unusual landscape of sinkholes, caves, and caverns.

Weathering weakens the Earth's surface rock, making it more prone to erosion by moving water, air, waves, and ice—all influenced by the ever-present pull of gravity. Some erosion takes place gradually, while in other areas, it occurs quickly with little prior warning, especially on steep slopes.

Key Concepts & Topics

1 How does weathering prepare landforms for erosion?

2 What unique landforms result when water erodes limestone?

3 How does mass movement alter landforms?

Resistant rock strata at the top of the structure helped preserve the arch beneath as surrounding rock eroded away to form Delicate Arch, Utah.

9.1 Weathering & Landforms

Key Learning Concepts

▶ **Explain** how the *angle of repose* influences the stability of slopes.

▶ **List** the factors that influence weathering and cause differential weathering.

A dizzying array of landforms blankets Earth's surface, including the arches, sinkholes, domes, and **karst** towers discussed in this chapter. The manner by which these landforms erode provides clues that help us reconstruct Earth's history. The study of **geomorphology** investigates the origin, evolution, and distribution of surface landforms. Once landforms are uplifted, **denudation** processes wear away or rearrange them. Denudation includes *weathering* and *mass movement*, discussed in this chapter, as well as *erosion* discussed in Chapter 10. **Weathering** breaks down rock on and just below Earth's surface, either disintegrating rock into mineral particles or dissolving it in water. *Erosion* includes the transport of materials to different locations by moving water, wind, waves, ice, and gravity. The interaction between the geologic forces that uplift land and the denudation processes that wear it down represent a continuing struggle between Earth's endogenic and exogenic processes.

Stable & Unstable Slopes

Slopes are inclined surfaces that vary with rock structure and climate (▼Fig. 9.1). On every slope, the weathering of rock loosens material for erosion. The slope angle, raindrops, freezing and thawing, moving animals, and wind also promote erosion. However, before moving downslope, material must overcome friction, inertia (resistance to movement), and the cohesion of particles to one another. The maximum incline at which **sediments** on a slope can remain at rest before gravity begins pulling them downward is called the *angle of repose* (▶Fig. 9.2). The angle commonly ranges between 33° and 37° for rock fragments and decreases with particle size. A slope is *stable* if its strength exceeds the combined force of gravity and the denudation processes, and *unstable* if the slope's materials are weaker, as we see in Figure 9.2.

geoCHECK ✓ Describe what forces rocks and soil must overcome before moving downslope.

▼9.1 **Anatomy of a slope** The directional forces acting on individual particles combine to move materials down a slope. Notice how different processes predominate on different parts of the slope.

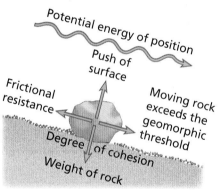

Potential energy of position

Push of surface

Frictional resistance

Moving rock exceeds the geomorphic threshold

Degree of cohesion

Weight of rock

(a) Directional forces act on materials along an inclined slope.

Angle of repose

35°

Fine sand

45°

Angular pebbles

(a) Angle of repose

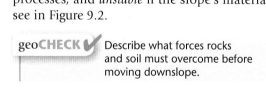

25°

(b) Arabian Desert, United Arab Emirates

▲9.2 **Slopes in disequilibrium** When particles on a slope exceed the angle of repose, movement downslope will occur.

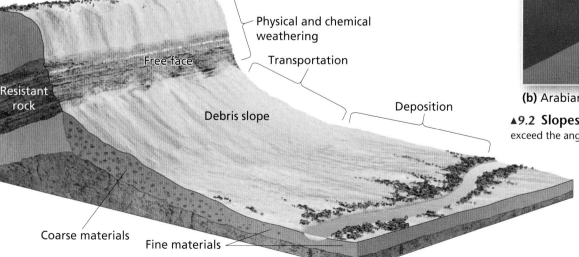

Soil processes

Physical and chemical weathering

Free face

Transportation

Resistant rock

Debris slope

Deposition

Coarse materials

Fine materials

(b) The principal elements of a slope.

Animation (MG)
Mass Movement

http://goo.gl/CWFyUr

Mobile Field Trip (MG)
The Critical Zone

https://goo.gl/iJe1NM

Weathering Processes

Weathering refers to the physical and chemical processes that break up or dissolve rock. Physical and chemical weathering frequently work together. Weathering does not transport material, but rather simply prepares it for erosion by water, wind, waves, and ice—all influenced by gravity.

Bedrock is the *parent rock* from which regolith and soil develop. The continual weathering of bedrock near the surface creates **regolith**—a layer of unconsolidated rocky material that often covers bedrock. As weathering proceeds, this regolith is transported and deposited downslope (▶Fig. 9.3). The unconsolidated fragments combine with other weathered rock to form the **parent material** from which soil evolves (▼Fig. 9.4a). In young soils, the parent rock is usually identifiable because of reduced erosion and mixing with parent material from other places.

In some areas regolith is missing or undeveloped, thus exposing an outcrop of unweathered bedrock. However, rock weathers and crumbles, landforms wear away, and soil covers much of the earth's land surface. Important factors that influence weathering include the following:

- Rock composition and structure, or jointing—**joints** are fractures in rock that occur without displacement of the sides of the fracture, as in faulting. Joints increase the surface area of rock exposed to physical and chemical weathering.
- Climate—precipitation, temperature, and freeze–thaw cycles all affect weathering.
- Subsurface water—the water table and water movement within soil and rock structures.
- Slope orientation—whether a slope faces north or south controls the slope's exposure to Sun, wind, and precipitation. Slopes facing away from the Sun's rays are cooler, moister, and more vegetated than those in direct sunlight.
- Organic processes—decaying organic matter from plants and animals produces acids that promote chemical weathering. Plant roots enter crevices and physically break up rock (▼Fig. 9.4b).
- Time—with all of these, *time* is a crucial factor, for these processes are slow to operate.

Differential Weathering Weathering rarely occurs evenly within a single landform. Subtle inconsistencies in rock composition, slight depressions in the land surface, and cracks in rock expose the rock surfaces to various weathering processes. Thus even rocks of the same composition and structure, found in the same place, may end up over time eroding at uneven rates. The resulting **differential weathering** often creates spectacular landscapes such as Delicate Arch shown in the photograph that opens this chapter.

geoCHECK ✔ Explain why rocks of similar composition may erode at very different rates.

Physical weathering of the upper canyon bedrock loosens rocks for their eventual movement downslope.

The vegetation adds an element of chemical weathering.

Gravity pulls this regolith downward.

At the canyon bottom, farmers till the soil particles that remain behind.

▲9.3 Bedrock to soil, Afghanistan

geoQUIZ

1. Explain how tectonic activity and the processes of weathering and erosion represent a continuing struggle between Earth's endogenic and exogenic forces.
2. Define the angle of repose.
3. List six important factors that influence weathering.

Animation (MG)
Physical Weathering

http://goo.gl/9qHlwK

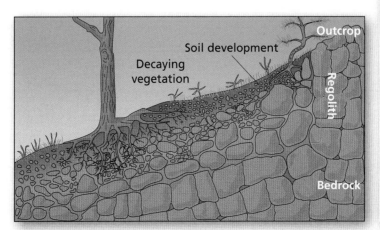

Soil development
Decaying vegetation
Outcrop
Regolith
Bedrock

(a) A cross section of a typical hillside.

(b) Tree roots force

▲9.4 Regolith, soil, and parent materials

9.2 Physical Weathering

Key Learning Concepts

▶ **Describe** how physical weathering breaks apart rock and soil.

Physical weathering (also called *mechanical* weathering) occurs when rock is broken and disintegrated without chemical alteration. By breaking up rock, physical weathering produces more surface area on which chemical weathering may operate. A single rock broken into eight pieces has doubled its surface area and is susceptible to three primary weathering processes: frost action, **salt-crystal growth**, and **pressure-release jointing**.

Frost Action

Freezing water expands in volume as much as 9% (see Chapter 5). The expansion of ice creates a powerful mechanical **frost action**, or *freeze–thaw action*, which can exceed the strength of rock. The repeated freezing (expanding) and thawing (contracting) of water breaks rocks apart—particularly in cold climates at high latitudes and high elevations (▶Figs. 9.5 and 9.7). This process is also called *frost wedging*. Ice initially works into small openings, gradually expanding until rocks split. Blocks of rock often separate along existing joints and fractures. Frost action then pushes the rock apart. Cracking and breaking create varied shapes, depending on the rock structure.

The Incas of South America and early pioneers of the American West used frost action as a force to quarry rock (▼Fig. 9.6). They chiseled and drilled holes in rock, then poured water into the holes, and over time let expanding ice break off large blocks to use for construction material. Today, frost action often damages road pavement and bursts water pipes.

Physical weathering in some mountain regions makes spring a risky time. As rising temperatures melt winter ice, newly fractured rock pieces fall without warning and often trigger rockslides. The falling rock may shatter on impact—another form of physical weathering. Over time, these rock fragments accumulate as a *talus slope*.

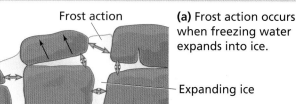

Frost action

(a) Frost action occurs when freezing water expands into ice.

Expanding ice

(b) Frost action breaks the layers apart in this shale outcrop.

▲**9.5 Frost action**

geoCHECK ✓ How does frost action break a rock apart?

▼**9.7 Closeup of frost action** Repeated freezing of water is wedging this granite slab apart.

▼**9.6 Humans using frost action** Four centuries ago, the Incas used frost action to split large stone blocks that still support buildings in Cuzco, Peru.

▼**9.8 Physical weathering in Canyon de Chelly, Arizona** Weathering caused by crystallization (salt-crystal growth) helped to form this niche in the canyon wall, where ancient Native Americans constructed cliff dwellings more than 700 years ago.

(a) The dark streaks are desert varnish which forms when bacteria oxidizes clays and other particles attached to the rock surface.

Salt-Crystal Growth

Especially in arid climates, dry weather draws moisture to the surface of rocks. As the water evaporates, dissolved mineral salts in the water form crystals—a process called *crystallization*. Over time, as the crystals grow and enlarge, they exert a force great enough to force apart individual mineral grains and begin breaking up the rock. This *salt-crystal growth*, or *salt weathering*, is a form of physical weathering.

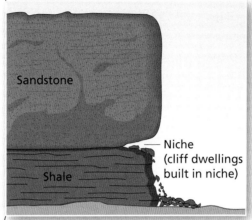

(b) Water and an impervious rock layer helped concentrate weathering processes in a niche in the overlying sandstone.

On the Colorado Plateau of the Southwest, salty water slowly flows from rock strata. As this water evaporates, crystallization loosens the sand grains. Subsequent erosion and transportation by water and wind complete the sculpturing process. Deep indentations develop in sandstone cliffs, especially where the sandstone lies above an impervious layer such as shale. More than 1000 years ago, Native Americans built entire villages in these weathered niches, including Mesa Verde in Colorado and Arizona's Canyon de Chelly (◄**Fig. 9.8**).

geoCHECK ✔ Describe how the growth of salt crystals contributes to the weathering of rocks in arid climates.

Pressure-Release Jointing

Recall from Chapter 8 how rising magma that is deeply buried and subjected to high pressure forms bodies of intrusive igneous rock called *plutons*. These plutons cool slowly and produce coarse-grained, crystalline, granitic rocks. During subsequent tectonic uplift, the overlying rock is weathered, eroded, and transported away, eventually exposing the pluton as a mountainous batholith (illustrated in Figure 8.7).

The removal of this overlying rock relieves pressure on the granite in the same way a person rising off a couch removes pressure on the seat cushions. Over millions of years, the granite slowly responds with an enormous physical heave. In a process known as *pressure-release jointing* (also called *exfoliation*), layer after layer of rock peels off in curved slabs or plates, thinner at the top of the rock structure and thicker at the sides. As the slabs weather, they slip off in large sheets. This *exfoliation process* creates arch-shaped and dome-shaped features on the exposed landscape, sometimes forming an **exfoliation dome** (◄**Fig. 9.9**). Just beneath the curved surfaces of large domes, gravity itself exerts a tension force that also promotes exfoliation.

geoCHECK ✔ How can weathering and erosion of overlying rock contribute to the formation of joints in granite?

geoQUIZ

1. Compare and contrast the three types of physical weathering described above. How are they similar? How are they different?
2. Explain why physical weathering is more prominent in Canada than in Florida.
3. Identify the physical weathering process that helped provide shelter to Native Americans in the Southwest.

▲**9.9 Exfoliation in granite** Pressure-release jointing formed the peeling slabs of rock on the surface of Half Dome in Yosemite National Park, California.

9.3 Chemical Weathering

Key Learning Concepts

▶ **Describe** chemical weathering processes, including hydration, hydrolysis, oxidation, and carbonation.

The chemical breakdown of minerals in rock, usually in the presence of water, is **chemical weathering**. The chemical decomposition and decay accelerate temperature and precipitation increase. Although individual minerals vary in susceptibility to weathering, all rock-forming minerals respond to some degree of chemical weathering. The most important types of chemical weathering are hydration and hydrolysis, oxidation, and carbonation.

Hydration & Hydrolysis

Although hydration and hydrolysis are different, we group these two processes together because they both use water to decompose rock—one by simply combining water with a mineral, and the other by a chemical reaction of water with a mineral.

Hydration, meaning "in combination with water," involves little chemical change. Water becomes part of the chemical composition of the mineral. When some minerals hydrate, they expand, forcing grains apart as in physical weathering. A cycle of hydration and dehydration can lead to the breakdown of rock grains and promote additional chemical weathering. Hydration works together with carbonation and oxidation to convert feldspar, a common mineral in many rocks, to clay

minerals and silica. Hydration also chemically weathers the sandstone niches shown in Figure 9.8a.

Hydrolysis occurs when minerals chemically react with water. This decomposition process breaks down silicate minerals in rocks. Compared with hydration, in which water combines with minerals in the rock, hydrolysis involves water and elements in chemical reactions that produce different compounds. For example, feldspar minerals in granite often break down by the mild acids dissolved in precipitation (▶Fig. 9.10).

The by-products of chemical weathering of feldspar in granite include clay and silica. As clay forms from some minerals in the granite, quartz particles are left behind. The resistant quartz may wash downstream to become sand on some distant beach. Clay minerals become a major component in soil and in shale, a common sedimentary rock.

When water-soluble minerals in rock are changed by hydrolysis, the sharp edges of rocks are rounded as the rock's individual minerals or cementing materials break down. The breakdown of rock grains makes the rock appear etched, corroded, and even crumbly. Such disintegration results in **spheroidal weathering** (▼Fig. 9.11). Joints in the rock offer more surfaces of opportunity for weathering. Eventually, the spherical shells of decayed rock resemble layers of an onion. Although this process resembles exfoliation, it does not result from physical pressure-release jointing.

▲9.10 **Hydration and hydrolysis of granite** These processes disintegrate the surface of this granite monolith that rises above the ruins of Machu Picchu, Peru.

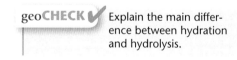 Explain the main difference between hydration and hydrolysis.

(a) Spheroidal weathering in Mexico's Sierra de Juárez.

(b) Steps in the process of spheroidal weathering.

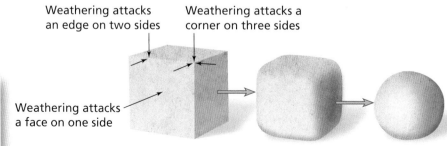

Weathering attacks an edge on two sides

Weathering attacks a corner on three sides

Weathering attacks a face on one side

▲9.11 **Spheroidal weathering**

▲9.12 **Oxidation** Oxidized sandstone rocks in the Red Rock Coulee Natural Area of Alberta, Canada

Oxidation

When certain metallic elements combine with oxygen to form oxides, **oxidation** occurs. The most familiar oxidation form is the "rusting" of iron in rocks or soil that produces a reddish-brown stain of iron oxide (▲Fig. 9.12). In much the same way, metal tools or nails left outside over time coat with iron oxide. The rusty color is visible on the surfaces of rock and in heavily oxidized soils such as those in the southeastern United States, southwestern U.S. deserts, and tropics. The removal of iron from the minerals within a rock disrupts the crystal structures, making the rock more susceptible to further chemical weathering.

geoCHECK ✔ Identify a common visible sign that rocks are oxidizing.

Carbonation

Since water is the universal solvent, it is capable of dissolving at least 57 of the natural elements and many of their compounds. **Carbonation** occurs when water vapor dissolves carbon dioxide, thereby yielding precipitation containing carbonic acid. In most areas carbonic acid is strong enough to dissolve many minerals. Over time and on level rock surfaces where the water and acid pools together, this process often creates small depressions called *solution pits* (▼Fig. 9.13).

Such chemical weathering transforms minerals that contain calcium, magnesium, potassium, and sodium. When rainwater attacks limestone (which is calcium carbonate, $CaCO_3$), the principal minerals dissolve and wash away with the mildly acidic rainwater.

The weathered limestone and marble in cemetery tombstones or in rock formations appear pitted and worn wherever adequate water is available for dissolution. Humans contribute to this form of chemical weathering as evident in Europe, North America, and increasingly in Asia, where acidic rains result from the burning of coal (▶Fig. 9.14).

geoCHECK ✔ Describe how a solution pit forms.

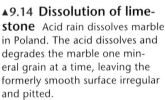

▲9.14 **Dissolution of limestone** Acid rain dissolves marble in Poland. The acid dissolves and degrades the marble one mineral grain at a time, leaving the formerly smooth surface irregular and pitted.

geoQUIZ

1. Explain how hydration and hydrolysis combine to break rocks apart.
2. How does spheroidal weathering change the shape of granitic rocks?
3. Describe how oxidation contributes to weathering of rocks.

▼9.13 **Solution pits** On the Grand Canyon rim, water combined with organic material from plants and animal droppings creates a weak carbonic acid that dissolves rock minerals.

Solution pits

9.4 Karst Topography

Key Learning Concepts

▶ ***Identify*** the processes and features associated with karst topography.

Limestone is abundant on Earth and composes many landscapes. The high calcium carbonate content of this gray-colored rock is easily dissolved in the acids produced by organic materials. This creates a unique landscape of pitted, bumpy topography, poor surface drainage, and well-developed underground caverns. Weathering and erosion caused by groundwater may also create remarkable underworld caverns.

The resulting **karst topography** is named for the Krs Plateau in Slovenia, where karst processes were first studied (▼Fig. 9.15). Approximately 15% of Earth's land area has some karst features, with outstanding examples found in southern China, Japan, Puerto Rico, Cuba, the Yucatán of Mexico, Kentucky, Indiana, New Mexico, and Florida.

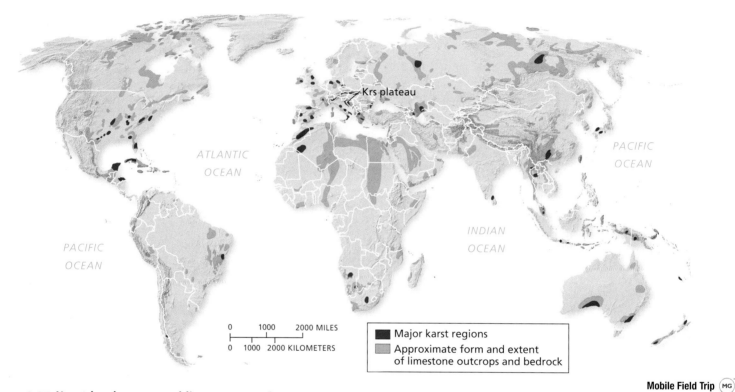

Krs plateau

ATLANTIC OCEAN

PACIFIC OCEAN

PACIFIC OCEAN

INDIAN OCEAN

| | Major karst regions |
| | Approximate form and extent of limestone outcrops and bedrock |

▲9.15 Karst landscapes and limestone regions

Mobile Field Trip (MG)
Mammoth Cave

https://goo.gl/2Fy46V

Formation of Karst

For a limestone landscape to develop into karst topography, several conditions are necessary:

- The limestone formation must contain 80% or more calcium carbonate for the dissolving processes to proceed.
- Complex joint patterns in the otherwise impermeable limestone are necessary for water to drain into subsurface channels.
- A zone containing air must exist in the soil between the ground surface and the water table.
- Vegetation cover is required to supply organic acids that enhance the dissolving process.

As with all weathering processes, the amount of time bedrock is exposed to weathering is a critical factor. However, karst development does not follow universal stages, but instead, is thought to be the result of unique combinations of local conditions. The importance of climate to karst processes remains under debate, although the amount and distribution of rainfall appear significant. Karst occurs in arid regions, but it is primarily due to former climatic conditions of greater humidity. Karst is rare in the Arctic and Antarctic regions, because the water is usually frozen.

 geoCHECK ✔ Identify at least three conditions that are necessary for the formation of karst topography.

Features of Karst Landscapes

Several landforms are typical of karst landscapes. Each results from the interaction between surface weathering processes, underground water movement, and processes occurring in subterranean cave networks described next. Because karst forms a fragile foundation to build upon, these changing landforms pose unique problems to human settlement.

Sinkholes & Towers Limestone weathering creates circular depressions called **sinkholes**. Rolling limestone plains are often pockmarked by slow subsidence of surface materials in *solution sinkholes* with depths of 2–100 m (7–330 ft) and diameters of 10–1000 m (33–3300 ft). Through continuing solution and collapse, sinkholes may coalesce to form a *karst valley*—an elongated depression up to several kilometers long.

A collapse sinkhole forms when a solution sinkhole collapses through the roof of an underground cavern, sometimes dramatically, such as in Guatemala City in 2010 (▶Fig. 9.16). A 91-m (300-ft) deep, 18-m (59-ft) wide sinkhole suddenly dropped, taking a three-story building and parts of

▲9.16 **Sinkhole in Guatemala City**

two streets with it. This disaster was probably years in the making and was likely triggered by heavy rains. In Florida, sinkholes make news when water tables lowered by municipal pumping begin to drop. As the water table lowers, dissolved limestone and other residue percolate further downward through the limestone, promoting additional erosion of subsurface limestone. Eventually, the surface collapses into underground solution caves, taking homes and businesses with them. In the humid tropics, karst topography forms in deeply jointed, thick limestone beds. Weathering leaves isolated resistant limestone towers above level plains. Examples include northern Puerto Rico and southern China (▼Fig. 9.17).

▲9.17 **Karst towers in southern China**

(a) Over time, a variety of "dripstone" features form in caverns through the deposition of calcium carbonate.

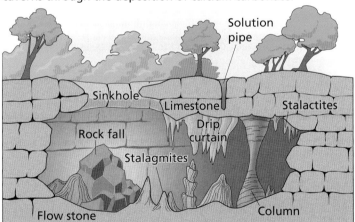

(b) A diver exploring a cave system, Dos Ojos, Mexico.

Caverns Caves form in limestone because it is so easily dissolved by carbonation. Caves form just beneath the water table. Eventual lowering of the water levels exposes the caves to further development. *Dripstones* form as water containing dissolved minerals slowly drips from the cave ceiling. Calcium carbonate slowly precipitates out of the evaporating solution, and accumulates on the cave floor. Over time, *stalactites* grow from the ceiling and *stalagmites* build from the floor, creating a dramatic subterranean world (◀Fig. 9.18).

The exploration and scientific study of caves is *speleology*. Many caves worldwide remain unexplored. Until recently, the Mammoth Cave system in Kentucky was the largest known cavern, but that distinction may now belong to the Son Doong Cave in Vietnam. Other large limestone caverns are found in Malaysia, Ukraine, Croatia, and Mexico. Karst landscapes blanket 21% of the United States, with the largest cave systems found in Kentucky, Tennessee, Missouri, Arkansas, and New Mexico.

geoCHECK Describe the difference between a stalactite and a stalagmite.

geoQUIZ

1. Explain why karst landscapes are susceptible to chemical weathering.
2. Identify three world regions where karst landscapes are common.
3. Explain the conditions that lead to the formation of a karst sinkhole.

◀9.18 **Cavern features** An underground cavern and related forms in limestone.

9.5 Mass Movement

Key Learning Concepts

▶ ***Explain*** the relationship between weathering and mass movement.
▶ ***Recognize*** how driving and resisting forces influence mass movement.
▶ ***Identify*** the mechanisms that trigger mass movement.

Physical and chemical weathering processes weaken surface rock, making it more susceptible to the pull of gravity. **Mass movement** is the movement of surface material—rocks, soil, vegetation—propelled downward by gravity. Mass movements can be surface processes or submarine landslides beneath the ocean. Mass-movement materials can range from dry to wet and involve particles ranging from small to large. Mass movement can be fast or slow, ranging from sudden free fall to gradual or intermittent flows. The term *mass movement* is often used interchangeably with *mass wasting*.

The Role of Slopes

All mass movements occur on slopes because of the downward-pulling force of gravity. Even grains of dry sand piled on a beach flow downslope until equilibrium is achieved. Before any sediment moves downslope, a slope must first exceed the *angle of repose*. As the *driving force* in mass movement, gravity works in conjunction with the weight, size, and shape of the surface material; the degree to which the slope is oversteepened (how far it exceeds the angle of repose); and the amount of ice or water moisture available. The greater the slope angle, the more susceptible the surface material is to mass-wasting processes. The *resisting force* is the *strength* of the slope material—that is, the cohesiveness of its individual particles that work against gravity. Over time, all slopes weaken until gravity overcomes friction and the slope begins to fail (◀Fig. 9.19).

 geoCHECK ✔ Explain the difference between driving and resisting forces, as they relate to hillslope environments.

▲**9.19 Gravity versus resistance** On this slope, physical and chemical weathering prepares rocks for their inevitable descent.

Mechanisms That Trigger Mass Movements

One factor that affects materials' resistance to mass movement is the presence of water. For example, clays, shale, and mudstones readily absorb water. If these water-saturated materials underlie rock strata in a slope, mass movement can occur more easily (▶Fig. 9.20). When clay surfaces are wet, they deform slowly in the direction of movement, and when saturated, they form a slowly moving mass that starts downhill with little gravitational pull. However, in less saturated slopes, the rock strata would be less likely to slip, and more driving force energy would be required, such as that generated by an earthquake. Examples in the next module demonstrate the mechanisms that trigger mass movements, including heavy rains, earthquakes, built-up pressure, and undercutting slopes.

▼**9.20 Hydration and slope failure** This slope failed when saturated clay soil overcame gravity in the Hoang Lien Mountains, Vietnam.

▲9.21 **Usoi Landslide and Lake Sarez, Tajikistan**

(a) Oso before the mudslide.

(b) Moving at an average speed of 0.8 k/hr (40 mph), this landslide buried 13 km² (0.5 mi²) and dammed the North Fork of the Stillaguamish River to a depth of 7.6 m (25 feet).

(c) A ground-level view of the destruction with the landslide scarp visible in the background. (A *scarp* is a steep, curved slope marking the surface where the landslide began.)

▲9.22 **Oso mudslide, Washington**

Mass Movement as a Natural Hazard Steep slopes and lots of sediment generate a high potential for large-scale mass movements. These geomorphic events also pose a serious risk to humans and their economic activities. Sarez Lake in the southeastern Pamir Mountains of Tajikistan is the largest natural sediment reservoir in Central Asia. The lake formed when a 1911 earthquake shook loose 6 billion metric tons of debris, damming the Murgab River (▲Fig. 9.21). The Usoi Dam—named after the village it buried—is the highest dam, natural or human-made, on Earth. The water level rose 240 m (787 ft) in three years and inundated 60 km (37 mi) of the narrow canyon before seepage through an underground outlet reestablished the river. Failure of the Usoi Dam is a potential cataclysmic disaster. An outburst flood from Lake Sarez would destroy the villages and roadways downstream. Close monitoring and an early warning system are currently in place.

A 1959 mass movement triggered by an earthquake also dammed the Madison River in Montana. The U.S. Army Corps of Engineers quickly excavated a channel to prevent a catastrophic flood in case the natural dam failed. In China, engineers also prevented many potential catastrophic floods by opening channels in 2008 on the Chaping River, when the magnitude 7.9 Wenchuan quake produced more than 100 of these landslide dams.

In 2014, mass movement near the small community of Oso, Washington, sent approximately 2.6 square kilometers (1 square mile) of sediment across the North Fork of the Stillaguamish River. The slide debris was 9–12 m (30–40 ft) deep and obliterated 30 homes and the adjacent highway (▶Fig. 9.22). Over 30 lives were lost. The blocked river caused flooding upstream until the river slowly carved out a new channel. Although scientists are still investigating this catastrophic mass-movement event, rainfall that averaged 175% of normal during the 45 days prior to the slide was the most likely cause.

geoCHECK ✔ Identify two natural events that can initiate catastrophic mass movement.

geoQUIZ

1. Explain what may happen to rocks on a slope when the resistance force weakens.
2. Suppose a building contractor keeps adding to a large pile of rocks and soil excavated from a construction site. What must the contractor guard against as the pile gets bigger? Explain.
3. Explain why heavy precipitation often triggers mass wasting events.

9.6 Types of Mass Movements

Key Learning Concepts

▶ **Classify** the types of mass movements according to moisture content and speed of movement.

▶ **Identify** examples of mass movement caused by human activities.

In any mass movement, gravity pulls on a slope until the failure point is reached—a **geomorphic threshold**. Moving slope material can *fall, slide, flow,* or *creep*—the four classes of mass movement (▼**Fig. 9.23**). As we have discussed above, physical and chemical weathering play a crucial role in preparing slope material for the types of mass movement discussed below.

▼9.23 Types of mass movement Variations in water content and rates of movement produce the different types of events shown here.

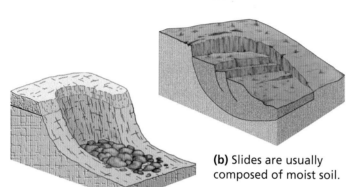

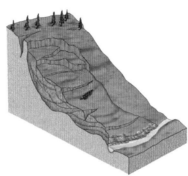

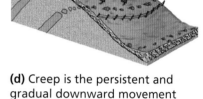

(b) Slides are usually composed of moist soil.

(d) Creep is the persistent and gradual downward movement of soil.

(a) Falls usually involve only rock, but a small amount of soil and plant debris may fall as well.

(c) Flows are much larger in volume than falls or slides and incorporate significant amounts of vegetation, rock and soil.

Falls & Avalanches

This class of mass movement includes rockfalls and debris avalanches. A **rockfall** is simply rock that falls through the air and hits a surface. The various weathering processes prepare the rock for erosion, which may occur one rock at a time. However, if an earthquake triggers the rockfall, enormous amounts of rocks may shake lose. The individual pieces characteristically form a cone-shaped pile of irregular broken rocks in a *talus slope* at the base of a steep incline (▶**Fig. 9.24**).

A **debris avalanche** is a mass of falling and tumbling rock, soil, and other materials. (*Debris* can consist of sediment of any kind or size, including boulders, trees, and even human-made objects). A debris avalanche is differentiated from a slower debris slide or landslide by the high velocity of the onrushing material. This speed often results from ice and water that fluidize the debris. The extreme danger of a debris avalanche results from its tremendous speed and consequent lack of warning.

▼9.24 Talus Slope Frost action breaks the rock apart, and then gravity pulls the fragments downslope in California's Sierra Nevada.

geoCHECK ✔ What is the difference between a rockfall and a debris avalanche?

Landslides

In 2014, weeks of heavy rain saturated the hillside above the village of Abi Barak in northeastern Afghanistan. The rains eventually triggered a massive mudslide that buried over 300 homes and killed an estimated 1000 people. A second phase of the slide buried scores of rescuers who had rushed to the scene to offer assistance (▶Fig. 9.25).

A sudden rapid movement of a cohesive mass of regolith or bedrock that is not saturated with moisture is a **landslide**—a large amount of material failing simultaneously. Landslides are dangerous because they occur without warning at the instant the downward pull of gravity wins the struggle for equilibrium. Slides occur in one of two basic forms: translational or rotational. In instances such as the massive Oso landslide portrayed in Figure 9.22, the mass movement may exhibit both translational and rotational movement.

Translational Slides In a translational slide, there is movement along a flat surface roughly parallel to the angle of the slope, with no rotation (▶Fig. 9.26). Translational slides may involve either rocks or soil. They may also grow into debris slides that incorporate a mix of both, along with vegetation and even human-built structures. The defining characteristic of a translational slide is that the top layer of material slides over a mostly level surface, much like a top layer of cake sliding off the bottom layer, with the frosting between them acting as the slip surface.

Rotational Slides When surface material moves along a concave surface, a rotational slide or slump occurs (▶Fig. 9.27). They are rotational because the slide material moves downward and outward, much like we might slip in a half-circular motion down and out of a leather chair. Frequently, underlying clay presents an impervious surface to percolating water. As a result, water flows along the clay surface, undermining the overlying block. As the slide occurs, the surface may rotate as a single unit, or it may develop a stepped appearance.

geoCHECK ✔ Explain the difference between a translational landslide and rotational landslide.

Mobile Field Trip Landslides
https://goo.gl/KUtdi6

Animation Mass Movements
http://goo.gl/CWFyUr

(a) (b)

▲9.25 **Abi-Barak landslide** Hundreds of people were killed when a landslide struck in northeastern Afghanistan following heavy rainfall.

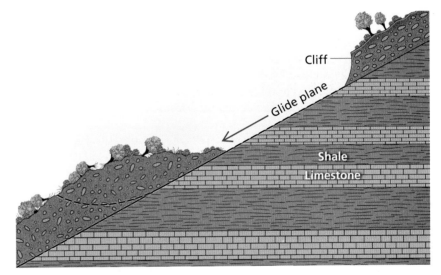

▲9.26 **Translational landslide—movement along a flat surface**

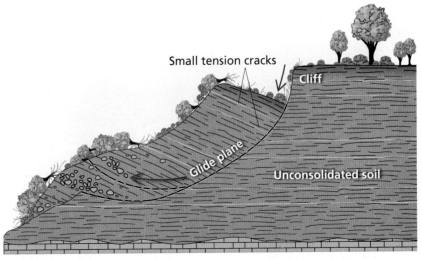

▲9.27 **Rotational landslide—movement along a concave surface**

9.6 (cont'd) Types of Mass Movements

Flows

When the moisture content of moving material is high, the suffix *-flow* is used, in place of *slide*. **Flows** include *earthflows* and more fluid *mudflows*. As with all mass movement, natural events, such as period of extraordinary rainfall or a seismic tremor, often trigger these flows. Flows can be fast or slow depending on the slope and the characteristics of the materials involved. For example, earthflows usually occur over longer periods of time than mudflows. They also frequently incorporate vegetation and human structures into the earthen mix, in which case the term *debris flow* is often applied (▶Fig. 9.28).

geoCHECK ✔ Explain what distinguishes an earthflow from a landslide.

Creep

A persistent and gradual mass movement of surface soil is **soil creep**. In creep, a number of different processes lift and disturb individual soil particles: the expansion of soil moisture as it freezes; cycles of moistness and dryness; diurnal temperature variations; or grazing livestock or digging animals.

In the freeze–thaw cycle, the freezing of soil moisture lifts particles at right angles to the slope (▼Fig. 9.29). When the ice melts, the particles fall straight downward in response to gravity. As the process repeats, the surface soil gradually creeps downslope. Creep may cover a wide area and cause fence posts, utility poles, and even trees to lean downslope.

▲9.28 **Debris flow** In December 1999, torrential rains triggered debris flows that killed thousands of people and destroyed buildings in the Vargas State of Venezuela.

geoCHECK ✔ Explain how both landslides and creep pose a threat to human structures.

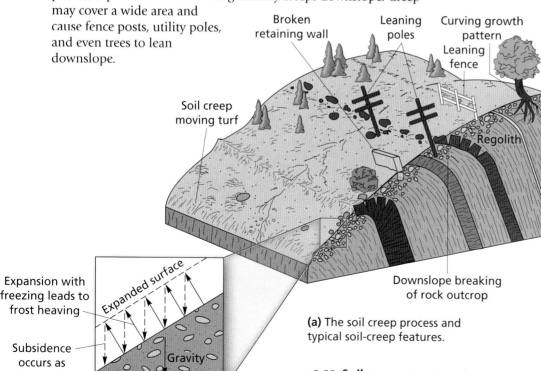

Broken retaining wall · Leaning poles · Curving growth pattern · Leaning fence · Soil creep moving turf · Regolith · Expansion with freezing leads to frost heaving · Expanded surface · Subsidence occurs as frost melts · Gravity · Downslope breaking of rock outcrop

(a) The soil creep process and typical soil-creep features.

(b) These ancient statues on Easter Island in the South Pacific Ocean are slowly moving downhill because of soil creep.

▲9.29 **Soil creep** Over time, the feeze-thaw cycle moves material downslope, producing distinctive changes in the surface that indicate soil creep has occurred.

Human-Induced Mass Movements

Every human disturbance of a slope—highway road cutting, surface mining, or construction of a shopping mall or housing development—can hasten mass movement. The newly destabilized and oversteepened surfaces, often now devoid of most vegetation, begin eroding to establish a new slope equilibrium. Scientists have estimated that humans annually move 40–45 billion tons (40–45 gross tons [Gt]) of Earth's surface, which greatly exceeds the natural movement of river sediment (14 Gt/year) or the movement of sediment by wave action and erosion along coastlines (1.25 Gt/year). Large open-pit surface mines—for copper, coal, and uranium, for example—are examples of human-induced mass movements, generally known as **scarification**. Such large excavations produce piles of unstable mine debris that are susceptible to further weathering, mass movement, or wind dispersal (▼Fig. 9.30). Where underground mining is common, particularly for coal in the Appalachians, land subsidence and collapse produce further mass movements. A controversial form of mining called *mountaintop removal* removes ridges and summits and then dumps the debris into streams—an extreme example of human-induced mass movement (▶Fig. 9.31).

 geoCHECK ✔ Identify at least three human activities that promote mass movement of rocks and soil.

geoQUIZ

1. What is a talus slope, and how do they form?
2. Identify and explain the difference between one rapid and one slow form of mass wasting.
3. Explain why landslides and debris flows are common in mountain areas.

(a) Large-scale scarification resulting from coal mining in Hayford Mountain, West Virginia.

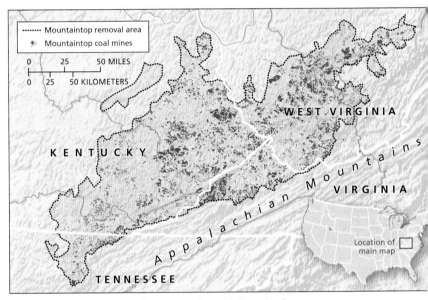

(b) Mountaintop removal areas and coal mines in the Appalachian Mountains of Kentucky and West Virginia.

▲9.31 **Mountaintop removal mining**

▼9.30 **Scarification by large-scale mining** Chile's Chuquicamata copper mine has produced a world record 29 million tons of copper.

GEOMORPHIC PROCESSES IMPACT HUMANS

- Chemical weathering processes break down carvings made by humans in rock, as on tombstones, cathedral facades, and bridges.
- Sudden sinkhole formation in populated areas can cause damage and human casualties.
- Mass movements cause human casualties and sometimes catastrophic damage, burying cities, damming rivers, and sending flood waves downstream.

HUMANS IMPACT GEOMORPHIC PROCESSES

- Mining causes scarification, often moving contaminated sediments into surface water systems and groundwater.
- Removal of vegetation on hillslopes may lead to slope failure, destabilizing streams and associated ecosystems.
- Lowering of water tables from groundwater pumping causes sinkhole collapse in population centers.

9a

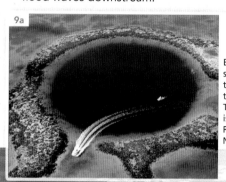

Blue holes are typical karst sinkholes. Lying offshore today, they formed on land during times when sea level was lower. The Great Blue Hole near Belize is in the Belize Barrier Reef Reserve System, a United Nations World Heritage site.

9b

The 71-m-tall (233-ft-tall) Grand Buddha at Leshan in southern China displays chemical weathering accelerated by air pollution. Over 1000 years old, the statue is now being corroded by acid rain from nearby industrial development.

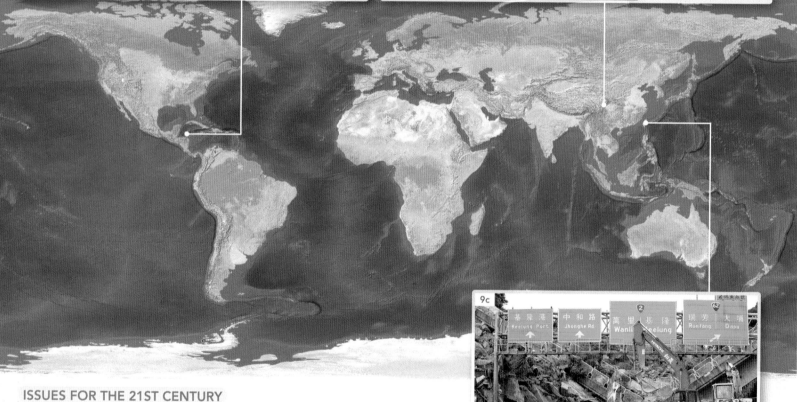

9c

In April 2010, a massive translational landslide covered parts of a highway near Taipei, Taiwan. The cause is uncertain but apparently not related to earthquake activity or excessive rainfall.

ISSUES FOR THE 21ST CENTURY

- Global climate change will affect forest health; declining forests (from disease or drought) will increase slope instability and mass-movement events.
- Open-pit mining worldwide will continue to move massive amounts of Earth materials, with associated impacts on ecosystems and water quality.
- Failures of containment ponds holding industrial by-products will spread toxic materials onto landscapes and into downstream areas.

Looking Ahead

In the next chapter we investigate how river systems drain the continents and transport nutrients and sediment downstream, and their role in shaping numerous landforms. Humans use rivers to water crops, power industry, and quench thirsty settlements. They also form vital transportation networks.

How does weathering prepare landforms for erosion?

9.1 Weathering & Landforms

Explain how the *angle of repose* influences the stability of slopes.

List the factors that influence weathering and cause differential weathering.

- The interaction between the geologic forces that uplift land and the denudation processes that wear it down represent a continuing struggle between Earth's endogenic and exogenic processes. Physical and chemical weathering of rock loosens material for erosion and transportation. Gravity exerts a constant force on all slopes, but before downward movement of rock and soil, they must overcome the geomorphic threshold.

1. Given all the interacting variables, do you think a landscape ever reaches a stable and unchanging condition? Explain your answer.

9.2 Physical Weathering

Describe how physical weathering breaks apart rock and soil.

- Physical weathering (also called *mechanical weathering*) breaks and disintegrates rocks without chemical alteration. In cold climates this occurs through *frost action,* when water enters pores and joints in rocks, then freezes and expands. In arid climates the *salt-crystal growth* pries rocks apart. Physical weathering also occurs through *pressure-release jointing*—the release of pressure after layers of rock are removed through erosion.

2. Why is freezing water such an effective weathering agent?
3. What role do joints play in the weathering process?

9.3 Chemical Weathering

Describe chemical weathering processes, including hydration, hydrolysis, oxidation, and carbonation.

- Chemical weathering is the breakdown of individual minerals in rock by water. The process is more intense as temperature and precipitation increase. Hydrolysis is the breakdown of rock by acidic water to produce clay and soluble salts. Oxidation occurs during the breakdown of rock by oxygen and water, during which the iron in the rock becomes red or rust colored. Carbonization takes place when rock minerals react with carbonic acid removal of rock in solution by acidic rainwater. Carbonation takes place when water combines with carbon dioxide, forming a carbonic acid that dissolves or breaks down the rock minerals. In particular, limestone is weathered by rainwater containing dissolved CO_2.

4. Explain why chemical weathering would be greater in the Amazon rain forest than in the Gobi Desert.

What unique landforms result when water erodes limestone?

9.4 Karst Topography

Identify the processes and features associated with karst topography.

- Chemical weathering of limestone creates a karst landscape of pitted, bumpy surface topography, poor surface drainage, and underground caverns. Karst topography is most developed when the limestone contains 80% or more calcium carbonate. A complex joint pattern to aid drainage, good airflow between the ground surface and the water table, and dense vegetation are also important.

5. Approximately what percent of the world is comprised of karst landscapes?
6. What are some of the characteristic erosional and depositional features found in a karst landscape?

How does mass movement alter landforms?

9.5 Mass Movement

Explain the relationship between weathering and mass movement.

Recognize how driving and resisting forces influence mass movement.

Identify the mechanisms that trigger mass movement.

- Mass movement applies to any material propelled downward by the driving force of gravity. However, before this erosion takes place and materials move downslope, gravity must overcome the *resisting force* of cohesiveness and internal friction of the slope material. The weathering of rock and soil weakens their resisting force, or shearing strength, and makes them more susceptible to the driving force of gravity and thus mass movement.

7. Summarize the environmental changes leading to a mass movement event.

9.6 Types of Mass Movements

Classify the various types of mass movements.

Identify examples of each in relation to moisture content and speed of movement.

- The four classes of mass movements are *falls, slides, flows,* and *creep.* Of these, rock and debris falls are very rapid. Landslides can be rapid or slow and usually involve just rocks and soil. When the moisture content of moving material is high, the suffix *-flow* is used, in place of *slide.* As the name suggests, creep involves a gradual downslope movement of surface soil.

8. List the major classes of mass movement and describe each briefly.
9. Identify the difference between a translational slide and a rotational slide.

Critical Thinking Questions

1. Summarize the process of spheroidal weathering.

2. What forces help a large rock that is perched on a steep slope to overcome the geomorphic threshold?

3. Explain the difference between physical and chemical weathering, and provide two examples of each.

4. Describe what effect increasing moisture and rising temperature would have on mass movement.

5. Assuming limestone is present, would karst landscapes be most developed in polar regions, the humid tropics, or a warm desert? Explain your answer.

Visual Analysis

Thousands of phantom-like rocks composed of limestone, siltstone, dolomite, and mudstone form the amphitheaters of Bryce Canyon National Park, on the edge of the Paunsaugunt Plateau in southern Utah.

1. How do the many spires illustrate the concept of differential weathering?

2. Bryce Canyon is located between 8000 to 9000 ft (2400 to 2700 m) in an arid region. Identify two weathering processes that most likely dominate in this environment.

3. Looking closely at one of the spires, identify areas that appear especially susceptible to physical weathering. Explain your answer.

▲R9.1 Bryce Canyon, Utah

(MG) Interactive Mapping | Login to the **MasteringGeography** Study Area to access **MapMaster**.

Launch MapMaster in MasteringGeography™, and note the location of the Andes Mountain chain in Latin America. Select **Physical Environment**, and then select *Physical Features* and *Environmental Issues*. Clicking these categories reveals the location of the mountain areas where forests have been cleared to make way for crops. By also clicking *Climate*, you will notice that the Andes fall into the complex Köppen H mountain climate zone, which, in the Andes, includes at least one season of heavy rainfall. The maps now reveal three environmental characteristics.

1. What do the layered data tell us about the topography or slope angle, condition of the forest cover, and precipitation in the Andes?

2. Using the environmental characteristics from the preceding question, identify two types of mass movement that likely pose a threat to human life in this region.

Explore | Use **Google Earth** to explore **Guilin, China.**

Viewing Earth from space allows us to visualize how karst processes create a distinctive landscape. In Google Earth, check the "Photos" tool, and leave other categories unchecked. Next, fly to the *Lijiang River, Guilin, China.* The next step is to zoom in slightly, and then begin clicking on the many photographs. You should also scroll (or pan) around the image, clicking on an even wider range of photos to learn more about each location.

1. Of the karst landscape characteristics presented in Chapter 9, identify at least three that are evident from the Google Earth image and the ground photos.

2. This area formed as a level plateau in a shallow marine environment. Tectonic forces then slowly uplifted the plateau above sea level. How does this karst landscape of today, as portrayed in the Google image and the ground photographs, illustrate the principle of differential weathering?

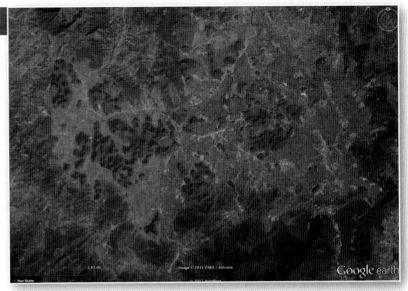

▲R9.2 Karst landscapes, China

Mastering Geography™

Looking for additional review and test prep materials? Visit the Study Area in MasteringGeography™ to enhance your geographic literacy, spatial reasoning skills, and understanding of this chapter's content by accessing a variety of resources, including MapMaster™ interactive maps, videos, *Mobile Field Trips, Project Condor* Quadcopter videos, *In the News* RSS feeds, flashcards, web links, self-study quizzes, and an eText version of *Geosystems Core.*

Key Terms

bedrock, p. 249
carbonation, p. 253
chemical weathering,
  p. 252
debris avalanche, p. 258
denudation, p. 248
differential weathering,
  p. 249

exfoliation dome, p. 251
flows, p. 260
frost action, p. 250
geomorphic threshold,
  p. 258
geomorphology, p. 248
hydration, p. 252
hydrolysis, p. 252

joints, p. 249
karst, p. 248
karst topography,
  p. 254
landslide, p. 259
mass movement, p. 256
oxidation, p. 253
parent material, p. 249

physical weathering,
  p. 250
pressure-release jointing,
  p. 250
regolith, p. 249
rockfall, p. 258
salt-crystal growth,
  p. 250

scarification, p. 261
sediment, p. 248
sinkholes, p. 255
slopes, p. 248
soil creep, p. 260
spheroidal weathering,
  p. 252
weathering, p. 248

Slip-Sliding Away:
Determining Slope Movement Variables

Mass wasting, the downward movement of rock, soil, and snow, occurs on slopes everywhere in the world. The many classes of mass wasting include enormous earthflows, free-falling rock, and the imperceptible movement of soil creep. Each is responding to the pull of gravity. Once the *angle of repose* is exceeded, the slope begins to fail. Although the average angle for dry materials falls between 30° and 35°, the composition of the material, temperature and moisture changes, earthquakes, and wind can all trigger the slope to exceed its critical angle of slope stability, resulting in slope failure.

Apply

As a land use planner working for a town in a region of hilly terrain, you are charged with calculating the slope angle of repose, so prospective home builders and buyers can properly assess the risks posed by mass wasting.

Objectives

- Examine the relationship between slope angles and the downward movement of two objects
- Calculate the angle of repose
- Analyze other factors that might influence the "mass wasting" of your objects, then relate these to the real mass wasting

Procedure

For this experiment, you will need a hard yet smooth surface about the size of your textbook (e.g., clipboard, kitchen cutting board, whiteboard), two hard objects about the size of a large marble (e.g., rocks, compacted soil, metal washers/hex nuts), a small amount of water and a cup, and a protractor to measure angles. If you don't have the protractor, trace the one provided below (**Fig. GL9.1**). Use these materials to perform the following experiment and answer the questions.

1. With your materials replicating a rock or soil clump positioned on a slope, use the protractor to calculate the slope angle indicated by angle α.
2. With your "slope" lying flat on a table and your "rock" positioned near the top, slowly and gently tilt the slope upward until rock exceeds its *angle of repose* and begins to move downward (**Fig. GL9.2**). Record that angle when the object first moves. Repeat this 10 times to simulate the variable conditions in the real world.

Object A Trial	1	2	3	4	5	6	7	8	9	10
Angle										

3. Calculate the average for the 10 trials.
   Average angle of repose for Object A =

4. Repeat the steps for Object B.

Object B Trial	1	2	3	4	5	6	7	8	9	10
Angle										

   Average angle of repose for Object B =

Analyze & Conclude

5. Was there any difference in the angle of repose for your two objects?
6. Explain any similarity or difference.

Place object A near its average angle of repose (but not allowing it to slip). With your fingers or a hard object (e.g., tool, rock, wooden stick), rap the middle of the board a few times.

7. What happens to the angle of repose?
8. What natural phenomena did your rapping simulate?

Moisten your "slope" and repeat the tilting procedure again with both objects. Then repeat this process after moistening both the slope and your object.

9. Did this alter the angle of repose? Explain.
10. **Figure GL 9.3** portrays the historic mining town of Telluride, Colorado. Using this photo, answer the following questions:

    a. Which slope appears to clearly exceed the angle of repose?
    b. Telluride sits at 2667 m (8750 ft) in the Rocky Mountains. Identify the climate factors that would most likely promote slope failure. Explain.
    c. Identify the tectonic phenomena that would most likely trigger slope failure.
    d. What visible natural material is contributing to slope stability? Why does this material not occur everywhere in the photo?
    e. What reasonable measures can humans take to reduce both catastrophic and minor slope failures?

11. Based your analysis of Telluride's natural setting, would you advise the town's planning board to approve a new housing development between the historic downtown and the nearby mountains? Why or why not?

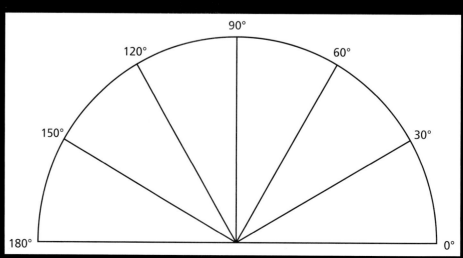

▲GL9.1

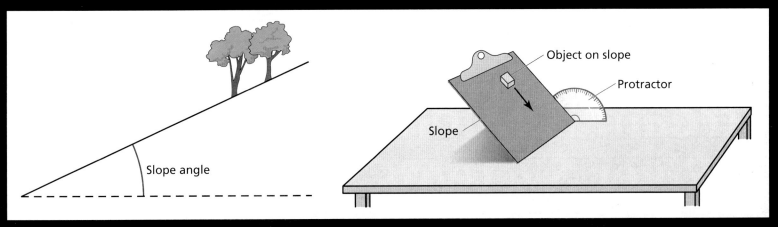

▲GL9.2

▲GL9.3 **Telluride, Colorado**

10

Stream Erosion & River Systems

Earth's waterways form vast arterial networks that drain the continents, redistribute mineral nutrients for soil formation and plant growth, and shape the landscape by transporting sediment downstream. They provide humans with essential water supplies for agriculture, industry, and settlements. They also form important transportation networks.

For example, the Joekulgilkvísl River, shown here, flows through south-central Iceland. The river begins as melting snow on the north side of the Bamur Range, in the Fjallabak Conservation Area. From these headwaters, numerous smaller tributary streams add to the water volume that eventually empties into the Atlantic Ocean.

Key Concepts & Topics

The Joekulgilkvísl River flows through the wild and mountainous Fjallabak Nature Reserve, Iceland.

10.1 Streams & Drainage Basins

Key Learning Concepts

▶ **Describe** how a stream forms.
▶ **Describe** the structure of a drainage basin.

Stream-related processes are **fluvial**, from the Latin *fluvius*, meaning "river." Fluvial systems have characteristic processes and produce recognizable landforms. Although the terms *river* and *stream* overlap, *river* is applied to a main stream. *Stream* is a more general term not necessarily related to size. A network of tributaries forms a *river system*.

Hydrology is the science of water and its global circulation, distribution, and properties. Insolation and gravity power the hydrologic cycle and drive fluvial systems. Individual streams vary greatly, depending on the climate, geology and soils, topography, plant cover. At any moment, approximately 1250 km³ (300 mi³) of water is flowing through Earth's waterways. Although flowing streams are just 0.003% of all freshwater, their energy is important for eroding, or wearing away, landscapes.

Stream Formation

Surface water initially moves downslope in a thin film of **sheetflow**. An area of high ground that separates two adjacent river valleys is called an *interfluve*. The interfluves direct sheetflow into small-scale downhill grooves called *rills* that may develop into deeper *gullies,* and then into a stream in the valley (▶Fig. 10.1). The point where two tributary streams join is called a *confluence*. As individual streams flow out of their valleys of origin, they merge with other streams to create a larger body of flowing water. As we will see below, as streams flow downhill, propelled by gravity, they can take on many different characteristics.

 geoCHECK ✔ Describe the movement of water and the features involved in stream formation.

Drainage Basins

Every stream has a **drainage basin**, or *watershed,* that collects water and sediment from many tributaries. Drainage basins may range in size from tiny to vast. All streams start small in the uppermost portions of their drainage basins—an area referred to as the stream's *headwaters*. Along with streams, drainage basins are the basic units of study in the science of surface water hydrology.

Drainage divides form on higher land separating drainage basins, often following along ridge tops. Drainage divides define the catchment (water-receiving) area of every drainage basin. These ridges form dividing lines that control into which basin runoff water from precipitation drains. Figure 10.1 illustrates a portion of a drainage basin. From this point, the stream will continue downslope to a confluence with another stream or until it empties into the ocean or another water body.

Every major drainage basin system is made up of many smaller drainage basins. Each basin gathers and delivers its runoff and sediment to a larger basin, concentrating the volume into the main stream. A good example from Figure 10.2 is the great Mississippi–Missouri–Ohio River system, draining some 3.1 million square kilometers (1.2 million square miles), or 41% of the continental United States.

 geoCHECK ✔ Where are the headwaters located within a drainage basin?

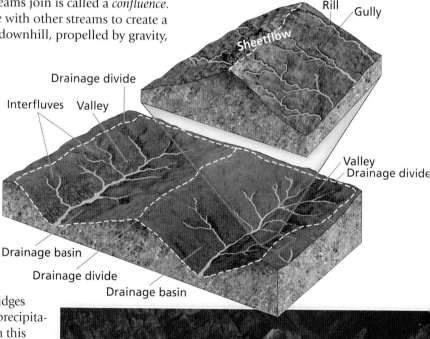

▼10.1 A drainage basin

Continental Divides On a continental scale, major drainage divides are called **continental divides**. In the United States and Canada, mountain ranges separate the major drainage basins and send flows to either the Pacific, the Gulf of Mexico, the Atlantic, Hudson Bay, or the Arctic Ocean (▶Fig. 10.2). In South America, the Andes Mountains form a continental divide that separates Amazonian waters that flow east into the Atlantic, from the many rivers that flow west into the Pacific Ocean. In Africa, a 2,000 km (1,200 mi) long continental divide separates the Congo and Nile River basins. On every continent, these divides provide a spatial framework for water-management planners who must balance the competing demands of urban, agricultural, and industrial water users.

International Drainage Basins Many of the world's major rivers flow across international boundaries. For example, the Danube River in Europe flows 2850 km (1770 mi) from Germany's Black Forest to the Black Sea. This important watershed exemplifies the complexity of an international drainage basin (▼Fig. 10.3). The river includes 300 tributaries and crosses or forms the borders between nine countries. The Danube is important for commercial transport, municipal water, agricultural irrigation, fishing, and hydroelectric power. An international struggle is under way to save the river from its burden of industrial and mining wastes, sewage, chemical discharge, agricultural runoff, and drainage from ships. The United Nations Environment Programme and the European Union, along with other organizations, are dedicated to fully restoring the environment of this valuable resource.

geoCHECK ✔ Explain the significance of a continental divide.

▲10.2 **Drainage basins and continental divides of North America** Continental divides (red lines) separate the major drainage basins that empty through the United States into the Pacific, Atlantic, and Gulf of Mexico and, to the north, through Canada into Hudson Bay and the Arctic Ocean. Subdividing these large-scale basins are major river basins.

geoQUIZ

1. Explain what the terms *river* and *stream* mean when applied to fluvial systems.
2. Explain the difference between an interfluve and drainage divide.
3. Explain why the Danube River is so important to humans.

(a) The Danube river flows through Budapest, Hungary.

(b) The Danube

(c) Often called the Everglades of Europe, the Danube delta provides wetland habitat for over 300 bird species and 45 species of freshwater fish.

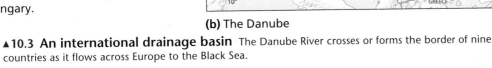

▲10.3 **An international drainage basin** The Danube River crosses or forms the border of nine countries as it flows across Europe to the Black Sea.

10.2 **Drainage Patterns**

Key Learning Concepts

▶ *Identify* different types of drainage patterns.
▶ *Describe* how the drainage density of a drainage basin is determined.

A **drainage pattern** is the arrangement of stream channels in an area. Patterns are quite distinctive and are determined by a combination of topography, variable rock resistance, climate, hydrology, and structural controls imposed by the underlying rocks. Consequently, the drainage pattern of any land area on Earth is a remarkable visual summary of every geologic and climatic characteristic of that region.

 Define drainage density.

Classifying Drainage Patterns

Drainage patterns are best viewed from the air, as shown in Figure 10.4, which portrays the dendritic and parallel drainage patterns. These patterns, and the other common types of drainage patterns, are shown in Figure 10.5. Dendritic drainage patterns are the most common. This branching, treelike pattern gets its name from the Greek word *dendron*, meaning "tree," and is similar to that of many natural systems, such as capillaries in the human circulatory system, the veins in leaves, or the roots in trees. Dendritic drainage patterns develop on a land surface where the rocks are of uniform resistance to erosion. The trellis drainage pattern (▶**Fig. 10.5b**) often develops in dipping or folded topography that varies in resistance to erosion. Parallel structures of more resistant rocks direct the main streams, while smaller tributary steams of less-resistant rocks join the main channels are right angles, as in a plant trellis. This pattern is clearly exposed along great anticlines and synclines of the Appalachian and Catskill Mountains in the eastern United States.

The remaining drainage patterns in Figure 10.5c–g are responses to other specific structural conditions.

- A *radial* drainage pattern (▶**Fig. 10.5c**) results when streams flow off a central peak or dome, such as occurs on a volcanic mountain.
- *Parallel* drainage (▶**Fig. 10.5d**) is associated with steep slope and high-velocity streams found on steep mountains.
- A *rectangular* pattern (▶**Fig. 10.5e**) is formed by a faulted and jointed landscape, which directs stream courses in patterns of right-angle turns.
- *Annular* patterns (▶**Fig. 10.5f**) are produced by structural domes or basins, with concentric patterns of rock strata guiding stream courses. Figure 10.5f provides an example of annular drainage on a dome structure.
- In areas having disrupted surface patterns, such as the glaciated regions of Canada, northern Europe, and some parts of Michigan and other states, a *deranged* pattern (▶**Fig. 10.5g**) is in evidence, with no clear geometry in the drainage and no true stream valley pattern.

A primary feature of any drainage basin is its **drainage density**. Drainage density is determined by dividing the total length of all stream channels in the basin by the area of the basin. The number and length of streams in a given area reflect the area's geology, topography, and climate. High drainage densities develop in areas where precipitation cannot infiltrate into the soil, and instead flows overland to form a dense network of stream channels. This is often the case in areas with sparse vegetation, such as deserts, and surface bedrock that is impermeable to water. Chile's Atacama desert, shown in Figure 10.5d, is one example.

◀**10.4 Two drainage patterns** Identify the two drainage patterns that developed in response to local topography and rock structure.

Dendritic

Trellis

Radial

Radial drainage flows off the Lower Klawasi mud volcano, Alaska. The flow from the center vent destroyed the forest in the upper third of the cone.

) Parallel

Parallel drainage in the Atacama desert, Chile.

e) Rectangular

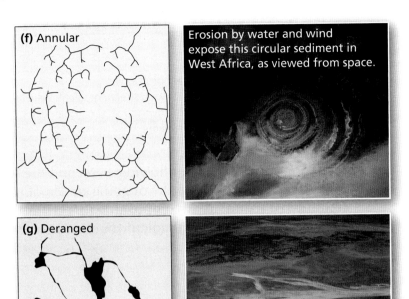

(f) Annular

Erosion by water and wind expose this circular sediment in West Africa, as viewed from space.

(g) Deranged

▲10.5 **Common drainage patterns** Each pattern is a visual summary of the geologic and climatic conditions of its region.

geoCHECK ✔ Use a pen or pencil to draw radial, dendritic, and rectangular drainage patterns, and indicate where you might find them.

geoQUIZ

1. Explain the characteristics that make drainage patterns distinct from one another.
2. Identify what a trellis drainage pattern reveals about the underlying topography.
3. Predict the soil and bedrock characteristics of a watershed with a low drainage density.

10.3 Stream Dynamics

Key Learning Concepts

▶ **Identify** the parts of a stream profile.
▶ **Describe** the relationship between stream gradient, velocity, depth, width, and discharge.
▶ **Identify** different types of streams.

Mobile Field Trip (MG)
Streams of the Great
Smoky Mountains

https://goo.gl/INIEzj

Project Condor Video (MG)
River Terraces
and Base Level

https://goo.gl/jsU1Kp

Streams are a mixture of flowing water and the sediment load they transport. Fluvial landscapes result from the ongoing *erosion* of rocks and soil, and the downstream *deposition* of sediment. Accomplishing this geomorphic work depends on a number of factors, including stream gradient (or slope), base level, and discharge. These concepts are discussed below.

Stream Gradient

Every stream has a degree of inclination or **gradient**, which is the rate of decline from its headwaters, where the stream originates, to its mouth, where it ends (▶**Fig. 10.6**). This gradient is calculated by dividing the difference in elevation between two points on a stream by the distance between the two points along the stream channel. The result is usually expressed in meters per kilometer or feet per mile. Since streams rarely maintain a straight path for long, determining the distance between two points must account for the many twists and turns.

Characteristically, the *longitudinal profile* (side view) of a stream has a steeper slope and V-shaped channel in the headwaters and upstream portions of the profile. Moving downstream, the gradient becomes more gradual, and the channel shape widens. Streams adjust their gradient and other channel characteristics such as width, depth, and velocity, to move sediment load. Where streams empty into a water body at their mouth, the gradient is typically gentle.

Streams erode the landscape in three ways: (1) headward erosion removes material upslope in the headwaters, (2) downcutting erodes material from the bottom of the stream channel, and (3) lateral erosion removes material from the sides of the river. An **ungraded stream** is actively eroding or depositing material in its channel so that its gradient is adjusted to carry the sediment load. In contrast, a **graded stream** has just enough energy to transport its sediment load.

Over time, the gradient adjusts to provide—for the available discharge and channel characteristics—just the velocity required to transport the load supplied from the drainage basin. This condition is a state of *dynamic equilibrium* among discharge, sediment load, and channel form. In streams of all gradients, the stream, channel, and landscape work together to maintain this balance within a drainage basin. An individual stream can have both graded and ungraded portions.

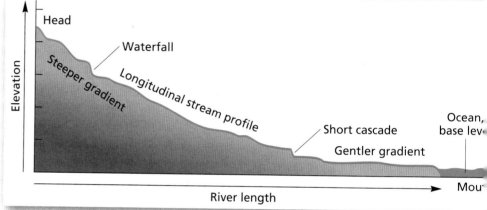

(a)

(b)

▲**10.6 An ideal longitudinal stream profile** The sloping profile of a stream is its gradient. The upstream portion usually has a steeper gradient. Downstream, the gradient usually lessens.

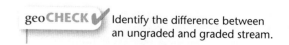

geoCHECK ✔ Identify the difference between an ungraded and graded stream.

Base Level

Base level is a level below which a stream cannot erode its valley. Sea level is the lowest practical level for all erosional processes in most areas (▶Fig. 10.7). However, not every landscape degrades all the way to sea level. For instance, a temporary *local base level* may control the lower limit of local streams for a region. The local base level may be a river, a lake, hard and resistant rock, or the reservoir formed by a human-made dam (▼Fig. 10.8). In arid landscapes, with their intermittent precipitation, valleys, plains, or other low points determine local base level.

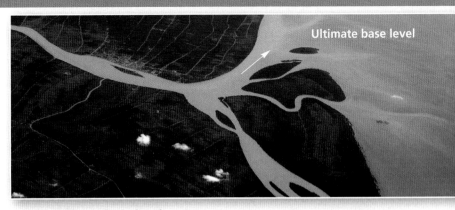

Ultimate base level

▲**10.7 Ultimate base level** Sea level is the ultimate base level destination for nearly all water, rock, and sediment on Earth.

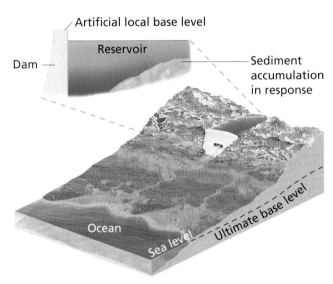

Artificial local base level

Reservoir

Dam

Sediment accumulation in response

Ocean

Sea level

Ultimate base level

(a) The ultimate base level is sea level. Note how base level curves gently upward from the sea as it is traced inland; this is the theoretical limit for stream erosion. The reservoir behind a dam is a local base level.

(b) The Srisailam Dam blocks India's Krishna River.

▲**10.8 Local base levels** Natural lakes are common natural local base levels, while a reservoir behind a dam represents an artificial local base level resulting from human activity.

As discussed in Chapter 5, there are presently 40,000 dams over 15 m (49 ft) high in the world. By temporarily blocking free-flowing rivers, dams create an artificial base level that also alters the downstream water temperatures, biota, sediment load, and flood regime, among other impacts. In Asia, Africa, and South America, large dams are still under planning and construction, following the example set in Europe and North America. However, since the early 1900s, the United States has removed nearly 1,150 dams. In 2014, explosives removed the final section of the 64 m (210 ft) high Glines Canyon Dam, thus freeing the Elwha River in Washington state. Dam removal improves water quality, revitalizes fish and riparian wildlife, and boosts local economies by providing more opportunities for outdoor recreation.

Tectonic uplift affects stream gradient by increasing the elevation of the stream relative to its base level. If tectonic forces slowly lift the landscape—increasing the potential energy of position—the stream gradient will increase and stimulate renewed erosion. Thus, a stream flowing through the uplifted landscape becomes *rejuvenated*. That is, the stream actively returns to downcutting (▶Fig. 10.9).

▲ **10.9 Tectonic uplift rejuvenated the Colorado River, providing energy for the stream to erode downward into this gorge at Horseshoe Bend, Arizona.**

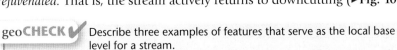

geoCHECK✔ Describe three examples of features that serve as the local base level for a stream.

10.3 (cont'd) Stream Dynamics

Stream Discharge

Stream water positioned above base level represents potential energy of position. As the water flows downhill, this becomes kinetic energy of motion. The rate of this conversion from potential to kinetic energy determines the ability of a stream to do geomorphic work, and this depends in part on the volume of water. **Stream discharge** is the volume of water moving past a point in a given unit of time (▼Fig. 10.10). Discharge is calculated using measurements of stream width, depth, and velocity of a given cross section of the channel. It is summarized in the equation

$$Q = wdv$$

Where Q = stream discharge, w = channel width, d = channel depth, and v = stream velocity.

Discharge is expressed in cubic meters per second (m^3/s) or cubic feet per second (ft^3/s), and is measured with a **hydrograph** at a stream gaging station.

Of the world's rivers, those with the greatest discharge are the Amazon of South America, the Congo of Africa, the Chang Jiang (Yangtze) of Asia, and the Orinoco of South America (▼Figs. 10.11 and 10.12). The Amazon's annual streamflow delivers millions of tons of sediment from a drainage basin that rivals the Australian continent in size. In contrast, the discharge of the Mississippi, the U.S. river with the highest discharge, averages only about one-tenth of the Amazon's discharge.

As the stream discharge (or volume) increases, some combination of channel width, channel depth, and stream velocity must also increase. As the water volume increases downstream, the gentler gradient typical of lower drainage basins often produces a smooth and quiet flowing river. However, the stream velocity may still be very high. In almost all watersheds, stream discharge fluctuates over time and is greatest during periods of heavy rainfall or when spring temperatures begin melting the winter snowpack. Greater discharge increases stream velocity. This increases the ability of the stream to downcut its bed and to transport the sediment downstream.

geoCHECK ✔ Describe how you would go about calculating the stream discharge of a river near your home.

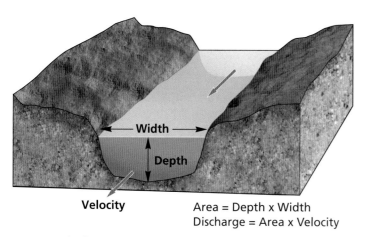

Area = Depth x Width
Discharge = Area x Velocity

▲10.10 **Calculation of stream discharge**

▼10.11 **Largest rivers on Earth ranked by discharge** In this cartogram, the sizes of the rivers' drainage basins are proportional to their discharge, which distorts the shapes of landmasses on the map.

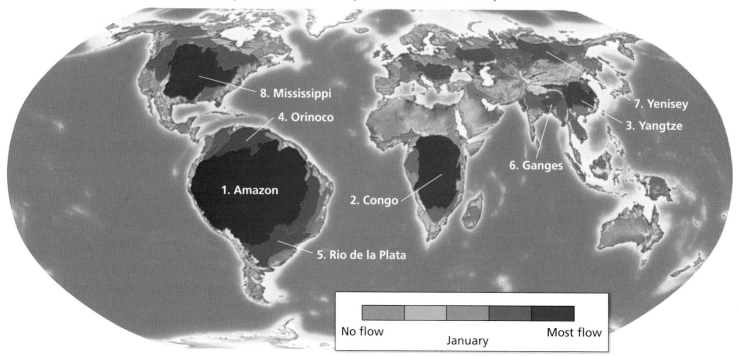

8. Mississippi
4. Orinoco
1. Amazon
2. Congo
5. Rio de la Plata
6. Ganges
3. Yangtze
7. Yenisey

No flow — January — Most flow

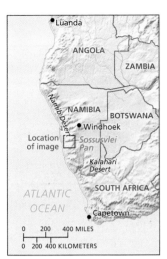

▲**10.12** Locations of rivers shown in Figures 10.13, 10.14, and 10.15.

▲**10.13 Ephemeral stream** Aerial view of the Sossusvlei region of the Namib Desert.

Stream Types

Discharge varies throughout the year for most streams, depending on precipitation and temperature. *Perennial streams* flow all year, which is typical for large rivers and in humid regions with ample precipitation. Rivers and streams in arid and semiarid regions may also have intermittent or ephemeral discharge. *Intermittent streams* flow for several weeks or months each year and may have some groundwater inputs. *Ephemeral streams* flow only after precipitation and are not connected to groundwater systems. Years may pass between flow events (▲**Fig. 10.13**).

Some river systems that begin in moist regions flow downstream into arid landscapes where high potential evapotranspiration and withdrawals for agriculture and settlements may reduce the discharge. Such a stream is an **exotic stream** of distant origin. The Nile River, one of Earth's longest rivers, originates in the more humid parts of northeastern Africa, but then flows through the deserts of Sudan and Egypt where it loses water because of evaporation and withdrawal for agriculture and urban uses (▼**Fig. 10.14**). The Colorado River in the United States is another exotic stream, which originates in Rocky Mountain headwaters and then flows across the parched Southwest (►**Fig. 10.15**).

geoCHECK ✔ Explain the difference between a perennial and ephemeral stream.

geoQUIZ

1. Explain why the ocean is considered the lowest practical base level for virtually all streams.
2. Identify three factors that influence stream discharge.
3. Explain why the Colorado and Nile Rivers are considered exotic streams.

◄**10.14 The Nile River flows through Cairo, Egypt,** here near its mouth in the Mediterranean Sea

►**10.15 Exotic stream** The Colorado River begins as melting snow in the Rocky Mountains, then flows through the arid southwestern United States.

10.4 Fluvial Processes

Key Learning Concepts

▶ *Explain* the processes of fluvial erosion and deposition.

The profile and character of streams change through time as erosion, transport, and deposition of sediment occurs. Streams maintain a dynamic equilibrium between discharge, slope, and sediment load. The ongoing *erosion* and *deposition* by flowing water produce the landscapes we see around us. Running water is an important erosional force. In desert landscapes, water is the most significant agent of erosion, greater than wind, even though precipitation is infrequent.

Stream Erosion

Streams carve and shape the landscape through hydraulic action and abrasion. **Hydraulic action** is the squeeze-and-release action of flowing water that loosens and lifts rocks. During transport, this sediment erodes the streambed further through the process of **abrasion**, with rock particles grinding and carving the streambed like liquid sandpaper.

Streams have two general types of flow, which vary with the shape, slope, and roughness of the channel. Flows are *turbulent* when water particles move randomly and cross paths in many directions. Turbulent flow increases erosion by hydraulic action, and is common in steep, shallow headwater streams and in sections with rapids on all streams (▶Fig. 10.16). Turbulent movement results from friction between the streamflow, the channel sides and bed, and obstacles such as boulders in the channel. Flows are laminar when water particles move in the same direction, in parallel paths. Laminar flow is common in downstream portions of rivers carrying large amounts of fine sediment particles, or in the calm water between rapids (▶Fig. 10.17). Both turbulent and laminar flow can occur within the same segment of a stream.

geoCHECK ✔ Describe two processes by which streams carry out erosion.

How Streams Transport Sediment

The amount of material carried by a stream depends on topographic relief, the nature of rock and soil, climate, vegetation, and human activity in a drainage basin. *Competence*, which is a stream's ability to move particles of a specific size, is a function of stream velocity and the energy available to move materials. *Capacity* is the total possible sediment load that a stream can transport. Streams with high competence and high capacity are able to transport large rocks and other types of sediment described ahead. The water in these streams runs a thick, dark color, while streams carrying a small sediment load run much clearer. Four processes transport eroded materials: solution, suspension, traction, and saltation (▶Fig. 10.18).

Solution *Solution* is the **dissolved load**, especially the chemical solution derived from minerals such as limestone, dolomite, or soluble salts. Other dissolved load originates as organic acid from biologic sources such as decaying vegetation, animal feces, and agricultural runoff. Material in solution is mostly derived from chemical weathering. In some rivers, such as the San Juan and Little Colorado, salt from dissolved rock and springs is sufficient to hinder human use.

Suspension A stream's **suspended load** consists of fine-grained, clastic particles (pieces of rock) held aloft in the stream, with the finest particles deposited when the stream velocity slows nearly to zero. These sediments wash into the stream and are carried along at the same speed. Turbulence in the water, with random upward motion, is an important mechanical factor holding sediment in suspension. Streams with a heavy suspended load run a muddy or cloudy color, similar to the color of the sediment they transport.

▲10.16 **The Magdalena River in Colombia is a *turbulent* stream**

▼10.17 **The Sepik River in Papua New Guinea demonstrates *laminar* flow**

Traction and Saltation Bed load refers to coarser materials that are moved down the streambed by **traction** or **saltation**. In traction, the current literally drags the materials along the streambed. When materials move downstream by saltation, the particles bounce along in short hops, because they are too large to be rolled or picked up to remain in suspension. Most bed load is moved downstream during floods (►Figs. 10.18 and 10.19).

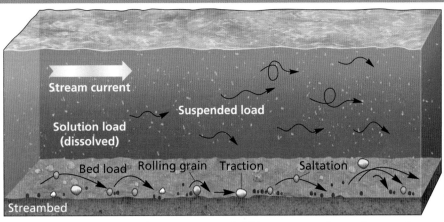

(a) A stream transports sediment in several different ways, depending on the size of the particles.

(b) Suspended load in this turbulent stream turns the water a brownish hue.

▲10.18 **Fluvial transportation of eroded materials**

geoCHECK ✔ Explain how dissolved load in a stream is different from bed load.

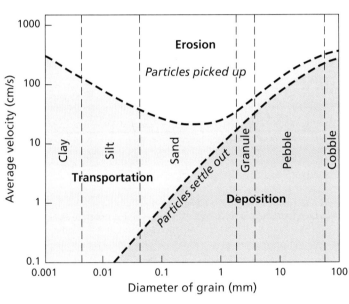

◄10.19 **Stream velocity, erosion, and deposition** As the average stream velocity (left axis) increases, larger diameter sediment particles are picked up and transported downstream (bottom axis). For example, an average stream velocity of 10 cm/s will transport clay and silt particles, but larger sand particles begin to settle out into the streambed, and few pebbles or cobbles are transported.

Deposition of Sediment

During a flood, a river may carry an enormous sediment load as traction and saltation move the larger bed load downstream. As flood flows ebb, stream energy is reduced, sediment transport slows or stops, and deposition increases. If the bed load and suspended load exceed a stream's capacity, sediment accumulates and the stream channel builds up through deposition—the process of **aggradation**. The general term for the clay, silt, sand, gravel, and mineral fragments deposited by running water is **alluvium**, which may be sorted or semisorted sediment on a floodplain, delta, or streambed.

When the stream's capacity to carry sediment has been exceeded, deposition of sediment begins. Deposition in **braided channels** occurs when reduced discharge lowers a stream's transporting ability, such as after flooding or near the mouth of a stream. Braided channels can also occur when a landslide occurs upstream or when loads increase where weak banks of sand or gravel exist (▲Fig. 10.20). Braided rivers commonly occur in glacial environments, which have abundant supplies of sediment and steep channel slopes, as is the case with the Brahmaputra River in Tibet. Glacial action and erosion produced the sediments that exceed stream capacity in this river channel.

▲10.20 **Interconnected and braided channels of the Rees River, South Island, New Zealand**

geoCHECK ✔ Describe the fluvial conditions that produce braided stream channels.

geoQUIZ

1. Explain why turbulent streamflow is more common than laminar streamflow in mountain environments.
2. Describe the difference between stream competence and stream capacity.
3. Describe the fluvial conditions that lead to stream aggradation.

10.5 Changes in Stream Channels

Key Learning Concepts

▶ **Develop** a model of a meandering stream.

▶ **Explain** how a stream erodes a nickpoint.

As discussed in Section 10.3, a stream gradient is the rate of decline from its headwaters, where the stream originates, to its mouth, where it ends (see Fig. 10.6). In most streams, the headwater and upstream slopes give way to a more gentle profile downstream. Accordingly, streams adjust their gradient, velocity, and channel shape to move the sediment load.

Features Formed by Stream Erosion

When the channel slope is gradual, streams **meander** through the landscape. Meandering streams migrate across their valley bottoms in a sinuous snake-like pattern, with distinct flow and channel characteristics. Rivers in this stage operate near equilibrium in that their gradient, velocity, and sediment load do not change much. These slow-moving waters exemplify a low-energy stream environment.

Flow characteristics of a meandering stream are best seen in a cross-sectional view (▶ **Fig. 10.21**). The greatest velocities are near the surface at the center of the channel, corresponding to the deepest part of the stream channel. Velocities decrease closer to the sides and bottom of the channel because of the frictional drag on the water flow. The portion of the stream flowing at maximum velocity moves diagonally across the stream from bend to bend in a meandering stream.

The outer portion of each meander receives the fastest water velocity, and therefore, the greatest scouring action. The outer curve can be the site of a steep **undercut bank**, or *cutbank* (Fig. 10.21). In contrast, the inner portion of a meander experiences the slowest water velocity, and thus, receives deposition, forming a **point bar** deposit. As meanders develop, these scour-and-fill processes gradually rework the stream banks, leaving residual deposits from prior channels. Figure 10.22 brings these cross-sectional views into real life. A looping meander often becomes isolated from the rest of the river as erosion and deposition configure a new channel through the narrow end of the meander. The resulting **oxbow lake** may gradually fill with organic debris and silt or may again become part of the river when it floods (▶ **Fig. 10.22**).

geoCHECK ✔ Explain the difference between a point bar and an undercut bank.

(a) Meandering stream channel

(b)

▲ **10.21 Meandering stream** The maximum flow velocity, indicated by the dark blue line, shifts from the center along a straight stretch of the stream channel to the outside bend of a meander, where erosion occurs. Lower stream velocity on the inside of the bend is where deposition of sediment occurs.

Animation (MG)
Meandering Streams
http://goo.gl/ySMJp5

Condor Video (MG)
Meandering Rivers
https://goo.gl/yboSd3

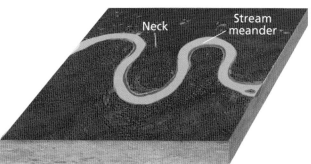

(a) Step 1: A neck forms where a lengthening meander loops back on itself.

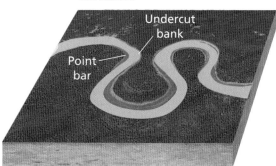

(b) Step 2: Over time, the neck narrows as erosion undercuts the banks.

▲ **10.22 Stream Meanders** Meandering stream channels increase the distance the water must travel through the drainage basin.

Stream Nickpoints

When the stream channel characteristics change, an ungraded stream works to restore its dynamic equilibrium. Stream **nickpoints** occur when the longitudinal profile of a stream experiences an abrupt change in gradient, such as at a waterfall or area of rapids (▶Fig. 10.23). Nickpoints can result when a stream flows across a zone of hard, resistant rock or from various tectonic uplift episodes, such as along a fault line. Temporary blockages in a channel caused by a landslide or a logjam, are also nickpoints. Over time, erosion by the stream eliminates the nickpoint and restores the stream's dynamic equilibrium.

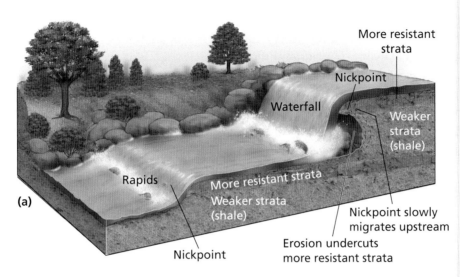

(a)

More resistant strata

Nickpoint

Waterfall

Weaker strata (shale)

Rapids

More resistant strata

Weaker strata (shale)

Nickpoint

Nickpoint slowly migrates upstream

Erosion undercuts more resistant strata

▲10.23 (a) **Stream nickpoint** (b) The Salto de Bordones waterfall in Colombia flows across a resistant layer of granite, causing a sudden and dramatic change in stream gradient.

Waterfalls are actually large stream nickpoints. In their most spectacular fashion, the water literally falls unimpeded down a large drop in the stream profile. The free-falling water moves at high velocity under the acceleration of gravity, causing increased abrasion and hydraulic action in the channel below. This action undercuts the waterfall. Eventually, the excavation causes the rock ledge at the lip of the fall to collapse, shifting the waterfall a bit farther upstream. The height of the waterfall is gradually reduced as debris accumulates at its base.

(b)

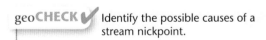 geoCHECK ✔ Identify the possible causes of a stream nickpoint.

geoQUIZ

1. Explain why stream velocity decreases closer to the sides and bottom of the stream channel.
2. Imagine that you are a drop of water in a stream. Describe how your path and velocity would change as you flow down a meandering stream.
3. Explain how the hydraulic action of a waterfall eventually works to reduce the actual height of the free-falling water.

(c) **Step 3:** Eventually, the stream erodes through the neck, forming a cutoff.

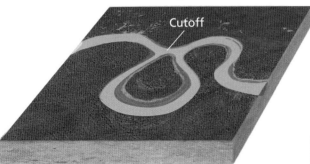

Cutoff

(d) **Step 4:** An oxbow lake forms as sediment fills the area between the new stream channel and its old meander.

Oxbow lake

10.6 Stream Deposition

Key Learning Concepts

▶ **Define** a floodplain and its characteristic landforms.
▶ **Describe** alluvial fan and delta environments.

After weathering, mass movement, erosion, and transportation, deposition is the next logical event. The general term for the unconsolidated clay, silt, sand, gravel, and mineral fragments deposited by running water is *alluvium*, which may accumulate as sorted or semisorted sediment. The process of *fluvial deposition* occurs when a stream deposits alluvium, thereby creating depositional landforms, such as bars, floodplains, terraces, and deltas.

Features of Floodplains

Flanking many stream channels are flat, low-lying areas called **floodplains**. Floodplains are formed by erosion, but their surfaces reflect centuries of recurrent flooding and deposition of sediment (▶Fig. 10.24). When the floodwater recedes, it leaves behind new deposits that cover the underlying rock with alluvium. Over time, the landscape near a meandering river develops meander scars made of residual deposits from abandoned channels. Meander scars may include traces of meander bends, oxbow lakes, natural levees, point bars, and undercut banks. Wetlands called backswamps often form in the poorly drained, fine sediments deposited when the river floods surrounding areas.

Natural Levees When a river overflows its banks, it loses stream competence and capacity. The water spreads out, dropping some of its sediment load to form **natural levees**. A *levee* is a mound of sediment that parallels both sides of the stream (▼Fig. 10.25). Successive floods increase the height of the levees. Deposition on the levees and within the river channel may elevate the river itself above the surrounding floodplain.

(b) Riparian marshes, or backswamps, are floodplain wetlands that store floodwaters and provide habitat for wildlife. Humans filled many such wetlands for development during the 20th century; restoration is now a priority, since wetland water storage feeds streamflow during drought conditions.

Meander scar   Oxbow lake   Meander scars
Floodplain
Cutoff
Yazoo stream   Point bar
Bluffs
Undercut bank
Alluvial deposits   Natural levees   Backswamp

(a) Typical floodplain landscape and related landscape features.

▲10.24 **A floodplain environment**

Animation (MG)
Oxbow Lake Formati
http://goo.gl/Axr01

Animation (MG)
Stream Processes,
Floodplains
http://goo.gl/ldctHL

Natural levee

▲10.25 **Stream levees**

Floodplains and Agriculture Since antiquity, humans have farmed the rich alluvial deposits on floodplains (▶Fig. 10.26). Examples include the Nile River (Egypt), the Tigris and Euphrates Rivers (Iraq), and the Mississippi River (USA).

Mobile Field Trip (MG)
Mississippi River Delta

https://goo.gl/2rbhF0

▼10.26 **Floodplain farming** For centuries, rice farmers have reworked the rich alluvial soils of this floodplain in Vietnam.

Alluvial Stream Terraces Forces other than flowing water can reshape the alluvial deposits on a floodplain. Tectonic uplift or a lowering of base level rejuvenates stream energy so that a stream again scours downward. The resulting entrenchment of the river into its own floodplain produces **alluvial stream terraces** on both sides of the valley, which look like topographic steps above the river (▶Fig. 10.27).

geoCHECK ✔ Describe the sequence of events that creates a floodplain.

Other Depositional Features

Deposition can occur in any area along a stream where the stream slows down, losing competence and capacity. Alluvial fans often occur along a mountain front not far from the headwaters of a stream. Deltas, on the other hand, form where a stream reaches its base level in a lake or the ocean.

Alluvial Fans In arid and semiarid climates **alluvial fans** form at the mouth of a canyon where an ephemeral stream channel exits into a wider valley (▼Fig. 10.28). Alluvial fans are produced when flowing water (such as a **flash flood**) abruptly loses velocity as it leaves the constricted channel at the canyon mouth. The stream gradient suddenly decreases, dropping layers of sediment along the base of the mountain block. The heavier coarse materials are deposited first, grading slowly to pebbles and sand with distance away from the mouth.

▼10.28 Alluvial fan in Death Valley

Project Condor Video (MG)
Characteristics
of Alluvial Fans

https://goo.gl/RA838E

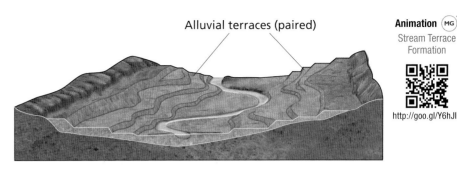

Alluvial terraces (paired)

Animation (MG)
Stream Terrace
Formation

http://goo.gl/Y6hJIZ

▲10.27 **Alluvial stream terraces** As a stream cuts into a valley, alluvial terraces form.

River Deltas The level or nearly level depositional plain that forms at the mouth of a river is a **delta**, after its characteristic triangular shape (▼Fig. 10.29). Deltas occur where a river's velocity rapidly decelerates as it enters a larger, standing body of water and the sediment load is deposited. The deposits are sorted, with coarse sand and gravel dropping out first closest to the river's mouth. Finer clays are carried farther and form the extreme end of the deposit, which may be underwater.

As with floodplains, humans settle deltas for their level terrain, proximity to water, and fertile soils. The Ganges–Brahmaputra River (South Asia), Zambezi River (Mozambique), and Mekong River (Vietnam) all have large deltas. Some rivers do not form deltas. For example, the Amazon lacks a true delta. Instead, its mouth has formed an underwater deposit on a sloping continental shelf.

geoCHECK ✔ List the reasons why river deltas are often favored for human settlement.

geoQUIZ

1. Describe the landforms created by alluvial deposition.
2. Explain how alluvial stream terraces are formed.
3. Identify the primary differences between an alluvial fan and a river delta.

▼10.29 The Nile River Delta

10.7 Floodplains & Human Settlement

Key Learning Concepts

▶ *Describe* recent examples of flood hazards.

▶ *Explain* flood probability estimates.

A flood is high water that overflows the natural bank along any portion of a stream. Throughout history, civilizations have settled floodplains and deltas, especially since the agricultural revolution of 10,000 years ago when the fertility of floodplain soils was discovered. Early villages were usually built away from the area of flooding, or on alluvial stream terraces, because the floodplain was dedicated exclusively to farming. However, as commerce grew, competition for sites near rivers increased, because these locations were important for transportation. Port and dock facilities were built, as were river bridges. Because water is a basic industrial raw material used for cooling and for diluting and removing wastes, waterside industrial sites became desirable. These competing human activities on vulnerable flood-prone lands place lives and property at risk during floods.

Flood Hazards

Today, with a world population of more than 7 billion, it is not surprising that hundreds of millions of people live on floodplains and deltas. These areas are subject to flooding under a range of circumstances. The catastrophic floods along the Mississippi River and its tributaries in 1993 and again in 2011 produced damage that exceeded $30 billion in each occurrence. In 2001, tropical storm Allison alone left $6 billion in damage in its wandering visit to Texas in June. In 2013, over 17 inches of rain fell during eight days in Boulder County, Colorado. The unprecedented flooding damaged or destroyed over 900 homes and commercial properties. In 2014, the wettest winter in 250 years triggered widespread flooding in England and Wales.

Catastrophic floods especially threaten less-developed regions of the world. Flood-prone Bangladesh provides a perennial example: It is one of the most densely populated countries on Earth, and more than *three-fourths* of its land area is a floodplain (▶Figs. **10.30** and **10.31**). The country's vast alluvial plain sprawls over an area the size of Alabama (130,000 km², or 50,000 mi²). The flooding severity in Bangladesh is a consequence of human economic activities, along with heavy precipitation episodes. Forest harvesting in the upstream portions of the Ganges–Brahmaputra River watersheds may also have augmented runoff. Over time, the increased sediment load carried by the Ganges was deposited in the Bay of Bengal, creating new islands. These islands, barely above sea level, became sites for new farming villages. As a result of the 1988 and 1991 floods and storm surges, about 150,000 people perished.

The Indus River, which flows through Pakistan into the Arabian Sea (▶Fig. **10.32**), is another recent example of flooding in a developing part of the world. Heavy monsoon rains in August 2010 swelled the Indus and its many tributaries beyond anything in recorded history. Destruction was greater than that from the 2004 Indian Ocean tsunami. More than 20 million Pakistanis were left homeless. High water returned in 2014, displacing over 1.3 million people (▶Fig. **10.33**).

▲**10.30 Bangladesh flood risk** Three of Asia's largest rivers —the Brahmaputra, the Ganges, and the Meghna—converge with lesser streams in this enormous lowland floodplain "of many mouths" that is subject to repeated flooding.

▲**10.31 Flooded fields and homes in Bangladesh** The river channel is normally confined by levees.

geoCHECK  Identify the main cause of the 2010 floods that inundated Pakistan's Indus River plain.

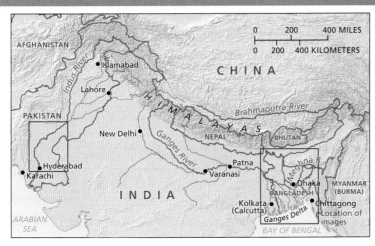

▲10.32 **The Indus, Ganges, Brahmaputra, and Meghna Rivers drain the high mountains of Asia** These rivers periodically flood their densely settled and heavily farmed floodplains.

Rating Floodplain Risk

Flood patterns in a drainage basin are as complex as the weather, for floods and weather are equally variable and both include a level of unpredictability. Measuring and analyzing how water progresses down large watersheds enables engineers to develop the best possible flood-management strategy. Unfortunately, reliable data often are not available for small basins or for the changing landscapes of urban areas, where paved areas and roofs prevent runoff from sinking into the soil. In the United States, the Federal Emergency Management Agency (FEMA) produces flood hazard maps to help Americans assess their flood hazard risk (see https://msc.fema.gov/portal).

Floods are rated statistically for the probability that a flood will occur sometime during a given time interval, based on historical data. For example, a *10-year flood* has a 10% likelihood of occurring in any one year and is likely to occur about 10 times each century, which indicates a moderate threat. A 50-year or 100-year flood is of greater and often catastrophic consequence, but is also less likely to occur in a given year. These probability ratings of flood levels are mapped for an area, and each defined floodplain that results is then labeled a "50-year floodplain" or

▼10.34 **Flooding in central Texas follows unusually heavy spring rains**

(a) A man and cattle search for higher ground.

(b) Indus River floodwaters inundate fields and villages in this rich agricultural region.

▲10.33 **Indus River flooding, 2014**

a "100-year floodplain." These statistical estimates are probabilities that floods will occur randomly during any single year of the specified period. Of course, two decades might pass without a 50-year flood, or a 50-year level of flooding could occur three years in a row. The record-breaking Mississippi River Valley floods in 1993 and again in 2011 (▼Fig. 10.34) easily exceeded a 1000-year flood probability, as did the 2005 Hurricane Katrina catastrophe.

geoCHECK ✔ Explain what is meant by a *200-year flood* rating.

geoQUIZ

1. Describe why, throughout history, floodplain environments have attracted human settlement.
2. Explain why the country of Bangladesh is so susceptible to devastating floods.
3. Why are urban areas so susceptible to large flood events?

RIVER SYSTEMS IMPACT HUMANS

• Humans use rivers for recreation and have farmed fertile floodplain soils for centuries.
• Flooding affects human settlements on floodplains and deltas.
• Rivers are transportation corridors and provide water for municipal and industrial use.

HUMANS IMPACT RIVER SYSTEMS

• Dams and diversions alter river flows and sediment loads, affecting river ecosystems and habitat. River restoration efforts include dam removal to restore ecosystems and threatened species.
• Urbanization, deforestation, and other human activities in watersheds alter runoff, peak flows, and sediment loads in streams.
• Levee construction affects floodplain ecosystems; levee failures cause destructive flooding.

10a

In June 2013, floodwaters following days of heavy rainfall inundated Germany, Austria, Slovakia, Hungary, and the Czech Republic. According to local residents, water levels in Passau, Germany, were higher than any recorded in the past 500 years.

10c

In 2011, Americans spent $42 billion on fishing-related activities. Streams in Montana, Missouri, Michigan, Utah, and Wisconsin are designated "blue ribbon fisheries" based on sustainability criteria such as water quality and quantity, accessibility, and the presence of certain species.

10b

Monsoon rains in August 2013 caused flooding across Pakistan, damaging over 80,000 homes, affecting over a million people, and causing more than 200 fatalities. The Swat Valley in northern Pakistan, pictured here, had the worst flooding in over a decade.

ISSUES FOR THE 21ST CENTURY

• Increasing population will intensify human settlement on floodplains and deltas worldwide, especially in developing countries, making more people vulnerable to flood impacts.
• Stream restoration will continue, including dam decommissioning and removal, flow restoration, vegetation reestablishment, and restoration of stream geomorphology.
• Global climate change may intensify storm systems, including hurricanes, increasing runoff and flooding in affected regions. Rising sea level will make delta areas more vulnerable to flooding.

Looking Ahead

In the next chapter we investigate how the oceans, coastal systems, and wind processes produce distinctive landforms. We will identify the components of the coastal environment, and explain coastal phenomena such as mean sea level, tides, barrier beaches, and coastal wetlands. We will also describe the erosion and deposition processes, and landforms created by wind.

How are drainage basins organized?

10.1 Streams & Drainage Basins

Describe how a stream forms.

Describe the structure of a drainage basin.

- Although the terms river and stream overlap, *river* is applied to a main stream. A network of tributaries forms a *river system* that empties a *drainage basin*, or *watershed*. Drainage divides and interfluves partition water into separate drainage basins. Smaller drainage basins deliver runoff and sediment to the main stream.

1. Explain the significance of the Continental Divide as it pertains to drainage basins in the United States.

2. On a map of your home state, the United States, or the world, identify a stream confluence.

10.2 Drainage Patterns

Identify different types of drainage patterns.

Describe how a stream's discharge is determined.

- Most streams follow one of seven distinctive drainage patterns. Each pattern results from a unique combination of topography, rock resistance, climate, hydrology, and underlying rocks. These factors all influence the stream basin discharge—the total streamflow past a point in a given unit of time. The drainage density of a drainage basin is determined by dividing the total length of all stream channels in the basin by the area of the basin.

3. Explain why a deranged drainage pattern is different from the other patterns.

4. Describe the factors that result in a high-density drainage basin.

What unique landforms result from moving water?

10.3 Stream Dynamics

Identify the parts of a stream profile.

Describe the relation between stream velocity, depth, width, and discharge.

Identify different types of streams.

- The stream gradient is the rate of decline from the headwaters, where the stream originates, to the mouth where the stream ends. The base level is the point below which a stream cannot further erode its valley. Base levels may be temporary, such as a pond or lake blocked behind a landslide, or the ultimate base level of the ocean. The discharge for most streams varies by season and by year. Perennial streams flow throughout the year. Intermittent streams flow seasonally, while ephemeral streams flow only after storms.

5. Describe the conditions under which stream discharge may actually decrease with distance downstream.

6. Identity at least one perennial and one intermittent stream in your home state.

7. Identify the five largest rivers on Earth in terms of discharge.

10.4 Fluvial Processes

Explain the processes of fluvial erosion and deposition.

- Fluvial landscapes result from erosion and deposition by streams. The hydraulic action of flowing water loosens rocks for movement downstream. The abrasion of this debris then erodes and carves the streambed. The ability to transport sediment—or competence—depends on stream velocity. Fluvial sediments are transported as dissolved load, as suspended load, or as bed load by traction or saltation.

8. Differentiate between the three types of stream load.

9. Describe the processes at work as a stream erodes its channel.

How do streams transport & deposit sediment?

10.5 Changes in Stream Channels

Develop a model of a meandering stream.

Explain the role of stream gradient in a stream's adjustment to changing conditions.

- Meandering streams migrate across valleys by the erosion of cutbanks and the deposition of material in point bars. Tectonic uplift can rejuvenate streams by increasing their gradient. Stream nickpoints, such as a cliff or temporary obstruction, create a sudden change in stream gradient and velocity.

10. Explain how streams usually adjust to a lessening gradient near the end of their profile.

11. Describe the formation of an oxbow lake.

10.6 Stream Deposition

Define a floodplain and its characteristic landforms.

Describe alluvial fan and delta environments.

- Streams deposit unconsolidated alluvial sediment on level floodplains. Natural levees form in these environments when overflowing streams spread out and deposit their alluvium both on the levee and in the adjacent floodplain. Other fluvial depositional landforms include alluvial stream terraces, oxbow lakes, meander scars, alluvial fans, and deltas.

12. Describe the process by which an alluvial fan forms.

10.7 Floodplains & Human Settlement

Describe recent examples of flood hazards.

Explain flood probability estimates.

- A *flood* is a high water flow that overspills the natural bank along any portion of a stream. Humans have farmed these rich alluvial soils since antiquity. However, repeated flooding in these lowland environments also makes them risky for human settlement. The keys to good flood management is reliable stream discharge data and understanding how the drainage basin's topography, vegetation, and rock type respond during periods of heavy precipitation.

13. Identify the factors used to calculate stream discharge.

Critical Thinking Questions

1. What role is played by rivers in the hydrologic and geologic cycles?

2. Explain what happens to a stream's base level when a reservoir is constructed.

3. Identify the drainage patterns that exist in your hometown and where you attend school. You may need to consult Google Earth or search for online maps and satellite images by the name of your watershed, or click on the EPA's "Surf Your Watershed" website: http://cfpub.epa.gov/surf/locate/index.cfm, which shows the network of streams in each watershed and is searchable by ZIP code.

4. Describe the main features of a floodplain, and explain the role played by natural levees, oxbow lakes, and riparian marshes.

5. Explain how tectonic uplift would influence the landforms and stream pattern of a drainage basin.

Visual Analysis

This photograph portrays a stream descending through the rainforest of the Cordillera Central in Puerto Rico.

1. Identify the sections of the stream that demonstrate laminar flow and turbulent flow.

2. Identify the small nickpoint, and hypothesize on its origin.

3. Explain why the stream color is brown and not clear like tap water.

▲R10.1 A stream runs through the Cordillera Central, Puerto Rico

Key Terms

abrasion, p. 278
aggradation, p. 279
alluvial fan, p. 283
alluvial stream terrace, p. 283
alluvium, p. 279
base level, p. 275
bed load, p. 279

braided channel, p. 279
continental divides, p. 271
delta, p. 283
dissolved load, p. 278
drainage basin, p. 270

drainage pattern, p. 272
exotic stream, p. 277
flash flood, p. 283
flood, p. 18
floodplain, p. 282
fluvial, p. 270
graded stream, p. 274
gradient, p. 274

hydraulic action, p. 278
hydrograph, p. 276
hydrology, p. 270
meander, p. 280
natural levee, p. 282
nickpoint, p. 281
oxbow lake, p. 280
point bar, p. 280

saltation, p. 279
sheetflow, p. 270
stream discharge, p. 276
suspended load, p. 278
traction, p. 279
undercut bank, p. 280
ungraded stream, p. 274
waterfalls, p. 281

(MG) Interactive Mapping | Login to the **MasteringGeography** Study Area to access **MapMaster**.

- Open: Launch MapMaster in MasteringGeography™.
- Select: the *Physical Environment*, and then *Physical Features*, which includes drainage basins of Russian rivers. Study the rivers that drain from them into the many seas of the Arctic, Pacific, and Atlantic Oceans.
- From the *Tools* feature, use the *Pencil* to trace one river each, from its headwaters to its mouth in the following seas: Barents Sea, Kara Sea, Bering Sea, Sea of Okhotsk, Caspian Sea, Black

Sea, and the Baltic Sea. Note: clicking *Physical Features* on/off will help reveal the rivers.

- Turn off all data layers, then open the *Population* category and select *Population Density*. Under *Physical Environment* select *Environmental Issues*. With both layers displayed, hypothesize why so many polluted rivers flow through sparsely populated drainage basins. Opening the *Economic* category and selecting *Major Natural Resources* will help guide you to the answer.

Explore | Use **Google Earth** to explore the **Amazon River**.

Viewing the Earth from space allows us to visualize several fluvial concepts. In Google Earth, fly to the *Amazon River, Brazil*. Once there, zoom out until the full sweep of this enormous drainage basin is visible.

1. Zoom back in on the Amazon headwaters, then identify individual drainage basins, interfluves, and stream nickpoints.

2. Zoom out slightly, and move downstream to identify stream confluences, meanders, cut banks, point bars, and oxbow lakes (or ones about to form).

3. Zoom in where the Amazon enters the Atlantic Ocean. Explain the relationship between stream velocity and the change in the suspended sediment load that colors the river.

▲R10.2 **Mouth of the Amazon River**

MasteringGeography™

Looking for additional review and test prep materials? Visit the Study Area in MasteringGeography™ to enhance your geographic literacy, spatial reasoning skills, and understanding of this chapter's content by accessing a variety of resources, including

MapMaster™ interactive maps, videos, *Mobile Field Trips, Project Condor* Quadcopter Videos, *In the News* RSS feeds, flashcards, web links, self-study quizzes, and an eText version of *Geosystems Core*.

Risky Waters: Determining Flood Risk

All streams generate an average annual flow, but over time, most streams exhibit a wide variation in streamflow from very low stages to large floods. On September 12, 2013, heavy precipitation in Rocky Mountain National Park sent floodwaters roaring down the Big Thompson River and its tributaries (Fig. GL 10.1). The epic flooding damaged or destroyed the major roadways along with hundreds of homes and businesses in this narrow canyon (Figure GL 10.2).

   The average number of years between floods of a certain size is the recurrence interval. These intervals are based on the magnitude of the annual peak flow, which occurs once a year. However, the actual number of years between floods of any given size varies because of highly variable weather patterns, plus human impacts such as deforestation, urbanization, dam building, and the planting of crops can affect runoff. The goal of this lab is to understand the concept of a stream flood recurrence interval.

GeoLab10 (MG)
Pre-Lab Video

https://goo.gl/HPS02J

Apply

As a watershed manager working for a county at the base of the Rocky Mountains, you are charged with calculating a flood recurrence interval, so that county residents who live near streams have some understanding of the potential frequency of large flood events.

Objectives

- Outline a drainage basin
- Calculate a recurrence interval using actual stream data
- Apply these data to predict flood risk

Procedure

1. Your first chore is to outline, with a colored pen or pencil, the drainage basin for the Big Thompson River (Figure GL 10.3). Note that in a few places (red marker), some of this has been done for you. Next, mark the following:
   a. headwaters
   b. interfluve
   c. mouth, or location where the river flows onto the Great Plains
2. The next step is to calculate the flood recurrence intervals and rank flood events by their maximum discharge and their recurrence intervals. From this data, it is possible to calculate the probability of the flood recurring. Table 10L.1 lists nine years of maximum daily streamflow stage (cfs) for the Big Thompson River.
3. In the third column, rank this data by year from the largest amount (rank of 1) to the smallest (rank of 9).
4. In the fourth column, calculate the probability, in percent, of the flood recurring using the formula:

rank/(n + 1) × 100; where n = the number of years with available data

Table 10.L1 Data for Calculating Flood Recurrence Interval (RI)

Year	Stage (cfs)	Rank	Probability (%)	RI (years)
2005	780	4	40	2.5
2006	357			
2007	284			
2008	240			
2009	400			
2010	1,800			
2011	627			
2012	1,600			
2013	19,000			
2016–2019				

Note: As an example, the calculations for 2005 are completed.
Data for additional years can be entered in the blank rows.

   In column five, rank the recurrence intervals. This is determined by the following equation:

$$RI = \frac{(N + 1)}{M}$$

Where:
- RI = the recurrence interval
- N is the number of years the floods are on record
- M is the rank of the flooding event (largest flood M = 1)

Because there are nine years of data, this calculation becomes 9 + 1, divided by the rank number you assigned.

Analyze & Conclude

5. In your completed data table, how is the recurrence interval related to the rank and probability of each year's maximum discharge?
6. Based on your answer to question 5, what maximum discharge might residents along the Big Thompson River typically expect? Explain.
7. Does knowing the recurrence interval enable you to predict floods on the Big Thompson River during the next nine-year period? Explain why or why not.
8. In many areas of the world, accurate steam recurrence interval data now extend over one century. After a record (rank 1 flood) in these areas, the media often refer to it as a "once in a century flood," or a "100-year flood." In your job as a watershed manager, does this mean that prospective home and business owners are safe from another event of that magnitude for more than the average human lifetime? Explain.
9. Looking at Figure GL 10.3 on the next page, describe the difference in flow characteristics discussed in module 10.5 for stream water flowing in the middle of the canyon and the same water as it flows out of the mountains and across the Great Plains just below the Rocky Mountains.
10. Hypothesize how and why the following landscape changes might influence future floods in the lower sections of the Big Thompson River watershed:
    a. More urbanization
    b. A large forest fire
    c. A series of upstream dams

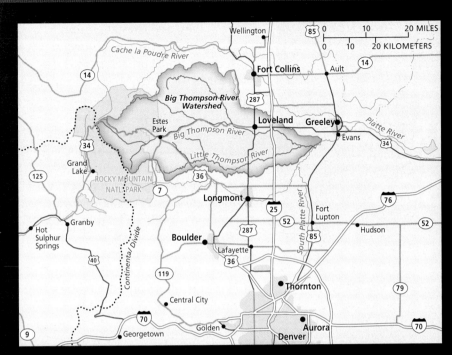

▲GL10.1 **Big Thompson River–Little Thompson River watershed in relation to cities and highways**

▲GL10.2 **Deadly topography** Steep slopes and sparse vegetation combined with heavy rain have lead to repeated flooding in the Big Thompson River canyon and communities downstream: A flash flood in 1976 killed 144. In 2013, 8 died as floods caused $2 billion in property damage.

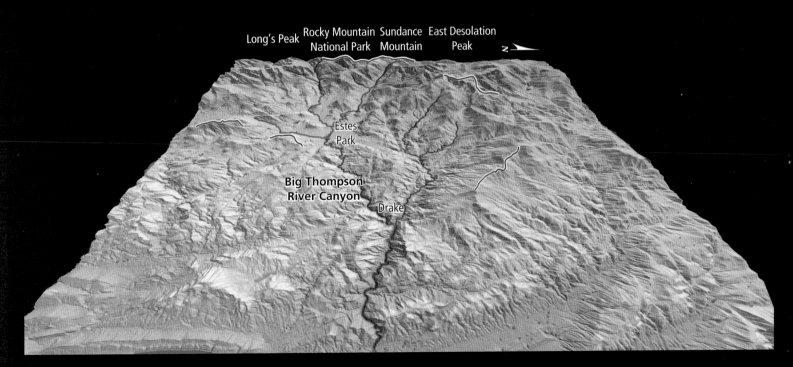

▲GL10.3 **Digital relief map of the Big Thompson watershed (NOAA)**

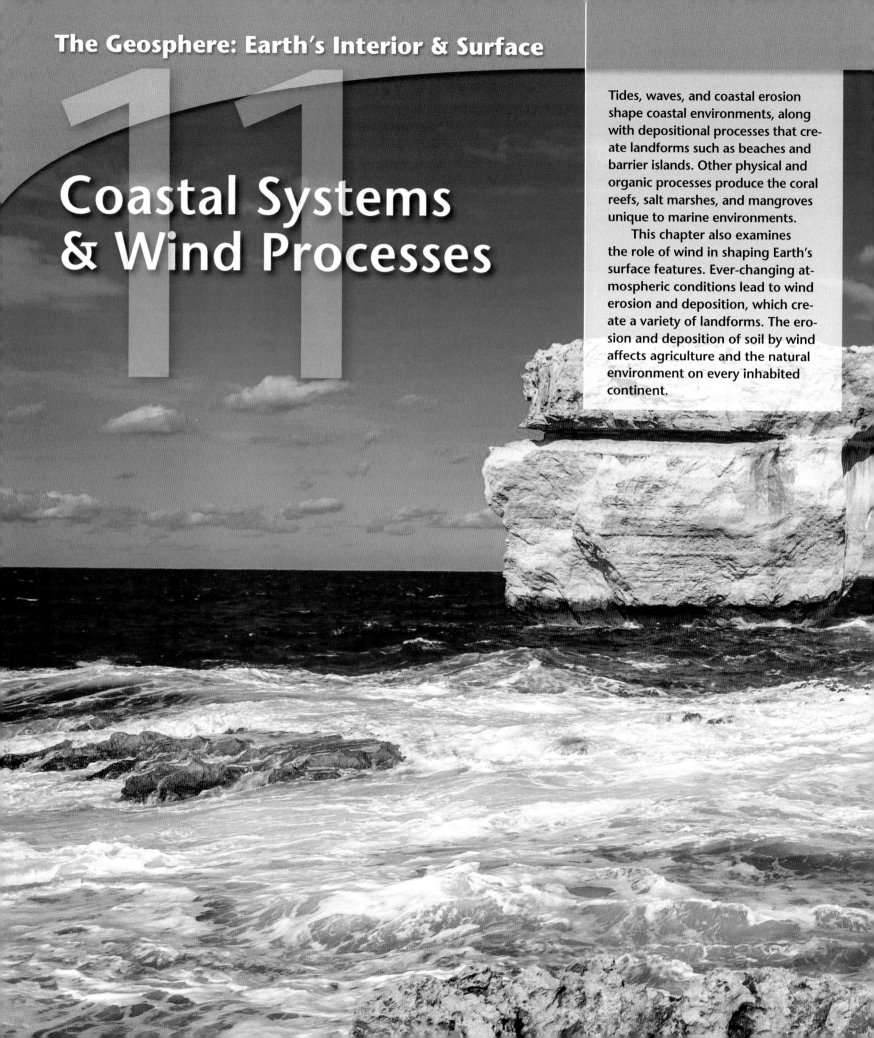

11 Coastal Systems & Wind Processes

Tides, waves, and coastal erosion shape coastal environments, along with depositional processes that create landforms such as beaches and barrier islands. Other physical and organic processes produce the coral reefs, salt marshes, and mangroves unique to marine environments.

This chapter also examines the role of wind in shaping Earth's surface features. Ever-changing atmospheric conditions lead to wind erosion and deposition, which create a variety of landforms. The erosion and deposition of soil by wind affects agriculture and the natural environment on every inhabited continent.

Key Concepts & Topics

Centuries of wave action produced the Azure Window, a sea arch on the Mediterranean Sea near Gozo, Malta.

11.1 Global Oceans & Seas

Key Learning Concepts

▶ *Analyze* the geographic distribution of oceans and seas.

▶ *Describe* the chemical composition of seawater.

▶ *Explain* how the ocean's physical and chemical properties determine its layered structure.

arth's vast oceans are intricately linked to life-sustaining systems in the atmosphere, the hydrosphere, and the lithosphere. Oceans act as a vast buffering system, absorbing excess atmospheric carbon dioxide and thermal energy. As one of Earth's last great scientific frontiers, oceans are of great interest to geographers. Remote sensing from satellites, surface vessels, and submersibles is providing a wealth of new data on the oceanic system.

Oceans & Seas

Figure 11.1 shows the locations of oceans and major seas. Although the terms *ocean* and *sea* are often used interchangeably, a sea is much smaller than an ocean and is usually enclosed, at least in part, by land (▶Fig. 11.2). Tectonic forces and uneven erosion of the coastlines may also produce a gulf, bay, or cove. Of these, gulfs are by far the largest, and while most have narrow openings, the Gulf of Alaska and Gulf of Mexico both feature large openings to their adjacent seas. Bays and coves exhibit similar geographic configurations, but they are much smaller in size.

geoCHECK ✔ Using precise definitions, what is the difference between an ocean and a sea?

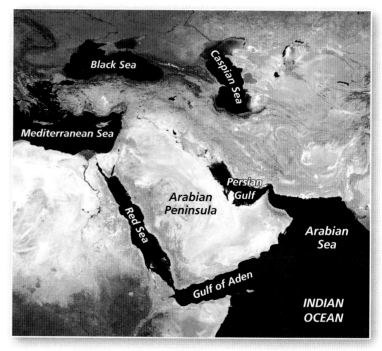

▲ **11.2 Five seas, two gulfs, and the Indian Ocean surround the Arabian Peninsula** The Caspian Sea is by area, Earth's largest inland body of water. Unlike the other seas that surround Arabia, the Caspian Sea does not connect to any ocean.

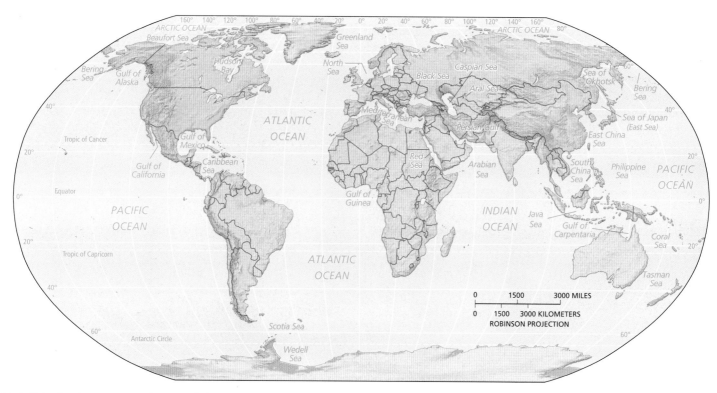

▲ **11.1 Principal oceans and seas of the world**

The Chemistry of Seawater

The physical and chemical properties of seawater change with latitude, water depth, proximity to land, and the addition of inflowing freshwater. The density of seawater generally increases with depth due to the weight of the overlying water. As depth increases, the temperature drops and salinity increases. As we shall see, these changing physical characteristics influence the overall layering and circulation of seawater, along with atmosphere–hydrosphere interactions.

Chemical Composition of Seawater Water is the "universal solvent," dissolving at least 57 of the 92 elements found in nature. Most natural elements and the compounds they form are found in the seas as dissolved solids, or *solutes*. Thus seawater is a solution composed of many ingredients (►Fig. 11.3).

Ocean chemistry results from complex exchanges among seawater, the atmosphere, minerals, bottom sediments, and living organisms. Significant flows of mineral-rich water also enter the ocean through hydrothermal (hot water) vents in the ocean floor. Seven elements account for more than 99% of the dissolved solids in seawater. These are chlorine, sodium, magnesium, sulfur, calcium, potassium, and bromine. Seawater also contains dissolved gases such as carbon dioxide, nitrogen, and oxygen, suspended and dissolved organic matter, along with many trace elements.

Salinity The concentration of dissolved solids is the solution's **salinity**. The worldwide average and most common notation for salinity is 35‰ (‰ indicates parts per thousand) or 3.5% (►Fig. 11.4). Variations in salinity are attributable to atmospheric conditions above the water, the volume of freshwater inflows, and water depth.

Seawater averages 3.5% salinity. The term **brine** is applied to water that exceeds this 35‰ average. **Brackish** applies to water that is less than 35‰ salinity. Oceans are usually lower in salinity near landmasses because of freshwater runoff and river discharges. Insolation also affects salinity because latitudes closer to the poles receive less insolation and have lower evaporation rates (and therefore lower salinity) than latitudes closer to the equator. Extreme examples include the Baltic Sea north of Poland and Germany, which averages 10‰ or less salinity because of heavy freshwater runoff and low evaporation rates. In contrast, the Persian Gulf has a salinity of 40‰. This briny composition results from high evaporation rates in a nearly enclosed basin in a subtropical desert climate.

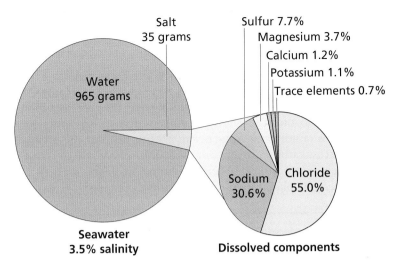

▲11.3 The chemistry of seawater The concentration of the dissolved solids (right) composes the salinity of seawater (left).

▼11.4 Ocean salinity Composite image of global ocean surface salinity from August 2011 to July 2012 using data from the Aquarius satellite, in orbit since 2011.

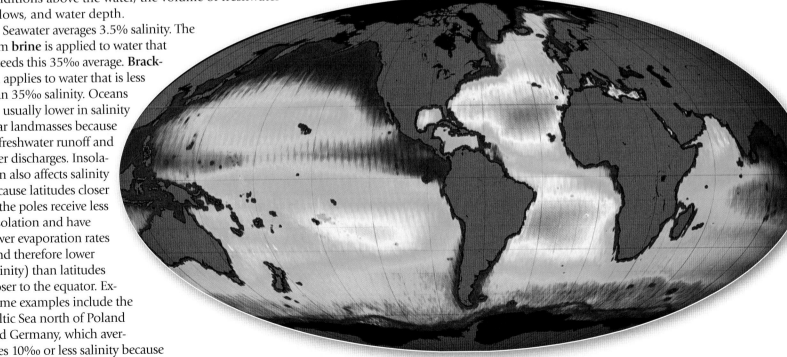

OCEAN SURFACE SALINITY

30   32   33   34   34.5   35   35.5   36   37   38   40   No data
Grams per kilogram

11.1 (cont'd) Global Oceans & Seas

Changes in Seawater Salinity & Acidity The ocean reflects conditions in Earth's environment. As an example, the salinity in high-latitude oceans has been declining over the past decade as melting ice contributes significant freshwater. Another change under way is rapidly increasing **ocean acidification**, as the ocean absorbs excess carbon dioxide from the atmosphere (▼Fig. 11.5a). A more acid ocean will cause certain marine organisms such as corals and some plankton to have difficulty maintaining external calcium carbonate shell structures (▼Fig. 11.5b). The overall pH could decrease by –0.4 to –0.5 units this century from the ocean's current average pH of 8.2. Because the scale is logarithmic, a decrease of 0.1 equals a 30% increase in acidity. Oceanic biodiversity and food webs will respond to this change in unknown ways.

geoCHECK ✔ Distinguish between brackish and brine water.

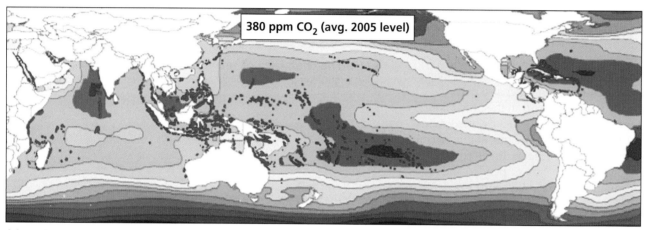

(a) As the ocean absorbs excess carbon dioxide from the atmosphere, the oceans become more acidic.

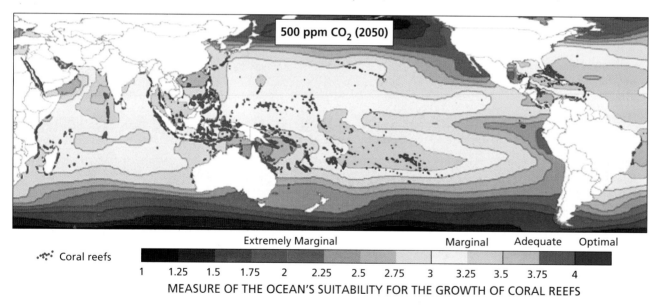

MEASURE OF THE OCEAN'S SUITABILITY FOR THE GROWTH OF CORAL REEFS

(b) Increased acidity reduces the levels of the calcium carbonate essential to building the shells and skeletons of many marine organisms, including the pteropod.

Animation (MG)
The Carbonate
Buffering System

http://goo.gl/nHTUpq

▲11.5 **Increasing ocean acidification**

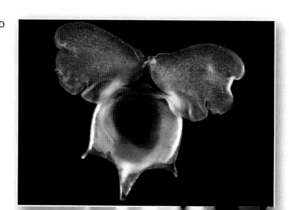

Ocean Layers

The physical structure of the ocean is layered like a cake. Increasing depth brings changes in average temperature, salinity, and density (▼Fig. 11.6). The Sun warms the ocean's surface layer, and the wind transfers energy to the water's surface, producing waves and currents. Variations in water temperature and solutes blend rapidly in a *mixing zone* that represents only 2% of the ocean. Below this is the *thermocline transition zone*, a region more than 1 km deep of decreasing temperature gradient that lacks the motion of the surface. Friction between these two layers can slow surface currents, and colder water temperatures also inhibit any convective mixing. From a depth of 1–1.5 km (0.6–0.9 mi) to the ocean floor is the *deep cold zone*, where water temperatures hover near 0°C (32°F). The water does not freeze because of its salinity and the intense pressures found at this depth. Surface seawater freezes at about –2°C (28.4°F).

geoCHECK ✔ Compare and contrast the three oceanic layers.

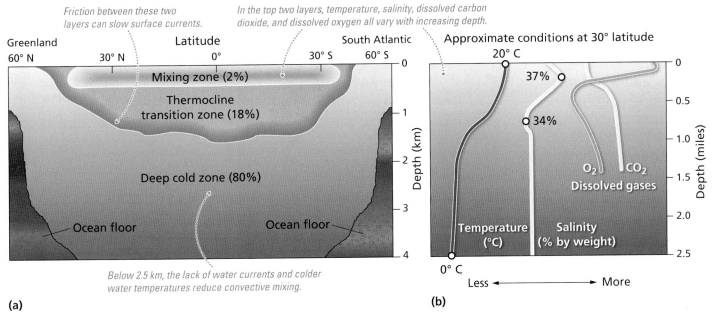

Friction between these two layers can slow surface currents.

In the top two layers, temperature, salinity, dissolved carbon dioxide, and dissolved oxygen all vary with increasing depth.

Below 2.5 km, the lack of water currents and colder water temperatures reduce convective mixing.

(a)

(b)

▲11.6 **The ocean's physical structure** The ocean consists of three horizontal layers.

Pollution & Oil Spills

The world's oceans have become repositories for much of the world's waste, whether discarded intentionally into the ocean or accidentally leaked or spilled. Marine and terrestrial oil pollution is a continuing problem in coastal regions as waste oil seeps and leaks into oceans from improper disposal and spills into oceans from offshore drilling and transpiration problems. On average, 27 oil-releasing accidents occur every day, totaling 10,000 a year worldwide and ranging from a few disastrous spills to numerous small ones (►Fig. 11.7).

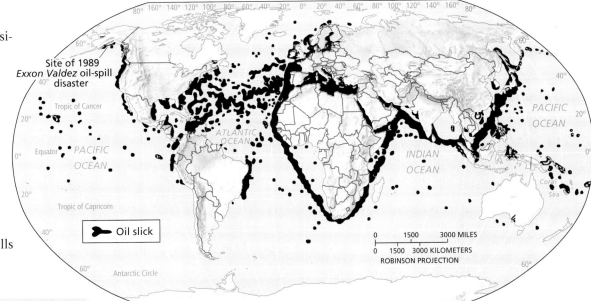

▲11.7 **Location of oil slicks worldwide in the 1990s**

geoQUIZ

1. Identify how mineral-rich water may enter the ocean.
2. Explain how and why the salinity of high-latitude oceans is changing.
3. What prevents seawater from freezing in the *deep cold zone*?

11.2 Coastal Environments

Key Learning Concepts

▶ *Identify* the main components of coastal systems.

▶ *Explain* how changes in sea level occur.

Although many of Earth's surface features, such as mountains and crustal plates, formed over millions of years, most of Earth's coastlines are relatively young. In coastal systems, the land, ocean, atmosphere, Sun, and Moon interact to produce the tides, currents, and waves that shape the shoreline's erosional and depositional features.

Coastal System Components

Inputs to the coastal environment include many elements discussed in previous chapters (▶ Fig. 11.8). All these inputs occur within the ever-present gravitational pull of Earth, Moon, and Sun. Gravity provides the potential energy of position (the energy an object has due to its position within a gravitational field) and produces the tides. A dynamic equilibrium among these components produces coastline features. The inputs include the following:

- *Solar energy* drives the atmosphere and hydrosphere. Conversion of insolation to kinetic energy produces prevailing winds, weather systems, and climate.
- *Atmospheric winds*, in turn, generate ocean currents and waves, key inputs to the coastal environment.
- *Climatic regions*, which result from insolation and moisture, strongly influence coastal geomorphic processes.
- Local characteristics of *coastal rock* and coastal geomorphology are important in determining rates of erosion and sediment production. Tectonic forces that move Earth's crust (Chapter 8) can cause coastlines to rise or sink in relation to sea level.
- *Human activities* are an increasingly significant input producing coastal change.

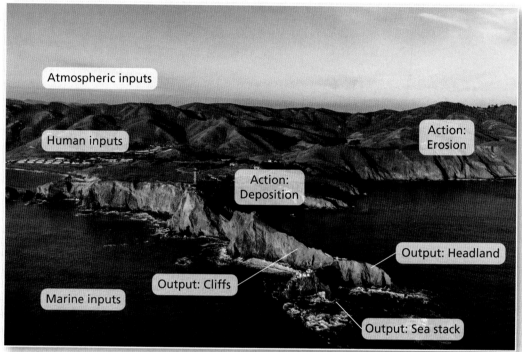

(a) Main inputs to coastal systems include waves and tides, atmospheric conditions, and human activities.

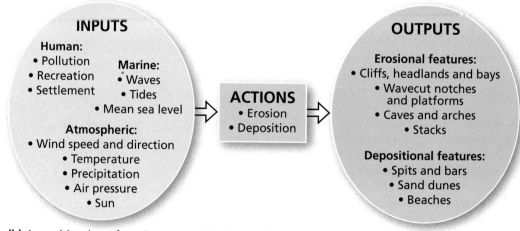

(b) A combination of marine, atmospheric, and human factors cause the erosion and deposition that shape coastlines.

▲ **11.8 Point Bonita, California** This rugged coastline can be described as as a system (a) with inputs and actions that lead to outputs—the varied coastal features visible in the photograph (b).

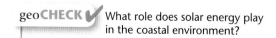 geoCHECK ✔ What role does solar energy play in the coastal environment?

The Coastal Environment & Sea Level

The inputs described above determine the nature of the coastal environment, along with another factor—the sea level itself. This environment is divided into distinct sections. The immediate coastal and shallow offshore areas make up the **littoral zone**, which extends to the highest waterline on shore during a storm (▶ **Fig. 11.9**). The littoral zone extends seaward to where water is too deep for storm waves to move sediments on the seafloor—about 60 m (200 ft) in depth. The *shoreline* is the line of contact between the sea and the land, although it shifts with tides, storms, and sea-level adjustments. The *coast* continues inland from high tide to the first major landform change. The littoral zone shifts over time as sea level varies. A rise in sea level submerges land, whereas a falling sea level exposes new coast.

Sea level is an important concept as it provides the reference point for every elevation on a map. **Mean sea level (MSL)** is a value based on average tidal levels recorded hourly at a given site over many years. The MSL varies daily across Earth's shorelines because of variations in ocean currents and waves, tides, air and ocean temperatures, air pressure, winds, gravity, and oceanic volume.

The MSL also varies over the long term with changing climate, tectonic movements, glaciation, and the reduction of liquid water storage on land. At the peak of Pleistocene glaciation about 18,000 years ago, sea level was 130 m (430 ft) lower than today. If the Antarctic and Greenland ice were to fully melt, sea level would rise 65 m (215 ft). Even with less extreme changes, sea-level fluctuations expose coastal landforms in many different areas to tidal and wave processes.

During the last 100 years, average sea level rose 10–20 cm (4–8 in.), a rate 10 times higher than the average during the last 3000 years. The present rise impacts local areas differently due to the ways sea, land, and ice interact globally. For instance, the rate of sea-level rise along Argentina's coast is 10 times that along the coast of France. The 2014 Intergovernmental Panel on Climate Change (IPCC) forecast for global MSL rise this century ranges from 0.26 to 0.55 m (10.2 to 21.6 in.). This estimate will be mostly attributable to worldwide losses in glacial ice and the thermal expansion of warming seawater. Based on the IPCC forecast, the global MSL rise will likely continue for many centuries beyond 2100, with the amount of rise dependent on future emissions of the greenhouse gases that cause climate change.

(a) The Chaves Beach littoral zone, Republic of Cabo Verde, Africa.

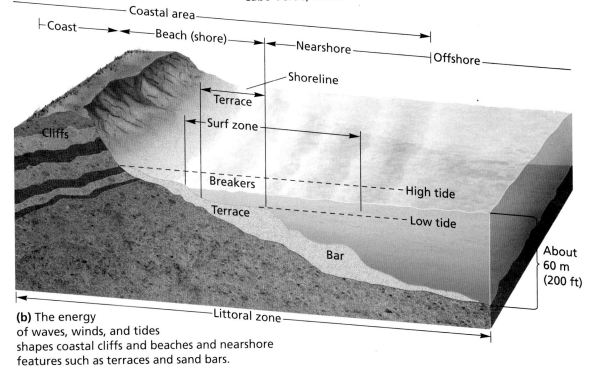

(b) The energy of waves, winds, and tides shapes coastal cliffs and beaches and nearshore features such as terraces and sand bars.

▲**11.9 The littoral zone** Spanning the coast, beach, nearshore, and part of the offshore environment, the littoral zone is dynamic and ever changing.

geoCHECK ✔ Describe how MSL is determined.

geoQUIZ

1. Describe how the main components of the coastal zone interact to form a dynamic environment.
2. Imagine you are standing on the shoreline, then identify the factors that could change this boundary within an hour.
3. Identify the short-term and long-term causes of sea level change.

11.3 Coastal System Actions

Key Learning Concepts

▶ ***Explain*** the cause and actions of tides.
▶ ***Describe*** wave motion at sea and near shore.

The coastal system experiences complex tidal fluctuation, winds, waves, ocean currents, and occasional storms. These forces shape landforms ranging from gentle beaches to steep cliffs, and they sustain delicate ecosystems. They also influence human settlement and commerce on coastlines throughout the world.

Tides

On every ocean shore, **tides** are twice-daily oscillations in sea level that range from barely noticeable to several meters. Tides also exist in large lakes, but there, the tides can be difficult to distinguish from the movements of windblown water. The gravitational pull of the Sun and the Moon produces tides. The Sun's influence is half that of the Moon's because of the Sun's greater distance from Earth (▶ Fig. 11.10a). Gravitational pull also raises *tidal bulges.* Because the gravitational pull on the oceans is different on opposite sides of Earth, two areas of raised water called *tidal bulges* form: one tidal bulge made up of water pulled directly toward the Moon and a tidal bulge on the side of Earth facing away from the Moon.

Tides appear to move in and out from the shoreline. However, it's actually Earth's surface that rotates in and out of the "fixed" tidal bulges as it rotates on its axis and changes its position relative to the Moon and Sun. Every 24 hours and 50 minutes, the Earth rotates through two bulges as a result of this rotational positioning (▶ Fig. 11.10b). Thus, every day most coastal locations experience two rising high tides and two falling low tides. The difference between consecutive high and low tides is the *tidal range.* For example, the narrow and high-latitude Bay of Fundy in Nova Scotia records the greatest tidal range on Earth, a difference of 16 m (52.5 ft) (▼ Fig. 11.11). In parts of the Bay of Fundy, the leading edge of incoming tidal waters resembles a large wave and is referred to as a **tidal bore**. Ocean basin characteristics (size, depth, and topography), latitude, and shoreline shape may also influence tides. The daily tides affect shoreline erosion and deposition and impact human navigation, fishing, and recreation.

geoCHECK ✔ Explain the difference between high tide and low tide.

Animation (MG)
Tidal Forces

https://goo.gl/IJuVhq

Animation (MG)
Monthly
Tidal Cycle

http://goo.gl/KYR2Pr

▼ 11.11 Tides in the Bay of Fundy, Nova Scotia, Canada

(a) Flood tide at Halls Harbor, Nova Scotia, Canada (near the Bay of Fundy).

(b) Ebb tide at Halls Harbor.

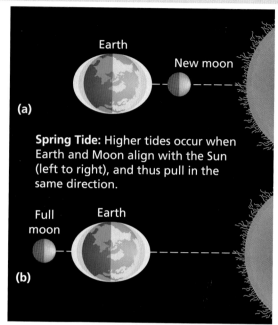

(a)

Spring Tide: Higher tides occur when Earth and Moon align with the Sun (left to right), and thus pull in the same direction.

(b)

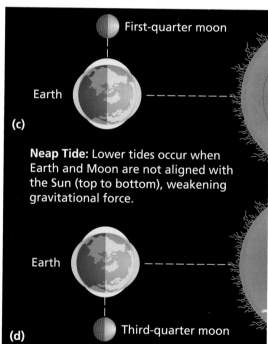

(c)

Neap Tide: Lower tides occur when Earth and Moon are not aligned with the Sun (top to bottom), weakening gravitational force.

(d)

▲ 11.10 The cause of tides

Waves

Friction between wind and the ocean surface generates **waves**. Waves vary in scale from small waves to large storm-generated *wave trains* that radiate outward in all directions across the ocean. Thus waves along a coast may result from a storm center far away, as well as from local winds. Regular patterns of smooth, rounded waves on the open ocean, are called **swells**.

Wave Formation In open water waves migrate in the direction of wave travel, but the water itself doesn't advance. It is *wave energy* that moves through the fluid medium of water (▶Fig. 11.12). Individual water particles move forward only slightly, following a vertically circular path. As a deep-ocean wave approaches the shoreline and enters shallower water, the water particles are vertically compressed. The slowdown begins at a depth equal to half the wavelength, and this friction with the ocean floor slows the wave, even as more waves continue arriving. The result is closer-spaced waves, growing in height and steepness, with sharper wave crests. As the crest of each wave rises, a

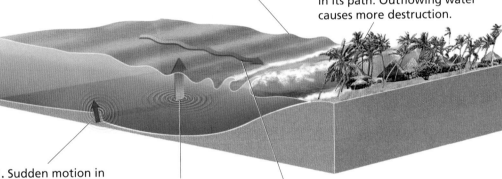

Path of water particles

Wavelength

Crest — Crest

Trough

Height

Movement of wave energy

Breaker

Beach

Surf

Deep water

Little motion below depth of one-half wavelength

Shallow depth shortens wavelength

Animation (MG) Wave Motion, Wave Refraction

http://goo.gl/RAeWk8

Animation (MG) Tsunami

http://goo.gl/clWb16

▲**11.12 Wave formation and breakers** Wavelength is the distance between corresponding points on any two successive waves. In open water, *wave energy* moves through the flexible medium of water, and not the water itself. When wave height exceeds its vertical stability, the wave *breaks* on the beach.

▼**11.13 How a tsunami forms**

4. Shallow water near a coast increases wave height.

5. The giant wave collides with the shore, leveling everything in its path. Outflowing water causes more destruction.

1. Sudden motion in seafloor (earthquake, landslide, or volcano) displaces a massive amount of water.

2. Energy from the seafloor motion creates waves at the surface.

3. The tsunami wave travels across open ocean at 600–800 kmph (375–500 mph).

(a)

(b) In 2011, a magnitude 9.0 earthquake triggered this catastrophic tsunami in northeastern Japan.

point is reached when its height exceeds its vertical stability, and the wave falls into a characteristic **breaker**, crashing onto the beach (Fig. 11.12).

Tsunamis Sudden motions in the seafloor, caused by earthquakes, submarine landslides, or eruptions of undersea volcanoes, produce **tsunamis**, or *seismic sea waves*. Japanese for "harbor wave," tsunamis have devastating effects in coves and harbors. These waves generally exceed 100 km (60 mi) in wavelength (crest to crest) but are only a meter (3 ft) or so in height in the open ocean. They travel at great speeds in deep-ocean water—velocities of 600–800 kmph (375–500 mph)—but often pass unnoticed on the open sea because their great wavelength makes the rise and fall of water hard to observe (▶Fig. 11.13). As a tsunami approaches a coast, the shallow water forces the wavelength to shorten, resulting in an increase in wave height up to 15 m (50 ft) or more. During the 20th century, records show 141 damaging tsunamis, with a total death toll of about 70,000. The 21st century has already seen two extremely deadly tsunamis: the Indian Ocean tsunami of 2004 (about 280,000 dead) and the tsunami that struck Japan after the 2011 earthquake (about 20,000 dead).

geoCHECK ✔ Explain why tsunamis often advance unnoticed in the open ocean.

geoQUIZ

1. Identify the interacting forces that generate the pattern of tides in oceans and large lakes.
2. Explain the energy source that generates typical ocean waves.
3. Describe what causes a tsunami wave to grow in height as it reaches the shoreline.

11.4 Coastal Erosion & Deposition

Key Learning Concepts

▶ **Explain** coastal straightening and coastal landforms.
▶ **Differentiate** between landforms of erosional and depositional coastlines.

Coastlines are active places, where the energy of moving water continuously delivers and removes sediment in a narrow environment. The action of tides, currents, wind, waves, and changing sea level produces a variety of erosional and depositional landforms.

Erosional coastlines (also called *emergent* coastlines) are zones of active coastal erosion. They result from a retreating shoreline that exposes land previously underwater. This occurs during periods of active tectonic uplift, or when the land rebounds upward when the weight of melting glacial ice is removed.

Depositional coastlines (also called *submergent* coastlines) are places of active deposition of sediments. They develop from relative sea-level rise that occurs with increased subsidence of land surface due to weight of sediment (such as the weight of large delta), decreased rates of erosion, or during periods of global sea-level rise from melting glaciers during global warming.

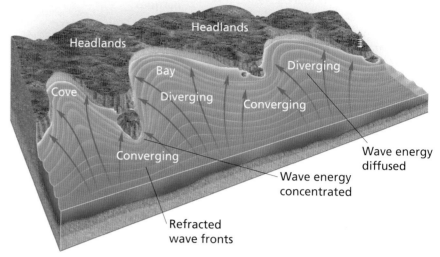

(a) Wave energy is concentrated as it converges on headlands and is diffused as it diverges in coves and bays.

(b) The Na Pali headland, Kauai, Hawai`i

▲ 11.14 **Wave refraction and coastal straightening**

Erosional Coastlines

The tectonically active margin of the Pacific Ocean along North and South America is a typical emergent coastline. High rates of erosion produce a rugged coastline with high relief, and wave action straightens the coastline, producing unique landforms. Protruding landforms of resistant rocks, called **headlands**, are formed where shallowing water depth causes waves to drag bottom, slowing them down and bending them toward the projecting land (▶ Fig. 11.14). The resulting **wave refraction**, concentrates erosional energy around headlands, but dissipates energy in coves and bays where sediment is deposited. Over time, headlands wear away and sediment builds up in bays and coves, gradually straightening an irregular coastline.

The sea's undercutting action along an emergent coastline forms sea *cliffs*. As indentations slowly grow at water level, a sea cliff becomes notched and eventually will collapse and retreat. Other erosional forms evolve along cliff-dominated coastlines, including *sea caves, sea arches,* and *sea stacks* (▶ Fig. 11.15). As erosion continues, arches may collapse, leaving isolated stacks. Wave action may also cut a horizontal **sea terrace** in the tidal zone, extending from a sea cliff out into the sea. If the relation between the land and sea

level changes due to tectonic uplift or lowering of sea level due to glaciation, multiple terraces may rise like stair steps back from the coast.

Longshore Drift Wave refraction is also involved in the formation of currents that move along shorelines. As incoming waves enter shallow water, they are refracted as the wave slows. The wave portion in deeper water moves faster in comparison, thus producing a current parallel to the coast (▶ Fig. 11.16). This **longshore current** depends on wind direction and wave direction. A longshore current is generated only in the surf zone and works with wave action to transport sand, gravel, sediment, and debris along the shore as **longshore drift**. This sediment-transportation process is important in the formation of depositional features along coastlines.

geoCHECK ✔ Summarize the causes and effects of the process of wave refraction.

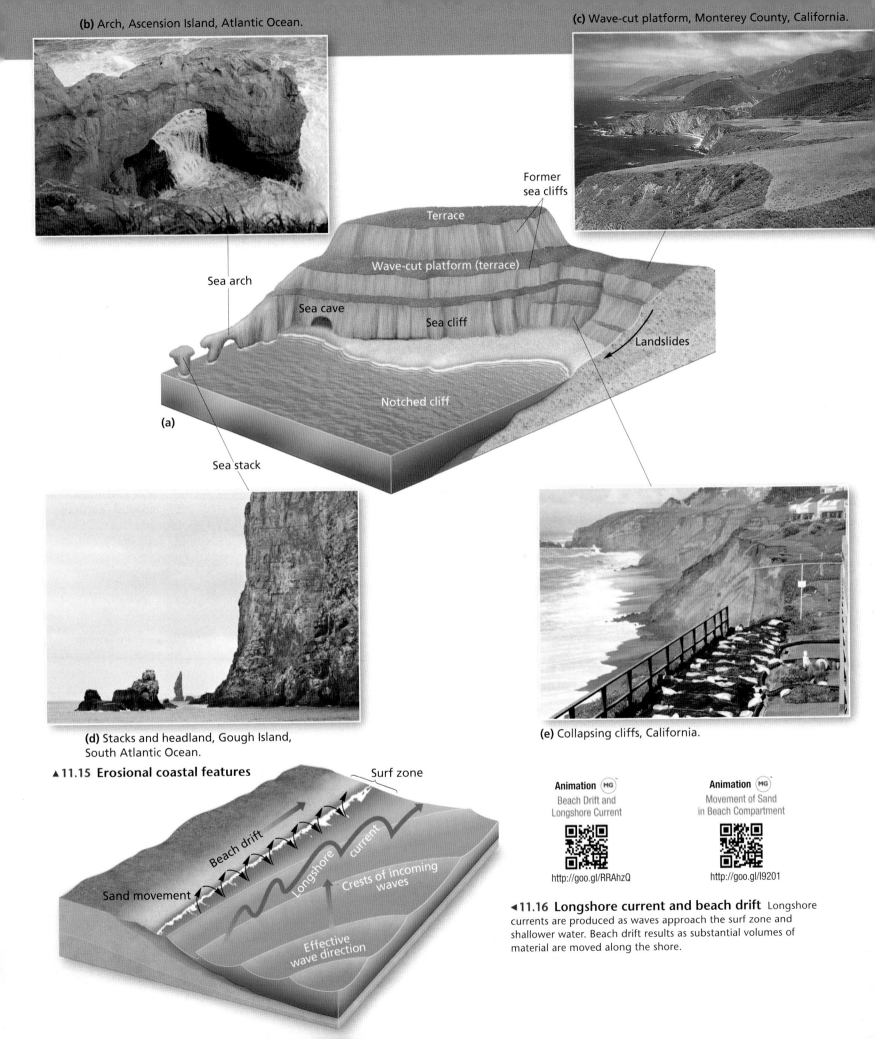

(b) Arch, Ascension Island, Atlantic Ocean.

(c) Wave-cut platform, Monterey County, California.

Terrace

Former sea cliffs

Wave-cut platform (terrace)

Sea arch

Sea cave

Sea cliff

Landslides

Notched cliff

(a)

Sea stack

(d) Stacks and headland, Gough Island, South Atlantic Ocean.

(e) Collapsing cliffs, California.

▲ **11.15 Erosional coastal features**

Surf zone

Beach drift

Longshore current

Sand movement

Crests of incoming waves

Effective wave direction

Animation (MG)
Beach Drift and Longshore Current

http://goo.gl/RRAhzQ

Animation (MG)
Movement of Sand in Beach Compartment

http://goo.gl/I9201

◀ **11.16 Longshore current and beach drift** Longshore currents are produced as waves approach the surf zone and shallower water. Beach drift results as substantial volumes of material are moved along the shore.

11.4 (cont'd) Coastal Erosion & Deposition

Depositional Coastlines

Depositional coasts feature gentle relief. Waves and currents deposit the characteristic landforms of these submergent coastlines. For example, the Atlantic and Gulf coastal plains of the United States, which lie far from an active plate tectonic boundary, are submergent coastlines.

Longshore drift (discussed above) along a depositional coastline can form a **barrier spit** (▼Fig. 11.17). The sand and silt is deposited in a long ridge extending out from a coast, partially blocking the mouth of a bay. Offshore currents must be weak for a spit to form, since strong currents transport material away. **Tidal flats** and **salt marshes** are coastal wetlands flooded and drained by tides. Both features occur on level terrain wherever tidal influence exceeds wave action. If these deposits completely cut off the bay from the ocean, an inland **lagoon** forms. A **tombolo** occurs when sediment deposits connect the shoreline with an offshore island or sea stack by accumulating on an underwater wave-built terrace.

Beaches vary in type and permanence, especially along coastlines dominated by wave action. Beaches often consist of sand particles that originated inland as rocks fragmented and weathered. Rivers transport the sand down to the ocean, where longshore currents and longshore drift then spread the sand along the coast. Most beaches are composed of sand that is rich in quartz, which resists weathering. In volcanic areas,

▲11.18 A black sand beach and the Bismarck Sea, Papua New Guinea

black sand beaches are derived from wave-crushed lava (▲Fig. 11.18). Other beaches are composed of gravel and shells or corals.

(b) A wave cut platform on England's North Devon coastline

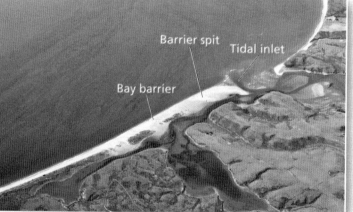
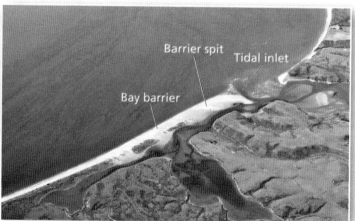

◄11.17 Depositional coastal features

Barrier spit

Tidal inlet

Bay barrier

(c) The Limantour barrier spit nearly blocks the entrance to Drakes Estero, along Point Reyes.

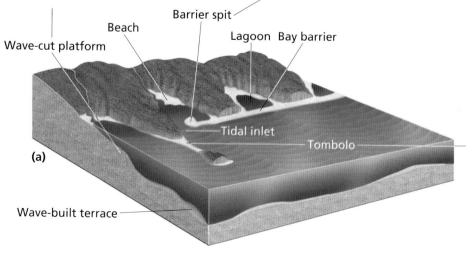

Barrier spit

Beach

Lagoon Bay barrier

Wave-cut platform

Tidal inlet

Tombolo

(a)

Wave-built terrace

(d) Tombolo Cluster, Koh Nang Yuan, Thailand

The beach zone usually stretches from about 5 m (16 ft) above high tide to 10m (33 ft) below low tide (see Fig. 11.9). Beaches stabilize the shoreline by absorbing wave energy, as is evident by the constantly moving material. Some beaches are seasonal accumulations in summer that disappear in winter when storm waves carry the sediments offshore.

Barrier beaches are long, narrow, depositional features, usually of sand, that form offshore parallel to the coast (▼Fig. 11.19). A broader, more extensive landform is a **barrier island**. These features form where tidal variation is usually low. On the landward side of the barrier island are tidal flats, marshes, coastal dunes, and beaches. Barrier beaches and islands line 10% of Earth's coastlines.

geoCHECK ✔ Why is quartz the most common component of beach sand?

▼11.19 Barrier island chain along North Carolina coast The Outer Banks of North Carolina. The area presently is designated as one of 10 national seashore reserves supervised by the National Park Service.

Mobile Field Trip (MG)
Gulf Coast Processes

https://goo.gl/0pYJkc

Animation (MG)
Movement of Barrier Island

http://goo.gl/KUPJcl

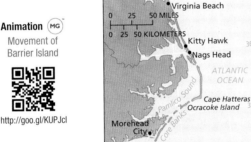

geoQUIZ

1. List the sequence of features that could form along a rocky, emergent coastline and describe how each feature is formed.
2. Explain the primary difference between erosional and depositional coastlines.
3. Identify the depositional characteristic that separates a tidal flat from a lagoon.

(a) Satellite photo of the Outer Banks.

(b) Flooding from Hurricane Irene cut through this road on Hatteras Island.

(c) Waves from a coastal storm batter homes in Buxton, North Carolina.

11.5 Wind Erosion

Key Learning Concepts

▶ **Differentiate** between eolian deflation and abrasion and the landforms they produce.

Wind blows over the entire surface of Earth but is an agent of geomorphic change primarily in coastal and desert environments. The work of wind is **eolian** (also spelled *aeolian*), for Aeolus, the Greek god of winds. The ability of wind to erode and transport sediment is actually small compared with that of water and ice because air is much less dense. Yet, over time, wind accomplishes enormous work. Wind can pile up sand dunes hundreds of meters high or sandblast a boulder, grinding away the solid rock into fine particles of dust. Consistent local wind can also prune and shape vegetation and sculpt ice and snow surfaces (▾Fig. 11.20).

(a) Wind-sculpted tree near South Point, Hawai`i. Nearly constant tradewinds keep this tree naturally pruned.

(b) Wind-eroded snow, called sastrugi, usually forms irregular grooves or ridges that are parallel to the wind direction.

▲ **11.20 The work of wind on vegetation (a) and snow (b)**

Eolian Erosion

Two principal wind-erosion processes are **deflation**, the removal and lifting of individual loose particles, and **abrasion**, the grinding of rock surfaces by the sandblasting action of particles transported in the air. Deflation and abrasion produce a variety of distinctive landforms and landscapes.

Deflation Deflation literally blows away loose or noncohesive sediment and works with rainwater to form a surface resembling a cobblestone street. The resulting **desert pavement** protects underlying sediment from further deflation and water erosion. Traditionally, deflation was regarded as a key formative process in all arid areas, eroding fine dust, clay, and sand that leave behind a concentration of pebbles and gravel as desert pavement (▶Fig. 11.21).

Another hypothesis that better explains some desert pavement surfaces states that deposition of windblown sediment, not removal, is the formative agent. Windblown particles settle between and below coarse rocks and pebbles that are gradually displaced upward. Rainwater is involved as wetting and drying episodes swell and shrink clay-sized particles. The gravel fragments are gradually lifted to surface positions to form the pavement. Desert pavements are so common that many provincial names are used for them: for example, *gibber plain* in Australia; *gobi* in China; and in Africa, *lag gravels* or *serir*, or *reg* desert if some fine particles remain.

Wherever wind encounters loose sediment, deflation may remove enough material to form basins known as **blowout depressions**. These depressions range from small indentations less than a meter wide up to areas hundreds of meters wide and many meters deep. Chemical weathering, the decomposition and decay of minerals in rock through chemical alteration(Chapter 9), although slow in arid regions owing to the lack of water, is important in the formation of a blowout, for it removes the cementing materials that give particles their cohesiveness. Large depressions in the Sahara Desert are at least partially formed by deflation. For example, the enormous Qattara Depression, which covers 18,000 km^2 (6950 mi^2) in the Western Desert of Egypt, is now about 130 m (427 ft) below sea level, partly due to deflation.

(a) A typical desert pavement.

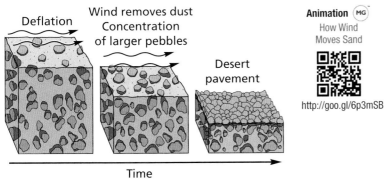

Deflation | Wind removes dust Concentration of larger pebbles | Desert pavement

Animation (MG)
How Wind
Moves Sand

http://goo.gl/6p3mSB

Time

(b) The deflation hypothesis: Wind removes fine particles, leaving larger materials to consolidate into pavement.

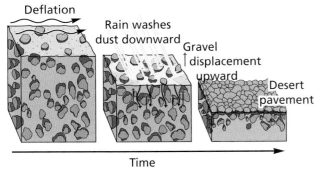

Deflation | Rain washes dust downward | Gravel displacement upward | Desert pavement

Time

(c) The sediment-accumulation hypothesis: Wind delivers fine particles that settle and wash downward as larger gravels migrate upward during cycles of swelling and shrinking, resulting in pavement.

▲11.21 Desert pavement

Mobile Field Trip (MG)
Desert Geomorphology

https://goo.gl/XngZ60

Abrasion You may have seen work crews sandblasting surfaces on buildings, bridges, or streets to clean them. Sandblasting blows compressed air filled with sand grains to quickly abrade a surface. Abrasion by windblown particles is nature's slower version of sandblasting, and it is especially effective at polishing exposed rocks when the abrading particles are hard and angular. Variables that affect the rate of abrasion include the hardness of surface rocks, wind velocity, and wind constancy. Abrasive action is restricted to the area immediately above the ground, usually no more than a meter or two in height, because sand grains are lifted only that high off the ground.

Rocks exposed to eolian abrasion appear pitted, grooved, or polished. They usually are aerodynamically shaped in a specific direction, according to the consistent flow of airborne particles carried by prevailing winds. On a larger scale, deflation and abrasion are capable of streamlining rock structures that are aligned parallel to the most effective wind direction, leaving behind distinctive, elongated ridges or formations called **yardangs**. These wind-sculpted features can range from meters to kilometers in length and up to many meters in height (►**Fig. 11.22**). Also in arid regions, **ventifacts** form when windblown ice and sand erode rocks into unusual shapes.

▼11.22 Sculpting by wind

(a) A yardang in Natural Bridges National Monument, Utah.

(b) A ventifact in the Altiplano, Bolivia.

geoCHECK ✔ Describe how abrasion by wind works to polish exposed rock surfaces.

geoQUIZ

1. Explain the two physical processes that scientists have proposed to explain how desert pavement is formed.
2. Compare and contrast the two erosional processes associated with moving air.
3. Identify the role that chemical weathering plays in forming a blowout depression.

11.6 Wind Transportation & Deposition

Key Learning Concepts

▶ *Describe* the factors that influence the different ways eolian transporation moves sediment particles.

▶ *Identify* eolian erosion and deposition and the resultant landforms.

The effects of wind as an agent of geomorphic change are most easily visualized in coastal and desert environments. Like water, moving air is a fluid, and like moving water, it transports materials such as dust, sand, and snow, creating erosional and depositional landforms.

Eolian Transportation

The distance that wind can transport particles varies greatly with particle size. Initially, wind exerts a frictional pull on surface particles until they become airborne, just as water in a stream picks up sediment. Intermediate-sized grains move most easily. The large particles are heavier and thus require stronger winds. The small particles bond with the surface, and they also present a smooth, aerodynamic shape that resists takeoff (▶ Fig. 11.23).

The term *saltation*, used in Chapter 10 to describe movement of particles by water, also explains the wind transport of grains larger than 0.2 mm (0.008 in.) along the ground. About 80% of wind transport of particles occurs through this skipping and bouncing action. Compared with fluvial transport, in which saltation occurs by hydraulic lift, eolian saltation results from aerodynamic lift, elastic bounce, and impact.

The collision of saltating particles knocks them loose, forward, and even airborne. This movement promotes **surface creep**, which slides and rolls particles too large for saltation—20% of eolian material, and similar to how bed load moves down a stream channel. Once in motion, lower wind velocities continue transporting material. In areas where moving sand threatens human settlement, introducing stabilizing native plants and building fences slows sand erosion from beaches.

geoCHECK ✔ 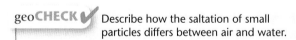 Describe how the saltation of small particles differs between air and water.

Eolian Deposition

Eolian deposition creates many distinctive landscapes. The most common are sand dunes in deserts and along coastlines. Sand dunes are also found in areas that were once more arid, with once active dunes that are now stabilized by grasses and other woody vegetation. Deep deposits of windblown **loess** (glacial silt) are also found in northern China, the Great Plains of North America, central Europe, and parts of Russia and Central Asia. Farmers across these regions have long valued the fertile soils formed from loess.

Humans have a history of accelerating wind erosion when they plow up semiarid grasslands to plant crops. While the term "dust bowl" originated during the 1930s in the American Midwest, the giant windstorms of fine dust set loose by plowing have since occurred

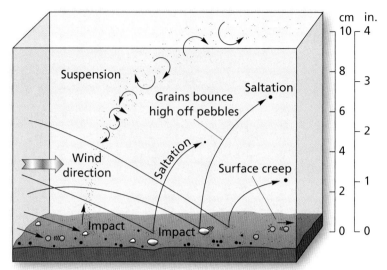

▲11.23 **How the wind moves sand** Eolian suspension, saltation, and surface creep all transport sand.

▲11.24 **Eolian ripples in Arabian Desert sand**

in Russia/Kazakhstan, northern China, southern Mongolia, and parts of Africa south of the Saharan Desert.

Distribution of Sand Dunes While desert pavements predominate across most subtropical arid landscapes, about 10% of desert areas are covered with sand. Sand grains are deposited as constantly moving ridges or hills called dunes. A **dune** is a wind-sculpted accumulation of sand. The smallest features shaped by individual saltating grains are ripples, which form in crests and troughs, positioned at a right angle to the direction of the wind (▲Fig. 11.24).

An extensive area of dunes, such as those found in North Africa, is characteristic of an **erg desert**, or **sand sea**. The Grand Erg Oriental in the central Sahara exceeds 1200 m (4000 ft) in depth and covers 192,000 km² (75,000 mi²), about the size of Nebraska. This sand sea has been in motion for more than 1.3 million years.

Stabilizing these giant sand seas would require a different climate—one with enough moisture to support grasslands.

Dune Movement and Form Dune fields in arid regions or along coastlines migrate in the direction of sand-transporting winds. Stronger seasonal winds or winds from a passing storm may prove more effective at moving sand than average prevailing winds. When saltating grains encounter small patches of sand, their kinetic energy of motion is dissipated, resulting in accumulation. As a dune's height increases above 30 cm (12 in.), a *slipface* and other characteristic dune features form.

Winds create a gently sloping *windward side*, with a more steeply sloped slipface on the *leeward side* (▲Fig. 11.25). The

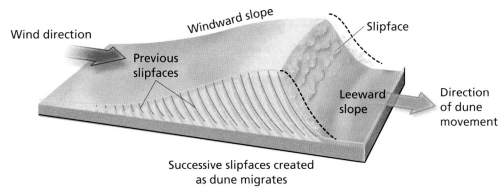

▲**11.25 Dune cross section** A dune grows as wind-borne particles accumulate on the gentler, windward slope and then cascade down the steep slipface of the leeward slope.

slipface angle is the steepest *angle of repose* at which loose material is stable. The constant flow of new material makes a slipface resemble an *avalanche slope*. Sand builds up as it moves over the crest of the dune to the brink. It then avalanches downwind as the slipface continually adjusts, seeking its angle of repose (usually 30°–34°). The many wind-shaped dune forms can be divided into five classes: *crescentic* (curved shape), *linear* (straight), *star*, *dome*, and *parabolic* (◀Fig. 11.26). Of these, crescentic dunes are the most common.

Dune fields are also present in humid climates, such as along coastal Oregon, the south shore of Lake Michigan, and the U.S. Gulf and Atlantic coastlines. Sand carried by continental ice sheets also created dunes in parts of northern Eurasia and the American Midwest. Most of these are today covered by grasslands. Large dune fields are also found in Antarctica and wherever large and meandering rivers have changed course, leaving piles of transported sand in their flood plains.

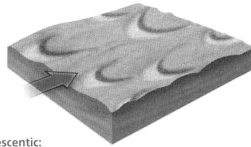

1. Crescentic:
Curved shape with horns pointed downwind; found in areas with constant winds and little directional variability, and where limited sand is available.

2. Linear:
Straight, slightly sinuous, ridge-shaped dune, aligned parallel with the wind direction.

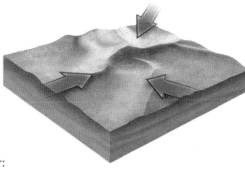

3. Star:
Pyramidal-shaped structure with three or more sinuous, radiating arms extending outward from a central peak; results from effective winds shifting in all directions.

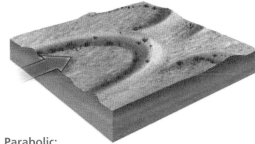

4. Parabolic:
Crescent-shaped dune with opening end facing upwind; U-shaped "blowout" and arms anchored by vegetation, which stabilizes dune form.

▲**11.26 Major dune forms** The different types of sand dunes vary in shape and size depending upon (1) wind speed and direction, (2) amount of sand present, and (3) presence or absence of vegetation that can anchor dunes and influence sand movement.

5. Dome:
Circular or elliptical dune with no slipface; sometimes modified into barchanoid forms, and sometimes stabilized by vegetation.

 geoCHECK ✔ Describe how a sand dune slope is similar to a slope covered with snow.

geoQUIZ
1. Why do intermediate-sized grains move more easily than small and large-sized particles?
2. Explain the process of surface creep.
3. Describe the difference between an erg desert and desert pavement.

11.7 Living Coastal Environments

Key Learning Concepts

▶ *Assess* living coastal environments: coral reefs and coastal wetlands.

▶ *Identify* the ecological requirements of reef-building corals.

Most coastlines form and change from the physical processes discussed in this chapter. However, some coastal environments, such as coral reefs and wetlands, form as the result of biological processes. These sensitive ecological zones are also important indicators of Earth's warming climate.

Corals

A **coral** is a simple marine animal with a small, cylindrical body called a *polyp*; it is related to other marine invertebrates such as anemones and jellyfish (▼Fig. 11.27). Corals secrete calcium carbonate ($CaCO_3$) from their lower bodies, forming a hard external skeleton. They obtain their energy and nutrients by trapping tiny organisms in passing currents. Corals live in a *symbiotic* relationship with algae, each dependent on the other for survival. Algae perform photosynthesis and convert solar energy to chemical energy, providing the coral with nutrients and assisting with the calcification process. In return, corals provide the hard exoskeleton structure that holds the algae.

▲11.27 **Coral polyps**

geoCHECK ✔ Summarize the characteristics of coral animals.

Coral Reefs

Most corals are "colonial" animals, that is, they live together in large colonies. Corals produce enormous structures, known as *coral reefs*, as their skeletons accumulate, forming coral rock. Coral reefs form through many generations as live corals near the ocean's surface build on the foundation of older coral skeletons, which, in turn, may rest upon a volcanic seamount or other submarine feature. Thus, a coral reef is a biologically derived sedimentary rock that can assume several distinctive shapes. Figure 11.28 portrays the examples of each reef stage: *fringing reefs* (platforms of surrounding coral rock, and the most common

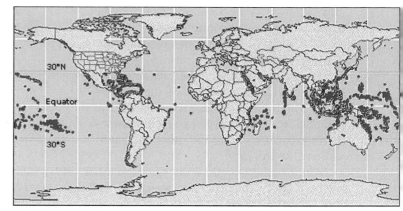

(a)

(b) Australia's Great Barrier Reef

▲11.29 **Worldwide distribution of coral reefs**

form), *barrier reefs* (reefs that enclose lagoons), and *atolls* (circular, ring-shaped reefs). Earth's largest barrier reef, the Great Barrier Reef along the shore of Queensland, Australia, exceeds 2025 km (1260 mi) in length and is 16–145 km (10–90 mi) wide.

Corals thrive in tropical oceans from about 30° N to 30° S, in shallow, clear, and warm eastern coastal currents (▲Fig. 11.29). The global distribution is determined by the narrow ecological requirements of the coral: 10–55 m

Fringing
Once a volcanic island forms, a fringing reef can grow.

Barrier
As the island erodes and subsides, reefs gradually encircle it.

Atoll
The island is completely submerged; a ring of reefs encloses a lagoon.

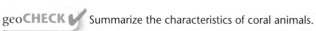

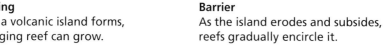

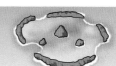  

▲11.28 **Formation of a coral atoll**

(30–180 ft) depth, 27‰–40‰ salinity, and 18°C–29°C (64°F–85°F) water temperature. Normally colorful corals can turn stark white when they expel their own nutrient-supplying algae in a process called **coral bleaching**. Exactly why they eject their symbiotic partner is unknown, for without algae the corals die. Possible causes include pollution, disease, sedimentation, changing ocean salinity, and increasing oceanic acidity.

geoCHECK ✔ What coral body part is the primary component of coral reefs?

Coastal Wetlands

Some coastal areas have great *biological productivity* represented by plant growth and spawning grounds for fish and shellfish. The high productivity stems from trapped organic matter and sediments, which make the areas fertile. In fact, these rich coastal marshes produce much more biomass than croplands. They also support rich wildlife habitats and buffer coastlines from hurricane storm surges. Unfortunately, these wetland ecosystems are fragile and threatened by human development.

Wetlands are saturated with water enough of the time to support plants that grow in water or wet soil. Wetlands usually occur on poorly drained soils. Geographically, they occur along coastlines and as northern bog peatlands, prairie land kettles, cypress swamps, river bottomlands and floodplains, and arctic and subarctic permafrost environments.

Salt marshes and mangrove swamps are two types of coastal wetlands. In the Northern Hemisphere, **salt marshes** form north of the 30th parallel, whereas **mangrove swamps** form within the tropics.

Salt Marshes Common along the East Coast of the United States, salt marshes form in **estuaries**, partly enclosed coastal bodies of brackish water, with one or more inflowing freshwater streams. With a connection to the open sea, estuaries form the transition zone between river and ocean environments. Salt marshes also form behind barrier beaches and spits. An accumulation of mud in these areas promotes the growth of salt-tolerant plants. This vegetation then traps additional alluvial sediments, which adds to the salt marsh area. Because salt marshes form in the intertidal zone, sinuous, branching channels are produced as tidal waters flood and ebb from the marsh (▶ Fig. 11.30).

Mangrove Swamps Made up of dense areas of mangrove trees whose long roots anchor the trees in shallow coastal waters, mangrove swamps are limited to the tropics because mangrove seedlings cannot survive freezing conditions. Sediment accumulation on tropical coastlines provides the site for mangrove trees and other vegetation of mangrove swamps. As conditions shift, the mangrove roots constantly find new anchorages. The roots are visible above the waterline, but reach below the water surface, providing a habitat for many specialized life forms (▶ Fig. 11.31). Mangrove swamps often accumulate enough material to form islands.

▲ 11.30 A coastal salt marsh, Arcata, California

▲ 11.31 Mangroves

geoCHECK ✔ Explain why coastal wetlands have high biological productivity.

geoQUIZ

1. Explain the symbiotic relationship that promotes coral growth.
2. Describe the sequential development of a coral reef.
3. Identify where coral reefs occur on Earth. What explains this distribution?
4. Identify the geographic range of salt marshes and mangrove swamps.

11.8 Coastal Hazards & Human Impacts

Key Learning Concepts

Identify the environmental risks of sea-level rise, storm surges, and coastal erosion.

Give examples of human adaptations to the changing shoreline environment in an era of rising seas.

Over 70% of the human population lives on or near coastal environments. Of the 15 largest cities on Earth, 11 of them are adjacent to an ocean. Humans favor these environments for their access to food resources, shipping lanes, moderate climates, and often the gentle topography found near the ocean. However, living adjacent to the sea can present many environmental challenges. The effects of climate change on coastal environments discussed below are making these challenges more acute.

Sea Level Rise

Recent satellite measurements combined with tide-gauge data demonstrate that global MSL has risen 10–20 cm (4–8 in.) during the last century. Scientists have determined that the rate of rise rapidly increased in the last 20 years. Although this rise may appear modest, the cumulative effect of the last century's rise is enough to increase coastal flooding and shoreline erosion. Rising oceans also push saltwater further inland, flood wetlands, contaminate aquifers and agricultural soils, and alter fish and wildlife habitats. If the current trajectory of sea level rise and population growth in large port cities around the world continues unchecked, by 2070 the homes and livelihoods of at least 14 million people will be at risk, according to the Organization for Economic Co-operation and Development (▶ Fig. 11.32). Unless the international community can implement policies to limit or reverse climate change, confronting this problem will require new and expensive infrastructure designed to hold back the rising seas. This includes barrier dikes, water filtration and pumping systems, and enormous gate-like storm barriers. In New York City, officials are considering a system of barriers and sea walls to protect the city's most vulnerable shorelines—a plan that could cost billions of dollars.

 Describe some of the problems caused by sea level rise in urban areas and wetlands.

(a) Southeastern United States and the Caribbean.

(b) Bangladesh. By 2070, rising sea level will alter the daily lives of at least 14 million people.

▲**11.32 Vulnerability to Sea-Level Rise** These maps represent the amounts of land that would be inundated by different amounts of sea-level rise. The IPCC projects up to about one meter of sea-level rise by 2100, but also predicts that sea levels will continue to rise for centuries in response to global warming.

Storm Surges & Erosion

The high winds, flooding, and erosion associated with coastal storms such as hurricanes (Atlantic and Northeast Pacific Oceans), cyclones (Northwest Pacific Ocean), typhoons (South Pacific and Indian Oceans), and nor'easters (eastern North America) pound coastal communities every year. The high winds and surf accompanying these storms produce a storm surge, which occurs when seawater rises above the expected tide line (▶ Fig. 11.33). Although a storm surge is not a single enormous wave, the water can rise 1–2 meters within a few minutes, and remain there for the duration of the storm.

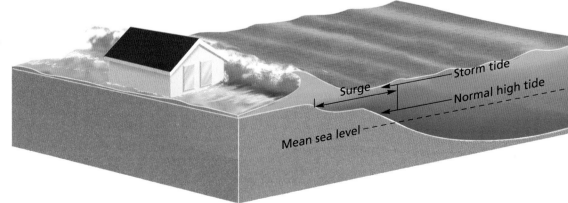

▲**11.33 A storm surge** Surges occur when high winds and surf accompanying tropical storms push seawater above the expected tide line.

The world record storm surge occurred in 1899, when Tropical Cyclone Mahina struck Australia's Bathurst Bay. The highest waves rose 12–14.6 m (43–48 ft). The U.S. record occurred in 2005 during Hurricane Katrina. The surge reached 8.2 m (27.8 ft) above sea level along the Gulf Coast of Mississippi. Just seven years later Hurricane Sandy pummeled the New York and New Jersey coastline. The accompanying storm surge broke records due to a full Moon that made tides higher than average. At the peak of the storm, a buoy in New York harbor measured a record-setting 10 m (32.5 ft) wave a full 2 m (6.5 ft) taller than the previous record.

While these examples are from low-lying submergent coastlines, structures built along coastal bluffs of emergent coastlines are also at risk as erosion whittles away the cliff face (▼Fig. 11.34). In these vertical environments, gravity, heavy rains, and earthquakes trigger large mass movements that threaten human structures. While the risk of landslides is not a new phenomenon for those living on coastal headlands, the warming climate is increasing both the frequency and intensity of storms, along with the waves that batter coastal cliffs.

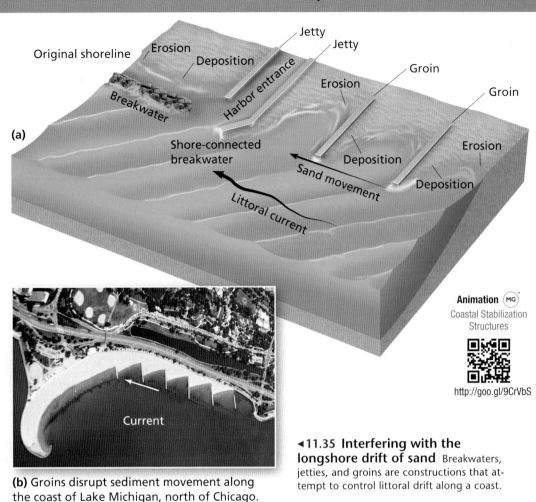

(a)

Animation (MG)
Coastal Stabilization Structures

http://goo.gl/9CrVbS

◄**11.35 Interfering with the longshore drift of sand** Breakwaters, jetties, and groins are constructions that attempt to control littoral drift along a coast.

(b) Groins disrupt sediment movement along the coast of Lake Michigan, north of Chicago.

▲**11.34 Sea-cliff erosion near Gleneden Beach, Oregon**

Beach Protection

Changes in coastal sediment transport can disrupt human activities as beaches are lost, harbors are closed, and coastal highways and beach houses are inundated with sediment. Thus people use various strategies to interrupt longshore drift, with the goal of either halting sand accumulation or forcing a more desirable type of accumulation, through construction of engineered structures

or through "hard" shoreline protection (▲Fig. 11.35). Common approaches include the building of *groins*, long and narrow structures built into the sea to slow drift action along the coast. Other protective structures include *jetties* to block material from harbor entrances and *breakwaters* to create zones of still water near the coastline. However, interrupting the longshore drift disrupts the natural beach replenishment process and may lead to unwanted changes in sediment distribution in areas nearby.

In contrast to "hard" protection, "soft" shoreline protection involves importing sand to replenish a beach. *Beach nourishment* refers to the artificial replacement of sand along a beach with sand from other areas. Theoretically, through such efforts a beach that normally experiences a net loss of sediment will be fortified with the new sand. However, a single storm can erase years of human effort and expense to build beaches. In addition, disruption of marine and longshore zone ecosystems may occur if the new sand does not physically and chemically match the beach's natural sand.

geoCHECK ✔ What is the role of groins and jetties?

geoQUIZ

1. Aside from the physical rise, explain other problems caused by rising sea levels.
2. How is climate change influencing the relationship between New Yorkers and the sea?
3. Describe the difference between "hard" and "soft" shoreline protection.

COASTAL SYSTEMS IMPACT HUMANS
- Rising sea level has the potential to inundate coastal communities.
- Tsunami cause damage and loss of life along vulnerable coastlines.
- Coastal erosion changes coastal landscapes, affecting developed areas; human development on depositional features such as barrier island chains is at risk from storms, especially hurricanes.

HUMANS IMPACT COASTAL SYSTEMS
- Rising ocean temperatures, pollution, and ocean acidification impact corals and reef ecosystems.
- Human development drains and fills coastal wetlands and mangrove swamps, thereby removing their buffering effect during storms.

A cargo vessel ran aground on Nightingale Island, Tristan da Cunha, in the South Atlantic in 2011, spilling an estimated 1500 tons of fuel, spilling tons of soybeans, and coating endangered Northern Rockhopper penguins with oil.

Dredgers pump sand through a hose to replenish beaches on Spain's Mediterranean coast, a popular tourist destination. Near Barcelona, pictured here, sand is frequently eroded during storms; natural replenishment is limited by structures that block longshore currents.

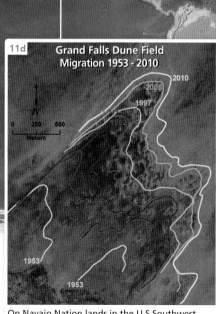

Grand Falls Dune Field Migration 1953 - 2010

2010
2005
1997
1953
1953

0   250   500
Meters

On Navajo Nation lands in the U.S Southwest, dune migration is threatening houses and transportation, and affecting human health. The Grand Falls dune field in northeast Arizona grew 70% in areal extent from 1997 to 2007. The increasingly dry climate of this region has accelerated dune migration and reactivated inactive dunes.

ISSUES FOR THE 21ST CENTURY
- Degradation and loss of coastal ecosystems—wetlands, corals, mangroves—will continue with coastal development and climate change.
- Continued building on vulnerable coastal landforms will necessitate expensive recovery efforts, especially as storm systems become more intense with climate change.

Mangrove planting: In Aceh, Indonesia, near the site of the 2004 Indian Ocean tsunami, authorities encourage local people to plant mangroves for protection against future tsunami.

Looking Ahead
In the next chapter we examine glacial and periglacial landscapes. We will investigate how glacial formation and movement sculpts the land and leaves behind many distinctive landforms. Changes in the Earth's total mass of glacial ice is also important evidence used to monitor our changing climate.

What unique physical traits characterize ocean & coastal environments?

11.1 Global Oceans & Seas

Analyze the geographic distribution of oceans and seas.

Describe the chemical composition of seawater.

Explain how the ocean's physical and chemical properties determine its layered structure.

- The ocean is divided by depth into the shallow mixing zone at the surface, the thermocline transition zone, and the deep cold zone. Seawater averages 35‰ (‰ parts per thousand) or 3.5% salinity. The term *brine* is applied to water that exceeds this average of 35‰. Brackish applies to water that is less than 35‰ salts. The dissolved solids remain in the ocean, but the water recycles endlessly through the hydrologic cycle, driven by energy from the Sun.

1. Analyze why salinity is less along the equator and greater in the higher latitudes.

2. Explain why ocean acidification might harm some marine life.

11.2 Coastal Environments

Identify the components of the coastal environment.

Explain how changes in sea level occur.

- The physical structure of the ocean is layered by density, as determined by average temperature, salinity, dissolved carbon dioxide, and dissolved oxygen. The littoral zone exists where the tide and wave-driven sea confront the land and is influenced by solar energy, winds, weather and climatic variation, geomorphology, and human activities. Mean sea level (MSL) variation results from ocean currents, waves, and tidal variations, and is rising worldwide due to to global warming of the atmosphere and oceans.

3. How is MSL determined? Is it constant or variable around the world?

How do erosion & deposition by oceans alter the landscape?

11.3 Coastal System Actions

Explain the cause and actions of tides.

Describe wave motion at sea and near shore.

- The coastal systems experience complex tidal fluctuation, winds, waves, ocean currents, and occasional storms. Tides are daily oscillations in sea level and are produced by the gravitational pull of both the Moon and the Sun. Waves result from friction between wind and the ocean surface. In open water, waves migrate in the direction of wave travel, but the water itself doesn't advance. It is *wave energy* that moves through the fluid medium of water. Wave refraction redistributes wave energy so that different sections of the coastline vary in erosion and in the abiltity to transport longshore drift.

4. Describe the refraction process that occurs when a wave reaches an irregular coastline.

5. Identify the interacting forces that generate the pattern of tides.

11.4 Coastal Erosion & Deposition

Explain coastal straightening and coastal landforms.

Differentiate between landforms of erosional and depositional coastlines.

- Erosional coastlines are zones of active coastal erosion in areas where tectonic uplift exposes land that was underwater. Depositional coastlines are places of active deposition of sediments that develop in areas of relative sea-level rise. Erosional coastlines produce sea terraces, headlands, arches, caves, and sea stacks. In contrast, depositional coasts have gentle relief, where sediments produce barrier spits, lagoons, marshes, and tidal flats.

6. Explain the physical conditions that lead to multiple sea terraces.

7. Identify the expected landforms of a depositional environment.

How do erosion & deposition by wind alter the landscape?

11.5 Wind Erosion

Differentiate between eolian deflation and abrasion and the landforms they produce.

- Eolian processes modify and move sand along coastal beaches and deserts. Deflation is the removal and lifting of individual loose particles. Abrasion is the "sandblasting" of rock surfaces with particles captured in the air. Deflation literally blows away loose or noncohesive sediment and works with rainwater to form a desert pavement that protects underlying sediment from further deflation and water erosion. On a larger scale, deflation and abrasion are capable of streamlining rock structures, leaving behind blowout depressions, yardangs, and ventifacts.

8. Describe the erosional processes associated with moving air.

9. How are yardangs formed by the wind?

11.6 Wind Transportation & Deposition

Describe the factors that influence the different ways eolian transportation moves sediment particles.

Identify eolian erosion and deposition and the resultant landforms.

- Wind exerts a frictional pull on surface particles until they become airborne. Only the finest dust particles travel far. Saltating particles crash into other particles, knocking them both loose and forward. Surface creep slides and rolls particles too large for saltation. A sand dune is a wind-sculpted accumulation of sand that forms in arid climates and along coastlines where sand is available. As height increases above 30 cm (12 in.), a slipface on the lee side forms. The slipface angle is the steepest angle of repose at which loose material is stable (usually 30°–34°). Dune forms are classified as crescentic, linear, and star crescent (barchan) and parabolic.

10. Explain how the angle of repose influences surface creep.

11. Describe how wind direction results in three classes of distinctive dune forms.

How do living things, including humans, interact with coastal systems?

11.7 Living Coastal Environments

Assess living coastal environments: coral reefs and coastal wetlands.

Identify the ecological requirements of reef-building corals.

- A coral is a simple marine invertebrate that forms a hard, calcified, external skeleton. Over generations, corals accumulate in large reef structures. Corals live in a symbiotic relationship with algae. Wetlands are lands saturated with water that support specific plants adapted to wet environments. Coastal wetlands form as salt marshes poleward of 30° latitude in each hemisphere and as mangrove swamps equatorward of these latitudes.

12. How are corals able to construct reefs and islands?

13. Describe a trend in corals that is troubling scientists and state some possible causes.

11.8 Coastal Hazards & Human Impacts

Identify the environmental risks of sea-level rise, storm surges, and coastal erosion.

Give examples of human adaptations to the changing shoreline environment in an era of rising seas.

- Over 70% of humanity lives on or near coastal environments. Global mean sea level has risen 10–20 cm (4–8 in.) during the last century, with the rate of rise increasing during the last 20 years. By 2070, rising seas will threaten the livelihoods of at least 14 million people. Rising seas also intensify storm surges, and the erosion of coastal headlands and beaches. A storm surge occurs when high winds and surf associated with tropical storms push seawater above the expected tide line.

13. Knowing that a tropical storm surge is coming, what Earth-Moon-Sun orientation would be advantageous to minimize potential damage? Explain your reasoning.

14. How do people modify longshore drift? Identify the positive and negative aspects of these actions.

Critical Thinking Questions

1. Characterize each of the three general physical zones of the ocean by temperature, salinity, dissolved oxygen, and dissolved carbon dioxide. How will a warming ocean influence the salinity and dissolved oxygen in the three zones?

2. Why are the coastal wetlands poleward of 30° N and 30° S latitude different from those that are equatorward?

3. What characteristic tides are expected during a new Moon and a full Moon? How might tides change if the distance between the Moon and Earth increased?

4. Compare and contrast eolian saltation and fluvial saltation.

5. Describe how three physical factors that influence tides might vary between the coast of Brazil and the English Channel. How will the differences in these factors affect tides at the two locations?

Visual Analysis

1. Identify the landforms that result from erosional processes and the landforms that result from depositional processes.

2. How will rising seas, caused by global warming, affect the coastal landforms in this scene?

3. Explain how the salinity content of the water might vary between the center of the image, and the open ocean on the right-hand side? Explain why this variation exists

▶ **R11.1** This photograph of Whitehaven Beach, in Australia's Whitsunday Islands (Richard L'Anson/LPI), portrays evidence of many landforms and processes discussed in this chapter.

(MG) Interactive Mapping | Login to the **MasteringGeography** Study Area to access **MapMaster**.

Oceans, Seas, & Refracting Waves

- Open MapMaster in MasteringGeography™.
- Select *Europe* from the *Physical Environment* Layer, and explore the map sublayers.

1. How many named seas appear in this watery region of the world?

2. Identify three cities that are particularly well protected from an inbound tsunami arriving from the Atlantic Ocean. Explain your reasoning.

3. Identify three locations in the United Kingdom where you would expect maximum wave refraction and an erosional coastline, and three areas where you would expect minimal wave refraction and a depositional coastline. Explain your choices.

Explore | Use **Google Earth** to explore Earth's **coral reefs**.

Reefer Madness

The combination of seawater and climate in the Pacific Ocean has produced thousands of coral reefs. In this exercise you will first visit the three common coral formations.

- In Google Earth, fly to *Oahu, Hawaii,* and zoom around to explore this fringing reef.
- In the next step, fly to *Bora Bora* in French Polynesia, and zoom around to explore this barrier reef.
- Finally, fly to *Kwajalein* in the U.S. Marshall Islands, and zoom around to explore this atoll.

Using your experience from above, what stage in the sequence of coral reef growth does each of the following three locations represent? Include a justification for each choice.

1. Providencia Island, Colombia (Caribbean Sea)
2. Funafuti, Tuvalu
3. Ningaloo Reef, Australia

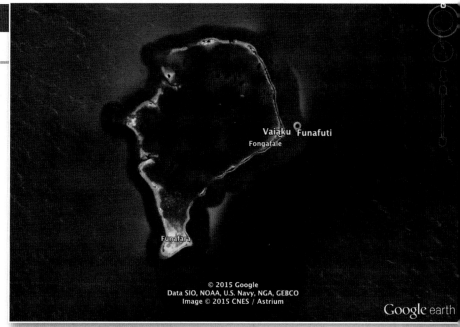

▲ R11.2

MasteringGeography™

Looking for additional review and test prep materials? Visit the Study Area in MasteringGeography™ to enhance your geographic literacy, spatial reasoning skills, and understanding of this chapter's content by accessing a variety of resources, including MapMaster™ interactive maps, videos, *Mobile Field Trips, Project Condor Quadcopter videos, In the News* RSS feeds, flashcards, web links, self-study quizzes, and an eText version of *Geosystems Core.*

Key Terms

abrasion, p. 306
barrier beach, p. 305
barrier island, p. 305
barrier spit, p. 304
beach, p. 304
blowout depression, p. 306
brackish, p. 295
breaker, p. 301
brine, p. 295

coral, p. 310
coral bleaching, p. 311
deflation, p. 306
depositional coastline, p. 302
desert pavement, p. 306
dune, p. 308
eolian, p. 306
erg desert/sand sea, p. 308

erosional coastline, p. 302
estuaries, p. 311
headland, p. 302
lagoon, p. 304
littoral zone, p. 299
loess, p. 308
longshore current, p. 302
longshore drift, p. 302
mangrove swamp, p. 311

mean sea level (MSL), p. 299
ocean acidification, p. 296
salinity, p. 295
salt marsh, p. 304
sand sea, p. 308
sea terrace, p. 302
surface creep, p. 308
swell, p. 300

tidal bore, p. 300
tidal flat, p. 300
tide, p. 300
tombolo, p. 304
tsunami, p. 301
ventifacts, p. 307
wave, p. 300
wave refraction, p. 302
wetland, p. 311
yardang, p. 307

Trash Talk: Can We Predict the Pathway & Decomposition of Trash in the Great North Pacific Garbage Patch?

GeoLab11 (MG)
Pre-Lab Video

https://goo.gl/WtlGHX

Ocean gyres are a system of circular ocean currents formed by Earth's wind patterns and the Coriolis force. Within the North Pacific Subtropical Ocean Gyre, the Great North Pacific Garbage Patch (GNPGP) circulates in an eddy-like pattern. The collective name for the garbage found within this gyre is *flotsam*, of which plastic components now compose 80% of the total amount. The flotsam eventually breaks into millions of small pieces called *microplastics* that over time sink down through the water column until gradually settling on the seafloor. Ocean currents and waves also deposit some flotsam on coastlines. While scientists identify many individual garbage patches within the GNPGP, they most commonly refer to a large eastern patch off the coast of Japan and a western patch between the states of Hawaii and California. Although often compared in size to Texas, the actual size of the GNPGP has not yet been scientifically calculated, in part, because so much of the microplastics circulate below the surface.

Apply

You are a NOAA scientist investigating the time and distance that different types of trash (plastic, wood, glass, etc.) will travel before decomposing or breaking into small pieces that sink below the ocean surface.

Objectives

- Calculate the time and distance it takes for trash to move into the center of the GNPGP
- Determine the distance different types of trash will circulate before decomposing into the ocean water

Procedure

Figure GL 11.1 portrays the wind-driven surface ocean currents that are part of the vast circular gyres that form a major part of the oceans' circulation. Figure GL 11.1 also shows the North Pacific Ocean, where the two main branches of the GNPGP exist. Scientists estimate that the upper level of this ocean gyre current flows at an average 3–4 km (1.8–2.5 mi) per hour. The distance between California and Japan is approximately 8851 km (5500 mi). Figure GL 11.2 presents the estimated the decomposition rates of common marine flotsam

b. In the following steps, compute the number of days that each type of trash travels before decomposing or, for microplastics, disappears beneath the surface. Multiply your result in (a) by the rate at which that particular type of trash decomposes (see Fig. GL 11.2) to arrive at the correct estimate.

Use your estimates from a and b above, to determine the following:

1. Calculate how many kilometers or miles a newspaper would travel before decomposing.
2. Calculate how many kilometers or miles a waxed carton container would travel before decomposing.
3. How many kilometers or miles would a plastic grocery bag travel before breaking into *microplastics*, at which point the small pieces would take years more to eventually settle on the ocean floor.
4. Compute how much further the aluminum can would travel within the gyre, compared to a large foam cup.
5. Would a plastic bottle tossed into the Pacific Ocean near Los Angeles break into microplastics before completing a full circuit around the gyre? To calculate this,

Analyze & Conclude

About 60 years ago, nearly all flotsam was biodegradable materials such as wood, hemp rope, and wool. Today, nonbiodegradable plastics compose 90% of the GNPGP. Furthermore, almost 80% of trash originates from land-based activities, with the remainder coming off ships (mostly fishing nets) and debris from offshore oil platforms.

6. Determine the gyres and ocean currents within which a plastic bottle discarded off the coastline of eastern India would circulate (see Figs. GL 11.1 and 11.2).
7. Estimate the cardinal direction, ocean currents, gyres, and the number of kilometers/miles a large piece of wood from an oil platform off the Florida Coast would travel until decomposing in the ocean waters (see figures and the prior calculations,
8. Given what you have learned about this issue, suggest two strategies that you and your classmates could follow that would prevent the GNPGP from expanding.

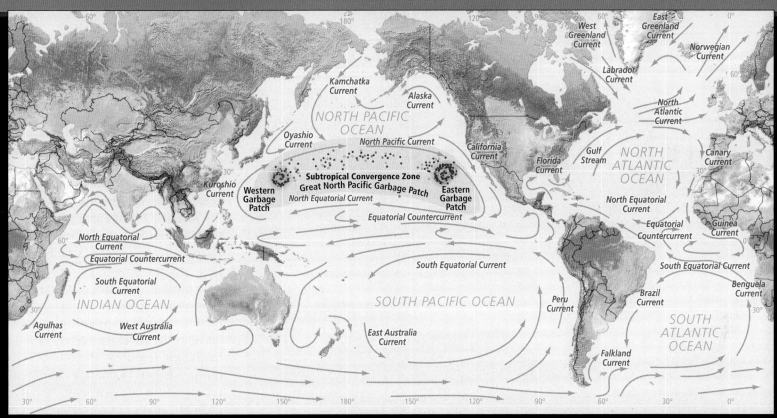

▲GL11.1 **Wind-driven surface currents and the Great North Pacific Garbage Patch**

GL11.2 Estimated decomposition rates of common marine flotsam	
Type of Debris	**Decomposition Rate**
Paper Towel	2–4 weeks
Newspaper	6 weeks
Apple Core	2 months
Cardboard Box	2 months
Cotton Shirt	2–5 months
Waxed Carton	3 months
Plywood	1–3 years
Wool Sock	1–5 years
Plastic Grocery Bag	10–20 years*
Foam Cup	50 years*
Tin Can	50 years
Aluminum Cans	200 years
Disposable Diaper	450 years*
Plastic Beverage Bottle	450 years*
Fishing Line	600 years*

*Items are made from a type of plastic. Although no one has lived for 450 or 600 years, many scientists believe plastics never entirely go away. These decomposition rates are estimates for the time it takes for these items to become microscopic and no longer be visible. Sources: EPA, Woods Hole Sea Grant.

Log in MasteringGeography™ to complete the online portion of this lab, view the Pre-Lab Video, and complete the Post-Lab Quiz.
www.masteringgeography.com

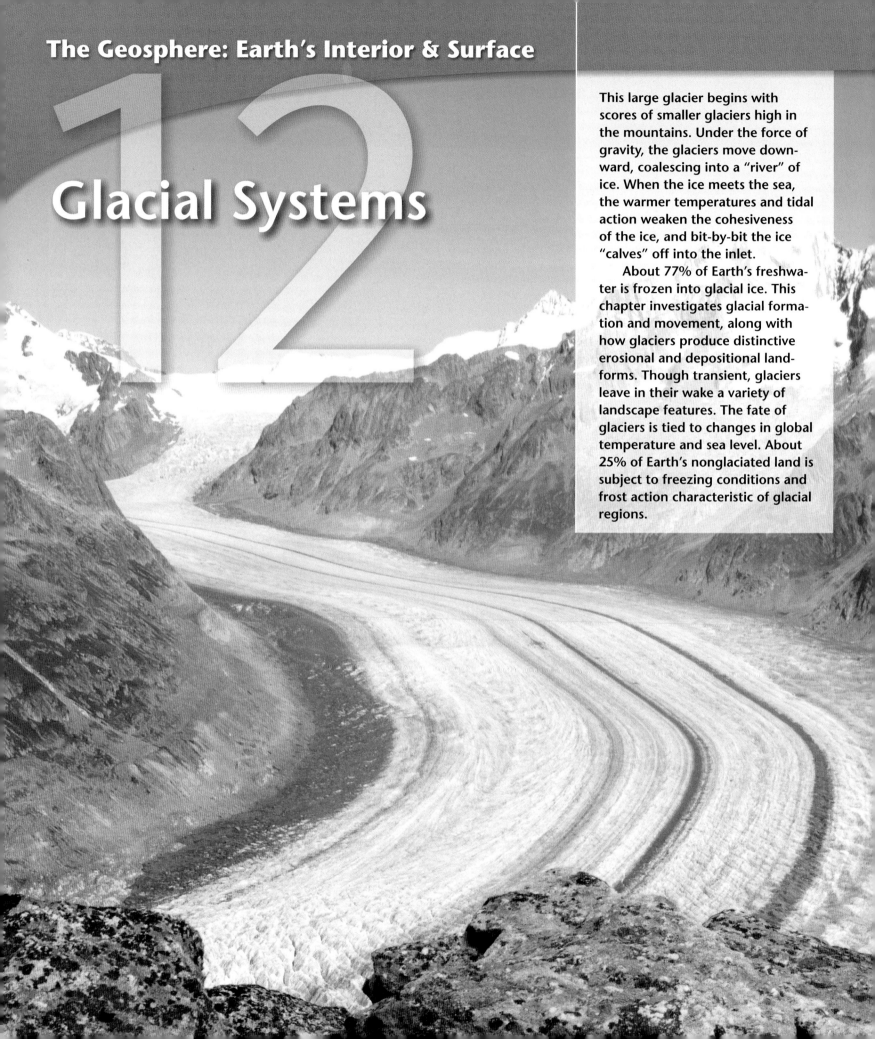

12

Glacial Systems

This large glacier begins with scores of smaller glaciers high in the mountains. Under the force of gravity, the glaciers move downward, coalescing into a "river" of ice. When the ice meets the sea, the warmer temperatures and tidal action weaken the cohesiveness of the ice, and bit-by-bit the ice "calves" off into the inlet.

About 77% of Earth's freshwater is frozen into glacial ice. This chapter investigates glacial formation and movement, along with how glaciers produce distinctive erosional and depositional landforms. Though transient, glaciers leave in their wake a variety of landscape features. The fate of glaciers is tied to changes in global temperature and sea level. About 25% of Earth's nonglaciated land is subject to freezing conditions and frost action characteristic of glacial regions.

Key Concepts & Topics

Above the Aletsch Glacier, Switzerland

12.1 Glacial Processes—Ice Formation & Mass Balance

Key Learning Concepts

▶ **Explain** how and where glaciers form.
▶ **Describe** the two main types of glaciers.

A glacier is a slow-moving river of ice. Glaciers form in areas of permanent snow, both at high latitudes and above the snowline in high mountains at any latitude. The *snowline* is the lowest elevation where snow survives year-round and is the lowest point where winter snow accumulation persists throughout the summer.

Glacial Formation

Glaciers form by the continual accumulation of snow that recrystallizes under its own weight into an ice mass. Glaciers then move slowly under the pressure of their own great weight and the pull of gravity. In fact, they move slowly in stream-like patterns, merging as tributaries into large rivers of ice (▶Fig. 12.1).

Today, glaciers and the much larger ice sheets of Greenland and Antarctica blanket 11% of Earth's land area. Glacial ice covered as much as 30% of continental land during colder episodes in the past. Through these "ice ages," below-freezing temperatures prevailed at lower latitudes more than they do today, allowing snow to accumulate year after year.

Glaciers are as varied as the landscape itself. They fall within two general groups, based on their form, size, and flow characteristics: alpine glaciers and continental glaciers.

Mobile Field Trip (MG)
The Glaciers
of Alaska

https://goo.gl/y4THpC

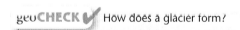 geoCHECK ✔ How does a glacier form?

Alpine Glaciers

A glacier in a mountain range is an **alpine glacier**. These form in several subtypes. Most alpine glaciers originate in a mountain *snowfield* that is confined in a **cirque**, a bowl-shaped erosional landform scooped out at the head of a valley (▶Fig. 12.2). Cirque glaciers often flow down to become **valley glaciers**, literally a river of ice confined within a valley that originally was formed by stream action.

Wherever several valley glaciers pour out of their confining valleys and merge at the base of a mountain range, a *piedmont glacier* is formed and spreads freely over the lowlands, such as the remnants of the Malaspina Glacier, which flows into Yakutat Bay, Alaska (▶Fig. 12.3). A *tidewater glacier* ends in the sea, calving to form floating pieces of ice known as **icebergs** (▶Fig. 12.4).

geoCHECK ✔ How does a piedmont glacier form?

▶**12.3 The Malaspina Glacier** Scores of smaller glaciers merge to form this enormous piedmont glacier at the base of the St. Elias Mountains, Alaska (USA) and Yukon (Canada).

▲**12.1 Rivers of ice** Glaciers flow from Alaska's Denali, the highest peak in North America.

Cirque Cirque Cirque

▲**12.2 Cirque glaciers**

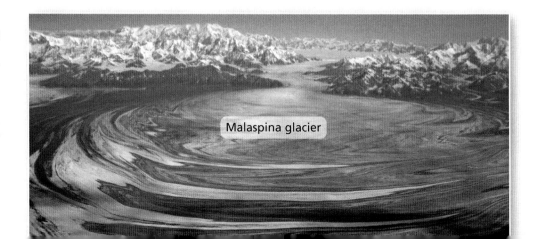

Malaspina glacier

▲12.4 Calving iceberg Ice calves off the end of the Hubbard Glacier, Alaska This large column will break into hundreds of small icebergs.

▼12.6 Greenland Ice Sheet Greenland is covered by an ice sheet and not the smaller dimension of ice defined as an ice cap.

▲12.5 Ice sheet

Continental Glaciers

On a much larger scale, a continuous mass of ice is a **continental glacier** or **ice sheet** (▲Fig. 12.5). Most of Earth's glacial ice exists in the ice sheets that blanket 81% of Greenland and 90% of Antarctica. Antarctica alone has 92% of all the glacial ice on the planet. These ice sheets have such enormous mass that large portions of each landmass beneath the ice are isostatically depressed (pressed down by weight) below sea level. The ice is more than 3000 m (9800 ft) thick, and often buries all but the highest peaks (▲Fig. 12.5).

Two additional types of continuous ice cover associated with mountain locations are *ice caps* and *ice fields*. An **ice cap** is roughly circular and, by definition, covers an area of less than 50,000 km² (19,300 mi²). Ice caps completely bury the underlying landscape (▼Fig. 12.6). An **ice field** is not extensive enough to form the characteristic dome of an ice cap. Instead, it extends in a characteristic elongated pattern in a mountainous region.

geoCHECK ✔ Where is most of the glacial ice located on Earth?

geoQUIZ

1. Explain why glaciers flow?
2. Place the following terms in the correct order to describe a tidewater glacier from top to bottom: iceberg, cirque, snowline, snowfield, and valley glacier.
3. How is an alpine glacier different from a continental glacier?

12.2 Glacial Mass Balance & Movement

Key Learning Concepts

▶ **Explain** what determines if a glacier grows or shrinks.
▶ **Describe** how a glacier moves.

A glacier collects and compresses snow into ice. Once formed, the outputs of a glacier include ice, meltwater runoff, and water vapor.

Glacial Mass Balance

Glaciers can be divided into two parts, as shown in Figure 12.7a. The upper *accumulation zone* is where snowfall and other moisture collects and compacts into ice. Furthermore, winter snow and ice survives the summer melting season. The lower portion of a glacier is the *ablation zone*, where internal and external melting of the ice occurs. Ice loss also occurs from:

- melting at the glacier's base,
- scouring from wind, which directly sublimates (evaporates) the ice from a solid to a gas,
- calving of ice blocks, and
- evaporating directly into the atmosphere—*sublimation*.

Geographers define the **equilibrium line** as the dividing line between the accumulation zone (above the line) and the ablation zone (below the line).

Whether a glacier grows larger and advances or shrinks and retreats depends on its *mass balance* (▼Fig. 12.7b). For a glacier to maintain a constant size, the accumulation of moisture must balance with the ice loss through ablation. If this mass balance is negative (more ablation than accumulation), the glacier will shrink in size. Conversely, the glacier increases in size when the mass balance is positive (more accumulation than ablation). A glacier grows larger during cold periods with adequate precipitation. In warmer times, the equilibrium line migrates up-glacier, and the glacier retreats—grows smaller. Whether a glacier is in positive or negative mass balance, gravity continues to move a glacier forward and downward, even though its lower end usually retreats owing to ablation.

Global warming is causing widespread reductions in middle- and lower-elevation glaciers. The present ice loss from alpine glaciers worldwide is thought to contribute over 25% to the measured rise in sea level.

(b) Accumulation Zone

Accumulation zone:
Snow builds up in this zone, is compressed by its own weight, and changes to glacial ice as the glacier increases in thickness.

(c) At least four tributary glaciers flowing into a valley glacier

geoCHECK ✔ Explain why the equilibrium line may migrate up or down a glacier.

Animation (MG)
Glacial Processes

http://goo.gl/OzhTkq

(d) Terminal moraine

Tributary glacier
Lateral moraine
Medial moraine
Snow
Plucking
Abrasion
Crevasses
Glacier ice
Melting and evaporation
Recessional moraine
Terminal moraine
Till
Meltwater stream
Outwash plain

Equilibrium line:
Accumulation and ablation are in balance

Ablation zone:
The glacier loses mass through melting and other processes.

(a)

◀12.7 A retreating alpine glacier and mass balance (a) Cross-section of a retreating alpine glacier. (b) The accumulation zone for a glacier in the Himalaya. (c) At least four tributary glaciers flow into a valley glacier in Greenland.

Glacial Movement

Glaciers are not rigid blocks that simply slide downhill. In contrast to the ice in your home freezer, the bottom of a glacier bends like plastic in response to weight and pressure from above and contours of the slope below. However, the glacier's upper portions are brittle. Rates of flow range from almost nothing to a kilometer or two per year on a steep slope. The rate of snow accumulation above the equilibrium line is critical to the speed.

The greatest movement within a valley glacier occurs *internally*, below the rigid surface. The ice often fractures as the underlying zone moves forward (▼Fig. 12.8). At the same time, the base slides along, varying its speed with temperature, friction, and the presence of any lubricating water beneath the ice. Because this *basal slip* is usually less rapid than the internal plastic flow, the upper portion of the glacier flows ahead of the lower portion.

▲12.9 A climber negotiates crevasses on Alaska's Matanuska Glacier

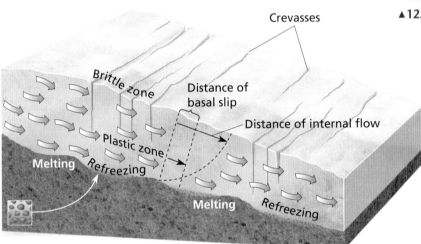

▲12.8 **Glacial movement** Cross section of a glacier, showing its forward motion, brittle cracking at the surface, and flow along its basal layer.

An uneven slope beneath the ice may also cause the pressure to vary, melting some of the basal ice by compression at one moment, only to have it refreeze downslope. This process may also incorporate rock debris into the glacier. Consequently, the basal ice layer has a much higher debris content than the ice above.

Crevasses All glaciers develop vertical cracks known as **crevasses** (▲Fig. 12.9). Crevasses result from friction with valley walls or from tension due to stretching as the glacier passes over slopes, or from compression as the glacier passes over concave slopes. Traversing a glacier is dangerous because a thin veneer of snow may mask the presence of a crevasse.

Glacier Surges Although glaciers flow plastically and predictably most of the time, some **surge** forward with little warning. A surge is a short-lived event where the ice may advance at velocities up to 100 times faster than normal. Thus in glacial terms, a surge can be tens of meters per day. The Jakobshavn Glacier on the western Greenland coast, for example, is one of the fastest moving at between 7 and 12 km (4.3 and 7.5 mi) a year. Global warming contributes to glacier surges by increasing the flow of lubricating meltwater into the basal layer.

geoCHECK ✔ What physical qualities make glacial ice different from ice in your freezer?

Animation (MG)
A Tour of the
Cryosphere

http://goo.gl/SjvDGd

Animation (MG)
Operation
IceBridge

http://goo.gl/qyNpOD

Animation (MG)
Flow of Ice
Within a Glacier

http://goo.gl/8TBHLV

geoQUIZ

1. Describe what happens to a glacier above *and* below the equilibrium line?
2. How can rock material become incorporated into the bottom of a glacier?
3. How does a glacial crevasse form?

12.3 Alpine Glacial Erosion & Landforms

Key Learning Concepts

▶ **Explain** how alpine glaciers erode the landscape.

▶ **Identify** the different landforms that alpine glaciers leave behind.

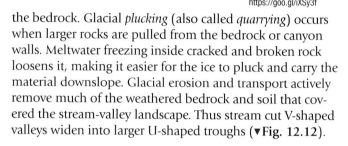
Glacial erosion and deposition produce distinctive landforms that differ greatly from the way the land looked before the ice came and went. Alpine and continental glaciers each generate their own characteristic landscapes.

How Glaciers Erode & Transport Sediment

Before glaciation, valleys eroded by streams have a prominent V shape. Figures 12.10 and 12.11 show how erosion changes landscapes during glaciation. As glacial ice moves downslope, it incorporates rocks and other sediment. This mixture then scours the canyon bottom and sidewalls in a process called glacial *abrasion*. The abrasion often scratches elongated *striations* into the bedrock. Glacial *plucking* (also called *quarrying*) occurs when larger rocks are pulled from the bedrock or canyon walls. Meltwater freezing inside cracked and broken rock loosens it, making it easier for the ice to pluck and carry the material downslope. Glacial erosion and transport actively remove much of the weathered bedrock and soil that covered the stream-valley landscape. Thus stream cut V-shaped valleys widen into larger U-shaped troughs (▼Fig. 12.12).

Glacial Landforms

Erosion by alpine glaciers produces spectacular mountain landscapes. Erosion in cirque basins is a fundamental part of this process.

Cirques, Arêtes, & Horns Recall that a cirque is a bowl-shaped basin carved by a glacier on the side of a mountain. As the *cirque* walls erode away, sharp ridges form, dividing adjacent cirque basins. These ridges, called **arêtes** ("knife-edge" in French) become the serrated (or "sawtooth") ridges in glaciated mountains. Two eroding cirques may reduce an arête to a saddle-like depression or pass, forming a *col* [(▼Fig. 12.12 (diagram) and Fig. 12.13 (photo)]. A **horn**, or pyramidal peak, results when several cirque glaciers gouge an individual mountain summit from all sides (Figs. 12.12 and 12.13). The most famous horn is the Matterhorn in the Swiss Alps, but many others occur worldwide.

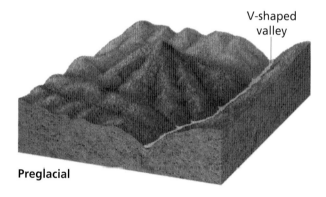

▲12.10 Preglacial landscape with V-shaped, stream-cut valleys

geoCHECK ✔ What does glacial ice pick up as it moves downslope?

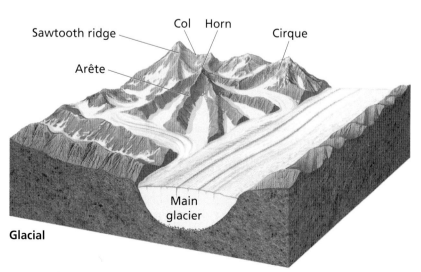

▲12.11 Alpine landscape filled with valley glaciers

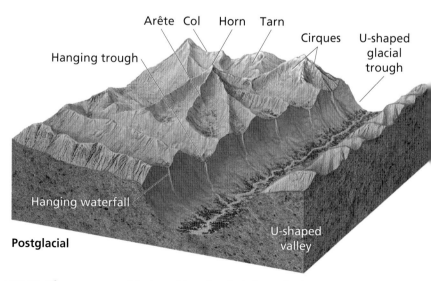

▲12.12 The geomorphic handiwork of alpine glaciers

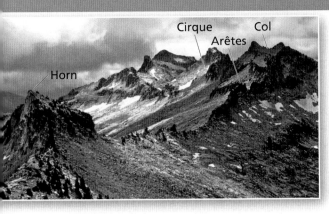

▲12.13 **Glacial horns, arêtes, cols, and cirques in the Trinity Alps, California**

U-Shaped Valleys & Talus Slopes Figure 12.12 shows the landscape during a warmer climate after the ice has retreated. The glaciated valleys now are U-shaped *troughs* greatly changed by erosion from their previous stream-cut V form. You can see the steep sides and the straightened course of the valleys. Physical weathering from the freeze–thaw cycle has loosened rock along the steep cliffs, where it has fallen to form *talus slopes* along the valley sides. Retreating ice leaves behind transported rocks as *erratics*.

Glacial Lakes Small mountain lakes called **tarns** often form in cirques where valley glaciers originate. One cirque contains a chain of small and circular lakes. These *paternoster lakes* form from the differing resistance of rock to glacial processes or from damming by glacial deposits.

Hanging Valleys The valleys carved by tributary glaciers are left stranded high above the valley floor because the primary glacier eroded the valley floor so deeply. These *hanging valleys* are the sites of spectacular waterfalls (▼Fig. 12.14). See how many of the erosional

forms (arête, col, horn, cirque, cirque glacier, U-shaped valley, erratic, and tarn, among others) you can identify in Figure 12.14.

Roche Moutonnée A *roche moutonnée* ("sheep rock") is an asymmetrical hill of exposed bedrock formed by glacial erosion. Its gently sloping upstream side has been polished smooth by glacial action, whereas its downstream side is abrupt and steep where the glacier plucked up rock pieces (▶Fig. 12.15).

Fjords When glacial ice pushes down a narrow U-shaped trough and into an ocean, the glacier may continue eroding the landscape, even below sea level. As the glacier retreats, ocean water floods the trough, forming a deep *fjord* in which the sea extends inland (▼Fig. 12.16). The fjord may be flooded further by rising sea level or when tectonic activity lowers the elevation of the coastal region. Along most of the glaciated coast of Alaska, retreating glaciers are opening many new fjords that previously were blocked by ice. Coastlines with notable fjords include those of Norway, Greenland, Chile, the South Island of New Zealand, Alaska, and British Columbia.

(a)

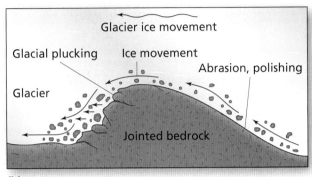

(b)

▲12.15 **Roche moutonnée** (a) Lembert Dome in Yosemite National Park, California, is a roche moutonnée. (b) Plucking by glacial ice splintered the rock on the downstream side of the roche moutonnée, while abrasion smoothed the upstream side.

▶12.16 **U-shaped valleys, Norway** The sea fills glacially carved U-shaped valleys in Norway. A glacial tarn, situated well above the ocean, fills a cirque in the lower left.

▲12.14 **Hanging valleys, Icy Bay, Alaska**

geoCHECK ✔ What is a hanging valley?

geoQUIZ

1. How does a V-shaped valley become transformed into a U-shaped trough?
2. Identify the common forces that result in arêtes, horns, cirques, and cols.
3. Explain how a glacial fjord is formed.

12.4 Glacial Deposition

Key Learning Concepts

▶ **Differentiate** between glacial drift and outwash.

As we see above, glacial abrasion and plucking incorporates tons of rock material and soil into the ice, which is then transported downslope. Like giant bulldozers, glaciers also push tons of sediment in front of them and along their sides. All of this sediment is eventually deposited within the mountains and valleys of alpine glaciers, or on the broad plains covered by continental glaciers.

Glacial Drift

The general term for all glacial deposits is **glacial drift**. Geographers distinguish two types of drift. The first is unsorted **till**—sediments of varying size that both moving and melting ice deposits in no particular order, often jumbled with the heaviest boulders on top of smaller rocks. As glaciers flow downslope and melt, this drift is deposited on the ground. The second type is *sorted* **outwash** sediments that occur when glacial meltwater deposits the heaviest rocks first, followed by the lighter gravels second, and the fine silt last. Thus sorted outwash deposits are layered or *stratified*, while more random and unsorted till deposits are *unstratified*.

geoCHECK ✔ What is outwash and how does it become sorted?

Landforms Deposited by Glaciers

Deposited glacial sediment produces a specific landform, called a **moraine**, composed of unsorted till. Several types of moraines are associated with alpine glaciation. A **lateral moraine** forms along each side of a glacier (▼Fig. 12.17). If two glaciers with lateral moraines join, a **medial moraine** may form (see ▼Fig. 12.18). Eroded debris that is dropped at the glacier's farthest extent is a **terminal moraine** (also called an *end moraine*). **Recessional moraines** form at other points where a glacier paused between advancing and retreating. These different types of moraines are associated with both alpine and continental glaciation.

All of these moraine types are unsorted and unstratified. In contrast, streams of glacial meltwater carry and deposit sorted and stratified glacial *outwash* beyond a terminal moraine. Distributary stream channels appear braided across its surface. The finely ground "rock flour" turns streams a brown chocolate color. When this suspended load enters more placid lakes, it produces a distinctive turquoise color. Meltwater comes from glaciers at all times, not just when they are retreating.

▲12.17 Lateral and terminal moraines A retreating glacier in Canada is depositing two different moraines. The long ridge on the left is a lateral moraine pushed along the edge of the retreating ice. The gravels between the glacier terminus and the meadow are terminal moraines dissected by glacial outwash.

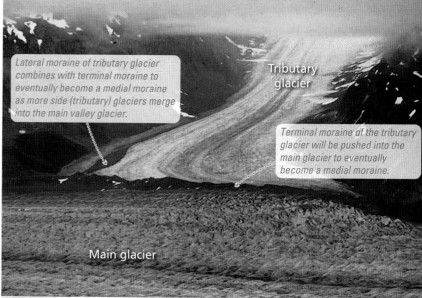

▲12.18 Terminal and medial moraines A smaller tributary glacier pushes a terminal moraine into a larger valley glacier, creating a new medial moraine.

In addition to moraines, glacial deposits form a variety of other landforms (▼Fig. 12.19). Although these landforms are common to both alpine and continental glaciers, the massive scale of continental ice sheets produce them on a much larger scale.

Eskers A sinuously curving, narrow ridge of coarse sand and gravel is an **esker**. It forms along the channel of a meltwater stream that flows in an ice tunnel inside a glacier. Retreating glaciers leave steep-sided eskers behind in a pattern parallel to path of the glacier. Sand and gravel are quarried from some eskers.

Kettles Sometimes an isolated block of ice, often a kilometer across on an outwash plain after a glacier has retreated. As much as 30 years is required for it to melt. In the interim, material continues to accumulate around the melting block. When the ice melts, it leaves behind a steep-sided **kettle** (hole) that often fills with water.

Kames Small hills of poorly sorted sand and gravel that are deposited by water or ice in crevasses, or in surface indentations, are called **kames**. Kames are also found shaped like a delta or as terraces along valley walls.

Drumlins Deposited till that was streamlined in the direction of continental ice movement is called a **drumlin**. Drumlins range from 100 to 5000 m (330 ft to 3.1 mi) in length up to 200 m (650 ft) in height. The streamlined shape of a drumlin often resembles an upside-down teaspoon bowl.

geoCHECK ✔ Describe the landforms common where glaciers deposit significant amounts of till.

geoQUIZ

1. What is glacial drift?
2. How does a medial moraine form?
3. Explain why some glacial sediment is stratified, and some is not stratified.

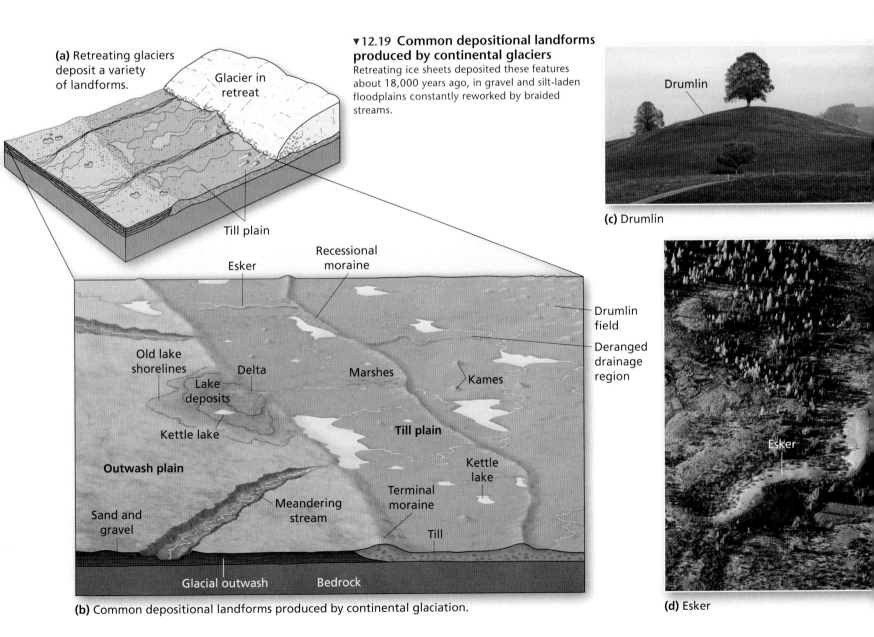

(a) Retreating glaciers deposit a variety of landforms.

Glacier in retreat

Till plain

▼12.19 Common depositional landforms produced by continental glaciers
Retreating ice sheets deposited these features about 18,000 years ago, in gravel and silt-laden floodplains constantly reworked by braided streams.

Drumlin

(c) Drumlin

Esker

Recessional moraine

Old lake shorelines

Delta

Lake deposits

Kettle lake

Marshes

Kames

Drumlin field

Deranged drainage region

Outwash plain

Till plain

Kettle lake

Sand and gravel

Meandering stream

Terminal moraine

Till

Esker

Glacial outwash   Bedrock

(b) Common depositional landforms produced by continental glaciation.

(d) Esker

12.5 Periglacial Landscapes

Key Learning Concepts

▶ *Explain* the cause and distribution of periglacial landscapes.

Periglacial environments are found in very cold climates, often around glaciated areas. When soil or rock temperatures remain below 0°C (32°F) for at least 2 years, **permafrost** develops. These areas of permanently frozen ground not covered by glaciers are considered periglacial. Permafrost occupies over 20% of Earth's land surface, with the largest extent of such lands in Russia (▶Fig. 12.20). Approximately 80% of Alaska has permafrost beneath its surface, as do large portions of Canada, China, Scandinavia, Greenland, Russia, and Antarctica. Many alpine mountain regions of the world also are affected.

The Geography of Permafrost

One main factor that defines permafrost regions is climate. This criterion is based solely on temperature and not the amount of water present. Permafrost regions are in *subarctic* and *polar* climates, especially *tundra* climate, either at high latitude (tundra and boreal forest environments) or at high elevation in lower-latitude mountains (alpine environments). Other factors that contribute to permafrost conditions are the presence of "fossil" permafrost from a previous ice age, the insulating effect of snow or vegetation that inhibits heat loss, and the slope and aspect that influences insolation. Permafrost regions are divided into two general categories, continuous and discontinuous, which merge along a transition zone.

Discontinuous permafrost zones are most susceptible to thawing with climate change. Peat-rich soils affected by thawing contain twice the amount of carbon that is in the atmosphere. The soil's high carbon content creates a powerful positive feedback as soils thaw and oxidation releases more carbon dioxide to the atmosphere.

Permafrost Behavior Figure 12.21 is a stylized cross-section of a permafrost landscape. The *active layer* is the zone of seasonally frozen ground that exists between the subsurface permafrost layer and the ground surface. It is subjected to consistent daily and seasonal freeze–thaw cycles. Higher temperatures reduce permafrost and increase the thickness of the active layer; lower temperatures gradually increase permafrost depth and reduce active-layer thickness.

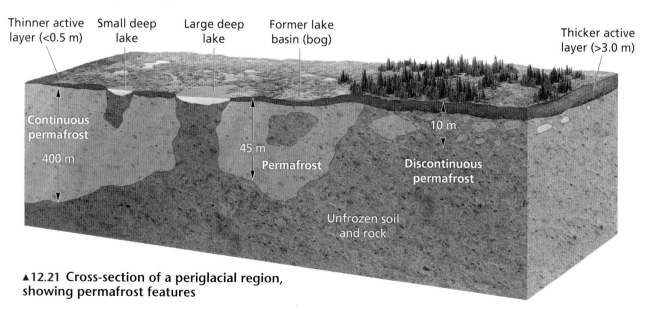

▲12.20 **Permafrost distribution** Most of Earth's continuous permafrost lies north of 60° N latitude.

Legend	
Subsea permafrost	Discontinuous permafrost
Glacial Ice	Sporadic permafrost
Continuous permafrost	Alpine permafrost

Thinner active layer (<0.5 m) — Small deep lake — Large deep lake — Former lake basin (bog) — Thicker active layer (>3.0 m)

Continuous permafrost
400 m
45 m
Permafrost
10 m
Discontinuous permafrost
Unfrozen soil and rock

▲12.21 **Cross-section of a periglacial region, showing permafrost features**

geoCHECK ✔ What is permafrost?

Periglacial Processes

In permafrost regions the frozen subsurface water varies from nearly none in drier regions to almost 100% in saturated soils. Since water expands 9% as it freezes, it can produce strong mechanical forces that fracture rock and disrupt soil at or below the surface. In periglacial regions this causes *frost heaving* of the soil upward and *frost thrusting* of the soil in a sideways direction. Boulders and rock slabs may be thrust to the surface. Soil horizons (layers) may appear stirred or churned.

An *ice wedge* develops when water enters a crack in the permafrost and freezes (▶Fig. 12.22). **Patterned ground** occurs when the expansion and contraction of frost action sorts surface stones and soil particles into separate areas. Greater slopes produce striped patterns, whereas lesser slopes result in sorted polygons.

Mass Movement Processes in Periglacial Landscapes

During the summer thaw cycle, the upper layer of soil regolith starts to flow downslope. This "current" is called *solifluction*, but since it may occur in all climates, in periglacial regions the more specific term *gelifluction* is applied. As gelifluction occurs, soil movement up to 5 cm (2 in.) per year is common. Over time, this landflow creates distinctive scalloped and lobed patterns on almost every slope. Periglacial mass movement may also include translational and rotational slides and rapid flows associated with melting ground ice.

Humans & Periglacial Landscapes

Permafrost continually challenges human settlement. Highways, rail, and utility lines warp or twist as thawing ground shifts. Buildings placed directly atop frozen ground eventually subside into the defrosting soil (▶Fig. 12.23). Thus structures in periglacial regions must sit above the ground to allow air circulation beneath. Water and sewer lines are built aboveground in "utilidors." Much of the trans-Alaska oil pipeline is above ground to avoid melting the frozen ground. Melting could cause the pipeline to shift and rupture. Warming annual temperatures in the Canadian and Siberian Arctic since 1990 are disrupting permafrost surfaces—leading to highway, railway, and building damage.

▲12.22 **Patterned ground on Nordaustlandet Island, Arctic Ocean** Over time, freezing and thawing produced this pattern of polygons and circles about a meter across in a stone-dominant area. Patterned ground also develops in soil-dominant areas.

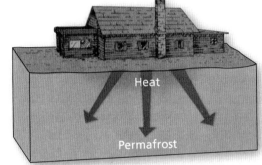

(a) Heat from a building can melt the permafrost below.

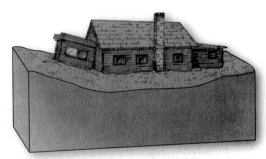

(b) As permafrost melts, the building collapses.

(c) A collapsed house sinks into permafrost, Alaska

▲12.23 **Permafrost as a natural hazard** Building failure due to improper construction and melting of permafrost, south of Fairbanks, Alaska.

 geoCHECK ✔ What is solifluction?

12.6 The Pleistocene Epoch

Key Learning Concepts

▶ **Describe** how Ice Ages come and go through geologic history.

Eon	Era		Period	Epoch
Phanerozoic	Cenozoic	Quaternary		Holocene
				Pleistocene
		Tertiary	Neogene	Pliocene
				Miocene
			Paleogene	Oligocene
				Eocene
				Paleocene
Phanerozoic	Mesozoic			Cretaceous
				Jurassic
				Triassic
Phanerozoic	Paleozoic			Permian
		Carboniferous		Pennsylvanian
				Mississippian
				Devonian
				Silurian
				Ordovician
				Cambrian
Prephanerozoic	Precambrian	Proterozoic		Neo-(late)
				Meso-(middle)
				Paleo-(early-)
				Archean

66.5 mya

251 mya

542 mya

Earth ~4600 Ma

▲ **12.24 The Pleistocene epoch in relation to geologic time** Time intervals not to scale

During the Pleistocene Epoch of the late Cenozoic Era, almost a third of Earth's land surface lay buried beneath ice sheets and glaciers—most of Canada, the northern Midwest, England, and northern Europe, and many mountain ranges further south. In addition, periglacial regions along the margins of the ice covered about twice their present area.

Ice-Age Temperatures & Landscapes

An **ice age**, or *glacial age*, is a cold period that may last several million years. The *glacial* periods of cold climate are interrupted by brief *interglacial* warm spells. Each carries a name usually based on the location where evidence of the episode is prominent—for example, the Wisconsinan glacial. The Pleistocene featured at least 18 advances and retreats of ice over Europe and North America.

Scientists determine Pleistocene temperatures by evidence from deep-sea cores—specifically, from oxygen-isotope fluctuations in fossil plankton, tiny marine organisms with a calcareous shell. The plankton shells are made up of layers of material in which differences in the oxygen–isotope ratios record historical changes in climate. Ice cores from Antarctica also provide evidence of climate conditions during the 300,000-year interval prior to our present Holocene Epoch. Figure 12.25 illustrates the continental ice sheets that covered portions of Canada, the United States, Europe, and Asia about 18,000 years ago. Ice sheets ranged in thickness to more than 2 km (1.2 mi). The ice sheet disappeared by 7000 years ago.

The Pleistocene Epoch began about 2.5 million years ago (◀**Fig. 12.24**) and is one of the more prolonged cold periods in Earth's history. It featured at least 18 expansions of ice over Europe and North America, each obliterating and confusing the evidence from the one before. Glaciation lasts about 100,000 years, whereas deglaciation requires less than 10,000 years to melt away the accumulation.

As both alpine and continental glaciers retreated, they exposed a drastically altered landscape: the drumlins and eskers of New England; the polished bedrock of Canada's Atlantic Provinces; the sharp crests of the Alps, Rocky Mountain, and the Sierra–Cascade chains; the Great Lakes of the United States and Canada; and much more. In the Southern Hemisphere, evidence of this ice age appears in the fjords and sculpted mountains in New Zealand and Chile.

geoCHECK ✔ How do geologists define the Pleistocene?

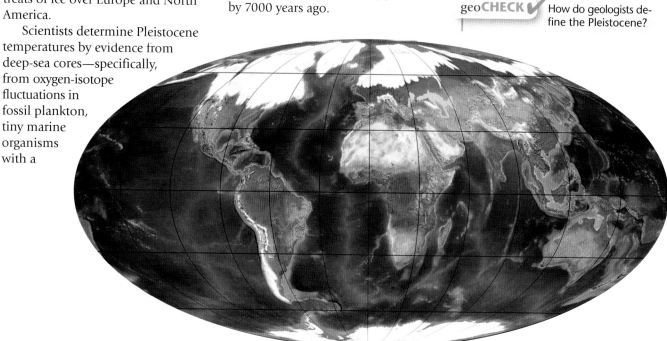

▲**12.25 Extent of Pleistocene glaciation, about 18,000 years ago**

Paleolakes

About 12,000–30,000 years ago, the American West was dotted with large lakes. Except for the Great Salt Lake in Utah, itself a remnant of the former Lake Bonneville noted on the map, and a few smaller lakes, only dry basins, ancient shorelines, and lake sediments remain today (▼Fig. 12.26). These ancient **pale-olakes** are also called **pluvial** lakes. *Pluvial*, derived from the Latin word for "rain," describes any period of wet conditions, such as occurred during the Pleistocene Epoch. During pluvial periods, the lake levels increase in arid regions. The drier *interpluvial* periods were marked by lake sediments that form terraces along former shorelines.

Scientists are trying to correlate pluvial periods and glacial ages to see whether they coincided during the Pleistocene. However, few paleolakes demonstrate such a simple relation. For example, in the western United States, the estimated volume of melted ice from glaciers is only a small portion of the water that filled paleolakes. Also, these lakes often predate glacial times and are correlated instead with periods of wetter climate or lower evaporation rates.

Paleolakes existed in every continent except Antarctica. In North America the two largest late-Pleistocene paleolakes were Lake Bonneville and Lake Lahontan, located in the western United States. Both were much larger than their present-day remnants. The Great Salt Lake—today the fourth largest saline lake in the world—and the Bonneville Salt Flats in western Utah are remnants of Lake Bonneville. Lake levels continue to decline as climate change produces drier conditions.

New evidence suggests that the occurrence of these pluvial lakes in North America was related to specific changes in the polar jet stream that steered storm tracks across the region. The continental ice sheet evidently influenced changes in jet-stream position.

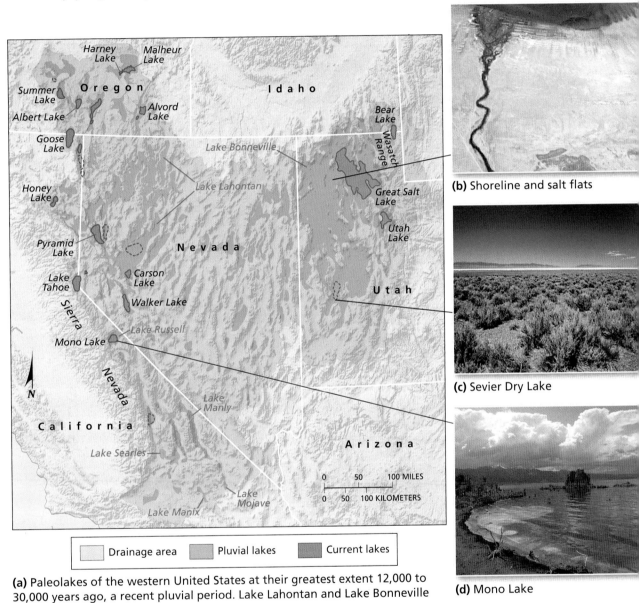

(a) Paleolakes of the western United States at their greatest extent 12,000 to 30,000 years ago, a recent pluvial period. Lake Lahontan and Lake Bonneville were the largest.

(b) Shoreline and salt flats

(c) Sevier Dry Lake

(d) Mono Lake

Drainage area   Pluvial lakes   Current lakes

▲12.26 **Paleolakes in the western United States**

geoCHECK ✔ What is a paleolake?

geoQUIZ

1. Describe the defining geographic characteristic of the Pleistocene.
2. What is the Latin origin and geographic meaning of the word *pluvial*?
3. How do changes in the polar jet stream help create pluvial lakes?

12.7 Ice-Age Climate Patterns

Key Learning Concepts

▶ **Summarize** how scientists interpret Earth's climate history and recent episodes of warmer and cooler climates.

▶ **Identify** natural mechanisms that cause fluctuations in climate.

The study of Earth's past climates is the science of *paleoclimatology*. Glacials and interglacials occur because Earth's climate alternates between cooler and warmer periods. Evidence for this is found in ice cores from Greenland and Antarctica, in layered ocean deposits of silts and clays, in the extensive pollen record from ancient plants, and in the relation of past coral productivity to sea level. This evidence is analyzed with radioactive dating methods and other techniques. The human historical record that appears in cave paintings and rock art (petroglyphs and pictographs) also provides clues to how our ancestors survived in climates of long ago. As it turns out, we humans of the last 200,000 thousand years have never experienced Earth's normal (more moderate, less extreme) climate, most characteristic of Earth's entire 4.6-billion-year span.

The climate pattern of ice ages and interglacial periods that characterized the Pleistocene Epoch began in earnest 1.65 million years ago. Although the Holocene Epoch began approximately 10,000 years ago, when average temperatures abruptly increased 6°C (11°F), today's climate may represent an end to the Pleistocene climate pattern, or merely a mild interglacial period (▾**Fig. 12.27**).

Medieval Warm Period & Little Ice Age

Climate has not remained constant within the current interglacial period. Two episodes of climate change that occurred over the last 1,200 years are the Medieval Warm Period and the Little Ice Age. Both time periods demonstrate that climate change influences human settlement, food production, and transportation routes.

An episode of warming around A.D. 1000 favored the Vikings as they sailed the North Atlantic to settle Iceland and Greenland (▾**Fig. 12.28**). This *Medieval Warm Period* lasted from about A.D. 800 to 1200. During the warmth, oats and barley were planted in Iceland, and wheat was sown in northern Norway. The shift to warmer, wetter weather pushed migration northward in Europe, and Asia, and likely influenced Native American settlement in North America

However, from approximately 1200 to 1900, a *Little Ice Age* took place. Parts of the North Atlantic froze, and expanding glaciers blocked key mountain passes in Europe. Snowlines in Europe lowered about 200 m (650 ft) in the coldest years. Greenland and other northern colonies were deserted.

geoCHECK ✔ How did the Medieval Warm Period influence human activity in Europe?

▲**12.27 Recent climates determined from ice cores** The graph shows the relationship between temperature (red line) and snow accumulation (blue line).

▶**12.28 Reconstructed Viking settlement, Iceland**
The Vikings colonized Iceland and Greenland during the Medieval Warm Period. The Iceland settlements survived the return of colder conditions, but people abandoned the Greenland settlements.

Mechanisms of Climate Fluctuation

Earth's short- and long-term climatic changes are receiving unprecedented attention. Ice and deep-sea cores now provide detailed records of historical weather and climate patterns, volcanic eruptions, and biosphere trends. These data provide insights on how past climate has changed, and how it might change again—especially during our present era of forced anthropogenic warming. Researchers have identified a complicated mix of interacting variables that appear to influence long-term climatic trends. The most important ones are discussed below.

Orbital Variations, Milankovitch Cycles, & Ice Ages The Earth's elliptical orbit about the Sun varies by more than 17.7 million kilometers (11 million miles) during a 100,000-year cycle (▶ **Fig. 12.29a**). In addition, Earth's axis "wobbles" through a 26,000-year cycle, called *precession*, which alters Earth's orientation to the Sun (▶ **Fig. 12.29b**). Earth's present 23.5° axial tilt varies from 21.5° to 24.5° during a 41,000-year period (▶ **Fig. 12.29c**). These variations affect the amount of incoming insolation the Earth receives. Serbian astronomer Milutin Milankovitch (1879–1958) was the first to hypothesize that reduced insolation due to these cycles periodically causes cooling that correlates with ice ages (▶ **Fig. 12.30**).

Solar Variability If the Sun's output varies over time, as the output of some other stars does, it could trigger global cooling. However, to date this hypothesis has little supporting evidence.

Climate & Tectonics Major glaciations are associated with plate tectonics, because plate movements shift some landmasses to higher, cooler latitudes. Mountain-building episodes over the past billion years also push summits above the line of summer melt, allowing alpine glaciers to form.

Atmospheric Factors Volcanic eruptions could temporarily reduce solar insolation by ejecting particles of ash and other substances into the upper atmosphere (▼ **Fig. 12.31**). These particles reflect more incoming solar radiation back to space, temporarily cooling the atmosphere and thus allowing more long-term snow cover at high latitudes. These high-albedo surfaces would then reflect more insolation away from Earth to augment cooling.

geoCHECK ✔ What are Milankovitch cycles, and how are they thought to affect climate?

geoQUIZ

1. How did the Medieval Warming Period favor Viking settlement in the New World?
2. Explain how plate tectonic movements could influence climate.
3. How can volcanic eruptions influence climate? Explain.

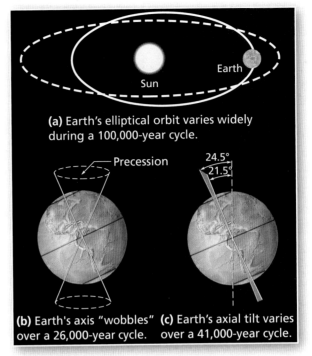

(a) Earth's elliptical orbit varies widely during a 100,000-year cycle.

Precession

24.5°
21.5°

(b) Earth's axis "wobbles" over a 26,000-year cycle. (c) Earth's axial tilt varies over a 41,000-year cycle.

▲12.29 Astronomical factors affecting climate cycles

Animation (MG)
End of the
Last Ice Age

http://goo.gl/XB2LMA

Animation (MG)
Orbital Variations &
Climate Change

http://goo.gl/p08UPy

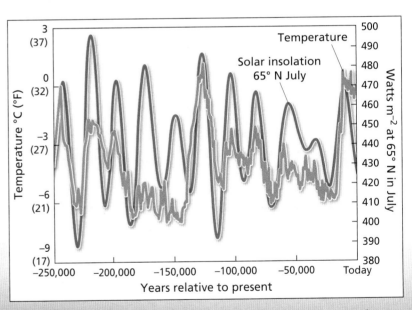

▲12.30 Milankovitch cycles and temperature from an ice core in Vostok, Russia

▼12.31 Eruption of the Volcano Holuhraun, Iceland

12.8 Shrinking Glaciers on a Warming Planet

Key Learning Concepts

▶ **List** the criteria climatologists use to define the polar regions.

▶ **Describe** recent changes in the Arctic and Antarctic linked to climate change.

As the earth–atmosphere warms, glacial ice in polar and mountain regions is shrinking almost everywhere. For example, if Peru's Quelccaya ice cap continues shrinking 600 ft (182.8 m) a year, it will vanish by 2100. The glaciers atop Africa's Mt. Kilimanjaro have retreated more than 80% in the last century. Glaciers in the central and eastern Himalaya may also disappear in the next three decades. When Montana's Glacier National Park was created in 1910, over 150 glaciers cloaked the cirques and narrow valleys of this Rocky Mountain landscape. Today, fewer than 30 exist, and these may disappear within 40 years (▼**Fig. 12.32**), and other shrinking glaciers from around the world will reduce the amount of freshwater available to downstream human populations.

(a) 1938 **(b)** 1998 **(c)** 2009

▲**12.32 The retreating Grinnell Glacier, Montana** Repeat photography clearly shows the glacier's retreat.

Polar Environments

Climatologists use environmental criteria to define the Arctic and the Antarctic regions. In the Northern Hemisphere, the 10°C (50°F) isotherm for July defines the *Arctic region* (green line in ▶**Fig. 12.33a**). This line coincides with the visible tree line—the boundary between the northern forests and tundra. The Arctic Ocean is covered by *floating sea ice* (frozen seawater) and *glacier ice* (frozen freshwater). This sea ice thins in the summer months and sometimes breaks up.

In the Southern Hemisphere, the Antarctic convergence defines the *Antarctic region*. The Antarctic convergence is a narrow zone that extends around the continent of Antarctica as a boundary between colder Antarctic water and warmer water at lower latitudes. This boundary follows roughly the 10°C (50°F) isotherm for February, the Southern Hemisphere summer, and is located near 60° S latitude (green line in ▶**Fig. 12.33b**). The Antarctic region covered with just sea ice represents an area greater than North America, Greenland, and Western Europe combined.

The Antarctic landmass is surrounded by ocean and is much colder overall than the Arctic, which is an ocean surrounded by land. In simplest terms, Antarctica is a continent covered by a single enormous glacier, although it contains distinct regions that respond differently to slight climatic variations. These ice sheets are in constant motion, as indicated by the arrows in Figure 12.33b.

geoCHECK ✔ Explain the difference between floating sea ice and glacial ice.

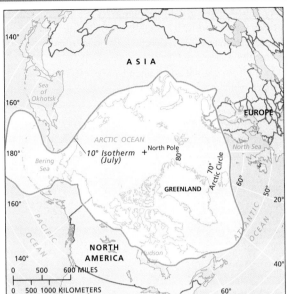

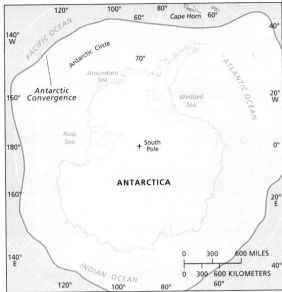

(a) Note the 10°C (50°F) isotherm in midsummer, which designates the Arctic region, dominated by pack ice.

(b) The Antarctic convergence designates the Antarctic region.

▲12.33 **The polar regions** (a) Arctic and (b) Antarctica.

Recent Polar Region Changes

Almost half of the Arctic ice pack by volume has disappeared since 1970 due to regional-scale warming. The year 2012 broke the record for lowest sea-ice extent. The fabled Northwest Passage, by which ships travel an ice-free route across the Arctic from the Atlantic to the Pacific was ice-free in September 2007. The Northeast Passage, north of Russia, has been ice-free for several years. Geographers investigating the Greenland Ice Sheet found that in the last decade the rate of ice melting is accelerating, and the melt zone is moving upward in elevation. However, the precise path of how this meltwater moves through the vast ice sheet and into the ocean—which contributes to rising sea levels—remains unclear (▶Fig. 12.34).

An important indicator of changing surface conditions is an increase in **meltponds** instead of ice across the polar regions. The darker meltponds represent positive feedback because they absorb more insolation and become warmer, which melts more ice, making more meltponds. Streams that flow and even melt through the ice sheet form a *moulin* (channel) that drains to the

▼12.35 **Melt stream** Jostedal Glacier, Norway.

base of the glacier (▼Fig. 12.35). This glacial water can flow through the basal layers of fine clays, thus lubricating and accelerating glacial flow rates.

The numerous ice shelves surrounding the margins of Antarctica constitute 11% of its surface area. Although ice shelves constantly break up into icebergs, more large sections are breaking free than expected. Since 1993, seven ice shelves have disintegrated in Antarctica. More than 8000 km² (3090 mi²) of ice shelf are gone, changing maps, freeing up islands to circumnavigation, and creating thousands of icebergs.

This and other ice loss during the last 50 years is likely a result of warmer water and the 2.5°C (4.5°F) air temperature increase in the Antarctic Peninsula, the northernmost part of Antartica and the only part of the continent to extend north of the Antarctic Circle. In response, the peninsula is experiencing previously unseen vegetation growth; reduced sea ice; and disruption of penguin feeding, nesting, and fledging activities. Ticks are a new problem for penguins.

▲12.34 **Surface melt over Greenland's ice sheet** Satellite measurements portray how quickly ice near the surface thaws during a four-day heat wave from (a) July 8, to (b) July 12, 2012.

geoCHECK✔ What is a moulin?

geoQUIZ

1. What environmental criteria do scientists use to define the Antarctic and Arctic regions?
2. How do meltponds create a positive feedback that promotes melting ice that creates more meltponds?
3. How are warming temperatures affecting the environment of the Antarctic Peninsula?

GLACIAL ENVIRONMENTS IMPACT HUMANS

• Glacial ice is a freshwater resource; ice masses affect sea level, which is linked to security of human population centers along coastlines.

• Snow avalanches are a significant natural hazard in mountain environments.

• Permafrost soils are a carbon sink, estimated to contain half the pool of global carbon.

HUMANS IMPACT GLACIAL ENVIRONMENTS

• Rising temperatures associated with human-caused climate change are accelerating ice sheet losses and glacial melting, and hastening permafrost thaw.

• Particulates in the air from natural and human sources darken snow and ice surfaces, which accelerates melting.

12a

A 675-km (420-mi) section of the trans-Alaska oil pipeline was constructed above the ground to prevent permafrost thaw. The pipeline is 1.2 m (4 ft) in diameter, and throughout this distance is supported on racks that average 1.5 to 3.0 m (5 to 10 ft) high.

12b

A USGS geospatial scientist ground truths the retreat of the Muir Glacier, Alaska

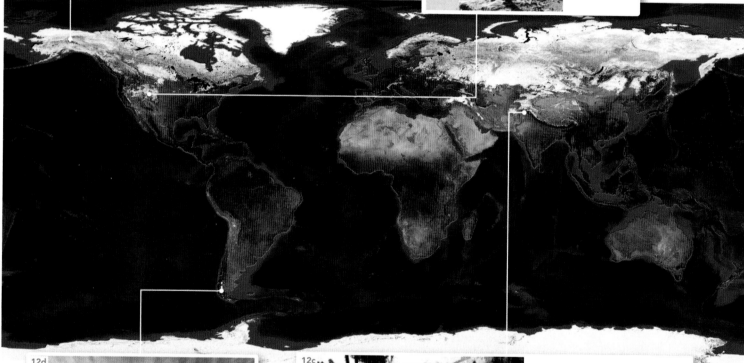

12d

Argentina's Perito Moreno Glacier on the South Patagonian Ice Field in Los Glaciares National Park is a premier tourist destination. Recent research found that glaciers in this region are thinning at a rate of about 1.8 m (5.9 ft) per year. These freshwater losses will impact regional water supplies.

12c

In the Indian-controlled region of Kashmir, avalanches buried a military camp and killed 16 Indian soldiers in February 2012. Three months later, 100 Pakistani soldiers were killed nearby. Large snowfall volumes and frequent winds combine with steep Himalayan slopes to create dangerous avalanche conditions in this area.

ISSUES FOR THE 21ST CENTURY

• Melting of glaciers and ice sheets will continue to raise sea level, with potentially devastating consequences for coastal communities and low-lying island nations.

• Thawing of permafrost in response to climate change will release vast amounts of carbon into the atmosphere, accelerating global warming.

Looking Ahead

In the next chapter we investigate the biosphere. Specifically, how weather and climate combine with the endogenic and exogenic forces to create the conditions in which life on Earth evolved and on which biologic systems depend upon for their survival.

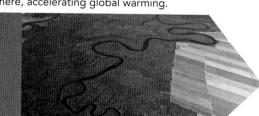

How do glaciers form, move, & erode Earth's surface?

12.1 Glacial Processes—Ice Formation & Mass Balance

Understand how and where glaciers form.

Describe the two main types of glaciers.

- About 77% of Earth's freshwater is frozen into glacial ice. Glaciers are slow-moving bodies found in high latitude and mountains where snow survives year-round. Glaciers form by the continual accumulation of snow that recrystallizes under its own weight into an ice mass. Alpine glaciers form in mountain ranges, while much larger continental glaciers or ice sheets are found in Greenland and Antarctica.

1. What environmental conditions lead to the formation of glaciers?
2. Differentiate between an alpine glacier, continental glacier, ice cap, and ice sheet.

12.2 Glacial Mass Balance, Movement, & Erosion

Explain what determines if a glacier grows or shrinks.

Describe how a glacier moves.

- The upper *accumulation* zone of a glacier is where snowfall collects and compacts into ice. The lower *ablation* zone is where most of the melting and sublimation occurs. The division between these two zones is called the equilibrium line. Glaciers flow downhill and outward under the force of gravity. The internal bottom of a glacier slides along, varying its speed with temperature, friction, and the presence of any lubricating water beneath the ice. The upper portions of a glacier are brittle and crack into many crevasses.

3. What line divides a glacier's accumulation and ablation zones, and how does glacial mass balance differ above and below that line?

How does glacial erosion & deposition change landscapes?

12.3 Alpine Glacial Erosion & Landforms

Explain how alpine glaciers erode the landscape.

Identify the different landforms that alpine glaciers leave behind.

- Alpine glaciers erode through abrasion (scraping) and plucking small and large rocks from the surface bedrock and canyon sidewalls. This scouring leaves behind a distinctive landscape of U-shaped troughs and hanging valleys, bowl-shaped cirques and tarns, along with pointed horns and sawtooth *arêtes*.

4. How does a glacier accomplish erosion, and what landforms result from these processes?
5. Describe the unique landform features present after the glacier retreats.

12.4 Alpine & Continental Glacial Deposition

Differentiate between different types of glacial deposition.

- Glaciers transport tons of sediment in front of them and along their sides. They deposit this *glacial drift* in the mountains and valleys of alpine glaciers, or on the broad plains of continental glaciers. Most drift is unsorted because moving ice both transports and deposits sediments randomly. However, *outwash* sediments deposited by running meltwater are stratified. The most common glacial depositional landform is a *moraine*, but kames, eskers, and drumlins are also common.

6. What are some common depositional features encountered on a glacial till plain?

How does climate influence glacial processes?

12.5 Periglacial Landscapes

Explain the cause and distribution of periglacial landscapes.

- *Periglacial* environments are found in very cold climates, often around glaciated areas. Areas of permafrost develop when soil or rock temperatures remain below 0°C (32°F) for at least 2 years. Permafrost covers 20% of Earth's surface and is found in subarctic and polar regions, and in some lower-latitude alpine environments. The seasonal melting of permafrost soil results in unique processes such as solifluction, patterned ground, and frost heaving.

7. What is permafrost, and in what climate types is it located?
8. How does the presence of permafrost affect human structures?

12.6 The Pleistocene Epoch

Describe how Ice Ages come and go through geologic history.

- An *ice age* is a cold period which may last several million years. Shorter *interglacial* warm spells interrupt these glacial periods. Scientists use evidence from deep-sea cores to determine ice age temperatures. Retreating ice age glaciers leave behind distinctive postglacial landscapes, including small remnant portions of paleolakes (or pluvial lakes).

9. Describe the role of ice cores in determining past climates.
10. What is a pluvial lake, and where do we find them?

12.7 Ice-Age Climate Patterns

Summarize how scientists interpret Earth's climate history and recent episodes of warmer and cooler climates.

Identify natural mechanisms that cause fluctuations in climate.

- *Paleoclimatology* investigates Earth's past climates. The most recent climate pattern of ice ages and interglacial periods that characterized the Pleistocene Epoch began in earnest 1.65 million years ago. During the last 1200 years, both the Medieval Warm Period and the Little Ice Age influenced human settlement, crops, and transportation routes. Scientists reason Milankovitch Cycles, solar variability, continental drift, volcanic eruptions, and changing ocean circulation could all promote climate change.

11. Describe how Milankovitch Cycles influence the conditions necessary for another ice age.

12.8 Arctic & Antarctic Regions

List the criteria climatologists use to define the polar regions.

Describe recent changes in the Arctic and Antarctic linked to climate change.

- Climatologists use environmental criteria to define the Arctic and the Antarctic regions. The 10°C (50°F) isotherm for July defines the *Arctic region*. The Antarctic convergence defines the *Antarctic region* in a narrow zone that extends around the continent as a boundary between colder Antarctic water and warmer water at lower latitudes. Almost half of the Arctic sea ice, and significant portions of the Antarctic ice shelf, have disappeared since 1970 due to regional-scale warming.

12. Explain the criteria used to define the Arctic and Antarctic regions.

Critical Thinking Questions

1. What can we learn about existing climate patterns from the conditions in glacial regions and glacial mass balances? Explain.

2. Describe the transformation of a V-shaped stream valley into a U-shaped glaciated valley.

3. Name three common depositional features found on a till plain and describe how they form.

4. Explain what is meant by the terms *glacial* and *interglacial* as they relate to ice ages.

5. How will human life be different if the Pleistocene conditions returned? How could these changes affect life in New York City, Chicago, Denver, Miami, or Los Angeles?

Visual Analysis

This photograph portrays the Casement Glacier and the Takinsha Range, Alaska.

1. What physical force shaped the many pointed peaks and sawtooth ridges above the glacier?

2. In the U-shaped trough filled by the Casement Glacier, what caused the broken and twisted ice in the lower part of the photo? What is the name for these glacial features?

3. What caused the black stripes running down the middle of the glacier?

▲R12.1 **Casement Glacier and the Takinsha Range, Alaska**

Key Terms

alpine glacier, p. 322
arête, p. 326
cirque, p. 322
continental glacier, p. 323
crevasse, p. 325
drumlin, p. 329
equilibrium line, p. 324
esker, p. 329

fjord, p. 327
glacial drift, p. 328
glacial lake, p. 327
glacier, p. 322
glacier surge, p. 325
hanging valley, p. 327
horn, p. 8
ice age, p. 332
ice cap, p. 323

ice field, p. 323
ice sheet, p. 323
iceberg, p. 322
kame, p. 329
kettle, p. 329
lateral moraine, p. 328
medial moraine, p. 328

meltpond, p. 337
moraine, p. 328
outwash, p. 328
paleolakes, p. 333
patterned ground, p. 331
periglacial, p. 330
permafrost, p. 330
pluvial, p. 333

recessional moraine, p. 328
surge, p. 7
talus slopes, p. 327
tarn, p. 327
terminal moraine, p. 328
till, p. 328
U-shaped valley, p. 327
valley glacier, p. 322

- Launch MapMaster in MasteringGeography. In the *Physical Environment* categories, select *Satellite* (Blue Marble). Clicking this category reveals the location of the mountain areas where glaciers occur. Of course, not every white speck is a glacier, but the white coloration does indicate the Köppen H climate zone where glaciers are located in cirques, shaded canyons, and in larger icecaps.
- Zoom in two clicks to enlarge India and Tibet so this part of the world occupies much of the map.

- Return to *Physical Environment* and click the *Global Surface Warming, Worst-Case Projections* category and note this worst-case scenario.

1. Summarize what impact warming will likely have on glaciers in the Himalaya and Tibet Plateau.

2. Meltwater from Himalayan glaciers feeds major rivers of China, Pakistan, and India. How will the effects of changing temperatures affect the two billion people who live downstream of these glaciers?

Explore | Use **Google Earth** to explore glaciers.

Viewing Earth from space allows us to visualize how glaciers occupy and modify the landscape. In Google Earth, check the *Borders and Labels* tool, and leave other categories unchecked. After arriving at each destination, zoom in and out to gain perspective. You may also check the *Photo* tool to learn more about each location. The destinations are: *Baltoro Glacier, Pakistan; Malaspina Glacier, Alaska;* and Palisade *Glacier, California.*

1. Of these three glaciers, identify which one is the valley glacier, which one is the piedmont glacier, and which one is the cirque glacier. Explain your reasoning for each choice.

2. If the Baltoro Glacier melts away in the next 100 years, identify the type and likely location of three landforms the receding ice will leave behind.

▲R12.2 **Baltoro Glacier, Karakoram Mountains, Pakistan**

MasteringGeography™

Looking for additional review and test prep materials? Visit the Study Area in MasteringGeography™ to enhance your geographic literacy, spatial reasoning skills, and understanding of this chapter's content by accessing a variety of resources, including MapMaster™ interactive maps, videos, *Mobile Field Trips, Project Condor* Quadcopter videos, *In the News* RSS feeds, flashcards, web links, self-study quizzes, and an eText version of *Geosystems Core.*

GeoLab12

Big Relief: Using Topographic Maps to Determine Elevation Profiles

Although abrasion by glacial ice does not create canyons, it does carve and usually enlarge them into a distinctive shape that is different from the work of fluvial erosion alone. The goal of this assignment is to construct an elevation profile across the wild glacial landscape surrounding both sides of Half Dome, in Yosemite National Park.

Apply

The ability to use elevation data from topographic maps is a core skill used by scientists, planners, and recreation specialists. As a park ranger, you are charged with providing an elevation profile across the two canyons that drain into Yosemite Valley. Helicopter pilots who fly search and rescue missions, fire suppression crews, and recreation planners who build and maintain hiking trails will use your measurements.

Objectives

- Plot elevation points from a topographic map on the graph.
- Construct a topographic profile.
- Calculate average vertical relief between two points.
- Hypothesize what geologic processes shaped this landscape profile.

Procedure

1. Using the graph paper provided below (GL 12.3), plot the elevation from the North Dome summit (Point A) to the summit of Half Dome (Point B). Plot your measurements to scale in feet using the map in GL 12.2.
2. Connect your plotted elevation points into an elevation profile.
3. Describe the shape of the profile.
4. Calculate the total decrease *and* increase in elevation between:
   a. North Dome to Tenaya Creek
   b. Tenaya Creek to the west shoulder of Half Dome
   c. the shoulder of Half Dome to its summit.
   If you were to hike this route (which would also involve significant rock climbing!), the total vertical relief would be _____ (feet and meters).
5. Calculate the difference in vertical relief between the two canyons: _____ (feet and meters).
6. Using the scale at the bottom of the map, calculate the distance a bird would fly from North Dome to the summit of Half Dome _____ (feet and meters).
7. Use the two values in Questions 5 and 6 above to determine the average elevation change for every 1000 feet (330 meters): _____

Analyze & Conclude

8. Describe how the profile you described in Question 3 above is different from a valley cut by fluvial erosion only.
9. Apart from the profile you just constructed, does the map in Figure GL 12.2 contain other evidence of glaciation? Explain.
10. Where in this picture do you think the large glaciers that sculpted Half Dome originated?
11. What geomorphic process caused waterfalls in Yosemite Valley, and what is the proper geomorphic name for the type of feature from which these waterfalls emerge?

▲GL12.1 **View of Half Dome from Sentinel Dome** Features visible in this photograph include (from left to right) North Dome (a), (left of the arrow) Tenaya Canyon, and Half Dome, (b).

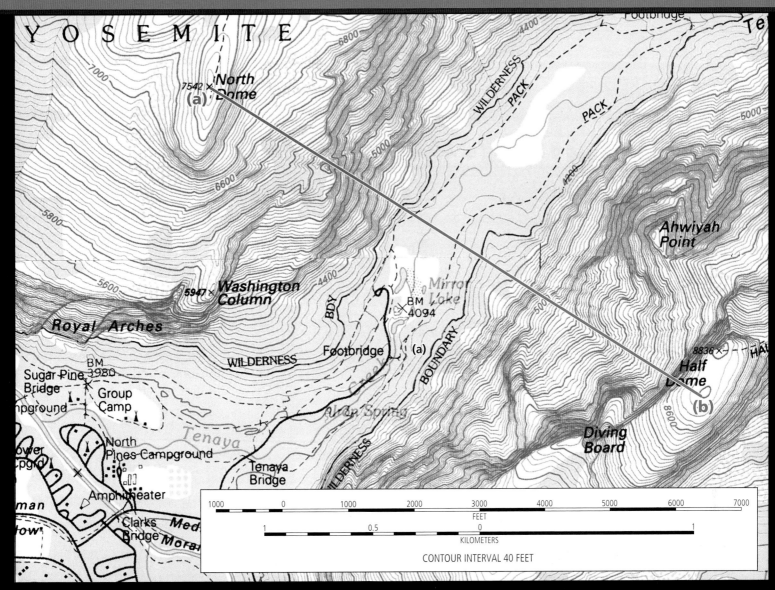

▲GL12.2 **Topographic map of Half Dome**

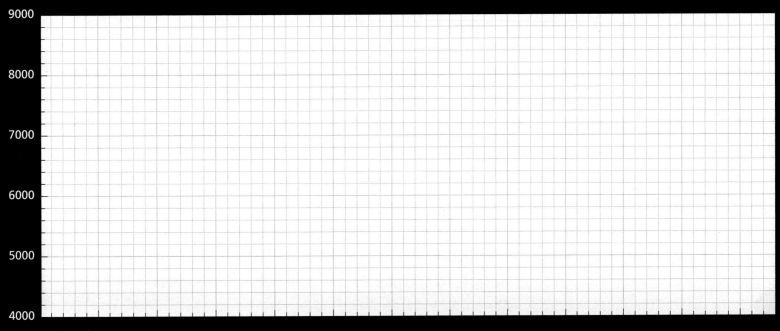

▲GL12.3 **Graph paper**

Ecosystems & Soils

13

Earth is the home of the solar system's only known biosphere—a uniquely complex system of interacting nonliving and living components working together to sustain a tremendous diversity of life. Most energy enters the biosphere through conversion of solar energy by photosynthesis in the leaves of plants. Energy then moves from plants to the animals that eat them or eventually decompose them. Together, these varied organisms, and the physical environments they inhabit, make up Earth's living communities. Groups of distinctive plants and animals are called ecosystems. Soil is the essential link connecting the living world to the lithosphere and the rest of Earth's physical systems.

Key Concepts & Topics

Cleared fields on the edge of rain forest in Queensland, Australia provide a stark contrast between the rich biodiversity of the rain forest and the industrial monocropping of the fields.

13.1 Energy Flows & Nutrient Cycles

Key Learning Concepts

▶ *Outline* the basic components of ecosystems.
▶ *Explain* photosynthesis and respiration.
▶ *Describe* the world pattern of net primary productivity.

The number of different species on Earth, our planet's **biodiversity**, reflects the interaction of living organisms and the nonliving components of the biosphere over time. The biosphere extends from below the ocean floor to an altitude of about 8 km (5 mi) into the atmosphere. The study of the relationships between organisms and their environment is **ecology**, which means "study of the place we live." **Biogeography** is the study of the distribution of plants and animals in the biosphere over space and time. The biosphere is composed of many *ecosystems*, which can range in size from a pond or city park to a large forest or desert. In this chapter we follow the flow of energy from the Sun through plants and animals and back to plants, as solar energy is converted into usable forms to energize life. Nonliving systems interact with living organisms and form a web that supports living ecosystems. Soil is an important medium for sustaining plant growth. Interwoven relationships among organisms as they consume or are consumed focus the energy and nutrients in forms useful to life. These relationships are the basis for community and species interactions. Over the span of the last 3.6+ billion years, evolution produced a vast biodiversity of living organisms. Understanding the resilience and stability of communities of plants and animals and how these living landscapes change over space and time is an essential concern of biogeography. Given the present state of global climate change, the challenge is to assess the impacts of these changes on living systems.

Ecosystem Structure & Function

An ecosystem is a self-sustaining association of living plants and animals and their nonliving physical environment (▶ Fig. 13.1). An ecosystem's **abiotic** components are environmental factors of the lithosphere, atmosphere, and hydrosphere, such as temperature and precipitation. The main abiotic component is solar energy, which powers nearly all ecosystems, except for a few ecosystems in caves or on the ocean floor. An ecosystem's **biotic** components are the organisms themselves. Different

▶ **13.1 Examples of ecosystems** Scientists classify the distinctive combinations of plants and animals found in different environments into ecosystems such as those shown here.

(a) Tropical savanna

(b) Temperate grassland

(c) Arctic tundra

(d) Temperate marsh

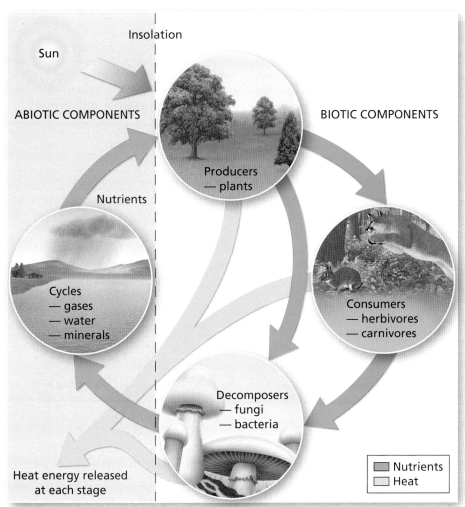

(b) Biotic and abiotic ingredients operate together to form this rain forest floor ecosystem in Costa Rica.

(c) Lichen and moss live in extreme Arctic climate conditions on rocks, Galindes Island, Antarctica.

(d) Brain coral off the island of St. Eustatius, Caribbean Netherlands, lives in a symbiotic relationship with algae.

(a) Solar energy input drives biotic and abiotic ecosystem processes. Heat energy and biomass are the outputs from the biosphere.

▲ **13.2 Essential ecosystem components** Nutrients flow from stage to stage in ecosystems, but heat flows in from the Sun and is released to the environment at each stage.

groups of organisms have particular roles in an ecosystem: for example, the ecosystem's plants, or **primary producers**, provide food for animals, or **consumers**, which in turn provide food for other consumers. **Decomposers** consume and recycle the remains of dead organisms. Figure 13.2 illustrates the essential parts of an ecosystem and how they work together.

Ecosystems are open systems with regard to both solar energy and matter. An ecosystem's abiotic processes include gaseous, hydrologic, and mineral cycles linked to larger Earth systems. As part of these cycles, the biotic processes of living organisms transfer matter and energy throughout the ecosystem.

geoCHECK ✔ Describe the roles of primary producers, consumers, and decomposers in ecosystems. Give an example of each from an ecosystem with which you are familiar.

13.1 (cont'd) Energy Flows & Nutrient Cycles

Energy in Ecosystems

The energy that powers the biosphere comes almost entirely from the Sun. Solar energy enters the energy flow of the biosphere by *photosynthesis*, the chemical process (described below) that occurs in the leaves of green plants. Plants convert about 1% of the energy that reaches Earth's surface into carbohydrates, which are the source of energy for the rest of the ecosystem.

Plants and algae link solar energy with the biosphere. Organisms that produce their own food by photosynthesis are primary producers. Nearly all the organisms in the biosphere, including humans, depend upon these organisms to convert light into food. The oxygen in our atmosphere is a by-product of photosynthesis. Some organisms in environments without access to solar energy are chemosynthetic, they generate energy by breaking down chemical compounds. Researchers found entire ecosystems around hot water vents at the bottom of the Pacific Ocean with *chemosynthetic* bacteria as their foundation.

The first organisms to produce oxygen appeared in the oceans about 2.7 billion years ago. These cyanobacteria helped create our modern atmosphere. Plants with veins to conduct fluids and roots to absorb nutrients and water are called vascular plants. In vascular plants, leaves are solar-powered chemical factories where photochemical reactions happen (▼Fig. 13.3). Veins in the leaf bring in water and nutrients,

carry off sugars produced by photosynthesis, and connect the stems and branches to the main circulation system.

Photosynthesis Powered by wavelengths of visible light, **photosynthesis** converts CO_2, water, and light into carbohydrates and oxygen.

$$H_2O + CO_2 + light \rightarrow C_6H_{12}O_6 + O_2$$

The light-sensitive structures in cells are chloroplasts, which contain the green pigment chlorophyll. Light stimulates the chlorophyll inside the chloroplasts and drives photosynthesis, which each year removes approximately 91 billion metric tons (100 billion tons) of CO_2 from the atmosphere and converts it into carbohydrates and oxygen. Competition for light is a major factor in the height, orientation, distribution, and structure of plants. Only about one-quarter of the light that hits leaves is useful. Chlorophyll only absorbs orange-red and violet-blue wavelengths and reflects mainly green light, which is why vegetation looks green. Plants store energy as sugar and other carbohydrates for later use, such as at night. They consume carbohydrates through respiration, which converts carbohydrates back into energy for other uses.

Respiration In *respiration*, plants break down carbohydrates with oxygen, releasing CO_2, water, and energy.

$$C_6H_{12}O_6 + O_2 \rightarrow H_2O + CO_2 + Energy$$

The difference between photosynthetic production of carbohydrates and the loss of carbohydrates through respiration is *net photosynthesis*.

geoCHECK ✔ Why are plants green?

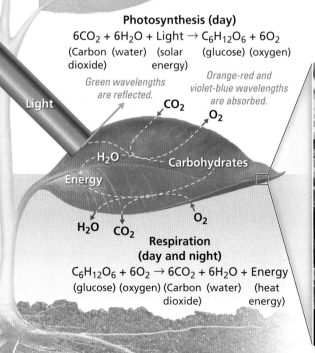

Photosynthesis (day)
$6CO_2 + 6H_2O + Light \rightarrow C_6H_{12}O_6 + 6O_2$
(Carbon (water) (solar (glucose) (oxygen)
dioxide) energy)

Green wavelengths are reflected.

Orange-red and violet-blue wavelengths are absorbed.

Light

CO_2

O_2

H_2O

Carbohydrates

Energy

H_2O CO_2

O_2

Respiration (day and night)
$C_6H_{12}O_6 + 6O_2 \rightarrow 6CO_2 + 6H_2O + Energy$
(glucose) (oxygen) (Carbon (water) (heat
dioxide) energy)

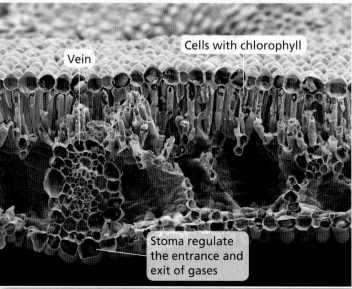

Vein

Cells with chlorophyll

Stoma regulate the entrance and exit of gases

Scanning electron microscope view of a leaf.

◄**13.3 Photosynthesis and respiration**
In photosynthesis, flows of CO_2, water, light, and oxygen enter and exit the surface of each leaf. Gases enter and exit a leaf through small pores, called stoma, which are more numerous on the underside of leaves. The plant can open and close the pores depending on its needs.

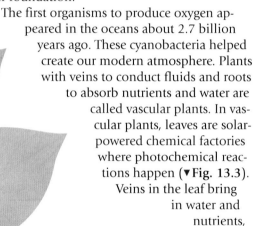

Net Primary Productivity

The net photosynthesis for an entire ecosystem is its **net primary productivity (NPP)**, the amount of energy captured and stored by an ecosystem. The total weight of the organic matter (plant and animal, living and dead) in an ecosystem is **biomass**. Net primary productivity determines the biomass available to animals, and the distribution of productivity is an important aspect of biogeography.

Net primary productivity is measured in carbon per square meter per year. Figure 13.4 shows that NPP tends to be highest between the tropics and decreases with higher latitudes and elevations. Productivity is tied to the main inputs of sunlight and rainfall, as shown by high productivity along the equator and low productivity in the subtropical deserts. NPP varies with the seasons, and is usually highest in high-sun months and lowest during low-sun months.

geoCHECK ✔ What is the initial source of the energy that powers the biosphere?

geoQUIZ

1. What are the main components of an ecosystem?
2. Explain the processes of photosynthesis and respiration.
3. Why does NPP vary seasonally in the mid- and high-latitude regions more than in the tropics?

Animation (MG)
Net Primary
Productivity

http://goo.gl/0YHzww

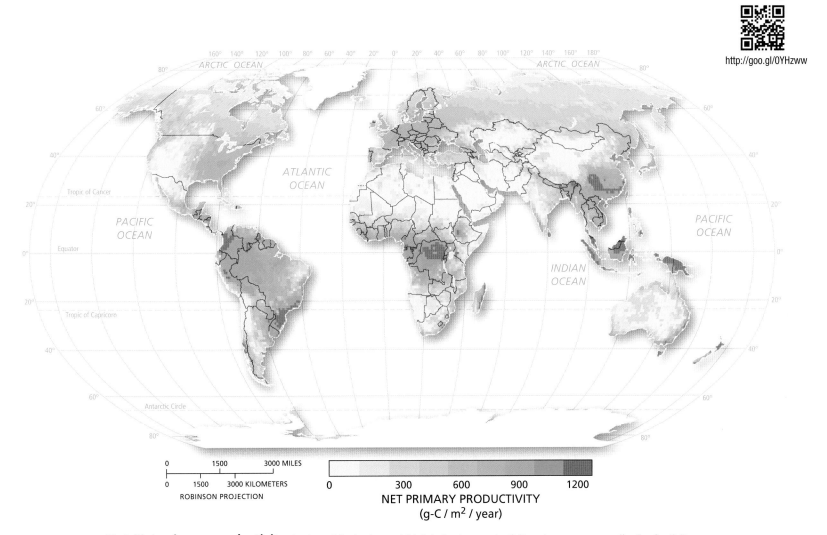

▲ **13.4 Net primary productivity** In the midlatitudes and high latitudes productivity rates vary seasonally. Productivity increases in spring and summer with more sun and precipitation, and decreases in winter. In the tropics, productivity is high throughout the year, due to constant amounts of sun and precipitation. In the oceans, nutrient levels affect productivity. Nutrient-rich upwelling currents along western coastlines are generally the most productive, while tropical oceans are quite low in productivity.

13.2 Biogeochemical Cycles

Key Learning Concepts

▶ *Summarize* the processes involved in the oxygen, carbon, and nitrogen cycles.

Many abiotic physical and chemical factors support the living organisms of each ecosystem. Abiotic factors include light, temperature, and water, as well as **nutrients,** chemicals essential for life. The Sun constantly supplies energy from outside the ecosystem, but most nutrients cycle within an ecosystem.

The most abundant elements in living matter are hydrogen, oxygen, and carbon. Together, these nutrients make up more than 99% of Earth's biomass. Nitrogen, calcium, potassium, magnesium, sulfur, and phosphorus are some other important nutrients.

These key chemicals flow through cycles called **biogeochemical cycles**. Oxygen, carbon, and nitrogen have gaseous cycles, parts of which take place in the atmosphere. The chemicals themselves cycle and recycle again and again in life processes.

Oxygen & Carbon Cycles

We consider the oxygen and carbon cycles together because they are so closely intertwined through photosynthesis and respiration (▼Fig. 13.5). Photosynthesis is the main source of atmospheric oxygen used by organisms. As you learned in Chapter 7, ocean water can absorb CO_2 from the atmosphere directly, causing ocean acidification. The ocean also absorbs CO_2 as a result of photosynthesis, and the carbon becomes part of living organisms, which can become carbonate minerals, such as limestone.

The atmosphere, which links carbon-fixing photosynthesis with carbon-releasing respiration, contains 800 billion metric tons of carbon as CO_2. This is far less carbon than is stored in fossil fuels and organic matter. CO_2 is released through respiration of plants and animals, volcanic activity, and humans' burning of fossil fuels.

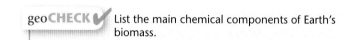 geoCHECK ✔ List the main chemical components of Earth's biomass.

▼13.5 **The oxygen and carbon cycles** Carbon is fixed through photosynthesis (orange arrows), with oxygen as a by-product. Respiration by living organisms, burning of fossil fuels, forests and grasslands release carbon to the atmosphere (blue arrows).

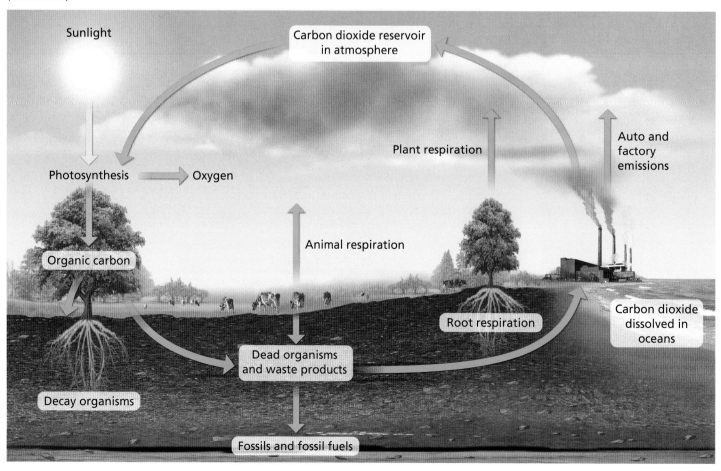

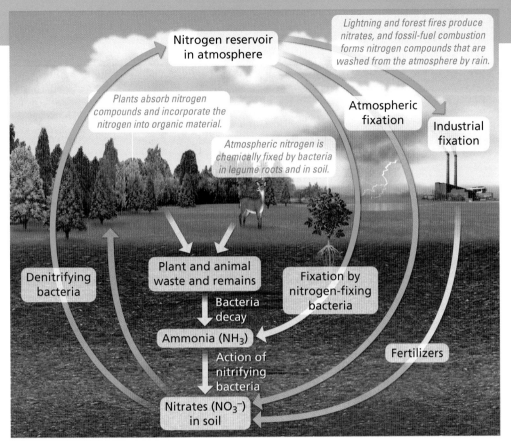

▲**13.6 The nitrogen cycle** Atmospheric nitrogen is fixed by bacteria, lightning, and industiral processes (yellow arrows). Fixed nitrogen is then added to soil (orange arrows), and then released back to the atmosphere by different bacteria (blue arrows).

Nitrogen Cycle

Nitrogen makes up over 78% of the atmosphere. Nitrogen is also an important component of organic molecules, especially proteins, and is essential to living processes. The nitrogen cycle is shown in Figure 13.6.

Nitrogen-fixing bacteria, which mainly live in the soil and the roots of certain plants, are crucial in taking nitrogen from the atmosphere and making it available to living organisms. Colonies of these bacteria live in nodules on the roots of legumes, such as peas, beans, clover, and peanuts, and chemically change the nitrogen from the air into nitrates and ammonia. Plants use these chemicals to produce their own organic matter. Consumers feed on the plants and incorporate the nitrogen into their tissues. Nitrogen in the waste of consumers is freed by other bacteria, which recycle it back to the atmosphere, beginning the cycle again.

Impact of Agriculture on the Nitrogen Cycle To improve crop yields, many farmers add synthetic inorganic fertilizers to their soil, rather than organic fertilizers that improve soil. The main process that creates synthetic nitrogen fertilizers uses large quantities of fossil fuels. It requires the equivalent of over 100,000,000 barrels of oil to produce the amount of nitrogen fertilizer used each year in the United States, 12,000,000 tons. Humans fix more atmospheric nitrogen to synthetic fertilizer than all the nitrogen in terrestrial sources.

The surplus of usable nitrogen accumulates in Earth's ecosystems. Excess nutrients from farmers overfertilizing their fields wash from soil into waterways and then into the ocean. The excess nitrogen begins a water pollution process that feeds algae and phytoplankton. When the excess algae and phytoplankton die, decomposers decrease the oxygen content of the water, creating *dead zones*.

The Mississippi River receives runoff from 41% of the area of the continental United States (▶**Fig. 13.7a**). It carries fertilizers, farm sewage, and other nitrogen-rich wastes to the Gulf of Mexico, causing huge spring blooms of phytoplankton. By summer, the decomposers feeding on the dead algae use so much oxygen that fish swimming into the low-oxygen areas are killed, resulting in a dead zone (▶**Fig. 13.7b**). Dead zones also occur in lakes.

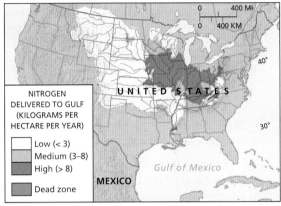

(a) Agricultural runoff from the Mississippi River watershed provides the nitrogen responsible for the Gulf of Mexico dead zone.

Bottom-water dissolved oxygen across the Louisiana shelf from July 22–28, 2013

(b) In 2013, the dead zone extended along much of the Louisiana coast.

▲**13.7 Dead zones** Nitrogen-rich runoff flows to the ocean, where the nitrogen nourishes phytoplankton blooms, producing the conditions that form a dead zone.

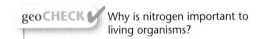

geoCHECK ✔ Why is nitrogen important to living organisms?

geo**QUIZ**

1. What is the main difference between the abiotic ecosystem components of solar energy and nutrients?

2. Explain how nitrogen gets from the atmosphere into living organisms.

3. What are dead zones? What factors create them? Why are they a concern?

13.3 Soil Development & Soil Profiles

Key Learning Concepts

▶ **List** the four major components of soil.
▶ **Describe** the five principal soil-formation factors.
▶ **Sequence** the horizons of a typical soil profile.

A mixture of weathered rock particles and organic material, **soil** covers most of Earth's land surface and is the medium in which plants grow. Soil is roughly 50% mineral and organic matter and 50% air and water. The organic matter is only about 5% of the mass of soil, but it is critical for soil function. Organic matter in soil includes living microorganisms and plant roots, dead and partially decomposed plant and animal matter, and fully decomposed plant matter called **humus**.

Soil is an open system with inputs of solar energy, water, rock and sediment, microorganisms, and outputs of plant ecosystems that sustain animal life and improve air and water quality. The five soil-forming factors are: parent material, climate, organisms, topography, and time. Soil is a good example of how Earth's spheres are related. Inputs from the atmosphere, hydrosphere, lithosphere, and biosphere must all be present and integrated to have fertile soil. Human activities, especially agriculture and ranching, also affect soil development. Soils are studied and classified using cross sections, usually extending from the surface down to bedrock.

Website (MG)
Soil Science
of America

http://goo.gl/mHHkcl

Animation (MG)
The Soil Moisture
(SMAP) Mission

http://goo.gl/h11xOB

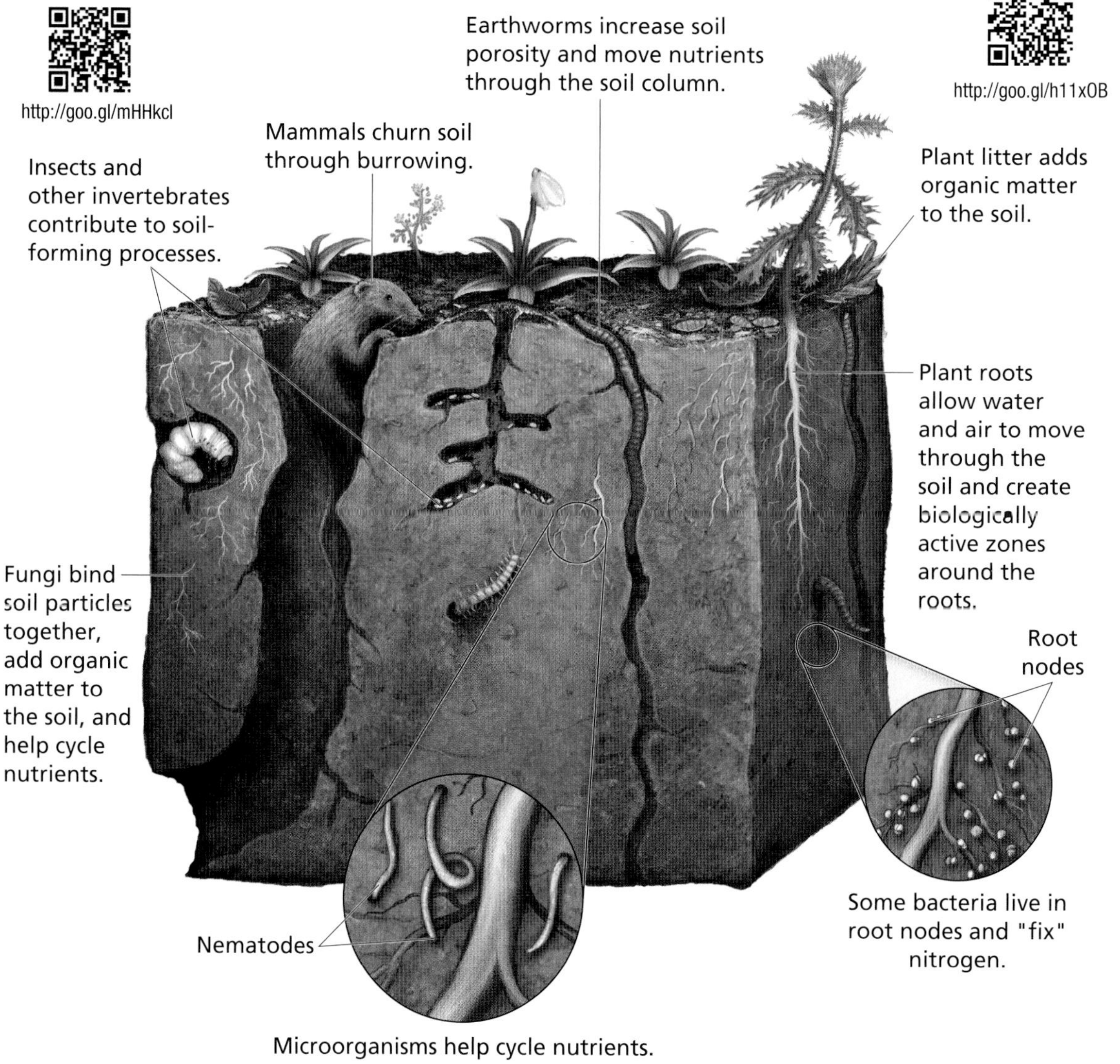

Insects and other invertebrates contribute to soil-forming processes.

Mammals churn soil through burrowing.

Earthworms increase soil porosity and move nutrients through the soil column.

Plant litter adds organic matter to the soil.

Plant roots allow water and air to move through the soil and create biologically active zones around the roots.

Fungi bind soil particles together, add organic matter to the soil, and help cycle nutrients.

Root nodes

Nematodes

Some bacteria live in root nodes and "fix" nitrogen.

Microorganisms help cycle nutrients.

▲**13.8 Soil organisms** Soil organisms range in size from land mammals that burrow into the ground, such as badgers, prairie dogs, and voles, to earthworms that ingest and secrete soil, to microscopic organisms that break down organic matter. The actions of these living organisms help maintain soil fertility.

Natural Soil-Formation Factors

As discussed in Chapter 9, weathering of rocks in the upper lithosphere provides the raw materials for soil formation. Bedrock, rock fragments, and sediments are the *parent material* and their texture and chemical make up help determine the resulting soil type. Climate is a major influence on soil development. The temperature and moisture characteristics of climates determine the chemical reactions, organic activity, and movement of water within soils. Past climates also leave their imprint on soils, often for thousands of years. For example, the heavily weathered soils formed from the Ione Formation in the Sierra Nevada are the result of tropical conditions that existed there approximately 60 million years ago.

Biological activity is an essential part of soil development. Vegetation, animals, and microorganisms contribute to the organic content of soil (◀Fig. 13.8). The chemical characteristics of vegetation and other life forms contribute to the soil's acidity or alkalinity. For example, broadleaf trees tend to increase alkalinity, while needleleaf trees tend to increase acidity.

Topography and slope affect soil formation as well. Gravity tends to remove soil from steep slopes. Level slopes typically have the thickest soils, but they may suffer from waterlogging. Slope orientation is important because it controls exposure to sunlight. In the Northern Hemisphere, north-facing slopes are cooler and provide more moisture than drier south-facing slopes (▼Fig. 13.9). More moisture means more vegetation, which acts to stabilize the soil and also adds organic matter to developing soil.

All of these factors require time to operate. The rate of soil development is related to the nature of the parent material (sediment makes soil more quickly than unbroken bedrock) and to climate (warm, moist climates develop soil more quickly). As plate tectonics has moved landmasses over geologic time, the surface features on those landmasses have experienced diverse conditions that affected soil-building processes.

geoCHECK ✔ What are the main components of soil?

Thicker soils develop on plateaus and in valleys

Thinner soils develop on steep slopes

Soils on north-facing slopes hold more moisture

Soils on south-facing slopes are drier

(a) Slope orientation affects the amount of sunlight a slope receives, and thus also affects soil moisture and vegetation.

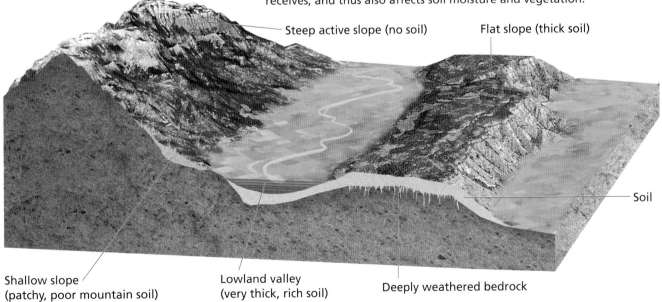

Steep active slope (no soil)

Flat slope (thick soil)

Soil

Shallow slope (patchy, poor mountain soil)

Lowland valley (very thick, rich soil)

Deeply weathered bedrock

(b) Deeper soils develop and accumulate on valley floors and flat plateaus.

▲13.9 **Slope orientation and soil moisture** The steepness of a slope and the slope's orientation both affect soil development.

13.3 (cont'd) Soil Development & Soil Profiles

Human Impacts on Soils

The soil-building process may take 500 years to create a few centimeters' thickness of soil. Because of poor farming practices, 35% of farmlands are losing soil more quickly than it is being formed (▶Fig. 13.10). Sustainable farming practices such as plowing on the contour and reducing plowing can dramatically reduce soil losses. More farmers are adopting *no-till agriculture,* which leaves crop stubble in the ground. No-till agriculture reduces moisture loss, soil erosion, and, because the soil is less compacted, encourages more vigorous root growth (▼Fig. 13.11). Poor farming practices in dry regions can result in the expansion of deserts, a process called *desertification.* Overgrazing, deforestation, and *salinization* (salts accumulate at the soil surface, discussed in Chapter 14) all contribute to desertification (▶Fig. 13.12). Global climate change is also contributing to desertification as temperature and precipitation patterns shift and the subtropical high-pressure cells strengthen.

geoCHECK ✔ What are the advantages of no-till agriculture?

▲ **13.10 Soil erosion** An example of soil loss through sheet and gully erosion on a northwest Iowa farm. One millimeter of soil lost from an acre weighs about 5 tons.

Mobile Field Trip (MG)
The Critical Zone

https://goo.gl/iJe1NM

(a) No-till farming leads to greater moisture retention and looser soil, allowing wheat to develop longer root systems that access moisture and nutrients farther below the surface

(b) Conventional farming, in which heavy tractors till (or plow) the soil to break it up, leads to soil compaction and shorter plant root systems.

▲ **13.11 No-till wheat and till wheat**

Soil Horizons

Scientists study soils using a **soil profile**, a vertical section of soil from the surface down to the deepest roots or bedrock. For soil classification, a three-dimensional soil profile called a **pedon** is used. A soil pedon is the smallest unit of soil that has all the characteristics used for classification. A soil profile is one side of a pedon (▶Fig. 13.13). When you look at a soil profile, you can see horizontal layers known as **soil horizons**. The horizons are roughly parallel to the surface and have different characteristics from the horizons above and below. The four main horizons are known as the O, A, B, and C horizons. The boundaries between horizons reflect differences between the layers of color, texture, structure, moisture, and other properties.

geoCHECK ✔ What characteristics distinguish soil horizons?

geoQUIZ

1. What are two ways that parent materials influence their soils?
2. What are two ways that biological activity influences soil development?
3. How do the A, E, and B horizons differ from the O, C, and R horizons?

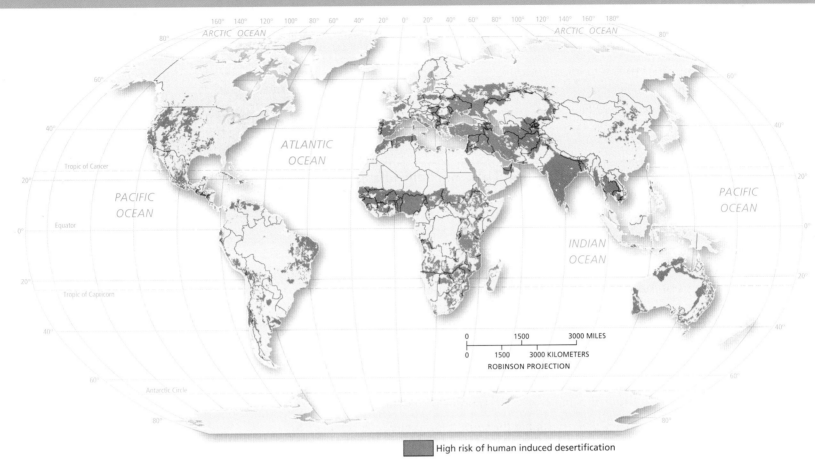

▲13.12 **Distribution of desertification** This map combines natural and anthropogenic factors that result in desertification.

Soil pedon

Soil horizons

Soil horizons

The A, E, and B horizons make up the true soil of the profile. The A, E, B, and C horizons extend below the O horizon to the R horizon, which is composed of sediment or bedrock. These middle layers are composed of sand, silt, clay, and other weathered by-products.

The R (rock) horizon is the bottom horizon.

O
A
E
B
C
R

True soil

The O horizon is made of organic material from living and former organisms. Plant and animal mattter from the surface is transformed into humus that is usually dark colored.

O and A — The A horizon is usually richer in organic content than lower horizons and is commonly called topsoil.

E ····· The E horizon is mainly made up of sand, silt, and leaching-resistant materials. Water carries clays, aluminum oxide, and iron oxide to lower horizons.

B ····· The B horizon is where clays, aluminum, and iron accumulate. The B horizon is often yellowish or reddish because of these deposited materials.

C ····· The C horizon is made of weathered bedrock and parent material.

(a) An idealized soil profile within a pedon.

Soil profile

(b) Profile of a well-drained soil with till as parental material (a Mollisol) in southeastern South Dakota. Carbonate nodules are visible in the lower B and upper C horizons.

▲13.13 **Soil profile** A typical soil profile within a pedon and example. [(b) Marbut Collection, Soil Science Society of America, Inc.]

13.4 Soil Characteristics

Key Learning Concepts

▶ **Describe** the physical properties used to classify soils: color, texture, structure, consistence, porosity, and soil moisture.

▶ **Explain** basic soil chemistry and relate this concept to soil fertility.

▲ **13.14 Soil colors**

The physical and chemical characteristics of soils not only differentiate them, but also affect soil fertility and resistance to erosion. **Soil fertility** is the ability of soil to grow plants. Farmers spend billions of dollars each year to create fertile soil conditions, but soil erosion increasingly threatens Earth's most fertile soils worldwide. While the following sections present the most widely applicable soil properties, the U.S. Department of Agriculture, Natural Resources Conservation Service (NRCS), *Soil Survey Manual* presents additional information on soil properties (http://soils.usda.gov/technical/manual/contents/chapter3.html).

Physical Properties

When you look at a soil profile, the physical properties that distinguish soils are readily observable: color, texture, structure, consistence, porosity, and moisture.

Color The chemical makeup of a soil determines the soil's color. Typical colors are the reds and yellows of iron oxides in the southeastern United States and the black richly organic soils of prairie soils (▶ Fig. 13.14). Colors can be deceptive: Soil with high amounts of organic matter is often dark, yet clays of the tropics with extremely low organic content are some of the world's blackest soils.

Texture The amounts of particles of different sizes in a soil gives the soil its texture—one of a soil's most permanent qualities. Scientists classify soil particles smaller than 2 mm (0.08 in.) as follows: *coarse sand* (2–0.5 mm), *silt* (down to 0.02 mm), and *clay* (under 0.02 mm).

▶ Figure 13.15 is a soil texture triangle showing the varying proportions of sand, silt, and clay in different soil types.

Loam is the designation for a balanced mixture of sand, silt, and clay. Farmers consider a sandy loam with clay content below 20% ideal because of its water-holding characteristics and ease of cultivation. Soil with too much clay retains too much water, is difficult to plow, and makes it more difficult for roots to grow. Soil texture is important in determining a soil's water-retention and water-transmission qualities.

Structure Structure refers to the size and shape of lumps of particles in the soil, not just particle

size, as is the case with texture. The shape of lumps determines the structural type of soils: crumb or granular, platy, blocky, or prismatic or columnar (▶ Fig. 13.16). Lumps have pores between them that are important for moisture storage and drainage. Rounded lumps have more space between them than do other shapes, which allows for better water permeability and plant growth.

Textural Analysis of Miami Silt Loam

Sample Points	% Sand	% Silt	% Clay
1 = A horizon	21.5	63.5	15.0
2 = B horizon	31.5	25.1	43.4
3 = C horizon	42.4	34.1	23.5

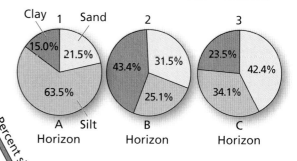

◀ **13.15 Soil texture triangle** The ratio of clay, silt, and sand determines soil texture. As an example, points 1 (horizon A), 2 (horizon B), and 3 (horizon C) designate samples taken from three different horizons in the Miami silt loam in Indiana. Note the ratios of sand to silt to clay shown in the three pie diagrams and table.

(a) Crumb or granular **(b)** Platy **(c)** Prismatic or columnar **(d)** Blocky

▲ **13.16 Soil structure** Structure is important because it controls drainage, rooting of plants, and how well the soil delivers nutrients to plants. The shape of individual peds, shown here, controls a soil's structure.

Consistence The term **consistence** describes the cohesion of soil particles. Consistence is a product of soil particle size and soil lump shape and reflects the soil's resistance to breaking under different moisture conditions. Typical consistence terms to describe dry soils are crumbly or brittle or hard, and for wetter soils they are sticky or moldable.

Porosity Pores in soil are spaces between soil particles that control the flow of water and air within soil. **Soil porosity** refers to the part of soil that is filled with air or water. Important factors are pore size, whether the pores are vertical or horizontal, and if they are located between or within lumps. Porosity is improved by plant roots and by tunneling animals such as worms. Much of the preparation work done before planting (such as tilling the soil) is to improve soil porosity so roots can penetrate the soil more easily.

Moisture Plants grow best when soil is at *field capacity*, which is the maximum water available for plant use after water has drained from the large pore spaces. Soil type determines field capacity.

geoCHECK ✔ Which soil property is the most permanent?

Chemical Properties

Soil chemistry involves soil particles, as well as the air and water in soil. The air in soil pores is similar to the air in the atmosphere, but with higher concentrations of carbon dioxide and lower concentrations of oxygen due to respiration processes in the ground. Water in soil pores is the medium for chemical reactions and a critical source of nutrients, which help create soil fertility.

Colloids The tiny particles of clay or organic matter in the soil solution are *soil colloids*. These colloids attract chemicals critical to plant growth (▼ **Fig. 13.17**). Without colloids, *leaching* would occur as water dissolves important chemicals from the soil, making them unavailable to plants. Organic colloids attract more nutrients than clay, but both organic and clay colloids attract more nutrients than silt and sand. Soil is fertile when it contains organic matter and clay that retains water and nutrients.

Acidity & Alkalinity Soils that are too acidic are not as fertile as more neutral soils, although each crop has its own optimal pH preference (▼ **Fig. 13.18**). If soil is too acidic, farmers may need to treat the soil by adding minerals that are alkaline.

geoCHECK ✔ What role do colloids play in soil fertility?

geoQUIZ

1. What are two possible reasons for black soil?
2. How does lump shape contribute to the qualities of a soil?
3. Explain how soil porosity affects soil qualities and plant growth.

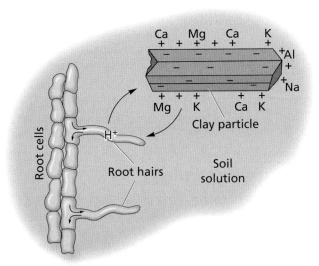

▲ **13.17 Soil colloids** Negatively charged colloids, such as this clay particle, hold on to positively charged mineral ions of calcium (Ca), potassium (K) until they are taken up by root hairs.

▼ **13.18 Soil pH** The pH scale measures acidity (lower pH) and alkalinity (higher pH). (The complete pH scale ranges between 0 and 14.)

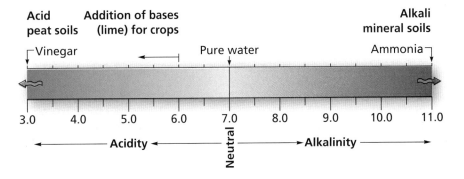

13.5 Energy Pathways

Key Learning Concepts

▶ ***Explain*** trophic relationships.
▶ ***Define*** the food chain, food web, and food pyramid concepts.
▶ ***Trace*** the flow of energy from primary producers to tertiary consumers.
▶ ***Analyze*** the cause and effects of biological amplification.

The feeding relationships between organisms make the energy pathways in an ecosystem. These trophic levels, or feeding levels, consist of food chains and food webs that range from simple to complex. A grassland food chain could consist of grasses that are eaten by mice, which are in turn eaten by coyotes, which will be decomposed by microorganisms in the soil.

Trophic Relationships

The producers in an ecosystem capture sunlight and convert it to chemical energy, fixing carbon, making new plant tissue, and liberating oxygen. Energy flows through an ecosystem from the producers along a pathway called a

food chain. Solar energy enters the system with the producers and flows to higher and higher levels of consumers. Organisms on the same feeding level are at the same **trophic level**. Food chains usually have between three and six levels, beginning with primary producers and ending with **detritivores**, which break down organic matter and free nutrients for recycling (▼Fig. 13.19).

The actual relationships between species in an ecosystem are more complex than the simple food chain model. Typically, an ecosystem's feeding relationships are made up of interrelated food chains called a **food web**. In a food web, consumers often are part of several different food chains. ▶Figure 13.20 shows a food web in the Antarctic that begins with microscopic algae. A consumer that feeds on both producers and consumers is an *omnivore*—a category that includes humans, among other animals.

Trophic Levels In a food web, who eats whom (or what) determines the trophic level:

- Producers represent the lowest trophic level.
- *Primary consumers* feed on producers. Producers are always plants, so primary consumers are herbivores.
- A *secondary* consumer mainly eats primary consumers and is therefore a carnivore.
- A *tertiary* consumer eats primary and secondary consumers and is referred to as the top carnivore of the food chain. The polar bear in the Arctic and the orca in the Antarctic are both examples of top carnivores.
- *Detritivores* and *decomposers* feed on *detritus*—dead organic matter produced by living organisms. Detritivores include worms, centipedes, snails, and slugs. Decomposers are primarily fungis and bacteria that also break down organic materials and release nutrients. Nutrient cycling is continuous within a food web, aided by detritivores and decomposers. Without detritivores and decomposers, nutrients would not be available to producers.

geoCHECK ✔ What was the top carnivore in your state 200 years ago? Today?

Energy Pyramids

The amount of energy moving through trophic levels decreases from lower to higher levels. You can see this pattern in an energy pyramid in which horizontal bars represent each trophic level (▶Fig. 13.21). At the bottom, are producers, which have the most energy and numbers of organisms, and usually the most biomass. The next level—primary consumers—has less energy available because energy is lost to metabolism and given off as heat as one organism eats another. Each higher trophic level has

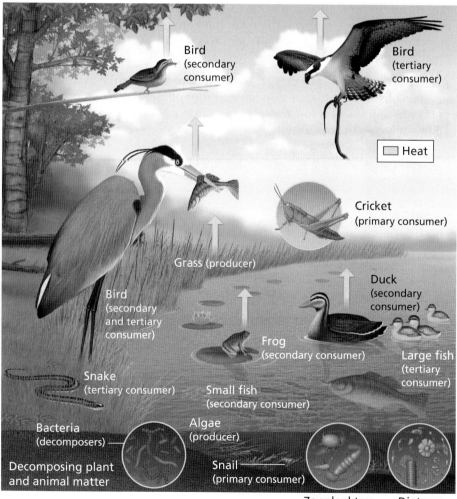

Bird (secondary consumer)

Bird (tertiary consumer)

☐ Heat

Cricket (primary consumer)

Grass (producer)

Duck (secondary consumer)

Bird (secondary and tertiary consumer)

Frog (secondary consumer)

Large fish (tertiary consumer)

Snake (tertiary consumer)

Small fish (secondary consumer)

Bacteria (decomposers)

Algae (producer)

Decomposing plant and animal matter

Snail (primary consumer)

Zooplankton (primary consumers)

Diatoms (producers)

▲13.19 **Trophic levels** The flow of energy, cycling of nutrients, and trophic (feeding) relationships portrayed for a generalized ecosystem. The operation is fueled by radiant energy supplied by sunlight and first captured by the plants.

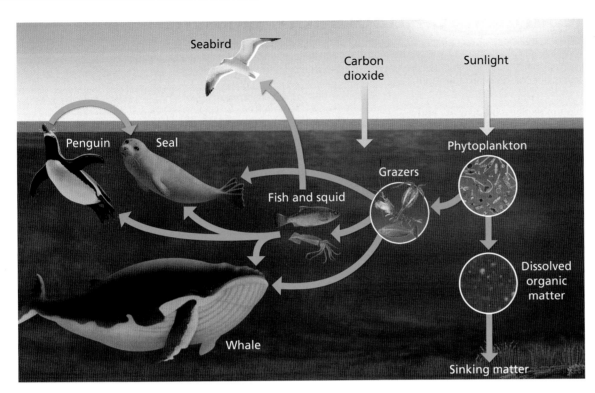

◄13.20 **13.20 Antarctic food web** Phytoplankton, the producers, use solar energy for photosynthesis. Krill and other herbivorous zooplankton eat the phytoplankton and are then consumed by organisms at the next trophic level.

less energy available, fewer organisms, and usually less biomass than the level underneath it.

Food Web Inefficiency Only 10% of the energy of each level is passed on to the level above it, because 90% of the energy is lost to heat. The most efficient use of resources occurs at the lower levels of the food pyramid by primary consumers. This concept applies to human eating habits and world food resources. If humans eat as carnivores, 90% of the food energy is lost to the primary consumers. The same amount of grain that would feed 1000 people as primary consumers would feed 100 people as secondary consumers. Today, roughly half of the cultivated land in the United States is planted for animal consumption. Much of our grain production goes to livestock feed rather than directly to human consumption.

geoCHECK ✔ Why is energy lost at each trophic level?

Biological Amplification

When chemical pesticides are applied to an ecosystem, the food web concentrates these chemicals—which are long-lived, stable, and soluble—in the fatty tissues of consumers. Pesticides become increasingly concentrated at each trophic level by the process of **biological amplification**. A famous example of this was the buildup of the pesticide DDT in birds. DDT had an especially serious effect on bald eagles, brown pelicans, and peregrine falcons. In these species, DDT caused a thinning of eggshells and increased hatchling deaths. The polar bears in the Barents Sea off of northern Europe have some of the highest levels of pollutants of any animal in the world, despite their distance from civilization. Many species are at risk of ingesting concentrated chemicals in this manner, including humans.

geoCHECK ✔ Summarize the process of biological amplification.

▼ **13.21 Energy and biomass pyramids** The higher the trophic level, the less energy is available to support the biomass of organisms at that level.

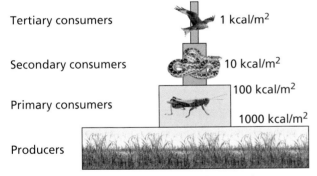

Tertiary consumers — 1 kcal/m²

Secondary consumers — 10 kcal/m²

— 100 kcal/m²

Primary consumers

— 1000 kcal/m²

Producers

(a) A pyramid shape illustrates the decrease in energy between lower and higher trophic levels. Kilocalorie amounts are idealized to show the general trend of the energy decrease.

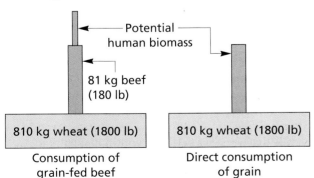

Potential human biomass

81 kg beef (180 lb)

810 kg wheat (1800 lb) | 810 kg wheat (1800 lb)

Consumption of grain-fed beef | Direct consumption of grain

(b) Biomass pyramids illustrate the difference in efficiency between direct and indirect consumption of grain.

geoQUIZ

1. How are detritivores and decomposers different? How are they similar?
2. What are two advantages of eating lower on the food chain?
3. Trace the amount of energy lost in a four-level food pyramid. How much of the initial energy makes it to the top level?
4. How would pesticide levels increase across four trophic levels if there was a 10-fold increase with each trophic level?

13.6 Communities: Habitat, Adaptation, & Niche

Key Learning Concepts

▶ *Relate* communities to the other levels of ecological organization in the biosphere.

▶ *Summarize* how the theory of evolution explains biodiversity and the adaptations of individual species.

▶ *Explain* the concept of the ecological niche.

The levels of organization within ecology and biogeography range from the biosphere, and all life on Earth, down to single living organisms. The biosphere is made up of ecosystems, which are made up of **communities**, which are interacting populations of plants and animals in a particular place. For example, in a forest ecosystem, one community may exist on the forest floor, while another community may function in the canopy of leaves high above. Similarly, within a lake ecosystem, the plants and animals living in the bottom sediments form one community, while those near the surface form another.

Each species has a **habitat**, the home of an organism. A habitat includes both the biotic and abiotic elements of the environment, and habitat size and character vary with each "species needs". For example, great blue herons are large wading birds that occupy shoreline habitats (river banks, freshwater marshes) throughout North America. They frequently nest in the tops of trees near their preferred feeding areas, keeping their young safe from predators on the ground (▶Fig. 13.22).

Evolution & Adaptation

The theory of evolution explains the origin of Earth's biodiversity. **Evolution** is the process in which the first single-celled organisms changed and diversified over time, eventually producing all the world's millions of species of organisms (▶Fig. 13.23). A **species** is a population of organisms that can reproduce sexually and produce fertile offspring.

British scientist Charles Darwin developed the theory of evolution in the mid-19th century to explain how a new species could evolve from an earlier species with different characteristics. In developing his theory, Darwin focused on a few key features of living things:

- More organisms are born than will survive.
- These numerous offspring are not identical, but inherit variations in a range of traits.
- Organisms engage in competition for survival.
- Those organisms that are better equipped to survive, reproduce and pass their genes on to the next generation.

Darwin reasoned that the successful organisms possessed traits, or **adaptations**, that enabled them to survive and reproduce in their environment. He theorized that through a process of **natural**

▲ **13.22 Aquatic bird habitat** Pelicans, cormorants, and herons perch on submerged trees at Prek Toal Bird Sanctuary on Tonlé Sap Lake in Cambodia.

selection, the organisms best adapted to an environment survived, passing on their favorable traits to their descendants. Over time, such traits become more frequent in a population, until the population is distinct from the species from which it arose. The result is a new species that is reproductively isolated from other species. Species that are unable to out-compete other species become extinct.

Today, evolution is understood in terms of genetics and DNA. Mutations in the DNA of organisms create new adaptations that are either beneficial or harmful to an organism's survival. Geography also affects natural selection. Species may disperse to new habitats where their traits are helpful in survival. Species may also have their habitat fragmented or reduced in size by human impacts, changes in sea level, warming or cooling temperatures, or by continental drift. If a population of organisms is split, the populations will continue to evolve, and over time the two groups may become different species.

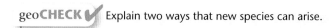 **geoCHECK** Explain two ways that new species can arise.

The Niche Concept

An ecological **niche** (from the French *nicher*, "to nest") is the "job" of an organism. A niche is determined by the physical and biological needs of the organism and is different from the concept of habitat. Many species can occupy the same habitat, but a niche is the unique role that a species performs within that habitat. An example is the white-breasted nuthatch and the brown creeper, two species of small bird that occur throughout North America in

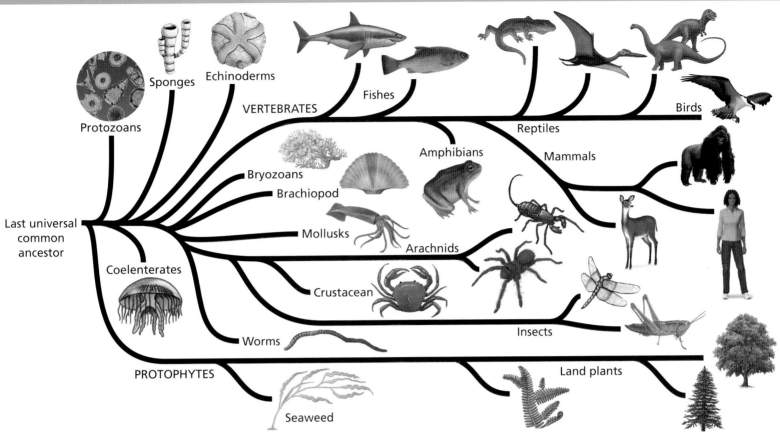

▲ **13.23 Brief history of life** This diagram shows the evolutionary relationships between species as they evolved from single-celled organisms over time. Note the dramatic decreases in biodiversity associated with mass extinction events, shown by the curved dotted lines, followed by the rise of new species.

forest habitats. The brown creeper and the white-breasted nuthatch both forage for insects on tree trunks, probing into the bark with their sharp bills. However, the white-breasted nuthatch travels up the trunks of trees while foraging, while the white-breasted nuthatch has feet that allow it to travel down the trunks of trees (▼ **Fig. 13.24**). This allows nuthatches to

occupy a different ecological niche and find insects overlooked by brown creepers. They also jam nuts and acorns into the bark and use their bills to extract the seeds. Although both nuthatches and brown creepers occupy the same habitat, the nuthatch's behavior places it in a different niche. The principle of *competitive exclusion* states that two different species will not occupy the same niche because one species will always out compete the other. Each species acts to reduce competition and maximize its reproductive success because species survival depends upon successful reproduction. This strategy leads to greater diversity as species shift and adapt to fill different niches— or become *extinct* if they fail to adapt to competition or environmental change.

▼ **13.24 White-breasted nuthatch and brown creeper**
These two birds occupy the same habitat, but exploit different niches.

(a) White-breasted nuthatch   **(b) Brown creeper**

geoCHECK ✔ Briefly describe your niche and your habitat.

geoQUIZ

1. Describe how a forest with tall trees could have different communities at different heights. How are environmental conditions different at different heights, from the forest floor to the tops of the tallest trees?
2. How do the white-breasted nuthatch and brown creeper exemplify the competitive exclusion principle?

13.7 Species Interactions & Distributions

Key Learning Concepts

▶ **Compare and contrast** the different forms of interaction between species.

▶ **List** several limiting factors on species distributions.

Living things compete for resources and space in the physical environment. In the process, they interact with other organisms (biotic factors) and with the environment (abiotic factors) in ways that influence their spatial distribution.

Species Interactions

Species interact in a wide variety of ways. You are probably familiar with predator–prey interactions, as when an owl catches and eats small rodents. Some species interact through **symbiosis**, a relationship in which two species are associated in a way that may benefit at least one of them. One type of symbiosis, **mutualism** occurs when each organism benefits from the relationship. For example, lichen is made up of algae and fungi living together (▶Fig. 13.25a). The fungi provides habitat, and the algae provides food. Their mutualism allows them to survive together in a niche where each one couldn't survive alone. Lichen evolved from a parasitic relationship in which fungi invaded algal cells. Today, they have a supportive and harmonious relationship.

Another form of symbiosis is **parasitism**, in which one species benefits and another is harmed by the relationship. This relationship often takes the form of a parasite living off a host organism, such as mistletoe growing on trees (▶Fig. 13.25b). A third form of symbiosis is **commensalism**, in which one species benefits, and the other species is neither helped nor harmed. Epiphytic plants, such as orchids and "air plants," grow on the branches of trees, using them for physical support (▶Fig. 13.25c).

geoCHECK ✔ Explain how mutualism and commensalism are similar and different.

Abiotic Influences on Distribution

A number of abiotic factors work together to determine the distributions of species, communities, and entire ecosystems. These factors include air and soil temperatures, precipitation and water availability, and water quality. Even day length can affect a species' distribution. For example, poinsettia needs at least 2 months of 14-hour nights to start flowering. Poinsettia cannot survive in equatorial regions with little day length variation and are restricted to latitudes with appropriate day lengths.

Early work in the study of species distribution was done by geographer and explorer Alexander von Humboldt (1769–1859), the first scientist to write about the zones of plant communities that change with elevation. After studying the Andes Mountains of Peru, von Humboldt hypothesized that plants and animals occur in characteristic groupings wherever similar climatic conditions occur. His ideas were the basis of the **life zone** concept, which describes the zonation of plants and animals at different elevations (▶Fig. 13.26). Each life zone possesses its own temperature and precipitation regime and its own biotic communities.

▼ **13.25 Three types of symbiosis**

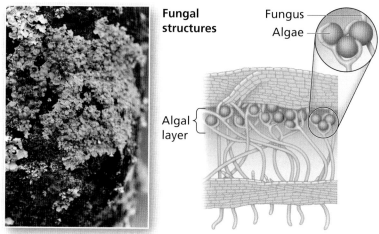

(a) Mutualism Both the fungus and the alga that make up the lichen benefit.

(b) Parasitism The mistletoe benefits, but the tree is harmed.

(c) Commensalism The orchid benefits, but the tree is unharmed.

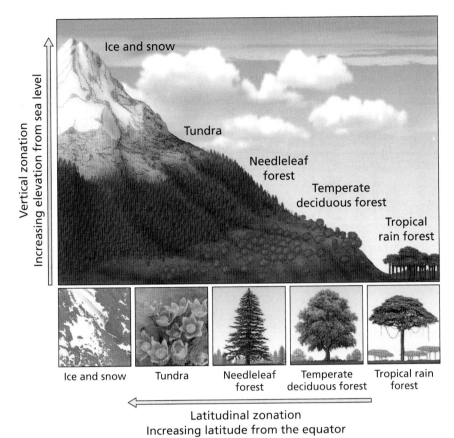

▲ **13.26 Life zones** Progression of plant community life zones with elevation or latitude.

Life Zones & Climate Change In the 1890s, C. Hart Merriam mapped 12 life zones in the mountains of northern Arizona. Merriam noticed that the pattern of life zones at progressively higher altitudes was similar to the sequence of life zones moving from lower to higher latitudes. For example, the alpine zone on an Arizona mountaintop had small, hardy plants similar to those of the tundra in northern Canada.

Recent studies show that climate change is causing plants and animals to move their ranges to higher elevations with more suitable climates. Some mountain species have run out of higher elevations, forcing them to migrate elsewhere or go extinct.

geoCHECK ✔ Give an example of how climate change could increase the range of some organisms while reducing the range of other organisms.

Limiting Factors Affecting Distribution

Physical, chemical, or biological factors that determine species distributions and population size are called **limiting factors**. For example, in some ecosystems, precipitation is a limiting factor on plant growth, either by too little or too much rainfall. Temperature, light levels, and soil nutrients all affect vegetation patterns and abundance:

- Low temperatures limit plant growth at high elevations.
- Lack of water limits growth in a desert; too much water limits growth in a bog.
- Lack of active chlorophyll above 6100 m (20,000 ft) limits plant growth.

For animals, limiting factors may be the number of predators, availability of food and habitat, availability of breeding sites, and disease. The snail kite, a tropical predatory bird with a small habitat in the Florida Everglades, is a *specialist* that feeds on only one type of snail. In contrast, the mallard duck is a *generalist*, eats a wide variety of food, and is found throughout most of North America.

For some species, one critical limiting factor determines their range. For other species, a combination of factors restricts their range, with no one factor being more important. Each organism has a range of tolerance for environmental characteristics. Within that range, the numbers of that species are high; at the edges of the range, the species is more rare; and outside the range limits, the species is absent. For example, the coast redwood is abundant in a narrow range along the California and Oregon coast where there is enough summer fog to provide the trees' water needs during the summer drought period. Redwoods at the limits of their range are shorter and smaller. In contrast, the red maple tree has a wide tolerance range and is distributed over a large area (▼ **Fig. 13.27**).

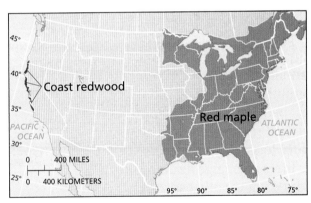

▲ **13.27 Ranges of red maples and redwoods** The coast redwood is limited by its need for fog as a moisture source; the red maple tolerates a variety of environmental conditions.

geoCHECK ✔ Describe an example of a limiting factor that affects animals.

geoQUIZ

1. What type of relationship do humans have with the biosphere? Mutualism? Parasitism? Commensalism? Explain your answer.
2. Why would higher elevations and higher latitudes have similar shifts in biotic communities?
3. What are the main limiting factors for plants where you live? Describe how those factors affect species numbers and distributions.
4. Recent studies show the California and Oregon coasts are having fewer foggy days. Explain how this could affect the distribution of coast redwood trees.

13.8 Ecosystem Disturbance & Succession

Key Learning Concepts

▶ *Discuss* wildfires as a natural hazard and disturbance factor in ecosystems.

▶ *Outline* the stages of ecological succession in both terrestrial and aquatic ecosystems.

Over time, ecosystems experience disturbances such as windstorms, flooding, wildfires, volcanic eruptions, logging of forests, and overgrazing of rangeland. These events damage or remove organisms, making way for different communities.

When an ecosystem is disturbed enough that most or all of its species are removed, the process of **ecological succession** begins, in which the cleared area undergoes changes in species composition as newer communities replace older ones, modifying the physical environment in a manner that favors a different community. During these transitions, species having an adaptive advantage, such as the ability to produce lots of seeds or to disperse them over large distances, will out compete other species for space, light, water, and nutrients. Succession occurs in both terrestrial and aquatic ecosystems.

Wildfire & Fire Ecology

Fire is one of Earth's natural hazards, and the economic impact from wildfires is increasing as more homes are built in areas that are becoming more fire-prone due to climate change. Wildfires burned a record 10 million acres in 2015, over 9 million acres in 2006, 2007, and 2012, and cost $2 billion in the United States in 2012. Scientists have linked record wildfires in the American West since 2000 to increased spring and summer temperatures and an earlier spring snowmelt. Perhaps more important is the fact that these climatic changes are occurring after 150 years or more of fire

(a) Fire-suppressed forest

In a forest where fires rarely happen, fuel builds up, such as: **surface fuel** (grass, logs, woody debris, brush); **ladder fuel** (shrubs, small trees, snags); and **tree crowns**.

1 Surface fires spread quickly through brush and woody debris.

2 Ladder fuels allow the fire to move up toward the forest canopy.

3 Tree crown fires are so intense, they're difficult to control.

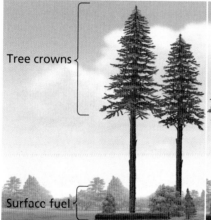

(b) Forests where fires are frequent

The greater the distance between **surface fuel** (grass, logs, twigs, fallen branches and low-lying foliage) and **tree crowns**, the more difficult it is for crown fires to start.

1 Periodic fires spread through surface fuel.

2 The surface fire cannot make the leap to the tree crowns.

3 The fire consumes small plants, but taller trees escape with scorched bark.

▲13.28 **Ground fires and canopy fires** Wildfires are a natural part of many ecosystems. In forests where fires are infrequent, fuel accumulates, leading to larger and more destructive fires. When fires are more frequent, fuel levels are lower and fires are smaller and less destructive.

Mobile Field Trip (MG)
Forest Fires in the West

https://goo.gl/vm6c9q

suppression across the United States, which has resulted in a buildup of undergrowth as fuel. Lightning-caused wildfire is a natural disturbance, with a dynamic role in community succession. Many ecosystems are *fire adapted*, meaning that plants in the ecosystem have traits that enable them to survive periodic fires. Over time, fire in these ecosystems creates a mosaic of different habitats, ranging

▲ **13.29 Primary succession** Plants establishing on recently cooled lava flows from the Kilauea volcano in Hawaii illustrate primary succession.

from totally burned, to slightly burned, to unburned. This patchwork increases the number of different habitats and prevents small fires from spreading to become catastrophic large fires. Small ground fires usually burn shrubs, young trees, and dead limbs on the ground, creating a healthier forest. Ground fires usually do not harm mature trees and are important in releasing nutrients back to the soil and thinning out small trees (◄Fig. 13.28a).

If fires are suppressed, however, dead limbs and trees build up, creating ladder fuel that allows ground fires to reach the canopy of trees. Canopy fires are hotter and more catastrophic and kill mature trees, small trees, shrubs, and dormant seeds in the soil. The hotter, larger canopy fires can also sterilize the soil by killing microorganisms and fungi and render it water-repellent, which leads to erosion and the loss of soil (◄Fig. 13.28b).

Species in ecosystems that experience wildfires range from fire-adapted to fire-dependent. Plants in grasslands, forests, and scrubland have evolved with fire and often have traits such as thick bark, a lack of lower branches to protect against ground fires, and energy stored in roots so they can resprout after a fire, which allows them to thrive in areas with wildfires.

Fire management policy has shifted away from fire prevention and efforts to immediately extinguish wildfires. Instead, land managers follow the lessons of fire ecology, using prescribed burns and wildfires to recreate a patchwork of habitats and reduce the likelihood of catastrophic wildfires.

geoCHECK ✔ How does wildfire increase the diversity of habitats?

Terrestrial Succession

An area of bare rock or a disturbed site without soil can be a site for **primary succession**, the beginning stage of an ecosystem. Primary succession can occur on new surfaces created by mass movement, glacial retreat, volcanic eruptions, surface mining, or clear-cut logging. Primary succession begins with the arrival of organisms that are adapted for colonizing new land, forming a *pioneer community*. For example, a pioneer community of lichens, mosses, and ferns may form on a new lava flow or bare rock (◄ **Fig. 13.29**). These early inhabitants prepare the way for future succession: Lichens break down rock, which begins the process of soil formation. As new organisms colonize soil surfaces, they bring nutrients that further change the habitat, eventually leading to the growth of grasses, shrubs, and trees.

More common is **secondary succession**, which occurs when soil, and possibly other parts of a former community, are still present, such as a disturbed area with the soil still intact. As secondary succession begins, new plants and animals having niches that differ from the previous community will colonize the area, and the groups of species will shift as soil develops and habitats and the communities change (▼Fig. 13.30).

▼ **13.30 Secondary succession after a fire** In fire-adapted ecosystems, a fire can stimulate plant growth by releasing nutrients into the soil, clearing space for new plants to establish themselves, and helping certain types of seeds to germinate.

(a) Immediately after fire

(b) One year later

13.8 (cont'd) Ecosystem Disturbance & Succession

Secondary Succession on Mount St. Helens Most of the areas affected by the eruption of Mount St. Helens in 1980 underwent secondary succession (▼Fig. 13.31). Some soils, young trees, and plants were protected from the blast, and community development began almost immediately after the event. The areas with soil and vegetation completely destroyed by the initial blast and landslide underwent primary succession.

Previously, communities were thought to pass through several succession stages, eventually reaching a mature state with a predictable "climax community"—a stable, self-sustaining assemblage of species that would exist until the next major disturbance. However, modern biogeography and ecology assume that disturbances constantly disrupt the successional sequence so communities may never reach a climax stage. Mature communities are in a state of constant adaptation—a dynamic equilibrium—to environmental changes. Natural and human-caused disturbances often create a patchwork of different habitats in different stages of succession.

The concept of *patch dynamics* refers to interactions between and within this mosaic of complex habitats. The overall biodiversity of an ecosystem is partially the result of such patch dynamics.

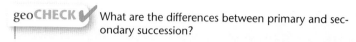 geoCHECK ✔ What are the differences between primary and secondary succession?

Aquatic Succession

Aquatic ecosystems occur in lakes, wetlands, and along shorelines, and the communities in these systems also undergo succession. For example, lakes and ponds exhibit successional stages as they fill with sediment and as aquatic plants take root and grow. The plant growth captures sediment and adds organic debris to the system (▶Fig. 13.32). This gradual enrichment in water bodies is known as **eutrophication** (Greek for "well nourished").

(a) 1980

◀13.31 **Secondary succession** After the eruption of Mount St. Helens, many locations had enough soil and vegetation for secondary succession to begin.

(b) Present

When viewed across geologic time, lakes and ponds are temporary features on the landscape that will fill with sediment over time. In most climates, a lake will develop a mat of floating vegetation that grows outward from the shore to form a bog. Cattails and other marsh plants become established, and decomposing organic matter accumulates in the basin, with additional vegetation bordering the lake shore. Vegetation and soil may fill in the marsh, creating a meadow, and willow or cottonwood trees may follow. Eventually the lake may evolve into a forest community. This process also makes oxbow lakes into meander scars.

geoCHECK ✔ Why are lakes and ponds temporary features?

geoQUIZ

1. Describe the process of primary succession on a new lava flow from bare rock to later communities.
2. How are terrestrial and aquatic succession similar and different?
3. Why do communities not evolve to a climax communities?

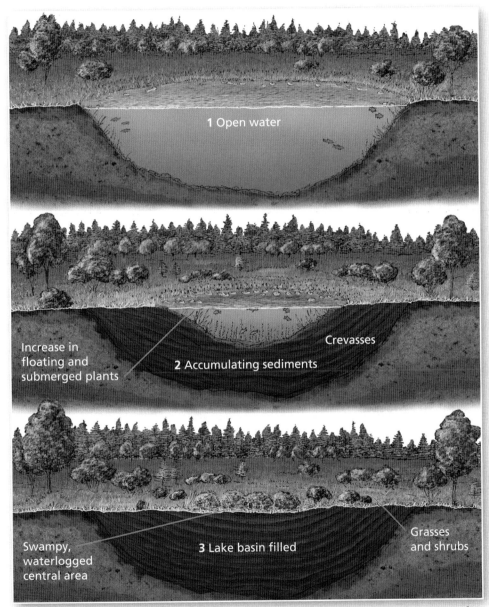

1 Open water

Increase in floating and submerged plants

Crevasses

2 Accumulating sediments

Swampy, waterlogged central area

3 Lake basin filled

Grasses and shrubs

(a) A lake gradually fills with organic and inorganic sediments, shrinking the area of open water. A bog forms, then a marsh, and finally a meadow, the last of the successional stages.

(b) Spring Mill Lake, Indiana

(c) Organic content increases as succession progresses in a western N.A. mountain lake.

(d) Peat bog with acidic soils, Richmond Nature Park, near Vancouver, British Columbia.

▲13.32 Aquatic succession In temperate climates, lake–bog–meadow succession could follow the idealized steps shown here, with examples of lakes at different stages of the process.

13.9 Ecosystem Stability & Biodiversity

Key Learning Concepts

▶ *Analyze* factors that contribute to the stability or resilience of ecosystems affected by disturbances.

▶ *Summarize* current threats to biodiversity and efforts to restore ecosystems.

Earth's ecosystems have been dynamic, and ever-changing from the beginning of life on the planet. Over time, communities of plants and animals have adapted and evolved to produce great diversity. Each ecosystem is constantly adjusting to changing conditions and disturbances. Ironically, the concept of change is key to understanding ecosystem "stability," or maintenance of dynamic equilibrium. A critical aspect of ecosystem stability is biodiversity, or the amount of biological diversity of an ecosystem.

Ecosystem Stability & Resilience

A stable ecosystem such as a redwood forest may be constantly changing, but not diverging greatly from its original state. **Resilience** is the ability of an ecosystem to recover from a disturbance. Higher biodiversity results in greater resilience. Studies of the prairie ecosystems of Minnesota confirmed that areas of prairie with higher biodiversity were more productive and more resilient than areas of prairie with lower biodiversity (▼ Fig. 13.33a). Ecosystems can also be stable, but not resilient. A tropical rain forest community is diverse and stable, but not resilient to large changes. A cleared tract of rain forest will regrow slowly, because nutrients are stored in the vegetation, not the soil (▼ Fig. 13.33b).

Agricultural systems are artificially created monoculture (one crop) communities that are neither stable nor resilient. Because they are kept in an arrested stage of succession, they are vulnerable to disturbances by invasive plants or insects and must be protected with weed killers and pesticides.

geoCHECK ✔ Briefly explain the differences between stability and resilience.

(a) A resilient ecosystem, such as the prairie grasslands, remains relatively stable in spite of disturbance—in this case, fire. After the fire, the grasses grow back vigorously.

◀ **13.33 Stability and resilience**
Ecosystems respond to disturbance in different ways.

Before disturbance (fire)

After disturbance

(b) A stable ecosystem that is not resilient, like this rain forest disturbed by logging, may not recover from disturbance for a very long time, if ever.

Before disturbance (fire)

After disturbance

Biodiversity on the Decline

Human activities have a great impact on global biodiversity. The present loss of species is irreversible and accelerating. We are facing a loss of genetic diversity that may be unprecedented in Earth's history, even compared with the six major mass extinction events (see Fig. 8.1, the geologic time scale) of the last 500 million years. The current wave of extinctions is the only mass extinction caused by living organisms. According to a 2014 study, the rate of species loss today is 1000 times faster than natural pre-human rates. A World Wildlife Fund study found that there are half as many wild animals alive today as were alive in 1970.

Scientists have identified 1.75 million species out of an estimated 13.6 million different species of plants, animals, fungi, and other organisms (**Table 13.1**) Globally, several areas stand out as having evolved ecosystems that are especially rich in biodiversity (▼**Fig. 13.34**).

Human Factors Affecting Biodiversity Estimates of species loss range from 1000 to 30,000 species, but the higher estimate may be too conservative. Harvard University biologist and founder of the Encyclopedia of Life website, E. O. Wilson, has calculated that at the present rate of extinction, half of Earth's present species will be extinct within 100 years. The total number of animals with backbones, not the number of different species, has been cut in half since 1970, a shocking statistic. Five categories of human impact are the greatest threat to biodiversity:

Table 13.1 Known and Estimated Species on Earth

Categories of Living Organisms	Number of Known Species	Estimated Number of Species High	Estimated Number of Species Low	Working Estimate	Accuracy
Viruses	4,000	1,000,000	50,000	400,000	Very poor
Bacteria	4,000	3,000,000	50,000	1,000,000	Very poor
Fungi	72,000	27,000,000	200,000	1,500,000	Moderate
Protozoa	40,000	200,000	60,000	200,000	Very poor
Algae	40,000	1,000,000	150,000	400,000	Very poor
Plants	270,000	500,000	300,000	320,000	Good
Nematodes	25,000	1,000,000	100,000	400,000	Poor
Crustaceans	40,000	200,000	75,000	150,000	Moderate
Arachnids	75,000	1,000,000	300,000	750,000	Moderate
Insects	950,000	100,000,000	2,000,000	8,000,000	Moderate
Mollusks	70,000	200,000	100,000	200,000	Moderate
Chordates	45,000	55,000	50,000	50,000	Good
Others	115,000	800,000	200,000	250,000	Moderate
Total	1,750,000	111,655,000	3,635,000	13,620,000	Very poor

- Habitat loss, degradation, and fragmentation as natural areas are converted to agricultural and urban uses
- Air, water, and soil pollution
- Unsustainable resource use and harvesting of plants and animals
- Human-induced climate change
- Human-introduced invasive species

▶ **13.34 Species richness (or biodiversity)** How does this map compare with the map of net primary productivity in Figure 13.4?

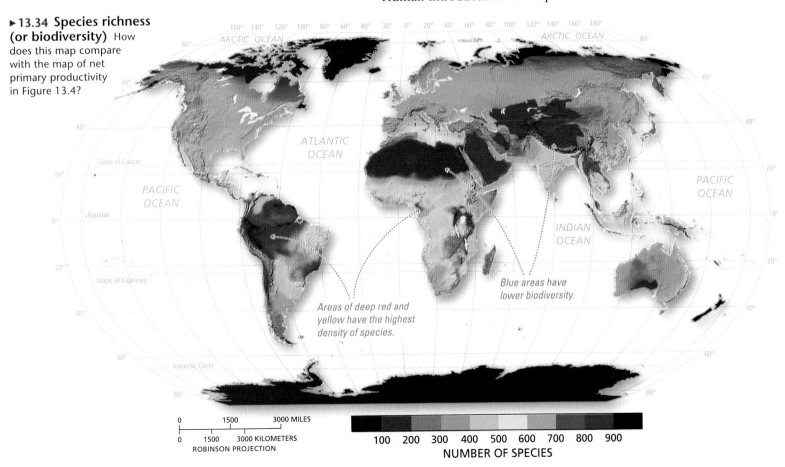

Areas of deep red and yellow have the highest density of species.

Blue areas have lower biodiversity.

100 200 300 400 500 600 700 800 900
NUMBER OF SPECIES

0   1500   3000 MILES
0   1500   3000 KILOMETERS
ROBINSON PROJECTION

13.9 (cont'd) Ecosystem Stability & Biodiversity

Amphibians are especially vulnerable to changes such as habitat destruction, pollution, invasive species, and climate change, putting them at a higher risk than mammals, fish, and birds (▼Fig. 13.35). A 2013 study found that vertebrates will have to evolve 10,000 times faster than their current rate of change. Genetic analysis of over 500 species of terrestrial vertebrates showed that species adapt to a 1 °C change in about a million years. The Intergovernmental Panel on Climate Change estimates that global temperatures could rise by 1.5 °C to 4.8 °C by 2100, depending upon emissions.

Species & Ecosystem Restoration Since the 1990s, species restoration efforts in North America have focused on returning predators such as wolves to parts of the American West and, recently, jaguars to the Southwest. Other efforts have resulted in increased numbers of black-footed ferrets and whooping cranes in the prairie regions. The preservation of large habitats has also played a critical role in restoring these, and other endangered species, worldwide. The reintroduction of wolves to Yellowstone National Park has been a spectacular success, with unexpected benefits across the park as a result of trophic cascades (▶Fig. 13.36).

Trophic cascades occur when the number of organisms at a trophic level change in response to a change in the number of organisms at another trophic level.

Trophic cascades are an example of the interconnections between organisms in a community. Changes to the numbers of organisms at the top of a trophic pyramid results in changes to the overall ecology of a region in surprising ways. The removal or reintroduction of large carnivores, such as wolves or bears, changes the numbers and behavior of the organisms at lower levels of a trophic pyramid. More humans in Yosemite National Park and Zion National Park resulted in a decline in mountain lions, which led to an increase in the number of deer. The increased numbers of deer resulted in a decline in California black oaks in Yosemite and cottonwoods in Zion, except in places inaccessible to deer. In Zion, just as in Yellowstone, the decline in stream side vegetation led to more erosion of stream banks, and a decline in the biodiversity of not just amphibians, but reptiles, butterflies, and wildflowers. Trophic cascades are important to understand, because humans have removed large carnivores from many regions through hunting and changes in land use.

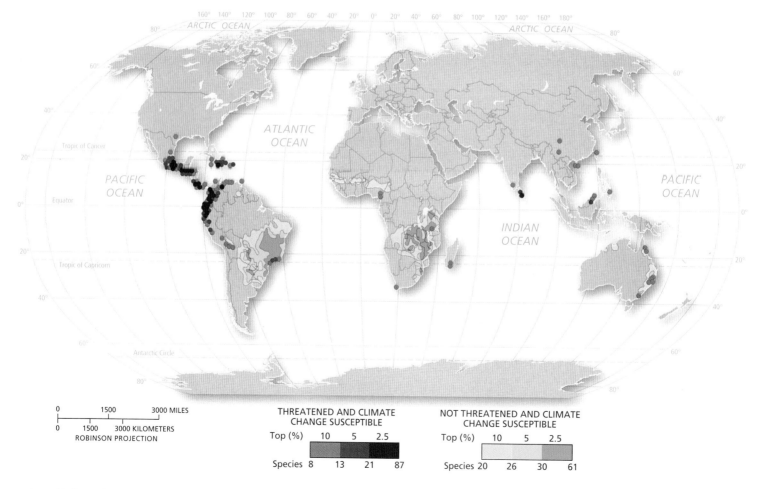

▲13.35 **Vulnerable amphibians** Areas of high concentration of amphibian species assessed as threatened and "climate change-susceptible" (reds) and not threatened but "climate-change-susceptible" (yellows).

Yellowstone without wolves (1926–1995)

Aspens

Coyote

Elk

Willow

Aspens were often unable to reach full height. New sprouts were usually eaten by elk.

Coyotes increased in numbers and ate small mammals, reducing food for foxes, badgers, and birds of prey.

Elk ate the streamside willows and shrubs, which decreased both fish and bird habitat and resulted in warmer water.

Yellowstone with wolves (1995–present)

Raven

Young aspens

Willow flycatcher

Willow

Wolves

Elk

Pronghorn

Grizzly bear

Beaver

Green-winged teal

Yellowstone trout

Boreal chorus frog

The elk population is half of its previous size. The presence of wolves keeps elk from streamsides where it is hard to escape from wolves.

The number of bison has also increased, probably due to fewer elk.

Pronghorn numbers have increased, possibly due to fewer coyotes.

Beavers have increased in number, creating ponds and better habitats for fish, amphibians, insects, birds, and small mammals.

Streamside vegetation has increased resulting in more stable stream banks, and better fish and bird habitat and cooler water.

Because wolves don't bury their kills, there is more food for scavengers such as golden eagles, ravens, magpies, and bears.

▲ **13.36 Trophic cascades** A change in one key species can trigger changes across an ecosystem. Wolves had been a top predator in Yellowstone National Park, but without wolves, the numbers of coyotes, elk, and deer increased. The reintroduction of gray wolves resulted in a decrease in those three species. The decrease in elk and deer increased streamside vegetation and habitat diversity. The decrease in coyotes led to an increase in small mammal populations, which led to increased numbers of hawks and eagles.

geoCHECK ✔ Which groups of species have the most accurate estimates of their numbers? Which groups have the least accurate estimates of their numbers? Explain why you think some groups are better known than other groups.

geoQUIZ

1. Give a hypothetical example of changes to an ecosystem that could lead to one species evolving into two separate species.
2. How can high biodiversity result in high ecosystems resilience?
3. Why is rapid climate change potentially so damaging to biodiversity?

SOILS IMPACT HUMANS

• Soils are the foundation of basic ecosystem function and are a critical resource for agriculture.
• Soils store carbon dioxide and other greenhouse gases in soil organic matter.

HUMANS IMPACT SOILS

• Humans have modified soils through agricultural activities. Recently, fertilizer use, nutrient depletion, and salinization have increased soil degradation.
• Poor land-use practices are combining with changing climate to cause desertification, soil erosion, and the loss of prime farmland.

13a

Grapevines grow in the Andisols of Lanzarote in the Canary Islands, producing wines from the fertile black ash of some of the most isolated vineyards in the world. Stone walls protect plants from the Atlantic winds.

13b

Nigerian women dig a trench to collect rainwater in the Sahel region of Africa. Although above-average rainfall in 2012 led to a successful harvest, the effects of desertification are ongoing throughout the region.

13c

1977 1998 2010

Desiccation of the Aral Sea began when rivers were diverted to irrigate cotton fields. The shrinking lake has accelerated desertification in the Aral basin of Kazakhstan and Uzbekistan and affected local climate, now hotter in summer without the water's moderating influence.

13d

Soybean fields are readied for planting in Rondônia, Brazil, surrounded by the Amazon rain forest. Oxisols, the soils of the tropics, are less fertile than most soils and obtain their red color from iron minerals.

ISSUES FOR THE 21ST CENTURY

• Continued soil erosion and degradation will cause lowered agricultural productivity worldwide and possible food shortages.
• Thawing of frozen soils (Gelisols) in the northern latitudes emits carbon dioxide and methane into the atmosphere, creating a positive feedback loop that leads to further warming.

Earth's biosphere is made up of abiotic and biotic components, all interacting through some 13.6 million species. Physical factors and living organisms—including humans—form soil, the medium for plant growth that bridges abiotic and biotic systems. Plants harvest sunlight through photosynthesis and thus begin vast food webs of energy and nutrition. Life on Earth evolved into this biologically diverse structure of organisms, communities, and ecosystems that gain strength through their biodiversity. We next move to the study of biomes, a topic that brings together all the book's topics to form a portrait of our planet.

What are the living & nonliving components of ecosystems?

13.1 Energy Flows & Nutrient Cycles

Outline the basic components of ecosystems.

Explain photosynthesis and respiration.

Describe the world pattern of net primary productivity.

- The biosphere, the sphere of life, extends from the ocean floor into the atmosphere. An ecosystem is a self-sustaining association of plants and animals and their nonliving physical environment. Ecology is the study of the relationships between organisms and their environment and between various ecosystems. Biogeography is the study of the distribution of plants and animals. Most ecosystems are powered by the Sun. Plants convert carbon dioxide and light energy to oxygen, sugars, and carbohydrates. The amount of carbon fixed by plants minus carbon used by plants in respiration is net primary productivity (NPP). The tropics have very high levels of NPP because of abundant rainfall and sunlight.

1. How are ecology and biogeography related? How are they different?

2. Why is NPP a good measure of the productivity of an ecosystem?

13.2 Biochemical Cycles

Discuss the oxygen, carbon, and nitrogen cycles.

- Abiotic factors such as light, temperature, nutrients, and water, are critical for ecosystem operation. Energy constantly flows in and out of ecosystems, but nutrients cycle within ecosystems. Oxygen and carbon are essential elements in ecosystems. Carbon is essential in photosynthesis and living organisms. Nitrogen is an essential nutrient for plants and is fixed by bacteria in the soil.

3. How are nutrient cycles different from energy inputs into ecosystems?

What is soil?

13.3 Soil Development & Soil Profiles

List the four major components of soil.

Describe the principal soil-formation factors.

Sequence the horizons of a typical soil profile.

- Soil is 50% minerals and organic matter and 50% air and water. Soil-forming factors are parent material, climate, organisms, topography, and time. Biological activity is important in soil formation, as is slope steepness and orientation. Steeper slopes have poorer soils than shallower slopes.

- Soils are studied in profile, from the surface down to bedrock. The main horizons are O, organic content; A, topsoil; and B, accumulation of clays and aluminum and iron oxides.

4. Briefly describe the main soil horizons. Why are the A, E, and B horizons considered the "true soil"?

13.4 Soil Characteristics

Describe the physical properties used to classify soils: color, texture, structure, consistence, porosity, and soil moisture.

Explain basic soil chemistry and relate this concept to soil fertility.

- Soil fertility refers to the ability to grow plants. The main characteristics of soil are color, texture, structure, consistence, porosity, and moisture. Soil chemistry involves air and water in the soil, as well as the chemicals that make up soil. Colloids are tiny particles that attract chemicals. Without colloids, nutrients would be leached from the soil and unavailable to plants.

5. How do colloids contribute to soil fertility?

How are ecosystems organized?

13.5 Energy Pathways

Explain trophic relationships.

Define the food chain, food web, and food pyramid concepts.

Trace the flow of energy from primary producers to tertiary consumers.

Analyze the cause and effects of biological amplification.

- The feeding relationships between organisms consist of food chains and food webs. Producers and consumers are on different trophic levels. Food chains usually have three to six levels. Producers are always plants, primary consumers are herbivores, and secondary consumers are carnivores, but omnivores both plants and animals. While only 10% of the energy at each trophic level is passed to the next higher level, pesticides and toxins accumulate and become more concentrated with each trophic level.

6. Describe the relationships between trophic levels in a typical ecosystem. Where do humans fit?

13.6 Communities: Habitat, Adaptation, & Niche

Relate communities to the other levels of ecological organization in the biosphere.

Summarize how the theory of evolution explains biodiversity and the adaptations of individual species.

Explain the concept of the ecological niche.

- Ecosystems are made of communities of plants and animals. Communities are identified by physical appearance, the species present, or trophic structure. Each species possesses traits, called adaptations, that contribute to its survival. As explained by Charles Darwin, different species and their adaptations arise as a result of evolution through natural selection. Each species has a home, its habitat, and a role or job, its niche.

7. Define community, habitat, adaptation, and niche.

13.7 Species Interactions & Distributions

Compare and contrast the different forms of interaction between species.

List several limiting factors on species distributions.

- Species interact with each other in symbiosis, which assists both; parasitism, when one benefits at the expense of the other; or commensalism, when one benefits and the other isn't helped or harmed. Abiotic factors such as temperature, precipitation, and sunlight influence species distributions, interactions, and growth. Alexander von Humboldt realized that plants and animals formed different communities at different elevations, which became the life zone concept. C. Hart Merriam found that altitude and latitude have similar effects on species.

8. What are life zones? Explain the relationships between elevation and latitude and the types of communities that develop.

How do ecosystems respond to change?

13.8 Ecosystem Disturbance & Succession

Discuss wildfires as a natural hazard and disturbance factor in ecosystems.

Outline the stages of ecological succession in both terrestrial and aquatic ecosystems.

- Communities experience disturbances from natural disasters and human actions. When most species are removed from a community, ecological succession begins. Wildfires are an important part of many natural forest communities. If the community is starting from bare rock, primary succession occurs. If soil remains, secondary succession occurs. Communities pass through succession stages in a dynamic and complex process with no clear climax community. Lakes and wetlands go through aquatic succession.

9. How are primary and secondary succession similar and different?

13.9 Ecosystem Stability & Biodiversity

Analyze factors that contribute to the stability or resilience of ecosystems affected by disturbances.

Summarize current threats to biodiversity and efforts to restore ecosystems.

- Earth's biodiversity is explained by the theory of evolution, the process by which species gradually change due to natural selection. Biodiversity influences ecosystem stability and resilience. Ecosystems can be stable but not resilient. Earth is losing biodiversity at a rate 1000 times faster than the natural pre-human rates. The current mass extinction event is the only event caused by living organisms.

10. Give some reasons why higher biodiversity is associated with higher productivity, stability, and resilience in ecosystems.

Key Terms

abiotic, p. 346	communities, p. 360	food web, p. 358	niche, p. 360	secondary succession,
adaptations, p. 360	consumers, p. 347	habitat, p. 360	nutrients, p. 350	p. 365
biodiversity, p. 346	decomposers, p. 347	humus, p. 352	parasitism, p. 362	soil, p. 352
biogeochemical, p. 350	detritivores, p. 358	life zone, p. 362	pedon, p. 354	soil fertility, p. 356
biogeography, p. 346	ecological succession,	limiting factor, p. 363	photosynthesis, p. 348	soil horizons,
biological amplification,	p. 364	mutualism, p. 362	primary producers,	p. 354
p. 359	ecology, p. 346	natural selection,	p. 347	soil profile, p. 354
biomass, p. 349	eutrophication, p. 366	p. 360	primary succession,	species, p. 360
biotic, p. 346	evolution, p. 360	net primary productivity,	p. 365	symbiosis, p. 362
commensalism, p. 362	food chain, p. 358	p. 349	resilience, p. 368	trophic level, p. 358

Critical Thinking

1. Trace two paths nitrogen might take from the atmosphere to the soil. Given that nutrients are crucial for soil fertility, what could be done to reduce fertilizer use and fertilizer loss.

2. Explain how plants and animals are connected through respiration and photosynthesis. What type of relationship do plants and animals have with regard to the exchange of gases? Across trophic levels?

3. What are the implications of energy loss across trophic levels and the bioaccumulation of pesticides for your diet? What dietary recommendations would you give to ensure adequate and healthier food for more people?

4. What are the implications of the ongoing loss of biodiversity? What steps should be taken to prevent irreversible species loss?

5. What are the unintended consequences of our past wildfire policies?

Visual Analysis

The Rim Fire outside of Yosemite National Park was the third largest fire in California's history. It burned from August 1, 2013, to October 24, 2013, and burned over 400 square miles (1000 square kilometers).

1. What are the different types of succession that will occur in this picture?

2. What are factors that are not favorable for soil development? How would soil develop differently on north-facing slopes from that on south-facing slopes? Steep slopes versus less steep slopes?

3. What are two factors regarding the ongoing drought that would increase the risk of catastrophic fires?

4. From the photo, does it look as though small fires had been allowed to burn? How can you tell?

5. If you were going to create a fire management policy for this region, what would you propose?

▲R13.1 **Effects of the Rim Fire, Yosemite National Park, 2013**

(MG) Interactive Mapping | Login to the **MasteringGeography** Study Area to access **MapMaster**.

- Open: MapMaster in MasteringGeography.
- Select: *Vegetation Zones* from the *Physical Environment* menu and *Countries, States, and Provinces* from the *Political* menu.

1. What is the main vegetation type in your state?

2. Where else does this vegetation type occur?

Turn off the *Vegetation Zones* layer, turn on the *Average Annual Precipitation* layer, and then turn on the *Vegetation Zones* layer.

3. Is there a good match between the outline of the main vegetation type from question 1 and precipitation amounts?

4. Find a vegetation zone that spans several precipitation zones. Which vegetation zone have you found? What appears to be the main factor that creates this range?

Explore | Use **Google Earth** to explore the relationship between **Net Primary Productivity & precipitation**.

Google Earth provides a cloud-free view of Earth's surface, which is useful in examining normally cloudy tropical regions. Pan and zoom so you can see all of Africa. Compare your view with Figure 13.4 Net Primary Productivity and Figure 6.1 Worldwide average annual precipitation.

1. Pan and zoom so you can see all of Africa. Compare your view with Figure GL13.1a Net Primary Productivity and 6.1 Precipitation. Do you notice a good correlation between the greenness on the Google Earth view and the NPP map? The Precipitation map?

2. Pan and zoom to the Amazon rain forest. Is the correlation stronger or weaker here? Pan and zoom to Indonesia and compare. Repeat with Australia, South Asia, Tibet and Mongolia, and finally with North America.

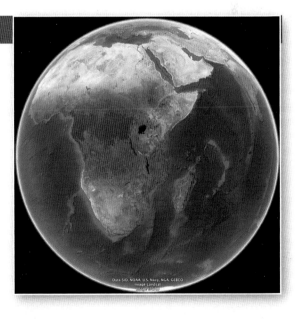

▶R13.2

MasteringGeography™

Looking for additional review and test prep materials? Visit the Study Area in MasteringGeography™ to enhance your geographic literacy, spatial reasoning skills, and understanding of this chapter's content by accessing a variety of resources, including MapMaster™ interactive maps, videos, *Mobile Field Trips*, *Project Condor* Quadcopter videos, *In the News* RSS feeds, flashcards, web links, self-study quizzes, and an eText version of *Geosystems Core*.

Getting Your Carbs: Net Primary Productivity

NPP is a measure of ecosystem productivity, measured in the net amount of carbon fixed per area per year. Recall that photosynthesis results in the formation of carbohydrates composed of carbon, hydrogen, and oxygen. NPP equals the amount of carbon from the atmosphere that is converted to sugar minus the carbon used during respiration. NPP shows clear patterns at the global and continental scales, and in this lab you will be exploring the relationship between NPP and the environmental factors of temperature and precipitation.

GeoLab13 (MG)
Pre-Lab Video

https://goo.gl/nbC3f1

Objectives

- Evaluate the effects of temperature and precipitation on NPP
- Identify regions of high and low NPP
- Predict NPP values for locations in North America

Apply

You have been tasked with identifying potential sites where ecosystems can store carbon that would otherwise enter the atmosphere. Vegetation converts atmospheric CO_2 into biomass. Higher NPP rates result in more CO_2 being converted into biomass.

Procedure

Figure GL 13.1 shows NPP, January and July temperatures are shown in Figures 2.29 and 2.30, and annual precipitation is shown in Figure 6.1.

1. Examine the global patterns of NPP, temperature, and precipitation. Identify regions of high and low NPP for South and Central America, Africa, Eurasia, and Australia either by region or by latitude and longitude.
2. Which region has the highest NPP? What factors are responsible for these values, temperature or precipitation?
3. Compare NPP, temperature, and precipitation for North Africa, India, and Indonesia.

Analyze & Conclude

4. Explain why Indonesia has approximately the same precipitation as India, yet Indonesia has much higher NPP. What other factors might be important in forming this pattern? Could locations with similar annual precipitation values have very different monthly precipitation amounts during the year? For example, would a tropical rain forest have the same NPP values as a region with a monsoonal or savannah precipitation regime? What data would you want to see to help you decide?

5. Why do you think Madagascar has the patterns of NPP and precipitation that it does? Which part of the island has higher NPP? Higher temperatures? Higher rainfall? How could prevailing wind direction explain the pattern of precipitation? Describe how temperature and precipitation combine to create the NPP shown on **Figure GL 13.1**.

6. Is temperature or precipitation more important in determining NPP values? Support your answer with a regional example.

7. Examine **Figure GL 13.2**, which shows the same data as Figure GL 13.1, but across North America. Compare the NPP, temperature, and precipitation values for the Pacific Northwest, the Southwest, the Southeast, and the Northeast.

8. How are these patterns similar to or different from the global patterns you discussed in previous questions? Provide at least two regional examples to support your answer.

9. The highest photosynthesis values in the world have been recorded in U.S. plains states in summer, yet the annual NPP values for these states are much lower than tropical regions. Explain what factors would result in midlatitude regions having high summer values, but moderate annual values.

10. Based upon the patterns the globe and North America, characterize the factors that would create high NPP values.

11. What factors create regions with low NPP?

12. Create a recipe for a region with high NPP. What combination of precipitation amounts, annual distribution of precipitation, annual average temperature and temperature range, and latitude would have the highest NPP values?

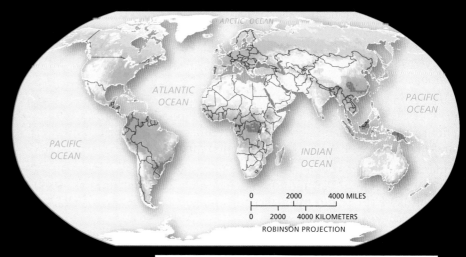

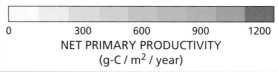

▲GL13.1 **World net primary productivity**

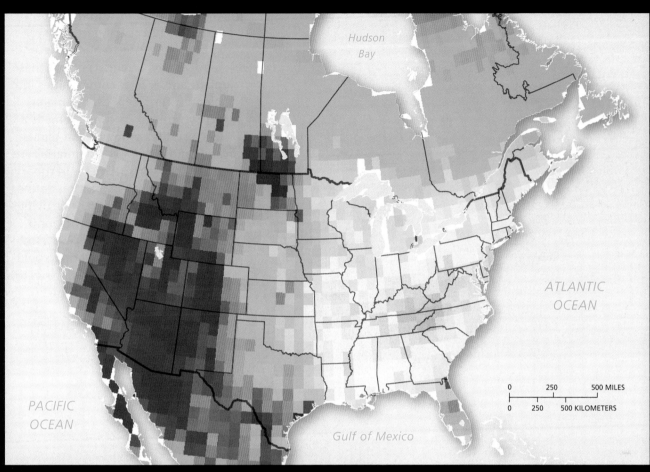

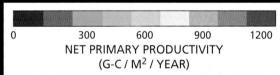

NET PRIMARY PRODUCTIVITY
(G-C / M² / YEAR)

▲GL13.2 **Net primary productivity: United States**

14

Terrestrial Biomes

Geology, climate, and the evolutionary history of particular species gave rise to our planet's biodiversity. Scientists group these plant and animal communities into biomes, representing the major ecosystems of Earth. A biome is a large stable community of plants and animals whose boundaries are closely linked to climate. In theory, mature, natural vegetation defines biomes. However, human activities have affected most biomes, and many are now experiencing accelerated rates of change that could produce dramatic alterations in the biosphere within our lifetime.

Key Concepts & Topics

The dramatic baobab trees of Morondava, Madagascar are an example of the amazing diversity of life on Earth.

14.1 **Biogeographic Divisions**

Key Learning Concepts

▶ *Identify* the world's biogeographic realms and explain how their boundaries are determined.
▶ *Explain* the basis for grouping plant communities into biomes.

Earth's biosphere has regions of similar plant and animal communities. Species' distributions and evolutionary history determine one class of region—the biogeographic realm. Plant communities' characteristics as they relate to climate and soils define another class of region—the biome.

Biogeographic Realms

Flora and *fauna* are general terms for the typical plants and animals of a region or ecosystem. Earth has eight distinct **biogeographic realms** made up of similar plants and animals (▶ **Fig. 14.1**). The organisms within each realm are the product of evolution and plate tectonics. As continents collided, different groups of organisms were brought together; as continents moved apart, populations of a species were separated. The boundaries of biogeographic realms are usually climatic and topographic barriers, such as deserts, rivers, mountain ranges, and oceans. Australia's unique native organisms, including 450 species of eucalyptus and 125 species of marsupials, are the result of its early isolation from the other continents. Australia moved away from Pangaea (see Fig. 8.14) and was never reconnected by a land bridge, even when sea levels were lower during glacial periods. Alfred Wallace (1823–1913) identified the differences between organisms in the Indo-Malay and Australasian realms, a boundary that is known today as "Wallace's line." The two realms on either side of Wallace's line remain distinct because of the deep ocean channel that separates them.

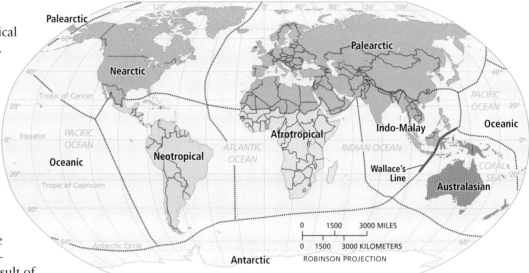

▲14.1 **The eight biogeographic Australasian realms, based on plant and animal associations and evolutionary relationships**

 geoCHECK ✔ Why would biogeographic realms correspond to continental plate boundaries?

Biomes

A **biome** is a large, stable ecosystem classified according to the main vegetation type and the adaptations of its organisms to that environment. Scientists study two main types of biomes: terrestrial biomes and aquatic biomes (the latter separated into freshwater and marine).

Biomes are defined by species that are native to a region. Today, most biomes have been greatly altered by humans, and few natural communities of plants and animals remain. The "native vegetation" identified on many biome maps reflects idealized potential mature vegetation. For example, in Norway, past old-growth forests are now a mix of second-growth forests, farmlands, and altered landscapes (▶ **Fig. 14.2**). However, the forest biome designation for this region remains, based on idealized conditions before human impacts.

▼14.2 **Boreal forest modified by human activity, rural Norway**
Species that tolerate disturbances often occupy the spaces where disturbed land meets natural habitat.

Vegetation Types Scientists determine biomes based on easily identifiable vegetation characteristics of the dominant plants, including size, woodiness, life span, and leaf traits (for example, whether trees are evergreen or deciduous). Biogeographers often designate six major groups of terrestrial vegetation: forest, savanna, shrubland, grassland, desert, and tundra. However, biome classifications are often more specific, with 10 to 16 total biomes, depending upon the classification system. The specific flora of each biome has a related fauna that helps define its geographic area.

For example, forests can be divided into several biomes—rain forests, seasonal forests, broadleaf mixed forests, and needleleaf forests—based on moisture regime, canopy structure, and leaf type. While rain forests occur in areas with high rainfall, they can be subdivided into tropical and temperate rain forests. Tropical rain forests are composed of mainly evergreen broadleaf trees (having broad leaves, as opposed to needles), while temperate rain forests are composed of both broadleaf and needleleaf trees (having needles as leaves).

The plants that grow in an area reflect the abiotic factors discussed in Chapter 13, such as the amount of sunlight, temperature and winds, the amount and seasonal timing of rain, and soils and nutrients. Because of the importance of these abiotic influences, plants growing in similar climates look similar, even if they have evolved on different continents or even different hemispheres. Biomes usually correspond to moisture and temperature regimes (▶Fig. 14.3), as well as the increasing influence of humans.

Ecotones

Boundaries between climates, biomes, or ecosystems are often zones of gradual transition, rather than abrupt borders. These boundary zones are called **ecotones**.

Ecotones are defined by physical factors, and they vary in width. Ecosystems separated by different climates usually have gradual ecotones, while those separated by differences in topography may have more abrupt boundaries. For example, the climatic boundary between grasslands and forests can occupy many kilometers of land, while a sharp boundary in the form of a river, or a landslide, or a lakeshore may occupy only a few meters. As human impacts cause ecosystem fragmentation, ecotones between ecosystems are becoming more numerous across the landscape (see Fig. 14.2).

▼14.3 **Vegetation patterns in relation to temperature and precipitation** This chart places Earth's biomes on a grid in which temperature and precipitation range from cold and dry (lower left) to hot and wet (upper right).

Temperate rain forest

Tropical rain forest

Broadleaf mixed forest

Tropical seasonal forest

Boreal forest

Midlatitude grassland

Tropical savanna

Arctic tundra

Cold desert

Warm desert

Wet — Dry — Precipitation

Cold ◄——————————————► Hot

Temperature

14.2 Conservation Biogeography

Key Learning Concepts

▶ **Describe** the impact of invasive species on biotic communities, using several examples.

▶ **Analyze** how the theory of island biogeography has led to new strategies for biodiversity conservation.

As humans have exerted more influence on natural species distributions, a new scientific field has emerged called *conservation biogeography*, which applies biogeographic tools to solve problems in biodiversity conservation. Conservation biogeographers research the impacts of rapid climate change on biodiversity (discussed in Chapter 7), the distribution and effects of invasive species, and the planning and establishment of protected areas.

Invasive Species

Native species inhabit a particular biome as a consequence of evolutionary and physical factors. However, communities, ecosystems, and biomes can also contain species that humans have introduced from elsewhere, either intentionally or accidentally.

Nonnative species that establish themselves in a new habitat often take over niches already occupied by native species, and become **invasive species**, leading to declines in native species. (Invasive species are also called *exotic species*). Invasive species often have an advantage over native species because invasive species often arrive without the organisms that keep their populations in check in their native community. Invasive species often do not provide shelter or food to native organisms, which further destabilizes communities. Prominent examples of invasive species are shown in ▶ **Fig. 14.4**.

Invasive species can alter the dynamics of entire biomes. For example, in southern California, the native Mediterranean shrubland vegetation is adapted to wildfire (see **Fig. 14.46b** ahead). Nonnative species often out-compete native species in burned areas, and

change the successional processes in this biome. Nonnatives can also create thick undergrowth, providing more fuel for the fires that are increasing in frequency. More frequent fires combine with the increasing numbers of nonnative species to cause the conversion of southern California's shrubland to grassland.

geoCHECK ✔ Why are introduced species disruptive to ecosystems and biomes?

▼ **14.4 Invasive species** These four invasive species from Eurasia now flourish in the United States.

(a) Zebra mussels cover most hard surfaces in the Great Lakes; they rapidly colonize on any surface, even sand, in freshwater environments.

(b) Kudzu, originally imported for cattle feed, spread from Texas to Pennsylvania; here, it overruns pasture and forest in western Georgia.

(c) Invasive Russian olive (green-gray color) and tamarisk (dark green color in the shade) along Chinle Wash, New Mexico, in streambank habitat formerly occupied by native cottonwoods.

(d) Purple loosestrife has invaded wetland habitats throughout much of the United States and Canada; shown here is southern Ontario.

Island Biogeography for Species Preservation

Organisms on islands often develop in isolation and evolve adaptations in response to the unique conditions of islands. However, new organisms introduced to islands often quickly replace the native species. For example, Europeans identified 43 species of birds when they first landed in the Hawaiian Islands in the late 1700s. Introduced nonnative species have driven 15 of those bird species to extinction and made 19 more species endangered or threatened. In most of Hawai'i, native birds no longer exist below elevations of 1220 m (4000 ft) because of an introduced avian virus (▼Fig. 14.5).

▲14.5 **The threatened Hawaiian I'iwi, a honeycreeper.** Once found throughout the Hawaiian Islands and prized for its striking plumage, the I'iwi is now extinct on Lana'i and extremely rare on O'ahu and Moloka'i. On the other islands, the bird is still relatively common above 1000 m (3280 ft), beyond the reach of mosquitos carrying disease.

The principles developed in the study of isolated species' evolution on islands and how isolated species decline after the introduction of nonnative species have become useful in global conservation efforts. In the 1980s, researchers discovered that a number of U.S. national parks had become isolated "islands" of biodiversity, and some species within them were declining or disappearing completely. Protected areas such as national parks appeared to be of limited value in species preservation because they were surrounded by human development and disconnected from other natural habitat. Such **habitat fragmentation** is problematic for species, such as bears and wolves in North America, requiring a large range for survival.

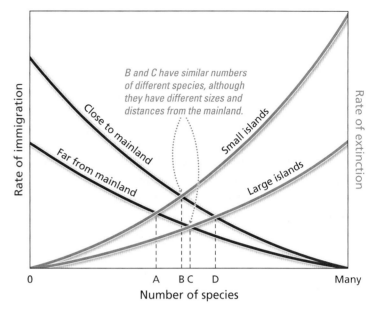

▲14.6 **Island biogeography species equilibrium curve.** The number of species in a given habitat is a function of the rates of immigration and extinction. Letters A, B, C, and D represent different numbers of species on different islands with different sizes and distances from the mainland. A shows the effects of a small, faraway island, and D shows the effects of a large island to the mainland.

A key model for understanding the effects of habitat fragmentation was Robert MacArthur and E. O. Wilson's theory of **island biogeography**. The theory, based on scientific work on small, isolated mangrove islands in the Florida Keys, as well as islands worldwide, links the number of species on an island to the island's size and distance from the mainland. The theory summarized three patterns of species distributions on islands: (1) The number of species increases with island area; (2) the number of species decreases with island isolation (distance from the mainland); and (3) the number of species on an island represents an equilibrium between the rates of immigration and extinction (▲Fig. 14.6). Larger islands have a wider variety of habitat and niches, and thus lower extinction rates. This theory helped scientists understand "islands" of fragmented habitat and led to a new approach to conservation: To prevent declines in biodiversity, parks and reserves needed to be larger and also preserve corridors of open space between protected areas.

geoCHECK ✔ Why are island organisms particularly vulnerable to extinction?

geoQUIZ
1. What factors make invasive species especially disruptive to existing communities?
2. How are parks and reserves like islands?
3. How do the main ideas of island biogeography help scientists design parks and reserves?

14.3 Soil Classification

Key Learning Concepts

▶ **Classify** the 12 soil orders of the Soil Taxonomy according to their characteristics.

▶ **Describe** the distribution of the soil orders across Earth and how soils affect plant cover and agriculture.

We have seen how Earth's biomes are large-scale patterns in the distribution of living things that result from many factors, including climate. The different types of soil are influenced by climate and geology, and they are another factor that influences the pattern of biomes. A number of soil classification systems are in use worldwide. Each system reflects the environment of the region in which it originated. Soils, like climates, are the product of interacting variables (▶Fig. 14.7). The interactions of geology, geomorphology, time, and climate have produced the pattern of soil orders we see today.

▲**14.7 Freshly plowed soil** Alfisol in Bradford on Avon, England.

Soil Taxonomy

The Natural Resource Conservation Service (NRCS) developed the U.S. soil classification system, called the **Soil Taxonomy**. The basis for the Soil Taxonomy system is field observation of soil appearance and structure. The smallest unit of soil used in soil surveys is a *pedon*. Recall that a pedon is a hexagonal column of soil measuring 1 to 10 m² in top surface area (see Figure 13.13).The soil profile within a pedon is used to evaluate its soil horizons. Pedons with similar characteristics are grouped into soil series, the most precise level of the classification system. The Soil Taxonomy is a hierarchy with six levels, or categories, beginning with 15,000 soil series classifications and culminating in the 12 soil orders discussed in this section (▶Fig. 14.8).

Diagnostic Soil Horizons Soil scientists use the soil horizons discussed in Chapter 13 to group soils into each soil series. A diagnostic horizon has distinctive physical properties (color, texture, structure, consistence, porosity, moisture).

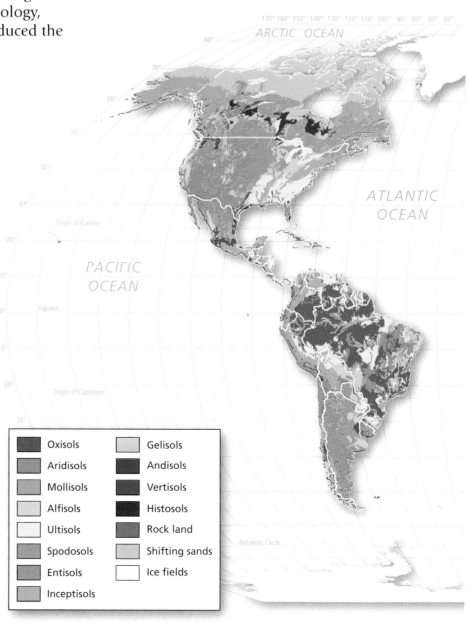

Oxisols

Aridisols

Mollisols

Alfisols

Ultisols

Spodosols

Entisols

Inceptisols

Gelisols

Andisols

Vertisols

Histosols

Rock land

Shifting sands

Ice fields

Pedogenic Regimes Specific soil-forming processes relate **pedogenic regimes** to climate regions. Prior to the U.S. Soil Taxonomy system, scientists used pedogenic regimes to describe soils. Although no longer used in classification, the five climate-based pedogenic regimes below are convenient for relating climate and soil processes:

- **Laterization** leaches nutrients and silica (SiO_2) in humid and warm climates (discussed with Oxisols).
- **Calcification** produces an accumulation of a horizon rich in calcium carbonates in arid and semiarid deserts and grasslands (discussed with Aridisols).

- **Salinization** concentrates salts in soils in climates with high potential evapotranspiration (POTET) rates (discussed with Aridisols).
- **Podzolization** is caused by the movement of iron and aluminum oxides in an acid regime, associated with forest soils in cool climates (discussed with Spodosols).
- **Gleization** results in an accumulation of humus and a thick, water-saturated gray layer of clay beneath, usually in cold, wet climates and poor drainage conditions (discussed with Gelisols).

geoCHECK ✔ What criteria are used to classify soils?

▼ **14.8 Soil Taxonomy** Worldwide distribution of the Soil Taxonomy's 12 soil orders, with each order listed and briefly described at left.

Website MG
Soil Science
of America

http://goo.gl/mHHkcl

14.3a Tropical, Arid, Grassland, & Forest Soils

Soil Orders of the Soil Taxonomy

The Soil Taxonomy includes 12 general soil orders, listed in **Figure 14.8** (left side). Their worldwide distribution is shown in **Figure 14.8** and in locator maps provided with the soil discussions that follow. Please consult the maps as you read the descriptions. As presented here, the descriptions form a progression arranged loosely by latitude; we begin, as with climates, along the equator (▶Fig. 14.9).

Oxisols The high temperature and rainfall of equatorial latitudes profoundly affect soils. These are among the most mature soils on Earth. They have been exposed to tropical conditions for millennia or hundreds of millennia, forming well-developed, heavily altered soils. Distinct horizons usually are lacking where these soils are well drained (▼Fig. 14.10a).

Oxisols (tropical soils) are named for their distinctive horizon of reddish (iron) and yellowish (aluminum) oxides. The concentration of oxides results from heavy precipitation, which leaches soluble minerals and soil constituents, leaving the soil with low fertility. **Laterization** is the leaching process that operates in well-drained soils in warm, humid climates. This process forms laterite, which can be quarried in blocks and used as a building material (▼Fig. 14.10b).

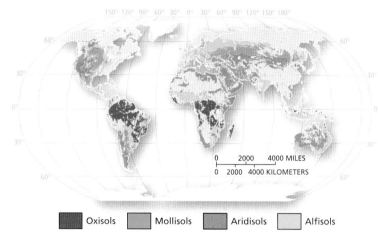

Oxisols ▪ Mollisols ▪ Aridisols ▪ Alfisols

▲**14.9 Worldwide Oxisol, Aridisol, Mollisol, and Alfisol distribution**

The world's lush rain forests are found in regions of Oxisols, even though these soils are poor in inorganic nutrients. This forest system relies on the recycling of nutrients from soil organic matter to sustain fertility. However, this nutrient-recycling ability is quickly lost when the ecosystem is disturbed, for example, when a forest is cleared to graze cattle.

▼**14.10 Oxisols**

(a) When Oxisols undergo laterization, leaching occurs as water moves downward through the soil.

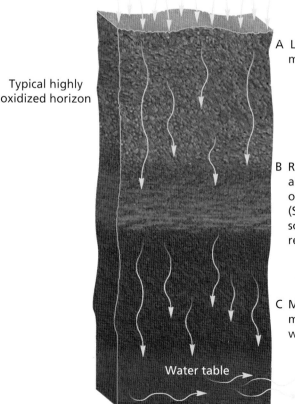

Warm and humid climates

Typical highly oxidized horizon

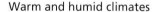

A LIttle organic matter

B Residual iron and aluminum oxides; silica (SiO2) and other soluble constituents removed

C Much soluble material to water table

Water table

To streams

(b) Laterization processes in Oxisols form plinthite, which was used to build this temple complex at Angkor Wat, Cambodia.

Aridisols The largest single soil order occurs in the world's dry regions. **Aridisols** (desert soils) occupy approximately 19% of Earth's land surface (see Fig. 14.9). A pale, light soil color near the surface is typical of Aridisols (▼Fig. 14.11).

Aridisol regions are characterized by shallow soil horizons, because they lack enough soil moisture for plant growth. Usually, these soils have adequate moisture for only 3 months a year. Lacking water and therefore lacking vegetation, Aridisols also lack organic matter of any consequence.

Calcification is a soil process characteristic of Aridisols and some Mollisols (discussed below) in which calcium carbonate or magnesium carbonate accumulates in the B and C horizons. Calcification by calcium carbonate ($CaCO_3$) forms its own characteristic layer in the soil.

Aridisols can be made productive for agriculture using irrigation. Two related problems common in irrigated lands are *salinization* and *waterlogging*. Salinization occurs as salts dissolved in soil water migrate to the surface and are deposited as the water evaporates. After more soil has developed on top of these deposits, they appear as subsurface salty horizons, which can kill plants if the horizons occur near the root zone. Waterlogging (saturation of the soil that interferes with plant growth) occurs with the introduction of irrigation water, especially in soils that are poorly drained.

Mollisols Some of Earth's most significant agricultural soils are **Mollisols** (grassland soils). The dominant horizon is a dark, organic surface layer some 25 cm (10 in.) thick (▼Fig. 14.12). Mollisols are soft, even when dry. These humus-rich soils have high fertility and are moderately moist. The B horizon can have clay accumulation and can be enriched in calcium carbonate in drier climates. The B horizon is thickest along the boundary between dry and humid climates.

Mollisols include soils of the steppes and prairies—the North American Great Plains, the Pampas of Argentina, and the "fertile triangle" of Ukraine and Russia. Agriculture in these areas ranges from large-scale commercial grain farming to grazing along the drier portions.

Alfisols Spatially, **Alfisols** (moderately weathered forest soils) are the most widespread of the soil orders, extending from near the equator to high latitudes. Representative Alfisol areas include interior western Africa; Fort Nelson, British Columbia; the states near the Great Lakes; and the valleys of central California. Most Alfisols are grayish brown to reddish and are considered moist versions of the Mollisol soil group. A subsurface horizon of clays is common because of leaching by precipitation (▼Fig. 14.13).

Alfisols are fertile, but their productivity depends on moisture and temperature. Alfisols usually are supplemented by a moderate application of lime and fertilizer in areas of active agriculture.

Some of the best U.S. farmland occurs in the humid continental hot-summer climates surrounding the Great Lakes. These Alfisols produce grains, hay, and dairy products. Mediterranean climates can also produce Alfisols. These naturally productive soils are farmed intensively for subtropical fruits, nuts, and special crops that grow in only a few locales worldwide—for example, California olives, grapes, citrus, artichokes, almonds, and figs.

In the Midwest of the United States, Alfisols, Mollisols, and Aridisols form a series. The soil order changes as precipitation decreases from east to west across the Great Plains (Fig. 14.14).

geoCHECK ✔ Why would regions with arid climates develop pale soils, while Mollisols are dark colored?

▼**14.11 Aridisols: A soil undergoing calcification** This process occurs in desert and grassland soils that have potential evapotranspiration equal to or greater than precipitation.

Potential evapotranspiration ≥ precipitation

Dark color, high in bases

O Dense sod cover of interlaced grasses and roots

A

E

Typical calcium enriched horizon

B Accumulation of excess calcium carbonate

C

▼**14.12 Mollisol profile in eastern Idaho, from loess that is high in calcium carbonate**

Organic layer

▼**14.13 Alfisol profile in northern Idaho loess**

Clay horizon due to leaching by precipitation

14.3b Weathered Forest Soils & Young Soils

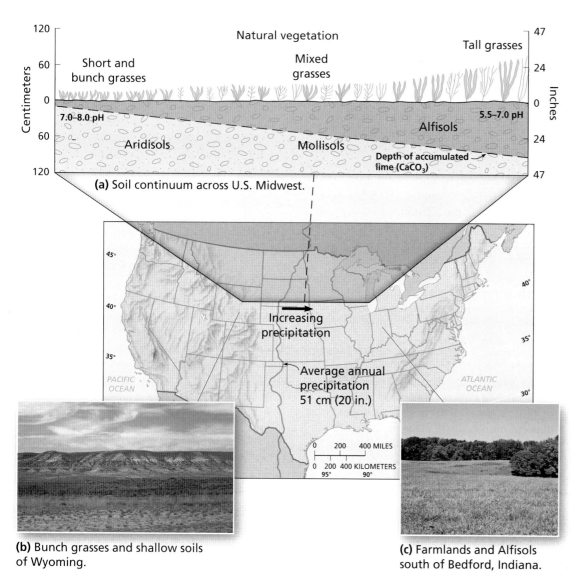

(a) Soil continuum across U.S. Midwest.

Natural vegetation

Tall grasses

Short and bunch grasses

Mixed grasses

7.0–8.0 pH

5.5–7.0 pH

Alfisols

Aridisols

Mollisols

Depth of accumulated lime (CaCO₃)

Increasing precipitation

Average annual precipitation 51 cm (20 in.)

PACIFIC OCEAN

ATLANTIC OCEAN

0   200   400 MILES

0   200   400 KILOMETERS

◄ **14.14 Soils of the Midwest** Aridisols (to the west), Mollisols (central), and Alfisols (to the east) are part of a soil continuum in the north-central United States and southern Canadian prairies. Note the changes that occur in soil pH and the depth of accumulated calcium carbonate.

Ultisols Farther south in the United States are the **Ultisols** (highly weathered forest soils) (▼Fig. 14.15). An Alfisol might evolve into an Ultisol, given time and exposure to increased weathering under moist conditions. These soils tend to be reddish because of residual iron and aluminum oxides in the A horizon (▼Fig. 14.16).

The relatively high precipitation in Ultisol regions causes greater mineral alteration and more leaching than in other soils. Therefore, soil fertility is lower. Fertility is further reduced by certain agricultural practices and the effect of soil-damaging crops such as cotton and tobacco, which deplete nitrogen and expose soil to erosion. These soils respond well to good management; for example, crop rotation restores nitrogen, and cultivation practices can prevent soil erosion. Peanut plantings assist in nitrogen restoration.

Spodosols The Spodosols (northern coniferous forest soils) occur generally to the north and east of the Alfisols, mainly in forested areas in the humid continental mild-summer climates of northern

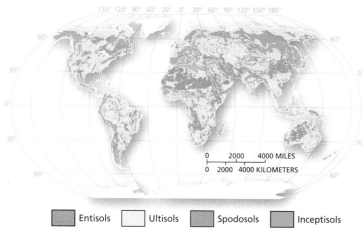

(b) Bunch grasses and shallow soils of Wyoming.

(c) Farmlands and Alfisols south of Bedford, Indiana.

▲**14.16 Ultisols planted with rows of peanuts in west-central Georgia have the characteristic reddish color**

150° 120° 90° 60° 30° 0° 30° 60° 90° 120° 150° 180°

0   2000   4000 MILES

0   2000   4000 KILOMETERS

☐ Entisols   ☐ Ultisols   ☐ Spodosols   ☐ Inceptisols

▲**14.15 Worldwide Ultisol, Spodosol, Entisol, and Inceptisol distribution**

North America and Eurasia, Denmark, the Netherlands, and southern England. Spodosols form from sandy parent materials, shaded under evergreen forests of spruce, fir, and pine. Mixed or deciduous forests have more moderate Spodosols. *Podzolization* refers to the leaching of organic material and minerals from the upper soil horizons and their deposition in lower soil horizons. Water percolating through the acidic needles of evergreen trees and sandy parent materials strips nutrients from the upper horizons and forms an ash gray layer (▶Fig. 14.17). The lack of nutrients and low pH make Spodosols difficult agricultural soils. Agricultural use of Spodosols often requires the addition of fertilizer and a soil amendment such as limestone to raise the pH of these acidic soils.

Entisols The **Entisols** (recent, undeveloped soils) lack vertical development of their horizons. The presence of Entisols is not climate dependent, for they occur in many climates worldwide. Entisols are true soils, but they have not had sufficient time to generate the usual horizons.

Entisols generally are poor agricultural soils, although those formed from river silt deposits are quite fertile. The same conditions that inhibit complete development also prevent adequate fertility—too much or too little water, poor structure, and insufficient accumulation of nutrients—also prevent adequate fertility. Entisols are characteristically found on active slopes, floodplains, poorly drained tundra, sand dunes, and plains of glacial outwash. ▼Figure 14.18 shows a landscape of Entisols in a desert climate where shales formed the parent material.

geoCHECK ✔ Would you expect Spodosols to become more acidic or less acidic as the soil develops? Explain your answer.

Inceptisols Although more developed than the Entisols, **Inceptisols** (weakly developed soils) are young, infertile soils. This order includes a wide variety of different soils, all having in common a lack of maturity, and most showing only the beginning stages of weathering.

Inceptisols are associated with moist soil regimes and demonstrate a loss of soil constituents throughout their profile; however, they do retain some weatherable minerals. This soil group has no distinct horizons. Most of the glacially derived till and outwash materials from New York down through the Appalachians are Inceptisols, as is the alluvium on the Mekong and Ganges floodplains.

▼14.18 **Entisols** A landscape showing Entisols—undeveloped soils forming in the shales of the Anza–Borrego Desert, California.

▼14.17 **Spodosols** These acidic soils occur in cool temperate, forested areas of North America and Eurasia.

Cool and moist climate

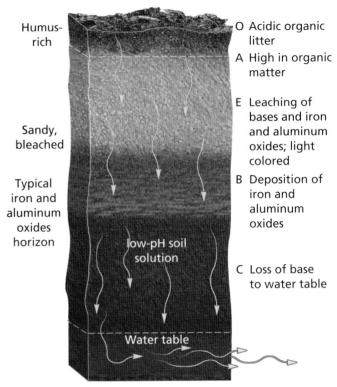

Humus-rich

Sandy, bleached

Typical iron and aluminum oxides horizon

low-pH soil solution

Water table

O Acidic organic litter

A High in organic matter

E Leaching of bases and iron and aluminum oxides; light colored

B Deposition of iron and aluminum oxides

C Loss of base to water table

(a) A soil undergoing podzolization This process is typical in cool and moist climatic regimes and Spodosols.

(b) Characteristic temperate forest and Spodosols in the cool, moist climate of central Vancouver Island.

14.3c Cold, Volcanic, & Organic Matter Soils

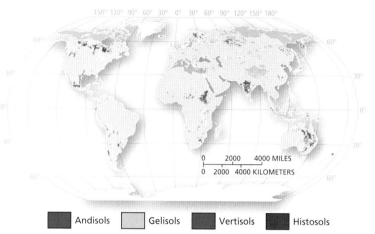

▼14.20 Gelisols

▲**14.19 Worldwide Gelisol, Andisol, Vertisol, and Histosol distribution**

Andisols Gelisols Vertisols Histosols

(a) The tundra is green in the brief summer season on Spitsbergen Island, northern Norway, as the active layer thaws.

Gelisols The **Gelisols** (cold and frozen soils) contain permafrost within 2 m (6.5 ft) of the surface and are found at high latitudes (Canada, Alaska, Russia, Arctic Ocean islands, and the Antarctic Peninsula) and high elevations (mountains) (▲**Fig. 14.19**). Temperatures in these regions are at or below 0°C (32°F), making soil development a slow process and disturbances to the soil long-lasting. Cold temperatures slow the decomposition of materials in the soil, so Gelisols can store large amounts of organic matter; thick O horizons are common (▶**Fig. 14.20**). Gelisols are associated with the pedogenic regime of *gleization*. The combination of cold temperatures and high water content in the soil results in high amounts of undecomposed organic matter and a gray-blue clay layer.

Gelisols contain about half of the pool of total global carbon, 1.7 trillion tons of carbon. As permafrost thaws, the decomposition of organic content releases enormous quantities of carbon dioxide and methane into the atmosphere.

Andisols The **Andisols** (volcanic soils) are derived from volcanic ash and volcanic glass and frequently bury previous soil horizons with materials from repeated eruptions. Volcanic soils are unique in their mineral content because they are recharged by eruptions.

(b) Fibrous organic content exposed on the underside of a soil clod, Spitsbergen.

In Hawai'i, fields of Andisols produce coffee, pineapples, macadamia nuts, and small amounts of sugar cane as important cash crops. Andisol soils are locally important in regions associated with the volcanic Pacific Rim of Fire, discussed in Chapter 8 (◀**Fig. 14.21**).

Vertisols Soils high in clays that swell when they get wet and shrink as they dry are **Vertisols**. Diagnostic horizons are usually absent, and an A horizon is

◀**14.21 Andisols in agricultural production**
Rows of pineapples on O'ahu.

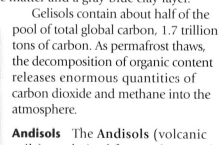

common (▼**Fig. 14.22**). Vertisols are located in regions experiencing highly variable soil-moisture balances through the seasons. Vertisols frequently form in savannas and grasslands of tropical and subtropical climates and are sometimes associated with a distinct dry season following a wet season.

Vertisols contain more than 30% clays that swell significantly when they absorb water. As they dry, vertical cracks may form. Loose material falls into these cracks, only to disappear when the soil again expands and the cracks close. After many such cycles, soil contents tend to mix vertically, bringing lower horizons to the surface.

Despite the fact that clay soils become heavy when wet and leave little soil moisture available for plants, Vertisols are high in nutrients and are some of the better farming soils. Vertisols often are planted with grain sorghums, corn, and cotton (**Fig. 14.22**b).

▼14.22 Vertisols

(a) Vertisol profile in the Lajas Valley of Puerto Rico.

Histosols   Accumulations of thick organic matter can form **Histosols** (organic soils). In the midlatitudes, when conditions are right, beds of former lakes may turn into Histosols, with water gradually replaced by organic material to form a bog (discussed in Chapter 13). Histosols also form in small, poorly drained depressions, where conditions can be ideal for significant deposits of sphagnum peat to form (▶**Fig. 14.23**).

(b) Commercial sorghum crop planted in Vertisols on the Texas coastal plain, northeast of Palacios. Note the characteristic dark soil color.

▼14.23 **Histosols** A Histosol profile on Mainland Island in the Orkneys, north of Scotland. The inset photo shows drying blocks of peat, used as fuel.

Note the fibrous texture of the sphagnum moss growing on the surface and the darkening layers with depth in the soil profile as the peat is compressed and chemically altered.

Peat beds, often more than 2 m thick, can be cut by hand with a spade into blocks, which are then dried, baled, and sold as a soil amendment. Once dried, the peat blocks burn hot and smoky. Peat is the first stage in the natural formation of lignite, an intermediate step toward coal. The Histosols that formed in lush swamp environments in the Carboniferous Period (359 to 299 million years ago) eventually became coal deposits.

Most Histosols are hydric soils, defined as soils that are saturated or flooded for long enough periods of time to develop anaerobic (oxygen-free) conditions during the growing season. The presence of hydric soils is the basis for the legal delineation of wetlands, which are protected from dredging, filling, and the discharge of pollutants under the U.S. Clean Water Act. Bogs are a type of wetland, and many Histosols develop in wetland environments.

geoCHECK ✔   Explain how Histosols can become an energy source.

geoQUIZ

1. Explain why climate is such an important factor in soil development.
2. Which soils are not related to specific climates? Why isn't climate a factor in these soils?
3. What are the characteristics of the soil order that is most common where you live? How is this soil used?

14.4 Earth's Terrestrial Biomes

Key Learning Concepts

▶ **Summarize** the characteristics of Earth's 10 major terrestrial biomes and locate them on a world map.

Climates and soils and many other factors have formed the patterns of distinctive natural landscapes across Earth. These distinctive landscapes, while often separated by ecotones, have been classified into biomes. In *Geosystems Core*, we describe 10 biomes: tropical rain forest, tropical seasonal forest and scrub, tropical savanna, midlatitude broadleaf forest, boreal and montane forest, temperate rain forest, Mediterranean shrubland, midlatitude grassland, desert, and arctic and alpine tundra.

The global distribution of these biomes is portrayed in ▼ **Figure 14.24**. The following descriptions of each biome synthesize all we have learned in previous chapters about the interactions of atmosphere, hydrosphere, lithosphere, and biosphere. Because plant distributions respond to environmental conditions and reflect variations in climate and soil, the world climate map in Chapter 6, Figure 6.6, is a helpful reference for this discussion.

Tropical Rain Forest

The lush biome covering Earth's equatorial regions is the **tropical rain forest**. In the tropical climates of these forests, with consistent year-round day length (12 hours), high insolation, average annual temperatures around 25°C (77°F), and plentiful moisture, live Earth's most diverse collection of plants and animals. Typical soils are Oxisols, formed under conditions of high precipitation and temperatures. Well-drained uplands may have Ultisols. Although this biome is stable in its natural state, undisturbed tracts of rain forest are becoming increasingly rare because of extensive deforestation.

Mobile Field Trip (MG)

Cloud Forest

https://goo.gl/bRb2R5

▶**14.24 The 10 major global terrestrial biomes** Note that this map subdivides midlatitude grasslands, deserts, and tundra and that *polar desert* is not counted as one of the 10 biomes.

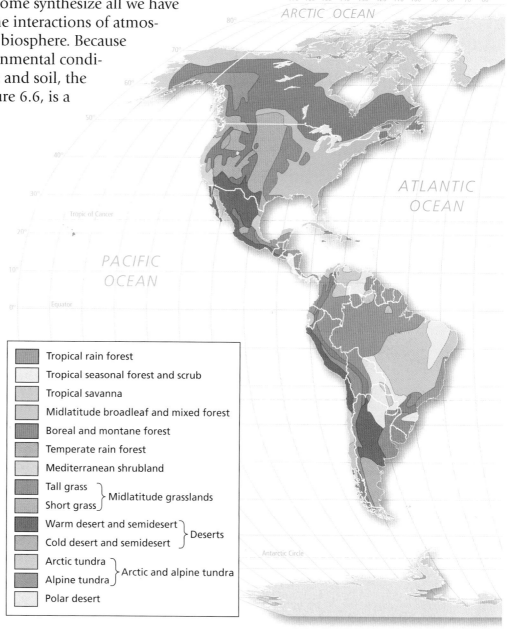

Tropical rain forest

Tropical seasonal forest and scrub

Tropical savanna

Midlatitude broadleaf and mixed forest

Boreal and montane forest

Temperate rain forest

Mediterranean shrubland

Tall grass ⎫
Short grass ⎭ Midlatitude grasslands

Warm desert and semidesert ⎫
Cold desert and semidesert ⎭ Deserts

Arctic tundra ⎫
Alpine tundra ⎭ Arctic and alpine tundra

Polar desert

The largest tropical rain forest occurs in the Amazon region, where it is called the *selva*. Tropical rain forests also cover equatorial Africa and parts of Indonesia and Southeast Asia, Madagascar, the Pacific coast of Ecuador and Colombia, and the east coast of Central America, with small patches elsewhere. Rain forests occupy about 7% of the world's total land area (▶Fig. 14.25) but are home to approximately 50% of Earth's species and about half of its remaining forests.

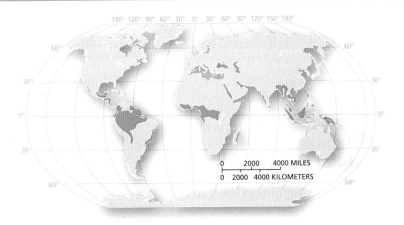

▲14.25 **Rain forest biomes**

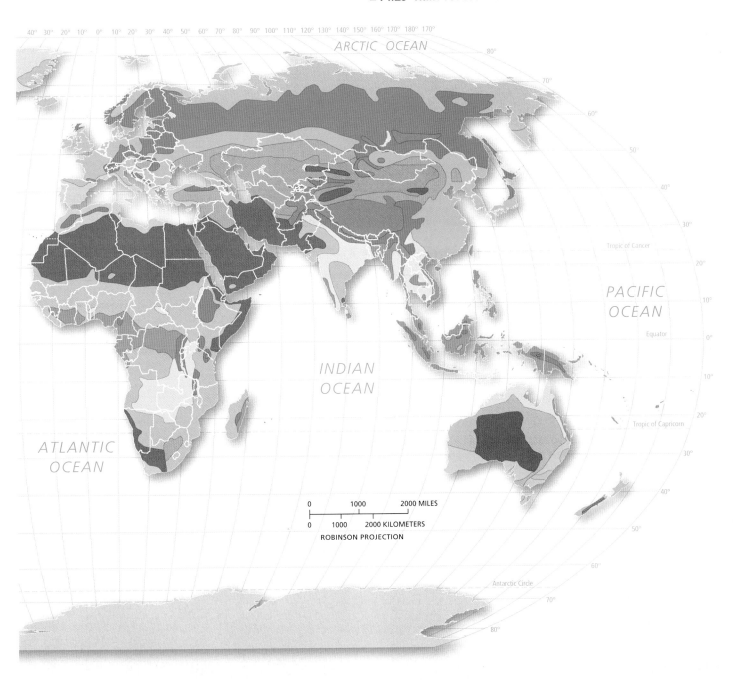

14.4a (cont'd) Tropics

Rain Forest Flora & Fauna The structure of a rain forest includes four levels, illustrated in **Figure 14.26**. Biomass is concentrated in the dense mass of leaves in the *overstory* and *middle canopy*. Beneath these levels, the *understory* is made up of shade-tolerant shrubs, herbs, and small trees. Only about 1% of the sunlight arriving at the canopy reaches the forest floor, which includes seedlings and ferns on a litter-strewn ground surface in deep shade. The high humidity, odors of mold and rotting vegetation, strings of thin roots and vines dropping down from above, windless air, and echoing sounds of life in the trees together create a unique environment.

Because of competition for light, ecological niches are distributed vertically rather than horizontally. The canopy is filled with a rich variety of plants and animals. Woody vines, called *lianas,* are rooted in the soil and stretch from tree to tree, entwining them with thick cords. Orchids, bromeliads, and ferns, are epiphytes—plants that live entirely above-ground, supported physically, but not nutritionally, by the structures of other plants. On the forest floor, the smooth, slender trunks of rain forest trees gain support from large, wall-like buttresses that grow out from and brace the trunks. These buttresses form angular hollows that provide habitat for various animals (▼**Fig. 14.27**). Branches are usually absent on at least the lower two-thirds of the tree trunks.

The rain forest's diverse animal life ranges from animals living exclusively in the upper stories of the trees to decomposers (bacteria) in the leaf litter and soil. Tree-dwelling animals include sloths, monkeys, lemurs, parrots, and snakes. Throughout the canopy are multicolored birds (▼**Fig. 14.28**), tree frogs, lizards, bats, and more than 500 species of butterflies. The rain forest's insect biodiversity is astounding: Biologists studying a tract of rain forest in Ecuador estimated that about 100,000 insect species inhabit one hectare (2.47 acres) of forest. On the forest floor, animals include pigs, small antelope, and mammalian predators (the tiger in Asia, jaguar in South America, and leopard in Africa and Asia).

▼**14.26 The triple canopy structure of the tropical rain forest**

60 m (200 ft)

50 m (165 ft)

Overstory, or Emergent Layer:
The overstory, or emergent layer, consists of the tallest trees, whose tops "emerge" from the main canopy, jutting above the surrounding forest.

40 m (130 ft)

Middle canopy:
Formed of the interlocking crowns of mature trees, the middle canopy is home to a variety of animals and plants.

20 m (65 ft)

15 m (50 ft)

Understory:
Between the middle canopy and the forest floor, the understory consists of shade-tolerant shrubs and trees and woody vines that climb up tree trunks toward sunlight.

5 m (15 ft)

Forest floor:
The forest floor is a deeply shaded area of ferns and litter of dead leaves and other plant material.

Animation (MG)
Plant Productivity in a Warming World

http://goo.gl/rwC5xC

▼**14.27 Thick buttresses stabilize rain forest trees** Leaf litter covers the rainforest floor, seen here with a typical buttressed tree trunk and lianas in the background.

▼**14.28 Immature green tree python in Queensland, Australia** This species mostly lives in trees, shrubs, or bushes in or near rain forests.

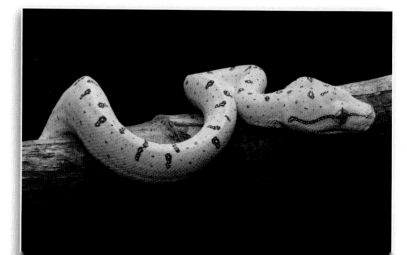

Deforestation of the Tropics For thousands of years, people have cleared Earth's tropical rain forest to make way for agriculture. Today, growing population, economic development, and globalization have increased the scale of deforestation. For several decades, tropical forests have been cleared at alarming rates for farming, fuel wood, cattle ranching, timber export, and most recently, palm oil production (▼Fig. 14.29). Total rain forest losses are now estimated at more than 50% in Africa, more than 40% in Asia, and 40% in Central and South America. Deforestation threatens native rain forest species, including potential sources of valuable pharmaceuticals and new foods—so much is still unknown and undiscovered. As discussed in Chapter 7, forest clearing and burning release millions of metric tons of carbon into the atmosphere each year.

Worldwide, an area nearly the size of Wisconsin is cleared each year (169,000 km², or 65,000 mi²). Selective cutting along the edges of deforested areas disrupts an additional 59,000 km² (22,000 mi²). Scientists are able to track tropical deforestation using satellite images, such as those in **Figure 14.30** of Rondônia in western Brazil. In this region, new roads and forest removal have significantly changed the landscape and disturbed adjacent uncut forest.

In Brazil, deforestation has decreased in recent years as the government has stepped up enforcement to preserve the forest: In 2012–2013, combined losses were down to approximately 10,000 km² (3800 mi²), which is slightly smaller than the big island of Hawai'i. This effort to reduce deforestation is controversial, especially for Brazil's growing cattle industry, which uses deforested lands for pasture.

geoCHECK ✔ Why is deforestation in the tropics such an important issue?

Animation MG
Amazon Deforestation

http://goo.gl/a5gKav

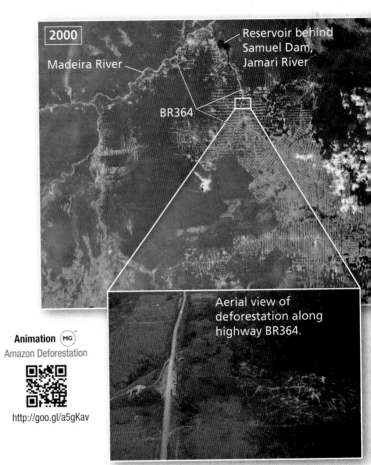

(a) True-color image of Rondônia in western Brazil in 2000 shows deforestation along highway BR364, the main artery of the region.

(b) The same region in 2009. Note the increased amount of deforested land and the branching pattern of feeder roads.

▲14.30 Amazon deforestation

Forest
Deforested by 2015
Projected deforestation by 2050

▲14.29 Deforestation in tropical Latin America

14.4a (cont'd) **Tropics**

Tropical Seasonal Forest & Scrub

At the margins of the world's rain forests are areas of tropical monsoon and tropical savanna climates, characterized by the **tropical seasonal forest and scrub** biome (▶**Fig. 14.31**). These regions have lower and more erratic rainfall than that which occurs in the equatorial zone. The biome includes tropical monsoon and tropical deciduous forest. The shifting intertropical convergence zone (ITCZ) creates a wet summer and dry winter pattern of precipitation with winter moisture deficits, which affects vegetation leaf loss and flowering. Many broadleaf trees lose some of their leaves during the dry season.

Thus, the tropical seasonal forest and scrub is a varied biome that occupies a transitional area from wetter to drier tropical climates. Natural vegetation ranges from monsoon forests to open woodlands to thorn forests to semiarid shrublands. The monsoonal forests have an average height of 15 m (50 ft) with no continuous canopy of leaves, transitioning into drier areas with open grassy areas or into areas choked by dense undergrowth. In the more-open tracts, a common tree is the acacia, with its flat top and usually thorny stems. Scrub vegetation consists of low shrubs and grasses with some adaptations to semiarid conditions.

Local names are given to these communities: the Caatinga of the Bahia State of northeastern Brazil; the Chaco (or Gran Chaco) of southeastern Brazil, Paraguay, and northern Argentina (▶**Fig. 14.32**); the brigalow scrub of Australia; and the dornveld of southern Africa. In Africa, this biome extends west to east from Angola through Zambia to Tanzania and Kenya. Tropical seasonal forests are also present in Southeast Asia and portions of India, from interior Myanmar through northeastern Thailand, and in parts of Indonesia.

The trees throughout most of this biome make poor lumber, but some, especially teak, may be valuable for fine cabinetry and furniture. In addition, some of the plants with dry-season adaptations produce usable waxes and gums, such as the carnauba wax produced by the Brazilian palm tree. Animal life includes the koalas and cockatoos of Australia (▶**Fig. 14.33**) and the elephants, large cats, anteaters, rodents, and ground-dwelling birds in other examples of this biome. Worldwide, humans use these areas for ranching; in Africa, this biome includes numerous wildlife parks and preserves.

geoCHECK ✔ What is the main climatic factor that creates the tropical seasonal forest and scrub?

▲**14.31 Tropical seasonal forest biome**

▲**14.32 Tropical seasonal forest and scrub** Bottle tree, Gran Chaco, Paraguay

▶**14.33 Koala in Australia** Koalas ferment food in their extremely long digestive tracts, which allows them to digest otherwise toxic eucalyptus leaves.

Tropical Savanna

The **tropical savanna** biome (▶Fig. 14.34) consists of large expanses of grassland, either treeless, or interrupted by scattered trees and shrubs (▶Fig. 14.35). The tropical savanna biome is associated with the tropical savanna climate, which receives precipitation during approximately 6 months of the year, when they are influenced by the shifting ITCZ. The rest of the year they are under the drier influence of shifting subtropical high-pressure cells. This is a transitional biome between the tropical seasonal forests and the semiarid tropical steppes and deserts. The soils of the tropical savanna include Alfisols, Ultisols, and Oxisols.

Shrubs and trees of the savanna biome are adapted to drought, grazing by large herbivores, and fire. Most plant species are drought resistant, with various adaptations to help them conserve moisture during the dry season, for example, small, thick leaves, or waxy leaf surfaces (other adaptations are discussed with the desert biomes).

Savanna vegetation is maintained by fire, both a natural and human-caused disturbance. During the wet season, grasses flourish, but as rainfall diminishes, this thick growth provides fuel for fires, which are often intentionally set to maintain the open grasslands. Hot-burning dry-season fires kill trees and seedlings and deposit a layer of nutrient-rich ash over the landscape (▶Fig. 14.36). These conditions foster the regrowth of grasses, which again grow vigorously as the wet season returns, sprouting from extensive underground root systems that are an adaptation for surviving fire disturbance. In northern Australia, the aboriginal people are credited with creating and maintaining many of the region's tropical savannas; as the traditional practice of setting annual fires declines, many savannas are reverting to forest.

Africa has the largest area of tropical savanna on Earth, including the famous Serengeti Plains of Tanzania and Kenya and the Sahel region, south of the Sahara. Portions of Australia, India, and South America also are part of the savanna biome. Local names for tropical savannas include the Llanos in Venezuela, stretching along the coast and inland east of Lake Maracaibo and the Andes; the Campo Cerrado of Brazil and Guiana; and the Pantanal of southwestern Brazil.

Particularly in Africa, savannas are the home of large land mammals—zebra, giraffe, buffalo, gazelle, wildebeest, antelope, rhinoceros, and elephant. These animals graze on savanna grasses, while others (lion, cheetah) feed upon the grazers themselves. Birds include the ostrich, martial eagle (largest of all eagles), and secretary bird. Many species of venomous snakes, as well as the crocodile, are present in this biome.

geoCHECK ✔ How does the tropical savanna differ from tropical seasonal forest and scrub?

▲14.34 Tropical savanna biome

▲14.35 Elephants and acacias in the tropical savanna of southern Africa

▼14.36 Fire on the Serengeti Plains, Tanzania

14.4b Midlatitudes

Midlatitude Broadleaf & Mixed Forest

Moist subtropical and continental climates support a mixed forest in areas of warm to hot summers and cool to cold winters. This **midlatitude broadleaf and mixed forest** biome includes several distinct communities in North America, Europe, and Asia (▶Fig. 14.37). Ultisols, and some Alfisols, are associated with the midlatitude broadleaf and mixed forest. In the United States, relatively lush evergreen broadleaf forests occur along the Gulf of Mexico. To the north are the mixed deciduous broadleaf and needleleaf trees associated with sandy soils and frequent fires—pines (longleaf, shortleaf, pitch, loblolly) predominate in the southeastern and Atlantic coastal plains. In areas of this region protected from fire, broadleaf trees are dominant. Into New England and westward in a narrow belt to the Great Lakes, white and red pines and eastern hemlock are the principal conifers, mixed with broadleaf deciduous oak, beech, hickory, maple, elm, chestnut, and many others (▶Fig. 14.38).

These mixed stands contain valuable timber, and logging has altered their distribution. Native stands of white pine in Michigan and Minnesota were removed before 1910, although reforestation sustains their presence today. In northern China, these forests have almost disappeared as a result of centuries of harvest. The forest species that once flourished in China are similar to species in eastern North America: oak, ash, walnut, elm, maple, and birch. This biome is quite consistent in appearance from continent to continent and at one time represented the principal vegetation of the humid subtropical (hot-summer) regions of North America, Europe, and Asia.

A wide assortment of mammals, birds, reptiles, and amphibians is distributed throughout this biome. Representative animals (some migratory) include southern flying squirrel (▶Fig. 14.39), white-tailed deer, red fox, opossum, bear, and a great variety of birds, including tanager and cardinal. To the west of this biome in North America are the rich soils and midlatitude climates that favor grasslands, and to the north is the gradual transition to the poorer soils and colder climates that favor the coniferous trees of the northern boreal forests.

geoCHECK ✔ How does forest composition change as you go poleward in the midlatitude broadleaf and mixed forest?

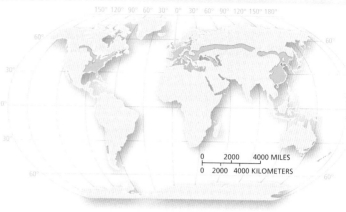

▲14.37 **Midlatitude broadleaf and mixed forest biome**

▲14.38 **Mixed broadleaf forest, southeastern United States**

▶14.39 **Flying squirrel** Found in the eastern half of North America, flying squirrels do not actually fly, but glide using a special membrane.

Boreal & Montane Forest

Stretching from the east coast of Canada and the Atlantic Provinces westward to the Canadian Rockies and portions of Alaska, and from Siberia across the entire extent of Russia to the European Plain is the **boreal forest biome** (▶ Fig. 14.40), also known as the northern needleleaf forest (▶ Fig. 14.41). The northern, less densely forested part of this biome, transitional to the arctic tundra biome, is called the *taiga*. This biome is characteristic of microthermal climates with cold winters and a summer of at least 3 months; the Southern Hemisphere has no such biome except in a few mountainous locales. The needleleaf forests at high elevations on mountains worldwide are called montane forests. The acid-rich needles of the trees found in the needleleaf forests form Spodosols, while wetter areas may have Histosols. Inceptisols and Alfisols are also found in this biome.

Boreal forests of pine, spruce, fir, and larch occur in most of the subarctic climates on Earth that are dominated by trees. The larch (*Larix*) is one of only a few needleleaf trees that drop needles in the winter months, perhaps as a defense against the extreme cold of its native Siberia. Larches also occur in North America (see the climograph for Churchill, Canada, in Figure 6.18a).

This biome also occurs at high elevations at lower latitudes, such as in the Sierra Nevada, Rocky Mountains, Alps, and Himalayas. Douglas fir and white fir grow in the western mountains of the United States and Canada. Economically, these forests are important for lumber in the southern margins of the biome and for pulpwood throughout the middle and northern portions (▶ Fig. 14.42). Present logging practices and the sustainability of these yields are issues of increasing controversy.

Representative fauna in this biome include wolf, elk, moose (the largest member of the deer family), bear, lynx, beaver, wolverine, marten, small rodents, and migratory birds during the brief summer season. Birds include hawks and eagles,

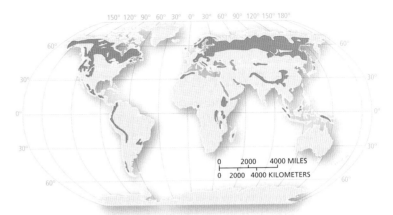

▲ 14.40 **Boreal and montane forest biome**

▲ 14.41 **Montane Forest** A hiker approaching a montane forest of southwest Montana.

several species of grouse, pine grosbeak, Clark's nutcracker, and several species of owls. About 50 species of insects particularly adapted to the presence of coniferous trees inhabit the biome.

geoCHECK ✔ How do the trees in the boreal and montane forest differ from those in the midlatitude broadleaf and mixed forest?

◀ 14.42 **A clear-cut hillside in British Columbia, Canada** Clear-cutting is profitable in the short term, but it destroys the soil microorganisms that coexist with vegetation, removes valuable habitat for animals, and exposes the soil to erosion.

14.4b (cont'd) Midlatitudes

Temperate Rain Forest

The lush forests in wet, humid regions make up the **temperate rain forest** biome (▶ Fig. 14.43). These forests of broadleaf and needleleaf trees, epiphytes, huge ferns, and thick undergrowth correspond generally to the marine west coast climates (occurring along middle- to high-latitude west coasts), with precipitation approaching 400 cm (160 in.) per year, moderate air temperatures, summer fog, and an overall maritime influence (▼ Fig. 14.44a). As with the boreal and montane forest, Spodosols are typical soils, but Inceptisols can be found in more mountainous regions. In North America, this biome occurs only along narrow margins of the Pacific Northwest. Similar temperate rain forests, too small to appear on the world map, exist in southern China, small portions of southern Japan, New Zealand, and a few areas of southern Chile.

The biome is home to bear, badger, deer, wild pig, wolf, bobcat, fox, and numerous bird species, including the northern spotted owl (▼ Fig. 14.44b). In the 1990s, this owl became a symbol for the conflict between species-preservation efforts and the use of resources to fuel local economies. In 1990, the U.S. Fish and Wildlife Service listed the owl as a "threatened" species under the U.S. Endangered Species Act, citing the loss of old-growth forest habitat as the primary cause for its decline. The next year, logging practices in areas with spotted owl habitat were halted by court order. The ensuing controversy pitted environmentalists against loggers and other forest users, with the end result being large-scale changes in forest management throughout the Pacific Northwest.

Later research by the U.S. Forest Service and independent scientists noted the failing health of temperate rain forests and suggested that timber-management plans balance resource use with ecosystem preservation. Sustainable forestry practices emphasize the continuing health and productivity of forests into the future

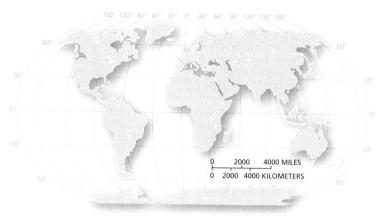

▲ **14.43 Temperate rain forest biomes**

and are increasingly based on a multiuse ethic that serves local, national, and global interests.

The tallest trees in the world occur in this biome—the coast redwoods (*Sequoia sempervirens*) of the California and Oregon coasts. These trees can exceed 1500 years in age and typically range in height from 60 to 90 m (200 to 300 ft), with some exceeding 100 m (330 ft). Old-growth stands of other representative trees, such as Douglas fir, spruce, cedar, and hemlock, have been reduced by timber harvests to a few remaining valleys in Oregon and Washington, less than 10% of the original forest that existed when Europeans first arrived. Most forests in this biome are secondary-growth forests, which have regrown from a major disturbance, usually human-caused.

 geoCHECK ✔ How does the temperate rain forest biome differ from the tropical rain forest biome?

Animation (MG)
End of the
Last Ice Age

http://goo.gl/XB2LMA

▼ **14.44 Temperate rain forest** [Actual Caption Depends Upon New Photo]

(a) Old-growth Douglas fir, redwoods, cedars, and a mix of deciduous trees, ferns, and mosses in the Gifford Pinchot National Forest, Washington. Only a small percentage of these old-growth forests remain in the Pacific Northwest.

(b) The northern spotted owl is an "indicator species" representing the health of the temperate rainforest ecosystem.

Mediterranean Shrubland

The **Mediterranean shrubland** biome, also referred to as a temperate shrubland, occupies temperate regions that have dry summers, generally corresponding to the Mediterranean climates (▶Fig. 14.45). Alfisols and Mollisols are common soils in this biome. The dominant shrub formations are low growing and able to withstand hot-summer drought. The vegetation is sclerophyllous (from *sclero*, for "hard," and *phyllos*, for "leaf"). Most shrubs average a meter or two in height, with deep, well-developed roots, leathery leaves, and uneven low branches.

Typically, the vegetation varies between woody shrubs covering more than 50% of the ground and grassy woodlands covering 25% to 60% of the ground. In California, the Spanish word *chaparro*, for "scrubby evergreen," gives us the name *chaparral* for this vegetation type (▶Fig. 14.46). This scrubland includes species such as manzanita, toyon, red bud, ceanothus, mountain mahogany, blue and live oaks, and the dreaded poison oak.

This biome is located poleward of the shifting subtropical high-pressure cells in both hemispheres. The stable high pressure produces the characteristic dry-summer climate and establishes conditions conducive to fire. The vegetation is adapted for rapid recovery after fire; many species are able to resprout from roots or burls after a burn or have seeds that require fire for germination.

A counterpart to the California chaparral in North America is the maquis of the Mediterranean region of Europe, which includes live and cork oak trees (the source of cork, ▼Fig. 14.47) as well as pine and olive trees. In Chile, this biome is known as the *mattoral*, and in southwestern Australia, it is *mallee scrub*. In Australia, the bulk of the eucalyptus species are sclerophyllous shrubs.

As described in Chapter 6, commercial agriculture of the Mediterranean climates includes subtropical fruits, vegetables, and nuts, with many food types (e.g., artichokes, olives, almonds) produced only in these climates. Animals include several types of deer, coyote, wolf, bobcat, various rodents and other small animals, and various birds. In Australia, this biome is home to malleefowl (*Leipoa ocellata*), a ground-dwelling bird, and numerous marsupials.

geoCHECK ✔ What are some adaptations to wildfire found in Mediterranean shrubland communities?

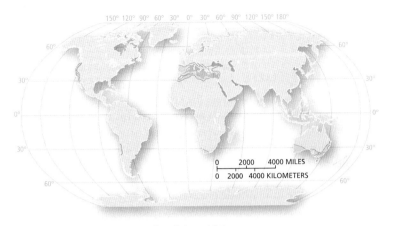

▲14.45 **Mediterranean shrubland biomes**

(a) Chaparral vegetation, southern California.

▲14.46 **Mediterranean chaparral and fire adaptations**

(b) Fire-adapted chaparral sends out sprouts from roots a few months after a wildfire in the San Jacinto Mountains, southern California.

◀14.47 **Recently stripped cork trees in Portugal** Cork oaks are a renewable resource since the outer bark of cork oak trees can be harvested every seven years without harming the tree. The bark is used to make cork stoppers for bottles, bulletin boards, and other products that make use of its insulating or cushioning properties.

14.4c Midlatitudes & Deserts

Midlatitude Grassland

Of all the natural biomes, the **midlatitude grassland** is the most modified by human activity. This biome includes the world's "breadbaskets"—regions that produce bountiful grain (wheat and corn), soybeans, and livestock (hogs and cattle). Mollisols are the most typical soil, but Aridisols are found in drier regions. The only naturally occurring trees found in the midlatitude grassland biome were deciduous broadleaf trees along streams and other limited sites. These regions are called grasslands because of the predominance of grasslike plants before human intervention (▶ Fig. 14.48 shows the natural location of this biome).

In North America, tallgrass prairie vegetation once grew to heights of 2 m (6.5 ft) and extended westward to about the 98th meridian, with shortgrass prairies in the drier lands farther west. As discussed in the previous section on Mollisols, wetter conditions are found to the east and drier conditions to the west of the 98th meridian (see Fig. 14.12).

Few patches of the original prairies (tall grassland) or steppes (short grassland) remain within this biome. One hundred million hectares of prairie is now down to a few areas of several hundred hectares each. Characteristic midlatitude grasslands outside North America are the Pampas of Argentina and Uruguay and the grassland of Ukraine. In most regions where these grasslands were the natural vegetation, human development of them was critical to territorial expansion.

This biome is the home of large grazing animals, including deer, pronghorn, and bison (▶ Fig. 14.49). Gophers, prairie dogs, ground squirrels, turkey vultures, grouse, and prairie chickens are common, as well as grasshoppers and other insects. Predators include the coyote, the black-footed ferret, badger, and birds of prey—hawks, eagles, and owls.

geoCHECK ✔ How are grasses at the far west range of the North American prairie different from the grasses at the far east range?

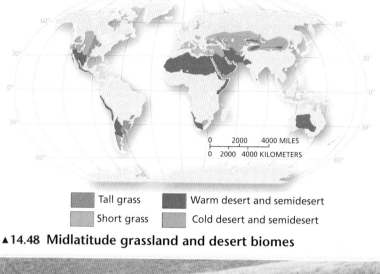

Tall grass	Warm desert and semidesert
Short grass	Cold desert and semidesert

▲14.48 **Midlatitude grassland and desert biomes**

▲14.49 **Bison** Grasslands and bison have evolved together. Areas with reintroduced bison have higher productivity and biodiversity than areas without bison.

Deserts

Earth's **desert** biomes cover more than one third of its land area (**Fig. 14.8**). We subdivide the desert biomes into warm desert and steppe, caused by the dry air and low precipitation of subtropical high-pressure cells, and cold desert and steppe, which tend toward higher latitudes, where subtropical high pressure affects climate for less than 6 months of the year. Typical desert biome soils are Aridisols.

Desert plants evolved many mechanisms to prevent water loss: storing water in thick, fleshy tissues (cacti and other succulents); long taproots to access groundwater (mesquite trees); shallow, spreading root systems to maximize water uptake (palo verde trees); small leaves to minimize surface area for water loss (acacia); waxy leaf coatings to retard water loss (creosote bush); and leaf drop during dry periods (ocotillo) (▶ Fig. 14.50). A number of desert plants developed spines, thorns, or bad-tasting tissue to discourage plant eaters. Finally, some desert wash plants produce seeds that require scratching of the surface for the seed to open and germinate. This can occur from the tumbling, churning action of a flash flood flowing down a desert wash, which also produces the moisture for seed germination.

Some desert plants are ephemeral, or short-lived, to take advantage of a short wet season or even a single rainfall event in desert environments. The seeds of desert ephemerals lie dormant until a rainfall stimulates the seed germination. Seedlings grow rapidly, mature, flower, and produce large

Thick fleshy tissue to store water (cacti and other succulents)

Waxy leaf coatings to retard water loss (creosote bush)

Leaf drop during dry periods (ocotillo)

Long taproots to access groundwater (mesquite trees)

Small leaves to minimize surface area for water loss (acacia)

Shallow, spreading root systems to maximize water uptake (palo verde trees)

▲14.50 Typical desert plant adaptations

numbers of new seeds, which are then dispersed long distances by wind or water and lay dormant until the next rains.

In cold deserts, where precipitation is greater and temperatures are colder, characteristic vegetation includes grasses and woody shrubs, such as sagebrush. Succulents that hold large amounts of water, such as the saguaro cactus (▶ **Fig. 14.51**), cannot survive in cold deserts with consecutive days or nights with freezing winter temperatures.

Because of the extreme conditions, there are only a few large resident animals in both warm and cold deserts. Camels, from the deserts of Africa and the Middle East, can lose up to 30% of their body weight in water without harm (for humans, a 10%–12% loss is dangerous). In western deserts of the United States, desert bighorn sheep populations declined precipitously from about 1850 to 1900 due to competition with livestock for food and water, as well as exposure to parasites and disease. Several states are transplanting the animals to their former ranges.

Other representative desert animals are the ring-tailed cat, kangaroo rat, lizards, scorpions, and snakes. Most of these animals become active only at night, when temperatures are lower. In addition, various birds have adapted to desert conditions and available food sources, for example, roadrunners, thrashers, ravens, wrens, hawks, grouse, and nighthawks.

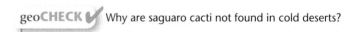 geoCHECK ✔ Why are saguaro cacti not found in cold deserts?

▼14.51 **Saguaro cacti in the Sonoran Desert**

14.4d Tundra & Anthropogenic Biomes

Arctic & Alpine Tundra

The **arctic tundra** biome is located in the extreme northern area of North America and Russia, bordering the Arctic Ocean and generally north of the 10°C (50°F) isotherm for the warmest month (▶Fig. 14.52). Day length varies greatly throughout the year, seasonally changing from almost continuous day to continuous night. The region, except for a few portions of Alaska and Siberia, was covered by ice during all of the Pleistocene glaciations.

This biome corresponds to the tundra climates; winters are cold and long, summers are cool and brief. A growing season of sorts lasts only 60 to 80 days, and even then frosts can occur at any time. Typical soils include Gelisols in colder regions, Histosols in wetter regions, and Entisols in regions of permafrost. In the summer months, thawing permafrost produces a mucky surface of poor drainage (▶Fig. 14.53). Roots can penetrate only to the depth of thawed ground, usually about a meter (3 ft). With recent climate change, these regions have been warming at more than twice the rate of the rest of the planet over the past few decades.

Arctic tundra vegetation consists of low, ground-hugging herbaceous plants such as sedges, mosses, arctic meadow grass, and snow lichen and some woody species such as dwarf willow. Owing to the short growing season, some perennials form flower buds one summer and open them for pollination the next. Animals of the tundra biome include musk ox, caribou, reindeer, rabbit, ptarmigan, lemming, and other small rodents, which are important food for the larger carnivores—the wolf, fox, weasel, snowy owl, polar bear, and, of course, mosquito. The tundra is an important breeding ground for geese, swans, and other waterfowl.

Alpine tundra is similar to arctic tundra, but it can occur at lower latitudes because it is associated with high elevations. This biome usually occurs above the tree line (the elevation above which trees cannot grow), which shifts to higher elevations closer to the equator. Alpine tundra communities occur in the Andes near the equator, the White Mountains and Sierra Nevada of California, the American and Canadian Rockies, the Alps, and Mount Kilimanjaro of equatorial Africa, as well as in mountains from the Middle East to Asia.

In many alpine locations, plants have forms that are shaped by frequent winds. Alpine tundra can also experience permafrost conditions. Characteristic fauna include mountain goats, Rocky Mountain bighorn sheep, elk, and voles (▶Fig. 14.54).

Vegetation of the tundra biome is slow growing, has low productivity, and is easily disturbed. Hydroelectric projects, mineral exploitation, and even tire tracks leave marks on the landscape that persist for hundreds of years. With rising population and energy demand, the region will face even greater challenges from the environmental impacts of petroleum resource development. The possibility of drilling for shale gas and methane hydrates is a new threat to this biome.

geoCHECK ✔ Why are arctic and alpine tundra biomes so sensitive to disturbance?

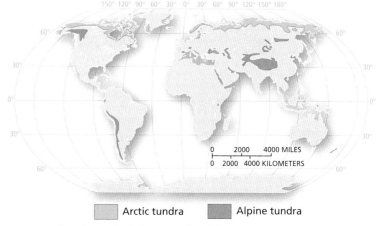

Arctic tundra ▢ Alpine tundra ▢

▲**14.52 Arctic and alpine tundra biomes**

▲**14.53 Arctic tundra** This treeless expanse of tundra in arctic Russia near the Barents Sea is typical of wide areas across northern Eurasia, Canada, and Alaska.

▶**14.54 Mountain goat family in Logan Pass, Montana** The largest mammals in their high altitude habitat, mountain goats rarely descend into valleys.

Anthropogenic Biomes

Even in many of the most pristine ecosystems on Earth, evidence of early human settlement exists. Today, we are the most powerful biotic agent on Earth, influencing all ecosystems on a planetary scale. Scientists are measuring ecosystem properties and building elaborate computer models to simulate the evolving human–environment experiment on our planet—in particular, the shifting patterns of environmental factors (temperatures and changing frost periods; precipitation timing and amounts; air, water, and soil chemistry; and nutrient redistribution) wrought by human activities.

In 2008, geographers presented the concept of **anthropogenic biomes**, based on today's human-altered ecosystems, as an updated and more accurate portrayal of the terrestrial biosphere than the "pristine" natural vegetation communities described in most biome classifications. Their map, shown in The Human Denominator shows five broad categories of human-modified landscapes: settlements, croplands, rangelands, forested lands, and wildlands. Within these categories, the scientists defined 21 biomes, which summarize the current mosaic of landscapes in terms of common combinations of land uses and land cover.

Anthropogenic biomes result from ongoing human interaction with ecosystems, linked to land-use practices such as agriculture, forestry, and urbanization (▼**Fig. 14.55**). The most extensive anthropogenic biome is rangelands, covering about 32% of Earth's ice-free land; followed by croplands, forested lands, and wildlands, each at about 20%; and settlements at about 7%.

The concept of anthropogenic biomes does not replace terrestrial biome classifications but instead presents another perspective. Understanding of the natural biomes presented in this chapter is essential for the advancement of basic and applied science as it applies to conservation biogeography and ecosystem and species restoration.

geoCHECK ✔ What is the most extensive anthropogenic biome?

geoQUIZ

1. Which biome do you live in? Which biome would you live in if the climate became warmer and wetter? Warmer and drier?
2. Which characteristics of rain forests produce their high biomass and biodiversity? Explain your answer.
3. Describe some adaptations by plants to climatic conditions in either the Mediterranean biome or the desert biome.

(a) Cattle ranch in Hawai'i.

(b) Rain forest land cleared for agriculture in Brazil.

▲**14.55 Anthropogenic biomes: Rangeland and cropland** Agriculture depends on preventing natural sucession. This results in a dramatic decrease in biodiversity, as well as resilience and stability. Farmers often use pesticides to prevent non-crop plants from reestablishing themselves.

BIOMES IMPACT HUMANS

• Natural plant and animal communities are linked to human cultures, providing resources for food and shelter.

• Earth's remaining undisturbed ecosystems are becoming a focus for tourism, recreation, and scientific attention.

HUMANS IMPACT BIOMES

• Invasive species, many introduced by humans, disrupt native ecosystems.

• Tropical deforestation is ongoing, with more than half of Earth's original rain forest already cleared.

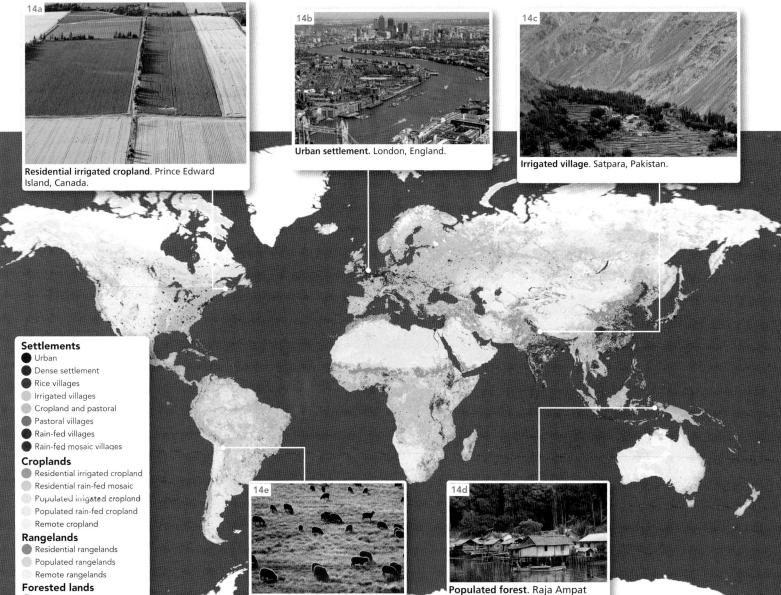

14a

Residential irrigated cropland. Prince Edward Island, Canada.

14b

Urban settlement. London, England.

14c

Irrigated village. Satpara, Pakistan.

Settlements
- ● Urban
- ● Dense settlement
- ● Rice villages
- ● Irrigated villages
- ● Cropland and pastoral
- ● Pastoral villages
- ● Rain-fed villages
- ● Rain-fed mosaic villages

Croplands
- ● Residential irrigated cropland
- ● Residential rain-fed mosaic
- ● Populated irrigated cropland
- ● Populated rain-fed cropland
- ● Remote cropland

Rangelands
- ● Residential rangelands
- ● Populated rangelands
- ● Remote rangelands

Forested lands
- ● Populated forest
- ● Remote forest

Wildlands
- ● Wild forest
- ● Sparse trees
- ● Barren or ice-covered

14e

Remote rangelands. Northern Chile.

14d

Populated forest. Raja Ampat Islands, Indonesia.

Map courtesy of Erle Ellis, University of Maryland, Baltimore County, and Navin Ramankutty, McGill University/NASA; available at: http://earthobservatory.nasa.gov/IOTD/view.php?id=40554

ISSUES FOR THE 21ST CENTURY

• Management of species and ecosystems must become a priority to avoid extinctions and loss of diversity.

• Shifting of species distributions in response to environmental factors will continue with ongoing climate change.

• Population control and global education (including education for women and disadvantaged minorities in all countries) are critical for sustaining natural and anthropogenic biomes.

Looking Ahead

All of Earth's physical systems combine to produce the biomes into which all regions on Earth can be classified. We see the patterns of these interactions throughout the biosphere and in the diversity of life. Today, we face the challenge of preserving the diversity of communities, ecosystems, and biomes amid climate change and other threats from human activities. Physical geography and Earth system science are important contributors to these efforts.

As our exploration of *Geosystems Core* concludes, your journey now leads onward into the 21st century. Study well and travel safely—with the wind at your back.

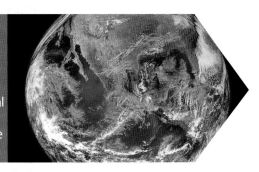

How are Earth's biomes organized?

14.1 Biogeographic Divisions

Identify the world's biogeographic realms and explain how their boundaries are determined.

Explain the basis for grouping plant communities into biomes.

- Plants and animals are grouped into biomes, representing the major ecosystems of Earth. Biomes are large, stable communities of plants and animals with boundaries related to climate. Biogeographic realms are determined by species distribution and evolutionary history. Today, eight biogeographic realms are recognized. Biomes are defined by vegetation characteristics, such as growth forms and vegetation types. Growth forms are based on size, woodiness, life span, leaf traits, and general shape. Biogeographers designate six major groups of terrestrial vegetation: forest, savanna, shrubland, grassland, desert, and tundra. These groups are further divided into 10 to 16 biomes. Biomes are separated by ecotones, which may be gradual transitions over a large area, such as the shortgrass-to-tallgrass transition of the American prairies, or abrupt transitions, such as the bank of a river. Because of the range of environmental conditions within an ecotone, they can have higher biodiversity than the habitats that are on either side of them.

1. How are biomes and biogeographic realms classified?

2. Give examples of wide and narrow ecotones.

What are geographic tools for conservation?

14.2 Conservation Biogeography

Describe the impact of invasive species on biotic communities, using several examples.

Analyze how the theory of island biogeography has led to new strategies for biodiversity conservation.

- Invasive species are those that have been introduced into a community from elsewhere, either by humans or by natural processes. Probably 90% of invasive species fail, but some have characteristics that allow them to flourish in their new location, such as a lack of predators. Invasive species can alter the dynamics of entire biomes. Invasive species often do not provide food or habitat for native organisms, and they can dramatically alter the fire dynamics of a biome. Species that have evolved on islands in isolation often are unable to compete with new, introduced nonnative species. In most of Hawai'i, native birds are no longer found below 1200 m (4000 ft) due to an introduced virus. Island biogeography relates habitat size to the number of species in that habitat. Just as an island in the ocean is one kind of habitat surrounded by a radically different one, parks and reserves act as islands. This makes the tool of island biogeography an important one in species preservation. Island biogeography compares the rate of immigration to the rate of extinction to determine species equilibrium numbers. Large islands close to the mainland will have higher species richness than will smaller, more distant islands.

3. Why are successful invasive species so devastating to native species and biomes?

4. What factors influence rates of immigration and extinction on islands?

How are soils classified?

14.3 Soil Classification

Classify the 12 soil orders of the Soil Taxonomy according to their characteristics.

Describe the distribution of the soil orders across Earth and how soils affect plant cover and agriculture.

- Soils in the United States are classified using the NRCS's Soil Taxonomy. The Soil Taxonomy classification system is based on soil appearance, form, and structure. The smallest unit of the taxonomy is the pedon, which is a hexagonal column from 1 to 10 m^2 in top surface area. The soil horizons within a pedon are used to group soils into soil series. The 15,000 soil series are grouped into families, subgroups, great groups, suborders, and finally 12 soil orders. In the past, soils were described by pedogenic regimes, which related climates to soil-forming processes. However, because one pedogenic regime can operate in many climates and soil orders, they are no longer used to classify soils. They are still useful in understanding soil formation processes. The five pedogenic regimes are laterization, found in tropical climates; calcification, found in arid climates; salinization, also found in arid climates; podzolization, found in forest soils with cool climates; and gleization, found in cold, wet climates in regions with poor drainage. The 12 soil orders are: the tropical Oxisols; the arid Aridisols; the grassland Mollisols; the forest Alfisols; the highly weathered forest Ultisols; the coniferous forest Spodosols; the recent Entisols; the weakly developed Inceptisols; the cold and frozen Gelisols; the volcanic Andisols; the swelling clay Vertisols; and the high organic content Histosols.

5. Briefly describe two pedogenic regimes. With which soil orders are they associated?

6. Why are the pedogenic regimes no longer used to classify soils? How are they useful in understanding soil formation processes?

What are the spatial patterns of Earth's biomes & soils?

14.4 Earth's Terrestrial Biomes

Summarize the characteristics of Earth's 10 major terrestrial biomes.

Locate the 10 biomes on a world map.

- *Geosystems Core* describes the following biomes: tropical rain forest, tropical seasonal forest and scrub, tropical savanna, midlatitude broadleaf forest, boreal and montane forest, temperate rain forest, Mediterranean shrubland, midlatitude grassland, desert, and arctic and alpine tundra. For an overview of the biomes and their vegetation characteristics, soil orders, climate type designations, annual precipitation ranges, temperature patterns, and water-balance characteristics, review Table 14.1 online at masteringgeography.com.

7. How do biomes differ from ecosystems?

8. Why does the boreal forest biome not exist in the Southern Hemisphere, except in mountainous regions? Where is this biome located in the Northern Hemisphere, and what is its relationship to climate type?

9. What are anthropogenic biomes? How do they differ from natural biomes?

Critical Thinking

1. Recent advances in genome analysis have resulted in major changes in the family trees of many kinds of organisms. If organisms are moved to new branches on their family trees, or even moved to new trees, would you expect biogeographic realm or biome boundaries to shift more dramatically? Explain your answer.

2. What are some possible reasons for Australia's unique native organisms?

3. Why are island organisms especially sensitive to disruption from invasive nonnative organisms?

4. How is the construction of a new road through a national park similar to a new river? Discuss possible consequences to species richness of new road construction in a park.

5. Some have proposed that all biomes should be considered anthropogenic biomes because of the extent of human impact on Earth. Explain why you agree or disagree with this statement.

Visual Analysis

In the United States and Canada, humans have established a new type of terrestrial plant community in developed areas. This new community, a mix of native and nonnative species used for landscaping, is somewhere between a grassland and a forest. Investments of water, energy, and capital are required to sustain the new community. This example from Arizona shows xeriscaping, landscaping that emphasizes drought-tolerant native plants.

1. Describe any observable characteristics of the plants in the photograph that represent adaptations to the desert environment.

2. What might be some environmental benefits of using native plants in landscaping?

▲R14.1

Interactive Mapping

Login to the **MasteringGeography** Study Area to access **MapMaster**.

- Open: MapMaster in MasteringGeography.
- Select: the *Soil Types/Taxonomy* layer from the *Physical Environment* menu. You should see the soil orders as different colors.

1. What is the main soil order in your state?

2. What is the main soil order in the southeastern United States?

3. What is the main soil order in the Amazon rain forest?

4. What is the main soil order across Norway, Sweden, Finland, from Connecticut north to Nova Scotia, and from the Great Lakes north to southern Ontario and Quebec?

5. What type of trees would you expect to find across the region in question 4? Explain your answer.

Explore

Use **Google Earth** to explore Earth's biomes.

Google Earth provides a cloud-free view of Earth that is useful in observing patterns of vegetation. Pan and zoom so that you can see all of Africa. The desert regions should be obvious, due to the light tan color of exposed rock and sand. The rain forests are also obvious, due to the dark green of forest vegetation. There are several climates and biomes between the Sahara Desert in the north and the central rain forests.

1. What is the latitudinal range (how far north and south) of the tropical rain forest in Africa? How far south does the Sahara Desert extend? What are the climates and biomes between the central tropical rain forest and the Sahara Desert?

2. Pan and zoom to southwestern Ethiopia. Why do you think southwestern Ethiopia has more vegetation than Somalia to the east, and Sudan and South Sudan to the west?

▲R14.2

MasteringGeography™

Looking for additional review and test prep materials? Visit the Study Area in MasteringGeography™ to enhance your geographic literacy, spatial reasoning skills, and understanding of this chapter's content by accessing a variety of resources, including MapMaster™ interactive maps, videos, *Mobile Field Trips, Project Condor* Quadcopter videos, *In the News* RSS feeds, flashcards, web links, self-study quizzes, and an eText version of *Geosystems Core.*

Key Terms

Alfisols, p. 378
alpine tundra, p. 404
Andisols, p. 390
anthropogenic biome, p. 405
arctic tundra, p. 404
Aridisols, p. 387
biogeographic realm, p. 380
biome, p. 380

boreal forest, p. 399
calcification, p. 385
desert, p. 402
ecotone, p. 381
Entisols, p. 389
Gelisols, p. 390
gleization, p. 385
habitat fragmentation, p. 383
Histosols, p. 391

Inceptisols, p. 389
invasive species, p. 382
island biogeography, p. 383
laterization, p. 385
Mediterranean shrubland, p. 401
midlatitude broadleaf and mixed forest, p. 398

midlatitude grassland, p. 402
Mollisols, p. 387
Oxisols, p. 386
pedogenic regime, p. 385
podzolization, p. 385
salinization, p. 385
Soil Taxonomy, p. 384
Spodosols, p. 388

temperate rain forest, p. 400
tropical rain forest, p. 392
tropical savanna, p. 397
tropical seasonal forest and scrub, p. 396
Ultisols, p. 388
Vertisols, p. 390

Hawaiian Habitats: Preserving the Big Island's Biomes

Hawai'i is a remarkable location in terms of biodiversity and climatic diversity. Because of the elevation range on the Big Island, most of the world's climates are found on this tropical island.

GeoLab14 (MG)
Pre-Lab Video

https://goo.gl/ONGDIW

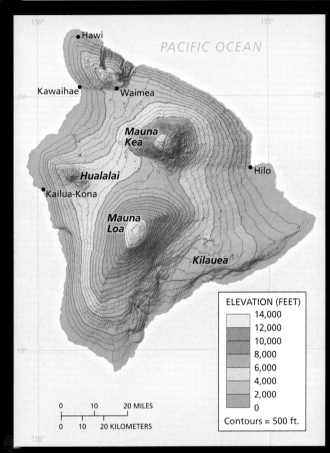

▲ GL14.1 **Hawai'i elevations**

ELEVATION (FEET)
14,000
12,000
10,000
8,000
6,000
4,000
2,000
0
Contours = 500 ft.

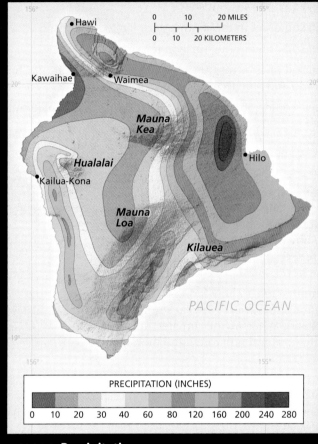

▲ GL14.2 **Precipitation on Hawai'i**

PRECIPITATION (INCHES)
0 10 20 30 40 60 80 120 160 200 240 280

Apply

Working for a nonprofit organization that sponsors research on preserving biodiversity, you are conservation planner looking for sites to create natural reserves for study. You will need to identify the climates and biomes found on Hawai'i, and identify sites that are representative for each of them.

Objectives

- Identify the climates and biomes on Hawai'i.
- Describe the factors responsible for creating the pattern of climates and biomes on Hawai'i.
- Relate the land cover to the climate.

Procedure

Use the maps of elevation, precipitation, climate, and land cover to complete your analysis and answer the questions.

1. What do you think are the main factors that create the pattern of precipitation?
2. The trade winds blow during most of the year. From which direction do they blow? Explain your answer and refer to the global patterns of wind.
3. How would the precipitation map help you determine the direction of the trade winds?
4. Identify two regions where the precipitation pattern is affected by orographic lifting processes.
5. What climate is at Kawaihae? What biome would you expect to find there, based on its climate?
6. What biome would be found at Kawaihae if the prevailing winds were from the opposite direction?
7. What climate is at Hilo? What biome would you expect to find there based on its climate?

8. What biome would be found at Hilo if the prevailing winds were from the opposite direction?
9. Examine Figure GL14.4 What are two reasons that bare ground would be on the Big Island, based on the locations of areas with bare ground?
10. Which climates are not represented? What would be required to have those climates?
11. How the Big Island have arctic tundra?

Analyze & Conclude

Review the maps in this GeoLab and if needed **Figure 14.24** and Table 14.1 online at masteringgeography.com.

12. List and describe the biomes that you would expect to find on Hawai'i.
13. Identify the biomes that you would classify as anthropogenic. Explain your answer.

14. How many reserves would be needed to capture the range of nonanthropogenic biomes of Hawai'i?

15. Describe the proposed locations for each of these biodiversity reserves, either by latitude/longitude or the nearest city.

15. Sketch and label the locations of your proposed biome reserves on **Figure GL 14.5**.

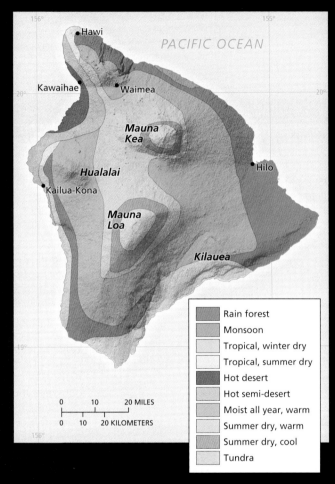

▲ GL14.3 **Climates of Hawai'i**

Legend:
- Rain forest
- Monsoon
- Tropical, winter dry
- Tropical, summer dry
- Hot desert
- Hot semi-desert
- Moist all year, warm
- Summer dry, warm
- Summer dry, cool
- Tundra

▲ GL14.4 **Land cover of Hawai'i**

Legend:
- Developed
- Cultivated crops
- Hay or pasture
- Evergreen forest
- Herbaceous
- Shrub or scrub
- Barren

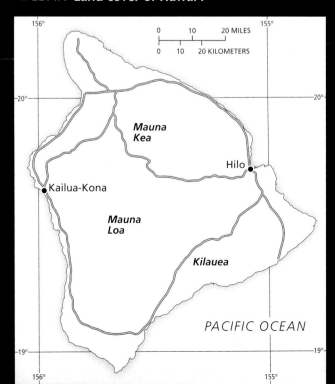

Common Conversions

Metric to English

Metric Measure	Multiply by	English Equivalent
Length		
Centimeters (cm)	0.3937	Inches (in.)
Meters (m)	3.2808	Feet (ft)
Meters (m)	1.0936	Yards (yd)
Kilometers (km)	0.6214	Miles (mi)
Nautical mile	1.15	Statute mile
Area		
Square centimeters (cm^2)	0.155	Square inches (in.^{2})
Square meters (m^2)	10.7639	Square feet (ft^2)
Square meters (m^2)	1.1960	Square yards (yd^2)
Square kilometers (km^2)	0.3831	Square miles (mi^2)
Hectare (ha) (10,000 m^2)	2.4710	Acres (a)
Volume		
Cubic centimeters (cm^3)	0.06	Cubic inches (in.^{3})
Cubic meters (m^3)	35.30	Cubic feet (ft^3)
Cubic meters (m^3)	1.3079	Cubic yards (yd^3)
Cubic kilometers (km^3)	0.24	Cubic miles (mi^3)
Liters (l)	1.0567	Quarts (qt), U.S.
Liters (l)	0.88	Quarts (qt), Imperial
Liters (l)	0.26	Gallons (gal), U.S.
Liters (l)	0.22	Gallons (gal), Imperial
Mass		
Grams (g)	0.03527	Ounces (oz)
Kilograms (kg)	2.2046	Pounds (lb)
Metric ton (tonne) (t)	1.10	Short ton (tn), U.S.
Velocity		
Meters/second (mps)	2.24	Miles/hour (mph)
Kilometers/hour (kmph)	0.62	Miles/hour (mph)
Knots (kn) (nautical mph)	1.15	Miles/hour (mph)
Temperature		
Degrees Celsius (°C)	1.80 (then add 32)	Degrees Fahrenheit (°F)
Celsius degree (C°)	1.80	Fahrenheit degree (F°)
Additional water measurements		
Gallon (Imperial)	1.201	Gallon (U.S.)
Gallons (gal)	0.000003	Acre-feet

1 cubic foot per second per day = 86,400 cubic feet = 1.98 acre-feet

Additional Energy and Power Measurements

1 watt (W) = 1 joule/s

1 joule = 0.239 calorie

1 calorie = 4.186 joules

1 W/m^2 = 0.001433 cal/min

697.8 W/m^2 = 1 cal/cm^{2}min^{-1}

1 W/m^2 = 2.064 cal/cm^{2}day^{-1}

1 W/m^2 = 61.91 cal/cm^{2}month^{-1}

1 W/m^2 = 753.4 cal/cm^{2}year^{-1}

100 W/m^2 = 75 kcal/cm^{2}year^{-1}

Solar constant:
1372 W/m^2
2 cal/cm^{2}min^{-1}

English to Metric

English Measure	Multiply by	Metric Equivalent
Length		
Inches (in.)	2.54	Centimeters (cm)
Feet (ft)	0.3048	Meters (m)
Yards (yd)	0.9144	Meters (m)
Miles (mi)	1.6094	Kilometers (km)
Statute mile	0.8684	Nautical mile
Area		
Square inches (in.^{2})	6.45	Square centimeters (cm^2)
Square feet (ft$^{2)}$)	0.0929	Square meters (m^2)
Square yards (yd^2)	0.8361	Square meters (m^2)
Square miles (mi^2)	2.5900	Square kilometers (km^2)
Acres (a)	0.4047	Hectare (ha)
Volume		
Cubic inches (in.^{3})	16.39	Cubic centimeters (cm^3)
Cubic feet (ft^3)	0.028	Cubic meters (m^3)
Cubic yards (yd^3)	0.765	Cubic meters (m^3)
Cubic miles (mi^3)	4.17	Cubic kilometers (km^3)
Quarts (qt), U.S.	0.9463	Liters (l)
Quarts (qt), Imperial	1.14	Liters (l)
Gallons (gal), U.S.	3.8	Liters (l)
Gallons (gal), Imperial	4.55	Liters (l)
Mass		
Ounces (oz)	28.3495	Grams (g)
Pounds (lb)	0.4536	Kilograms (kg)
Short ton (tn), U.S.	0.91	Metric ton (tonne) (t)
Velocity		
Miles/hour (mph)	0.448	Meters/second (mps)
Miles/hour (mph)	1.6094	Kilometers/hour (kmph)
Miles/hour (mph)	0.8684	Knots (kn) (nautical mph)
Temperature		
Degrees Fahrenheit (°F)	0.556 (after subtracting 32)	Degrees Celsius (°C)
Fahrenheit degree (F°)	0.556	Celsius degree (C°)
Additional water measurements		
Gallon (U.S.)	0.833	Gallons (Imperial)
Acre-feet	325,872	Gallons (gal)

Additional Notation

Multiples	Prefixes	
$1,000,000,000 = 10^9$	giga	G
$1,000,000 = 10^6$	mega	M
$1,000 = 10^3$	kilo	k
$100 = 10^2$	hecto	h
$10 = 10^1$	deka	da
$1 = 10^0$		
$0.1 = 10^{-1}$	deci	d
$0.01 = 10^{-2}$	centi	c
$0.001 = 10^{-3}$	milli	m
$0.000001 = 10^{-6}$	micro	µ

Glossary

The page number on which each term appears is **boldfaced** in parentheses and followed by a specific definition relevant to the term's usage in the chapter.

A

Aa (235) Rough, jagged, and clinkery basaltic lava with sharp edges. This texture is caused by the loss of trapped gases, a slow flow, and the development of a thick skin that cracks into a jagged surface.

abiotic (i-11, 346) Nonliving; Earth's nonliving systems of energy and materials.

Abrasion (278, 306) Mechanical wearing and erosion of bedrock accomplished by the rolling and grinding of particles and rocks carried in a stream, removed by wind in a "sandblasting" action, or imbedded in glacial ice.

absolute time (214) The actual number of years before the present, in contrast to relative time is the sequence of events.

absorption (37) Assimilation and conversion of radiation from one form to another in a medium. In the process, the temperature of the absorbing surface is raised, thereby affecting the rate and wavelength of radiation from that surface.

actual evapotranspiration (128) ACTET; the actual amount of evaporation and transpiration that occurs; derived in the water balance by subtracting the deficit (DEFIC) from potential evapotranspiration (POTET). (*Compare* Potential evapotranspiration.)

adaptations (360) Inherited characteristics that enhance an organisms survival and reproduction in specific environments.

adiabatic (94) Pertaining to the cooling of an ascending parcel of air through expansion or the warming of a descending parcel of air through compression, without any exchange of heat between the parcel and the surrounding environment.

advection fog (98) Active condensation formed when warm, moist air moves laterally over cooler water or land surfaces, causing the lower layers of the air to be chilled to the dew-point temperature.

aerosols (185) Small particles of dust, soot, and pollution suspended in the air.

aggradation (279) The general building of land surface because of deposition of material; opposite of degradation. When the sediment load of a stream exceeds the stream's capacity to carry it, the stream channel becomes filled through this process.

air (16) Earth's atmosphere, air is a mixture of gases that is odorless, colorless, taste-less, and blended so thoroughly that it behaves as if it were a single gas.

air mass (100) A distinctive, homogeneous body of air that has taken on the moisture and temperature characteristics of its source region.

air pressure (16) Pressure produced by the motion, size, and number of gas molecules in the air and exerted on surfaces in contact with the air; an average force at sea level of 1 kg/cm³ (14.7 lb/in²). Normal sea-level pressure, as measured by the height of a column of mercury (Hg), is expressed as 1013.2 millibars, 760 mm of Hg, or 29.92 inches of Hg. Air pressure can be measured with mercury or aneroid barometers (*see listings for both*).

albedo (37) The reflective quality of a surface, expressed as the percentage of reflected insolation to incoming insolation; a function of surface color, angle of incidence, and surface texture.

Alfisols (387) A soil order in the Soil Taxonomy. Moderately weathered forest soils that are moist versions of Mollisols, with productivity dependent on specific patterns of moisture and temperature; rich in organics. Most wide-ranging of the soil orders.

alluvial fan (283) Fan-shaped fluvial landform at the mouth of a canyon; generally occurs in arid landscapes where streams are intermittent. (*See* Bajada.)

alluvial stream terrace, (283) Level areas that appear as topographic steps above a stream, created by the stream as it scours with renewed downcutting into its floodplain; composed of unconsolidated alluvium. (*See* Alluvium.)

alluvium (279) General descriptive term for clay, silt, sand, gravel, or other unconsolidated rock and mineral fragments transported by running water and deposited as sorted or semisorted sediment on a floodplain, delta, or streambed.

alpine glacier (322) A glacier confined in a mountain valley or walled basin, consisting of three subtypes: valley glacier (within a valley), piedmont glacier (coalesced at the base of a mountain, spreading freely over nearby lowlands), and outlet glacier (flowing outward from a continental glacier; *compare* Ice sheet).

alpine tundra (404) A biome found above treeline at high elevation in mountains, featuring ground-hugging grasses, herbs, lichens, mosses, and low-growing woody shrubs. (*See* Arctic tundra.)

altitude (12) The angular distance between the horizon (a horizontal plane) and the Sun (or any point in the sky) .

altocumulus (97) Middle-level, puffy clouds that occur in several forms: patchy rows, wave patterns, a "mackerel sky," or lens-shaped "lenticular" clouds.

Andisols (390) A soil order in the Soil Taxonomy; derived from volcanic parent materials in areas of volcanic activity. A new order, created in 1990, of soils previously considered under Inceptisols and Entisols.

anemometer (65) A device that measures wind velocity.

aneroid barometer, (64) A device that measures air pressure using a partially evacuated, sealed cell. (*See* Air pressure.)

Antarctic Circle (14) This latitude (66.5° S) denotes the northernmost parallel (in the Southern Hemisphere) that experiences a 24-hour period of darkness in winter or daylight in summer.

Antarctic high (71) A consistent high-pressure region centered over Antarctica; source region for an intense polar air mass that is dry and associated with the lowest temperatures on Earth.

anthropogenic biome, (405) A recent conceptual term for large-scale, stable ecosystems that result from ongoing human interaction with natural environments. Human modifications are often linked to land-use practices such as agriculture, forestry, and urbanization.

anticline (228) Upfolded rock strata in which layers slope downward from the axis of the fold, or central ridge. (*Compare* Syncline.)

anticyclone (67) A dynamically or thermally caused area of high atmospheric pressure with descending and diverging airflows that rotate clockwise in the Northern Hemisphere and counterclockwise in the Southern Hemisphere. (*Compare* Cyclone.)

aquifer (134) A body of rock that conducts groundwater in usable amounts; a permeable rock layer.

Arctic Circle (14) This latitude (66.5° N) denotes the southernmost parallel (in the Northern Hemisphere) that experiences a 24-hour period of darkness in winter or daylight in summer.

arctic tundra (404) A biome in the northernmost portions of North America and northern Europe and Asia, featuring low, ground-level herbaceous plants as well as some woody plants. (*See* Alpine tundra.)

arête (326) A sharp ridge that divides two cirque basins. Derived from "knife edge" in French, these form sawtooth and serrated ridges in glaciated mountains.

Aridisols (387) A soil order in the Soil Taxonomy; largest soil order. Typical of dry climates; low in organic matter and dominated by calcification and salinization.

artesian well (134) Pressurized gro-undwater that rises in a well or a rock structure above the local water table; may flow out onto the ground without pumping.

auroras (7) A spectacular glowing light display in the ionosphere, stimulated by the interaction of the solar wind with principally oxygen and nitrogen gases at high latitudes; called *aurora borealis* in the Northern Hemisphere and *aurora australis* in the Southern Hemisphere.

autumnal equinox (27) The fall or September equinox in the northern hemisphere, and the March equinox in the southern hemisphere. See March equinox and September equinox.

axial parallelism (13) Earth's axis remains aligned the same throughout the year (it "remains parallel to itself"); thus, the axis extended from the North Pole points into space always near Polaris, the North Star.

axial tilt (13) Earth's axis tilts 23.5° from a perpendicular to the plane of the ecliptic (plane of Earth's orbit around the Sun).

axis (13) The imaginary line from the North Pole to the South Pole around which Earth rotates daily.

B

barrier beach (305) Narrow, long, depositional feature, generally composed of sand, that forms offshore roughly parallel to the coast; may appear as barrier islands and long chains of barrier beaches. (*See* Barrier island.)

barrier island (305) Generally, a broadened barrier beach offshore. (*See* Barrier beach.)

barrier spit (304) A depositional landform that develops when transported sand or gravel in a barrier beach or island is deposited in long ridges that are attached at one end to the mainland and partially cross the mouth of a bay.

base level (275) A hypothetical level below which a stream cannot erode its valley—and thus the lowest operative level for denudation processes; in an absolute sense, it is represented by sea level, extending under the landscape.

batholith (220) The largest plutonic form exposed at the surface; an irregular intrusive mass; it invades crustal rocks, cooling slowly so that large crystals develop. (*See* Pluton.)

beach (304) The portion of the coastline where an accumulation of sediment is in motion.

bed load (279) Coarse materials that are dragged along the bed of a stream by traction or by the rolling and bouncing motion of saltation; involves particles too large to remain in suspension. (*See* Traction, Saltation.)

bedrock (249) The rock of Earth's crust that is below the soil and is basically unweathered; such solid crust sometimes is exposed as an outcrop.

biodiversity (346) A principle of ecology and biogeography: The more diverse the species population in an ecosystem (in number of species, quantity of members in each species, and genetic content), the more risk is spread over the entire community, which results in greater overall stability, greater productivity, and increased use of nutrients, as compared to a monoculture of little or no diversity.

biogeochemical (350) One of several circuits of flowing elements and materials (carbon, oxygen, nitrogen, phosphorus, water) that combine Earth's biotic (living) and abiotic (nonliving) systems; the cycling of materials is continuous and renewed through the biosphere and the life processes.

biogeographic realm (380) One of eight regions of the biosphere, each representative of evolutionary core areas of related flora (plants) and fauna (animals); a broad geographical classification scheme.

biogeography (346) The study of the distribution of plants and animals and related ecosystems; the geographical relationships with their environments over time.

biological amplification (359) The process in which retained substances, often pesticides or other harmful chemicals, become more concentrated at each higher trophic level in a food web or chain.

biomass (349) The total mass of living organisms on Earth or per unit area of a landscape; also the weight of the living organisms in an ecosystem.

biome (380) A large-scale, stable, terrestrial or aquatic ecosystem classified according to the predominant vegetation type and the adaptations of particular organisms to that environment.

biotic (i-11, 346) Living; referring to Earth's living system of organisms.

blowout depression, (306) Wind eroded indentation on Earth's surface ranging in size from small indentations less than a meter wide up to areas hundreds of meters wide and many meters deep.

boreal forest (399) *See* Needleleaf forest.

brackish (295)   Descriptive of seawater with a salinity of less than 35%; for example, the Baltic Sea. (*Compare* Brine.)

braided channel (279)   A stream that becomes a maze of interconnected channels laced with excess sediment.

breaker (301)   The point where a wave's height exceeds its vertical stability and the wave breaks as it approaches the shore.

brine (295)   Seawater with a salinity of more than 35%; for example, the Persian Gulf. (*Compare* Brackish.)

C

calcification (385)   The illuviated (deposited) accumulation of calcium carbonate or magnesium carbonate in the B and C soil horizons.

caldera (235)   An interior sunken portion of a composite volcano's crater; usually steep-sided and circular, sometimes containing a lake; also can be found in conjunction with shield volcanoes.

carbonation (253)   A chemical weathering process in which weak carbonic acid (water and carbon dioxide) reacts with many minerals that contain calcium, magnesium, potassium, and sodium (especially limestone), transforming them into carbonates.

carbon sink (187)   An area in Earth's atmosphere, hydrosphere, lithosphere, or biosphere where carbon is stored; also called a *carbon reservoir*.

cartography (i-16)   The making of maps and charts; a specialized science and art that blends aspects of geography, engineering, mathematics, graphics, computer science, and artistic specialties.

chemical weathering, (252)   Decomposition and decay of the constituent minerals in rock through chemical alteration of those minerals. Water is essential, with rates keyed to temperature and precipitation values. Chemical reactions are active at microsites even in dry climates. Processes include hydrolysis, oxidation, carbonation, and solution.

cinder cone (235)   A volcanic landform of pyroclastics and scoria, usually small and cone-shaped and generally not more than 450 m (1500 ft) in height, with a truncated top.

circle of illumination (13)   The division between light and dark on Earth; a day–night great circle.

cirque (322)   A scooped-out, amphitheater-shaped basin at the head of an alpine glacier valley; an erosional landform.

cirrus (97)   Wispy, filamentous ice-crystal clouds that occur above 6000 m (20,000 ft); appear in a variety of forms, from feathery hairlike fibers to veils of fused sheets.

climate (152)   The consistent, long-term behavior of weather over time, including its variability, in contrast to weather, which is the condition of the atmosphere at any given place and time.

climate change science (176)   The interdisciplinary study of the causes and consequences of changing climate for all Earth systems and the sustainability of human societies.

climate feedback (186)   A process that either amplifies or reduces a climatic trend toward either warming or cooling.

climatic regions (152)   An area of homogenous climate that features characteristic regional weather and air mass patterns.

climatology (152)   The scientific study of climate and climatic patterns and the consistent behavior of weather, including its variability and extremes, over time in one place or region; includes the effects of climate change on human society and culture.

climograph (153)   A graph that plots daily, monthly, or annual temperature and precipitation values for a selected station; may also include additional weather information.

closed system (i-10)   A self-contained entity, shut off from the surrounding environment. Although rare in nature, Earth is a closed system in terms of physical matter and resources.

cloud (96)   An aggregate of tiny moisture droplets and ice crystals; classified by altitude of occurrence and shape.

cold front (104)   The leading edge of an advancing cold air mass; identified on a weather map as a line marked with triangular spikes pointing in the direction of frontal movement. (*Compare* Warm front.)

commensalism (362)   A symbiotic relationship where one organism benefits and one the other organism is neither helped nor harmed.

communities (360)   A convenient biotic subdivision within an ecosystem; formed by interacting populations of animals and plants in an area.

composite volcano (234)   A volcano formed by a sequence of explosive volcanic eruptions; steep-sided and conical in shape; sometimes referred to as a stratovolcano, although composite is the preferred term. (*Compare* Shield volcano.)

condensation nuclei (96)   Microscopic particles upon which water vapor condenses to form water droplets, can be sea salt, ash, dust, or smoke.

conduction (35)   The slow molecule-to-molecule transfer of heat through a medium, from warmer to cooler portions.

consumers (347)   An organism in an ecosystem that depends on producers (organisms that use carbon dioxide as their sole source of carbon) for its source of nutrients; also called a *heterotroph*. (*Compare* Producer.)

Coordinated Universal Time (UTC) (i-14) The official reference time in all countries, formerly known as Greenwich Mean Time; now measured by primary standard atomic clocks, the time calculations are collected in Paris at the International Bureau of Weights and Measures BIPM); the legal reference for time in all countries and broadcast worldwide.

continental divides (271)   A ridge or elevated area that separates drainage on a continental scale; specifically, that ridge in North America that separates drainage to the Pacific Ocean on the west side from drainage to the Atlantic Ocean and the Gulf of Mexico on the east side and to Hudson Bay and the Arctic Ocean in the north.

continental glacier (323)   A continuous mass of ice covering that blankets 81% of Greenland and 90% of Antarctica. The ice is more than 3000 m (9800 ft) thick, and often buries all but the highest peaks.

continentality (48)   Generally, old, low-elevation heartland regions of continental crust; various cratons (granitic cores) and ancient mountains are exposed at the surface.

convection (35)   Transfer of heat from one place to another through the physical movement of air; involves a strong vertical motion. (*Compare* Advection.)

convection current (216)   Flow of a fluid due to convection, for example in the atmosphere or lithosphere.

Convergent lifting (102) Air flowing from different directions forces lifting of air upward, initiating adiabatic processes.

convectional lifting (102) Air passing over warm surfaces gains buoyancy and lifts, initiating adiabatic processes.

coral (310) A simple, cylindrical marine animal with a saclike body that secretes calcium carbonate to form a hard external skeleton and, cumulatively, landforms called reefs; lives symbiotically with nutrient-producing algae; presently in a worldwide state of decline due to bleaching (loss of algae).

coral bleaching (311) The process by which normally colorful corals turn stark white when they expel their own nutrient-supplying algae.

Coriolis force (66) The apparent deflection of moving objects (wind, ocean currents, missiles) from traveling in a straight path, in proportion to the speed of Earth's rotation at different latitudes. Deflection is to the right in the Northern Hemisphere and to the left in the Southern Hemisphere; maximum at the poles and zero along the equator.

crater (235) A depression that usually forms on or near the summit of a volcano.

crevasse (325) A vertical crack that develops in a glacier as a result of friction between valley walls, or tension forces of extension on convex slopes, or compression forces on concave slopes.

cumulonimbus (97) A towering, precipitation-producing cumulus cloud that is vertically developed across altitudes associated with other clouds; frequently associated with lightning and thunder and thus sometimes called a *thunderhead*.

cumulus (97) Bright and puffy cumuliform clouds up to 2000 m (6500 ft) in altitude.

cyclone (67) A dynamically or thermally caused area of low atmospheric pressure with ascending and converging airflows that rotate counterclockwise in the Northern Hemisphere and clockwise in the Southern Hemisphere. (*Compare* Anticyclone; *see* Midlatitude cyclone, Tropical cyclone.)

D

day length (12) Duration of exposure to insolation, varying during the year depending on latitude; an important aspect of seasonality.

debris avalanche (258) A mass of falling and tumbling rock, debris, and soil; can be dangerous because of the tremendous velocities achieved by the onrushing materials.

December solstice (14) The time when the Sun's declination is at the Tropic of Capricorn, at 23.5 S latitude, on December 21–22 each year (also known as *winter solstice*). The day is 24 hours long south of the Antarctic Circle. The night is 24 hours long north of the Arctic Circle. (*Compare* June solstice.)

declination (12) The latitude that receives direct overhead (perpendicular) insolation on a particular day; the subsolar point migrates annually through 47° of latitude between the Tropics of Cancer (23.5° N) and Capricorn (23.5° S).

decomposers (346) Bacteria and fungi that digest organic debris outside their bodies and absorb and release nutrients in an ecosystem (See Detritivores.)

deep ocean trench (225) An elongated and narrow depression on the seafloor resulting from convergent plate boundaries. The denser basaltic ocean crust slides underneath the lighter continental crust in a process called subduction, creating a trench.

deflation (306) A process of wind erosion that removes and lifts individual particles, literally blowing away unconsolidated, dry, or noncohesive sediments.

delta (283) A depositional plain formed where a river enters a lake or an ocean; named after the triangular shape of the Greek letter delta, Δ.

denudation (248) A general term that refers to all processes that cause degradation of the landscape: weathering, mass movement, erosion, and transport.

depositional coastline (302) The process whereby weathered, wasted, and transported sediments are laid down by air, water, and ice.

derechos (97) Strong linear winds in excess of 26 m/sec (58 mph), associated with thunderstorms and bands of showers crossing a region.

desalination (143) In a water resources context, the removal of organics, debris, and salinity from seawater through distillation or reverse osmosis to produce potable water.

desert (403) The biome caused by a climate that recieves less than half the precipitation required to meet the needs of potential evapotranspiration.

desert pavement (306) On arid landscapes, a surface formed when wind deflation and sheetflow remove smaller particles, leaving residual pebbles and gravels to concentrate at the surface; an alternative sediment-accumulation hypothesis explains some desert pavements; resembles a cobblestone street. (*See* Deflation, Sheetflow.)

detritivores (358) Detritus feeders and decomposers that consume, digest, and destroy organic wastes and debris. *Detritus feeders*—worms, mites, termites, centipedes, snails, crabs, and even vultures, among others—consume detritus and excrete nutrients and simple inorganic compounds that fuel an ecosystem. (*Compare* Decomposers.)

dew-point temperature (90) The temperature at which a given mass of air becomes saturated, holding all the water it can hold. Any further cooling or addition of water vapor results in active condensation.

differential weathering, (249) The effect of different resistances in rock, coupled with variations in the intensity of physical and chemical weathering.

dissolved load (278) Materials carried in chemical solution in a stream, derived from minerals such as limestone and dolomite or from soluble salts.

drainage basin (270) The basic spatial geomorphic unit of a river system; distinguished from a neighboring basin by ridges and highlands that form divides, marking the limits of the catchment area of the drainage basin.

drainage density (272) Determined by dividing the total length of all stream channels in the basin by the area of the basin. The number and length of streams in a given area reflect the area's geology, topography, and climate.

drainage pattern (272)   A distinctive geometric arrangement of streams in a region, determined by slope, differing rock resistance to weathering and erosion, climatic and hydrologic variability, and structural controls of the landscape.

drought (129)   Does not have a simple water-budget definition; rather, it can occur in at least four forms: *meteorological drought, agricultural drought, hydrologic drought,* and/or *socioeconomic drought.*

drumlin (329)   A depositional landform related to glaciation that is composed of till (unstratified, unsorted) and is streamlined in the direction of continental ice movement—blunt end upstream and tapered end downstream with a rounded summit.

dry adiabatic rate (94)   The rate at which an unsaturated parcel of air cools (if ascending) or heats (if descending); a rate of 10 C° per 1000 m (5.5 F° per 1000 ft). (*See* Adiabatic; *compare* Moist adiabatic rate.)

dry climates (arid and semiarid steppe) (153, 167)   Regions where an annual moisture deficit normally develops when the potential evapotranspiration exceeds moisture supply. Deserts have greater moisture deficits than semiarid steppes—semiarid grasslands of Eastern Europe and Asia where the climate is too dry to support forest, but too moist to be a desert.

dune (308)   A depositional feature of sand grains deposited in transient mounds, ridges, and hills; extensive areas of sand dunes are called sand seas.

dust dome (50)   A dome of airborne pollution associated with every major city; may be blown by winds into elongated plumes downwind from the city.

E

earthquake (229)   A sharp release of energy that sends waves traveling through Earth's crust at the moment of rupture along a fault or in association with volcanic activity. The moment magnitude scale (formerly the Richter scale) estimates earthquake magnitude; intensity is described by the Mercalli scale.

easterly waves (114)   A trough of tropical low pressure that moves east to west with the trade winds, and can spawn tropical cyclones.

ecological succession, (364)   The process whereby different and usually more complex assemblages of plants and animals replace older and usually simpler communities; communities are in a constant state of change as each species adapts to changing conditions. Ecosystems do not exhibit a stable point or successional climax condition as previously thought. (*See* Primary succession, Secondary succession.)

ecology (346)   The science that studies the relations between organisms and their environment and among various ecosystems.

ecotone (381)   A boundary transition zone between adjoining ecosystems that may vary in width and represent areas of tension as similar species of plants and animals compete for the resources. (*See* Ecosystem.)

effusive eruption (234)   A volcanic eruption characterized by low-viscosity basaltic magma and low-gas content, which readily escapes. Lava pours forth onto the surface with relatively small explosions and few pyroclastics; tends to form shield volcanoes. (*See* Shield volcano, Lava, Pyroclastic; *compare* Explosive eruption.)

electromagnetic spectrum (8)   All the radiant energy produced by the Sun placed in an ordered range, divided according to wavelengths.

El Niño (78)   Southern Oscillation (ENSO) (4)   Sea-surface temperatures increase, sometimes more than 8 C° (14 F°) above normal in the central and eastern Pacific, replacing the normally cold, nutrient-rich water along Peru's coastline. Pressure patterns and surface ocean temperatures shift from their usual locations across the Pacific, forming the Southern Oscillation.

Entisols (389)   A soil order in the Soil Taxonomy. Specifically lacks vertical development of horizons; usually young or undeveloped. Found in active slopes, alluvial-filled floodplains, and poorly drained tundra.

environmental lapse rate (19, 94)   The actual rate of temperature decrease with increasing altitude in the lower atmosphere at any particular time under local weather conditions; may deviate above or below the normal lapse rate of 6.4 C° per km, or 1000 m (3.5 F° per 1000 ft). (*Compare* Normal lapse rate.)

eolian (306)   Caused by wind; refers to the erosion, transportation, and deposition of materials; spelled *aeolian* in some countries.

equal area (i-18)   A quality of a map projection; indicates the equivalence of all areas on the surface of the map, although shape is distorted.

equilibrium line (324)   On glaciers, the dividing line between the accumulation zone (above the line) and the ablation zone (below the line).

erg desert/sand sea (308)   An extensive area of sand and dunes; from the Arabic word for "dune field." (*Compare* Sand sea.)

erosional coastline (302)   Denudation by wind, water, or ice, which dislodges, dissolves, or removes surface material.

esker (329)   A sinuously curving, narrow deposit of coarse gravel that forms along a meltwater stream channel, developing in a tunnel beneath a glacier.

estuaries (311)   The point at which the mouth of a river enters the sea, where freshwater and seawater are mixed; a place where tides ebb and flow.

eustasy (126)   Refers to worldwide changes in sea level that are related not to movements of land, but rather to changes in the volume of water in the oceans.

eutrophication (366)   The gradual enrichment of water bodies that occurs with nutrient inputs, either natural or human-caused.

evaporation (128)   The movement of free water molecules away from a wet surface into air that is less than saturated; the phase change of water to water vapor.

evaporation fog (98)   A fog formed when cold air flows over the warm surface of a lake, ocean, or other body of water; forms as the water molecules evaporate from the water surface into the cold, overlying air; also known as steam fog or sea smoke.

evolution (360)   A theory that single-cell organisms adapted, modified, and passed along inherited changes to multicellular organisms. The genetic makeup of successive generations is shaped by environmental factors, physiological functions, and behaviors that created a greater rate of survival and reproduction and were passed along through natural selection.

exfoliation dome (251)   The physical weathering process that occurs as mechanical forces enlarge joints in rock into layers of curved slabs or plates, which peel or slip off in sheets; also called *sheeting*.

exosphere (16)   An extremely rarefied outer atmospheric halo beyond the thermopause at an altitude of 480 km (300 mi); probably composed of hydrogen and helium atoms, with some oxygen atoms and nitrogen molecules present near the thermopause.

exotic stream (277)   Earth's external surface system, powered by insolation, which energizes air, water, and ice and sets them in motion, under the influence of gravity. Includes all processes of landmass denudation. (*Compare* Endogenic system.)

explosive eruption (234)   A violent and unpredictable volcanic eruption, the result of magma that is thicker (more viscous), stickier, and higher in gas and silica content than that of an effusive eruption; tends to form blockages within a volcano; produces composite volcanic landforms. (*See* Composite volcano; *compare* Effusive eruption.)

extrusive igneous (219)   A rock such as basalt that solidifies and crystallizes from a molten state as it extrudes onto the surface.

F

faulting (229)   The process whereby displacement and fracturing occur between two portions of Earth's crust; usually associated with earthquake activity.

fjord (327)   A drowned glaciated valley, or glacial trough, along a seacoast.

flash flood (283)   A sudden and short-lived torrent of water that exceeds the capacity of a stream channel; associated with desert and semiarid washes.

flood (284)   A high water level that overflows the natural riverbank along any portion of a stream.

floodplain (282)   A flat, low-lying area along a stream channel, created by and subject to recurrent flooding; alluvial deposits generally mask underlying rock.

flows (260)   Pertains to mass wasting when the moisture content of moving material is high. Flows include earthflows and more fluid mudflows.

fluvial (270)   Stream-related processes; from the Latin *fluvius* for "river" or "running water."

fog (98)   A cloud, generally stratiform, in contact with the ground, with visibility usually reduced to less than 1 km (3300 ft).

folding (228)   The bending and deformation of beds of rock strata subjected to compressional forces.

food chain (358)   The circuit along which energy flows from producers (plants), which manufacture their own food, to consumers (animals); a one-directional flow of chemical energy, ending with decomposers.

food web (358)   A complex network of interconnected food chains. (*See* Food chain.)

friction force (67)   The effect of drag by the wind as it moves across a surface; may be operative through 500 m (1600 ft) of altitude. Surface friction slows the wind and therefore reduces the effectiveness of the Coriolis force.

frost action (250)   A form of physical weathering that occurs when repeated freezing (expanding) and thawing (contracting) of water that breaks rocks apart— particularly in cold climates at high latitudes and high elevations. Blocks of rock often separate along existing joints and fractures.

funnel clouds (112)   The visible swirl extending from the bottom side of a cloud, which may or may not develop into a tornado. A tornado is a funnel cloud that has extended all the way to the ground. (*See* Tornado.)

fusion (6)   The process of forcibly joining positively charged hydrogen and helium nuclei under extreme temperature and pressure; occurs naturally in thermonuclear reactions within stars, such as our Sun.

G

Gelisols (390)   A new soil order in the Soil Taxonomy, added in 1998, describing cold and frozen soils at high latitudes or high elevations; characteristic tundra vegetation.

general circulation model (198)   Complex, computer-based climate model that produces generalizations of reality and forecasts of future weather and climate conditions. Complex GCMs (three-dimensional models) are in use in the United States and in other countries.

geography (i-6)   The science that studies the interdependence and interaction among geographic areas, natural systems, processes, society, and cultural activities over space–a spatial science.

geographic information system (GIS) (i-22)   A computer based data processing tool or methodology used for gathering, manipulating, and analyzing geographic information to produce a holistic, interactive analysis.

geologic cycle (218)   A general term characterizing the vast cycling that proceeds in the lithosphere. It encompasses the hydrologic cycle, tectonic cycle, and rock cycle.

geologic time scale (214)   A depiction of eras, periods, and epochs that span Earth's history; shows both the sequence of rock strata and their absolute dates, as determined by methods such as radioactive isotopic dating.

geomagnetism (224)   A polarity change in Earth's magnetic field. With uneven regularity, the magnetic field fades to zero and then returns to full strength, but with the magnetic poles reversed. Reversals have been recorded nine times during the past 4 million years.

geomorphic threshold (258)   The threshold up to which landforms change before lurching to a new set of relationships, with rapid realignments of landscape materials and slopes.

geomorphology (248)   The science that analyzes and describes the origin, evolution, form, classification, and spatial distribution of landforms.

geostrophic winds (67)   A wind moving between areas of different pressure along a path that is parallel to the isobars. It is a product of the pressure gradient force and the Coriolis force. (*See* Isobar, Pressure gradient force, Coriolis force.)

glacial drift (328) The general term for all glacial deposits, both unsorted (till) and sorted (stratified drift)

glacial lake (327) Water bodies formed by glacial erosion or deposition. The smallest lakes are commonly called tarns.

glacier (322) A large mass of perennial ice resting on land or floating shelflike in the sea adjacent to the land; formed from the accumulation and recrystallization of snow, which then flows slowly under the pressure of its own weight and the pull of gravity.

glacier surge (325) The rapid, lurching, unexpected forward movement of a glacier.

gleization (385) A process of humus and clay accumulation in cold, wet climates with poor drainage.

global carbon budget (187) The exchange of carbon between sources and sinks in Earth's atmosphere, hydrosphere, lithosphere, and biosphere.

global dimming (39) The decline in sunlight reaching Earth's surface due to pollution, aerosols, and clouds.

graben (238) Pairs or groups of faults that produce downward-faulted blocks; characteristic of the basins of the interior western United States. (*Compare* Horst; *see* Basin and Range Province.)

graded stream (274) An idealized condition in which a stream's load and the landscape mutually adjust. This forms a dynamic equilibrium among erosion, transported load, deposition, and the stream's capacity.

gradient (274) The drop in elevation from a stream's headwaters to its mouth, ideally forming a concave slope.

gravity (4) The mutual force exerted by the masses of objects that are attracted one to another and produced in an amount proportional to each object's mass.

greenhouse effect (38) The process whereby radiatively active gases (carbon dioxide, water vapor, methane, and CFCs) absorb and emit energy at longer wavelengths, which are retained longer, delaying the loss of infrared to space. Thus, the lower troposphere is warmed through the radiation and re-radiation of infrared wavelengths. The approximate similarity between this process and that of a greenhouse explains the name.

greenhouse gases (38) Gases in the lower atmosphere that delay the passage of longwave radiation to space by absorbing and re-radiating specific wavelengths. Earth's primary greenhouse gases are carbon dioxide, water vapor, methane, nitrous oxide, and fluorinated gases, such as chlorofluorocarbons (CFCs).

groundwater mining (136) Pumping an aquifer beyond its capacity to flow and recharge; an overuse of the groundwater resource.

groundwater (p. 133) Water beneath the surface that is beyond the soil-root zone; a major source of potable water.

groundwater recharge (133) The subsurface water that is frozen in regions of permafrost. The moisture content of areas with ground ice may vary from nearly 0% in regions of drier permafrost to almost 100% in saturated soils.

gyres (76) Large, rotating ocean currents. They are driven by prevailing winds and the Coriolis force.

H

habitat (360) A physical location to which an organism is biologically suited. Most species have specific habitat parameters and limits. (*Compare* Ecological niche.)

habitat fragmentation, (383) The process in which habitat loss results in large, continuous habitats being reduced into smaller, isolated habitats.

Hadley cells (69) Two circulation cells on either side of the equator. Air rises in the ITCZ, flows toward the poles, sinks over subtropical high, and flows toward the ITCZ..

hail (110) A type of precipitation formed when a raindrop is repeatedly circulated above and below the freezing level in a cloud, with each cycle freezing more moisture onto the hailstone until it becomes too heavy to stay aloft.

hanging valley (327) Basins carved by tributary glaciers, that are stranded above the valley floor because the primary glacier eroded the valley floor so deeply.

Hawai'ian high (70) See Pacific high.

headland (302) A protruding coastal landform of resistant rocks.

heat (34) The flow of kinetic energy from one body to another because of a temperature difference between them.

heat wave (54) A prolonged period of abnormally high temperatures, usually, but not always, in association with humid weather.

heterosphere (17) A zone of the atmosphere above the mesopause, from 80 km (50 mi) to 480 km (300 mi) in altitude; composed of rarefied layers of oxygen atoms and nitrogen molecules; includes the ionosphere.

highland climates (153) A climate that occurs on Earth's mountain ranges. Even at low latitudes, the cooling effects of elevation can produce tundra, glaciers, and polar conditions.

Histosols (391) A soil order in the Soil Taxonomy. Formed from thick accumulations of organic matter, such as beds of former lakes, bogs, and layers of peat.

homosphere (17) A zone of the atmosphere from Earth's surface up to 80 km (50 mi), composed of an even mixture of gases, including nitrogen, oxygen, argon, carbon dioxide, and trace gases.

horn (326) A pyramidal, sharp-pointed peak that results when several cirque glaciers gouge an individual mountain summit from all sides.

horst (238) Upward-faulted blocks produced by pairs or groups of faults; characteristic of the mountain ranges of the interior of the western United States. (*See* Graben, Basin and Range Province.)

hot spot (p. 227) An individual point of upwelling material originating in the asthenosphere, or deeper in the mantle; tends to remain fixed relative to migrating plates; some 100 are identified worldwide, exemplified by Yellowstone National Park, Hawai'i, and Iceland.

humidity (90) Water vapor content of the air. The capacity of the air for water vapor is mostly a function of the temperature of the air and the water vapor

humus (352) A mixture of organic debris in the soil worked by consumers and decomposers in the humification process; characteristically formed from plant and animal litter deposited at the surface.

hurricane (114)   A tropical cyclone that is fully organized and intensified in inward-spiraling rainbands; ranges from 160 to 960 km (100 to 600 mi) in diameter, with wind speeds in excess of 119 kmph (65 knots, or 74 mph); a name used specifically in the Atlantic and eastern Pacific. (*Compare* Typhoon.)

hydration (252)   A chemical weathering process involving water that is added to a mineral, which initiates swelling and stress within the rock, mechanically forcing grains apart as the constituents expand. (*Compare* Hydrolysis.)

hydraulic action (278)   The erosive work accomplished by the turbulence of water; causes a squeezing and releasing action in joints in bedrock; capable of prying and lifting rocks.

hydrograph (276)   A graph of stream discharge (in m^3/s or ft^3/s) over a period of time (minutes, hours, days, years) at a specific place on a stream. The relationship between stream discharge and precipitation input is illustrated on the graph.

hydrology (270)   The science of water, including its global circulation, distribution, and properties—specifically water at and below Earth's surface.

hydrolysis (252)   A chemical weathering process in which minerals chemically combine with water; a decomposition process that causes silicate minerals in rocks to break down and become altered. (*Compare* Hydration.)

I

ice age (332)   A cold episode, with accompanying alpine and continental ice accumulations, that has repeated roughly every 200 to 300 million years since the late Precambrian Era (1.25 billion years ago); includes the most recent episode during the Pleistocene Ice Age, which began 1.65 million years ago.

ice cap (323)   A large, dome-shaped glacier, less extensive than an ice sheet, although it buries mountain peaks and the local landscape; generally, less than 50,000 km^2 (19,300 mi^2).

ice cap and ice sheet climates (165)   A climate characterized by dry, frigid air masses where all months average below freezing. This includes the Antarctic continent, most of Greenland, and the North Pole.

ice field (323)   The least extensive form of a glacier, with mountain ridges and peaks visible above the ice; less than an ice cap or ice sheet.

ice sheet (323)   A continuous mass of unconfined ice, covering at least 50,000 km^2 (19,500 mi^2). The bulk of glacial ice on Earth covers Antarctica and Greenland in two ice sheets (*Compare* Alpine glacier.)

iceberg (322)   A floating mass of ice that has calved (detached) from from tidewater glacier and is carried out to sea.

igneous rock (219)   One of the basic rock types; it has solidified and crystallized from a hot molten state (either magma or lava). (*Compare* Metamorphic rock, Sedimentary rock.)

Inceptisols (389)   A soil order in the Soil Taxonomy. Weakly developed soils that are inherently infertile; usually, young soils that are weakly developed, although they are more developed than Entisols.

industrial smog (24)   Air pollution associated with coal-burning industries; it may contain sulfur oxides, particulates, carbon dioxide, and exotics.

insolation (10)   Solar radiation that is incoming to Earth systems.

International Date Line (i-14)   The 180° meridian, an important corollary to the prime meridian on the opposite side of the planet; established by an 1884 treaty to mark the place where each day officially begins.

intertropical convergence zone (ITCZ) (68)   *See* Equatorial low.

intrusive igneous (219)   A rock that solidifies and crystallizes from a molten state as it intrudes into crustal rocks, cooling and hardening below the surface, such as granite.

invasive species (382)   Species that are brought, or introduced, from elsewhere by humans, either accidentally or intentionally. These non-native species are also known as *exotic species* or *alien species.*

ionosphere (20)   A layer in the atmosphere above 80 km (50 mi) where gamma rays, X-rays, and some ultraviolet radiation are absorbed and converted into longer wavelengths and where the solar wind stimulates the auroras.

island biogeography, (383)   Island communities are special places for study because of their spatial isolation and the relatively small number of species present. Islands resemble natural experiments because the impact of individual factors, such as civilization, can be more easily assessed on islands than over larger continental areas.

isobar (66)   An isoline connecting all points of equal atmospheric pressure.

isotherm (52)   An isoline connecting all points of equal temperature.

isotope analysis (176)   A technique for long-term climatic reconstruction that uses the atomic structure of chemical elements, specifically the relative amounts of their isotopes, to identify the chemical composition of past oceans and ice masses.

J

jet contrails (39)   Condensation trails produced by aircraft exhaust, particulates, and water vapor can form high cirrus clouds, sometimes called *false cirrus clouds.*

jet streams (73)   The most prominent movement in upper-level westerly wind flows; irregular, concentrated, sinuous bands of geostrophic wind, traveling at 300 kmph (190 mph).

joints (249)   A fracture or separation in rock that occurs without displacement of the sides; increases the surface area of rock exposed to weathering processes.

June solstice (14)   The time when the Sun's declination is at the Tropic of Cancer, at 23.5 N latitude, on June 20–21 each year (also known as *summer solstice*). The day is 24 hours long north of the Arctic Circle. The night is 24 hours long south of the Antarctic Circle. (*Compare* December solstice.).

K

kame (329)   A depositional feature of glaciation; a small hill of poorly sorted sand and gravel that accumulates in crevasses or in ice-caused indentations in the surface.

karst topography, (248, 254)   Distinctive topography formed in a region of chemically weathered limestone with poorly developed surface drainage and solution features that appear pitted and bumpy; originally named after the Krs Plateau in Slovenia.

kettle (329) Forms when an isolated block of ice persists in a ground moraine, an outwash plain, or a valley floor after a glacier retreats; as the block finally melts, it leaves behind a steep-sided hole that frequently fills with water.

kinetic energy (18) The energy of motion in a body; derived from the vibration of the body's own movement and stated as temperature.

Köppen climate classification (152) The most widespread system used to classify world climate, based on statistics or other data that measure the observed effects; developed by German climatologist Wladimir Köppen.

L

La Niña (79) The weaker, cold-water phase of the El Nino Southern Oscillation. Sea-surface temperatures decrease by at least 0.4° C (0.7° F) below normal in the central and eastern Pacific.

lagoon (304) An area of coastal seawater that is virtually cut off from the ocean by a bay barrier or barrier beach; also, the water surrounded and enclosed by an atoll.

landslide (259) A sudden rapid downslope movement of a cohesive mass of regolith and/or bedrock in a variety of mass-movement forms under the influence of gravity; a form of mass movement.

latent heat (34, 88) Heat energy is stored in one of three states—ice, water, or water vapor. The energy is absorbed or released in each phase change from one state to another. Heat energy is absorbed as the latent heat of melting, vaporization, or evaporation. Heat energy is released as the latent heat of condensation and freezing (or fusion).

latent heat of condensation (89) The heat energy released to the environment in a phase change from water vapor to liquid; under normal sea-level pressure, 540 calories are released from each gram of water vapor that changes phase to water at boiling, and 585 calories are released from each gram of water vapor that condenses at 20°C (68°F).

latent heat of evaporation (89) The heat energy absorbed from the environment in a phase change from liquid to water vapor at the boiling point; under normal sea-level pressure, 540 calories must be added to each gram of boiling water to achieve a phase change to water vapor.

latent heat of vaporization (89) See latent heat of evaporation.

lateral moraine (328) Debris transported by a glacier that accumulates along the sides of the glacier and is deposited along these margins.

laterization (385) A pedogenic process operating in well-drained soils that occurs in warm and humid regions; typical of Oxisols. Plentiful precipitation leaches soluble minerals and soil constituents. Resulting soils usually are reddish or yellowish.

latitude (i-12) The angular distance measured north or south of the equator from a point at the center of Earth. A line connecting all points of the same latitudinal angle is a parallel. (Compare Longitude.)

lava (220) Magma that issues from volcanic activity onto the surface; the extrusive rock that results when magma solidifies. (See Magma.)

LIDAR (i-21) A remote sensing technology that uses laser pulses to make a high-definition model of Earth's surface.

life zone (362) A zonation by altitude of plants and animals that form distinctive communities. Each life zone possesses its own temperature and precipitation relations.

lifting condensation level (94) The altitude at which a parcel of air, cooling at the dry adiabatic rate, will cool to dew point temperature.

lightning (110) Flashes of light caused by tens of millions of volts of electrical charge heating the air to temperatures of 15,000°–30,000°C (27,000°–54,000°F).

limiting factor (363) The physical or chemical factor that most inhibits biotic processes, through either lack or excess.

lithospheric plates (222) Enormous and unevenly shaped slabs of the outer crust and upper mantle (also called tectonic plates), of which the largest cover entire continents. Oceanic lithosphere is made up mostly of basalt, whereas continental lithosphere has a foundation of mostly granitic-type rocks.

littoral zone (299) A specific coastal environment; that region between the high-water line during a storm and a depth at which storm waves are unable to move seafloor sediments.

loess (308) Large quantities of fine-grained clays and silts left as glacial outwash deposits; subsequently blown by the wind great distances and redeposited as a generally unstratified, homogeneous blanket of material covering existing landscapes; in China, loess originated from desert lands.

longitude (i-12) The angular distance measured east or west of a prime meridian from a point at the center of Earth. A line connecting all points of the same longitude is a meridian. (Compare Latitude.)

longshore current (302) A current that forms parallel to a beach as waves arrive at an angle to the shore; generated in the surf zone by wave action, transporting large amounts of sand and sediment. (See Beach drift.)

longshore drift (302) The transport of sediments along a coast at an angle to the shoreline. Caused when incoming waves slow as they enter shallow water. The drift occurs only in the surf zone, and depends on wind direction and wave direction.

M

magma (220) Molten rock from beneath Earth's surface; fluid, gaseous, under tremendous pressure, and either intruded into existing crustal rock or extruded onto the surface as lava. (See Lava.)

magnetosphere (7) Earth's magnetic force field, which is generated by dynamo-like motions within the planet's outer core; deflects the solar wind flow toward the upper atmosphere above each pole.

mangrove swamp (311) A wetland ecosystem between 30° N and 30° S; tends to form a distinctive community of mangrove plants. (Compare Salt marsh.)

map (i-16) A generalized view of an area, usually some portion of Earth's surface, as seen from above at a greatly reduced size.

map projection (i-17) The reduction of a spherical globe onto a flat surface in some orderly and systematic realign-ment of the latitude and longitude grid.

March equinox (14)   The time around March 20–21 when the Sun's declination crosses the equatorial parallel (0° latitude) and all places on Earth experience days and nights of equal length (also known as *vernal equinox*). The Sun rises at the North Pole and sets at the South Pole. (*Compare* September equinox.)

marine effect (48)   A quality of regions that are dominated by the moderating effect of the ocean and that exhibit a smaller range of minimum and maximum temperatures, both daily and annually, than do continental stations. (*See* Continental effect, Land–water heating difference.)

mass movement (256)   All unit movements of materials propelled by gravity; can range from dry to wet, slow to fast, small to large, and free-falling to gradual or intermittent.

mean sea level (MSL) (299)   The average of tidal levels recorded hourly at a given site over a long period, which must be at least a full lunar tidal cycle.

meander (280)   The sinuous, curving pattern common to graded streams, with the energetic outer portion of each curve subjected to the greatest erosive action and the lower-energy inner portion receiving sediment deposits. (*See* Graded stream.)

medial moraine, (328)   Debris transported by a glacier that accumulates down the middle of the glacier, resulting from two glaciers merging their lateral moraines; forms a depositional feature following glacial retreat.

Mediterranean dry-summer climates (161)

Mediterranean shrubland (401)   A major biome dominated by the *Mediterranean* (dry summer) climate and characterized by sclerophyllous scrub and short, stunted, tough forests. (*See* Chaparral.)

meltpond (337)   Meltwater that pools on the surface of a glacier. Meltponds are darker than ice and thus absorb more insolation, which melts more ice, making more meltponds.

Mercator projection (i-18)   A true-shape projection, with meridians appearing as equally spaced straight lines and parallels appearing as straight lines that are spaced closer together near the equator. The poles are infinitely stretched, with the 84th north parallel and 84th south parallel fixed at the same length as that of the equator.

meridian (i-14)   A line designating an angle of longitude. (See Longitude.)

mercury barometer, (64)   A device that measures air pressure using a column of mercury in a tube; one end of the tube is sealed, and the other end is inserted in an open vessel of mercury. (*See* Air pressure.)

mesocyclone (112)   A large, rotating atmospheric circulation, initiated within a parent cumulonimbus cloud at midtroposphere elevation; generally produces heavy rain, large hail, blustery winds, and lightning; may lead to tornado activity.

mesosphere (18)   The upper region of the homosphere from 50 to 80 km (30 to 50 mi) above the ground; designated by temperature criteria; atmosphere extremely rarified.

mesothermal climates (153)   Climates of midlatitudes with mild winters. Mesothermal (middle) climates occur between the extreme temperatures of tropical and polar climates.

metamorphic rock (219)   One of three basic rock types, it is existing igneous and sedimentary rock that has undergone profound physical and chemical changes under increased pressure and temperature. Constituent mineral structures may exhibit foliated or nonfoliated textures. (*Compare* Igneous rock, Sedimentary rock.)

meteorology (86)   The scientific study of the atmosphere, including its physical characteristics and motions; related chemical, physical, and geological processes; the complex linkages of atmospheric systems; and weather forecasting.

microclimate (50)   The study of local climates at or near Earth's surface or up to that height above the Earth's surface where the effects of the surface are no longer determinative.

microthermal climates, (153)   Climates of mid- and high latitudes, with cold winters. The micro refers to climates with low temperatures.

midlatitude broadleaf and mixed forest (398)   A biome in moist *continental* climates in areas of warm-to-hot summers and cool-to-cold winters; relatively lush stands of broadleaf forests trend northward into needleleaf evergreen stands.

midlatitude cyclone (106)   The consistent orbital cycles—based on the irregularities in Earth's orbit around the Sun, its rotation on its axis, and its axial tilt—that relate to climatic patterns and may be an important cause of glacials and interglacials. Milutin Milankovitch (1879–1958), a Serbian astronomer, was the first to correlate these cycles to changes in insolation that affected temperatures on Earth.

midlatitude grassland (402)   A midlatitude biome with vegetation dominated by grasses, rather than shrubs or trees.

mid-ocean ridges (224)   An interconnected worldwide mountain chain on the ocean floor (also called spreading centers), found at divergent plate boundaries. As the plates move apart, magma rises from below to fill the gaps, creating small volcanoes of new seafloor along the ridge.

Milankovitch cycles (183)   The consistent orbital cycles–based around the variations in Earth's orbit aroud the Sun, its rotation on its axis, and its axial tilt--that relate to climatic patterns and may be in important cause of glacials and interglacials. Milutin Milankovitch (1879–1953), a Serbian astronomer, was the first to link these cycles with changes in insolation that affected temperatures on Earth.

Milky Way Galaxy (4)   A flattened, disk-shaped mass in space estimated to contain up to 400 billion stars; a barred-spiral galaxy; includes our Solar System.

mineral (219)   An element or combination of elements that forms an inorganic natural compound; described by a specific formula and crystal structure.

Moho (217)   The boundary between the crust and the rest of the lithospheric upper mantle, and a zone of sharp material and density contrasts. Also called the Mohorovičić discontinuity.

Mohorovičic´ discontinuity (217)   The boundary between the crust and the rest of the lithospheric upper mantle; named for the Yugoslavian seismologist Mohorovičić; a zone of sharp material and density contrasts.

moist adiabatic rate (94)   The rate at which a saturated parcel of air cools in ascent; a rate of 6 C° per 1000 m (3.3 F° per 1000 ft). This rate may vary, with moisture content and temperature, from 4 C° to 10 C° per 1000 m (2 F° to 6 F° per 1000 ft). (*See* Adiabatic; *compare* Dry adiabatic rate.)

moisture droplet (96)   A tiny water particle that constitutes the initial composition of clouds. Each droplet measures approximately 0.002 cm (0.0008 in.) in diameter and is invisible to the unaided eye.

Mollisols (387)   A soil order in the Soil Taxonomy. These have a mollic epipedon and a humus-rich organic content high in alkalinity. Some of the world's most significant agricultural soils are Mollisols.

moment magnitude scale (231)   An earthquake magnitude scale. Considers the amount of fault slippage, the size of the area that ruptured, and the nature of the materials that faulted in estimating the magnitude of an earthquake—an assessment of the seismic moment. Replaces the Richter scale (amplitude magnitude); especially valuable in assessing larger-magnitude events.

monsoons (75)   An annual cycle of dryness and wetness, with seasonally shifting winds produced by changing atmospheric pressure systems; affects India, Southeast Asia, Indonesia, northern Australia, and portions of Africa. From the Arabic word *mausim*, meaning "season."

moraine (328)   Marginal glacial deposits (lateral, medial, terminal, ground) of unsorted and unstratified material.

mutualism (362)   A symbiotic relationship where both organisms benefit.

N

natural levee (282)   A long, low ridge that forms on both sides of a stream in a developed floodplain; a depositional product (coarse gravels and sand) of river flooding.

natural selection (360)   A process in which organisms with more beneficial inherited characteristics are more likely to survive and reproduce than are organisms with other characteristics.

net primary productivity (349)   The net photosynthesis (photosynthesis minus respiration) for a given community; considers all growth and all reduction factors that affect the amount of useful chemical energy (biomass) fixed in an ecosystem.

net radiation (NET R) (42)   The net all-wave radiation available at Earth's surface; the final outcome of the radiation balance process between incoming shortwave insolation and outgoing longwave energy.

niche (360)   An organism's function in an ecosystem.

nickpoint (281)   The point at which the longitudinal profile of a stream is abruptly broken by a change in gradient; for example, a waterfall, rapids, or cascade.

nimbostratus (97)   Rain-producing, dark, grayish stratiform clouds characterized by gentle drizzle.

nitrogen dioxide (NO_2) (24)   A noxious (harmful) reddish-brown gas produced in combustion engines; can be damaging to human respiratory tracts and to plants; participates in photochemical reactions and acid deposition.

nonpoint source (139)   Pollution of water from non-fixed sources, such as runoff containing fertilizers, pesticides, animal wastes, and oil or road salt. Water moving downslope collects and transports these pollutants before depositing them into rivers, lakes, and wetlands.

normal fault (229)   A type of geologic fault in rocks. Tension produces strain that breaks a rock, with one side moving vertically relative to the other side along an inclined fault plane. (*Compare* Reverse fault.)

normal lapse rate (19, 94)   The average rate of temperature decrease with increasing altitude in the lower atmosphere; an average value of 6.4 C° per km, or 1000 m (3.5 F° per 1000 ft). (*Compare* Environmental lapse rate.)

northeast trade winds, (70)   The prevailing northeast winds flowing from the subtropical high toward the ITCZ.

nutrients (350)   An element or compound necessary for organisms use to survive and grow.

O

occluded front (107)   In a cyclonic circulation, the overrunning of a surface warm front by a cold front and the subsequent lifting of the warm air wedge off the ground; initial precipitation is moderate to heavy.

ocean acidification (296)   The increasing acid component of ocean water caused by the absorption of excess carbon dioxide from the atmosphere.

orogenesis (238)   The process of mountain building that occurs when large-scale compression leads to deformation and uplift of the crust; literally, the birth of mountains.

orographic lifting (102)   The uplift of a migrating air mass as it is forced to move upward over a mountain range—a topographic barrier. The lifted air cools adiabatically as it moves upslope; clouds may form and produce increased precipitation.

outgassing (126)   The release of trapped gases from rocks, forced out through cracks, fissures, and volcanoes from within Earth; the terrestrial source of Earth's water.

outwash (328)   Area of glacial stream deposits of stratified drift with meltwater-fed, braided, and overloaded streams; occurs beyond a glacier's morainal deposits.

overdraft (136)   The decline in the water table that occurs when pumping of groundwater exceeds the natural replenishment flow of water into the aquifer or the horizontal flow around the well.

oxbow lake (280)   A lake that was formerly part of the channel of a meandering stream; isolated when a stream eroded its outer bank, forming a cutoff through the neck of the looping meander (*see* Meandering stream). In Australia, known as a *billabong* (the Aboriginal word for "dead river").

oxidation (253)   A chemical weathering process in which oxygen dissolved in water oxidizes (combines with) certain metallic elements to form oxides; most familiar is the "rusting" of iron in a rock or soil (Ultisols, Oxisols), which produces a reddish-brown stain of iron oxide.

Oxisols (386)   A soil order in the Soil Taxonomy. Tropical soils that are old, deeply developed, and lacking in horizons wherever well drained; heavily weathered, low in cation-exchange capacity, and low in fertility.

oxygen (17)   The second most common gas in the atmosphere, critical for most forms of life on Earth.

ozone layer (20)   The region of the stratosphere that has higher levels of ozone (o3) that absorbs most of the Sun's ultraviolet radiation.

P

Pacific high (70)   A subtropical high-pressure cell, centered over the Pacific Ocean about 1600 km northeast of Hawai'i.

pahoehoe (222)   Basaltic lava that is more fluid than aa. Pahoehoe forms a thin crust that forms folds and appears "ropy," like coiled, twisted rope.

paleoclimatology (176)   The science that studies the climates, and the causes of variations in climate, of past ages, throughout historic and geologic time.

paleolakes (also called pluvial) (333)   An ancient lake, such as Lake Bonneville or Lake Lahonton, associated with former wet periods when the lake basins were filled to higher levels than today.

Pangaea (222)   The supercontinent formed by the collision of all continental masses approximately 225 million years ago; named in the continental drift theory by Wegener in 1912. (*See* Plate tectonics.)

parallel (i-12)   A line, parallel to the equator, that designates an angle of latitude. (See Latitude.)

parasitism (362)   A trophic relationship where one organisim benefits and another organism is harmed.

parent material (249)   The unconsolidated material, from both organic and mineral sources, that is the basis of soil development.

particulate matter (PM) (25)   Dust, dirt, soot, salt, sulfate aerosols, fugitive natural particles, or other material particles suspended in air.

patterned ground (331)   Areas in the periglacial environment where freezing and thawing of the ground create polygonal forms of arranged rocks at the surface; can be circles, polygons, stripes, nets, and steps.

pedogenic regime (384)   A specific soil-forming process keyed to a specific climatic regime: laterization, calcification, salinization, and podzolization, among others; not the basis for soil classification in the Soil Taxonomy.

pedon (354)   A soil profile extending from the surface to the lowest extent of plant roots or to the depth where regolith or bedrock is encountered; imagined as a hexagonal column; the basic soil sampling unit.

periglacial (330)   Cold-climate processes, landforms, and topographic features along the margins of glaciers, past and present; periglacial characteristics exist on more than 20% of Earth's land surface; includes permafrost, frost action, and ground ice.

permafrost (330)   Forms when soil or rock temperatures remain below 0°C (32°F) for at least 2 years in areas considered periglacial; criterion is based on temperature and not on whether water is present. (*See* Periglacial.)

phase change (88)   The change in phase, or state, among ice, water, and water vapor; involves the absorption or release of latent heat. (*See* Latent heat.)

photochemical smog (24)   Air pollution produced by the interaction of ultraviolet light, nitrogen dioxide, and hydrocarbons; produces ozone and PAN through a series of complex photochemical reactions. Automobiles are the major source of the contributive gases.

photosynthesis (348)   The process by which plants produce their own food from carbon dioxide and water, powered by solar energy. The joining of carbon dioxide and hydrogen in plants, under the influence of certain wavelengths of visible light; releases oxygen and produces energy-rich organic material, sugars, and starches. (*Compare* Respiration.)

physical weathering (also called mechanical weathering) (250)   The breaking apart and disintegration of rock without chemical alternation. These processes include frost action, salt crystal growth, and pressure-release jointing

plane of the ecliptic (13)   A plane (flat surface) intersecting all the points of Earth's orbit.

pluton (220)   A mass of intrusive igneous rock that has cooled slowly in the crust; forms in any size or shape. The largest partially exposed pluton is a batholith. (*See* Batholith.)

podzolization (385)   A pedogenic process in cool, moist climates; forms a highly leached soil with strong surface acidity because of humus from acid-rich trees.

point bar (280)   In a stream, the inner portion of a meander, where sediment fill is redeposited. (*Compare* Undercut bank.)

point source (139)   Pollution of water from fixed and easily identified sources such as a single pipe, factory, or ship.

polar climates (153)   Climates of high latitudes and polar regions, where cold temperatures occur year-round, with no true summer.

polar easterlies (71)   Variable, weak, cold, and dry winds moving away from the polar region; an anticyclonic circulation.

polar front (70)   A significant zone of contrast between cold and warm air masses; roughly situated between 50° and 60° N and S latitudes.

polar high-pressure cells (68)   Weak, anticyclonic, thermally produced pressure systems positioned roughly over each pole; that over the South Pole is the region of the lowest temperatures on Earth. (*See* Antarctic high.)

pollutants (22)   Natural or human-caused gases, particles, and other substances in the troposphere that accumulate in amounts harmful to humans or to the environment.

potential evapotranspiration (128)   POTET, or PE; the amount of moisture that would evaporate and transpire if adequate moisture were available; it is the amount lost under optimum moisture conditions, the moisture demand. (*Compare* Actual evapotranspiration.)

precipitation (96, 128)   Rain, snow, sleet, and hail—the moisture supply; called PRECIP, or P, in the water balance.

pressure-gradient force, (66)   Causes air to move from an area of higher barometric pressure to an area of lower barometric pressure due to the pressure difference.

pressure-release jointing (also called exfoliation) (250)   A form of mechanical weathering the occurs when slabs of weathered rock slip of in large sheets, removing pressure on the underlying rock. Gravity also exerts a tension force that also promotes exfoliation.

primary producers (346)   Organisms that manufacture their own food.

primary succession, (365) The process of ecological succession beginning without soil or organisms.

prime meridian (i-14) An arbitrary meridian designated as 0° longitude, the point from which longitudes are measured east or west; established at Greenwich, England, by international agreement in an 1884 treaty.

principle of superposition (214) The geological deposition of rock and sediment always are arranged with the youngest beds (or layers) "superposed" toward the top of a rock formation and the oldest at the base, if they have not been disturbed.

proxy method (176) Information about past environments that represent changes in climate, such as isotope analysis or tree ring dating; also called a *climate proxy*.

pyroclastic (234) An explosively ejected rock fragment launched by a volcanic eruption; sometimes described by the more general term *tephra*.

R

radiation fog (98) Formed by radiative cooling of a land surface, especially on clear nights in areas of moist ground; occurs when the air layer directly above the surface is chilled to the dew-point temperature, thereby producing saturated conditions.

radiative forcing (196) The amount by which some perturbation causes Earth's energy balance to deviate from zero; a positive forcing indicates a warming condition, while a negative forcing indicates cooling; also called *climate forcing*.

rain shadow (104) The area on the leeward slope of a mountain range where precipitation receipt is greatly reduced compared to the windward slope on the other side. (*See* Orographic lifting.)

recessional moraine (328) Unsorted glacial sediments at points where a glacier paused between advancing and retreating.

reflection (37) The portion of arriving insolation that is returned directly to space without being absorbed and converted into heat and without performing any work. (*See* Albedo.)

refraction (37) The bending effect on electromagnetic waves that occurs when insolation enters the atmosphere or another medium; the same process disperses the component colors of the light passing through a crystal or prism.

regolith (249) Partially weathered rock overlying bedrock, whether residual or transported.

relative humidity (90) The ratio of water vapor actually in the air (content) to the maximum water vapor possible in the air (capacity) at that temperature; expressed as a percentage. (*Compare* Vapor pressure, Specific humidity.)

relative time (214) The age of one feature with respect to another within a sequence of events and is deduced from the relative positions of rock strata above or below each other. (Compare to absolute time).

relief (i-19) Elevation differences in a local landscape; an expression of local height differences of landforms.

remote sensing (i-20) Information acquired from a distance, without physical contact with the subject—for example, photography, orbital imagery, and radar.

resilience (368) The ability of a community to recover and return to a pre-disturbance state.

reverse fault (229) Compressional forces produce strain that breaks a rock so that one side moves upward relative to the other side; also called a *thrust fault*. (*Compare* Normal fault.)

revolution (13) The annual orbital movement of Earth about the Sun; determines the length of the year and the seasons.

Richter scale (231) An open-ended, logarithmic scale that estimates earthquake magnitude based on measurement of the maximum seismic wave amplitude; designed by Charles Richter in 1935; now replaced by the moment magnitude scale. (*See* Moment magnitude scale.)

ridge push (231) The process by which the addition of new rock along a mid-ocean ridge pushes a tectonic plate away from the ridge. Because mid-ocean ridges lie at a higher elevation than the surrounding ocean floor, gravity causes the oceanic lithosphere to slide "downhill" toward a trench.

ridges (72) An elongated region of high air pressure, usually oriented north-south.

rock (219) An assemblage of minerals bound together, or sometimes a mass of a single mineral.

rock cycle (219) A model representing the interrelationships among the three rock-forming processes: igneous, sedimentary, and metamorphic; shows how each can be transformed into another rock type.

rockfall (258) Free-falling movement of debris from a cliff or steep slope, generally falling straight or bounding downslope.

Rossby waves (73) An undulating horizontal motion in the upper-air westerly circulation at middle and high latitudes.

rotation (13) The turning of Earth on its axis, averaging about 24 hours in duration; determines day–night relation; counterclockwise when viewed from above the North Pole and from west to east, or eastward, when viewed from above the equator.

S

salinity (295) The concentration of natural elements and compounds dissolved in solution, as solutes; measured by weight in parts per thousand (‰) in seawater.

salinization (385) A pedogenic process that results from high potential evapotranspiration rates in deserts and semiarid regions. Soil water is drawn to surface horizons, and dissolved salts are deposited as the water evaporates.

salt marsh (304) A wetland ecosystem characteristic of latitudes poleward of the 30th parallel. (*Compare* Mangrove swamp.)

saltation (279) The transport of sand grains (usually larger than 0.2 mm, or 0.008 in.) by stream or wind, bouncing the grains along the ground in asymmetrical paths.

salt-crystal growth (250) A form of physical weathering that occurs when dry weather draws moisture to the surface of rocks. As the water evaporates, dissolved mineral salts in the water form crystals that over time, grow and break up the rock.

sand sea (308) An extensive area of sand and dunes; characteristic of Earth's erg deserts. (*Compare* Erg.)

saturation (90) State of air that is holding all the water vapor that it can hold at a given temperature, known as the dew-point temperature.

scale (i-17) The ratio of the distance on a map to that in the real world; expressed as a representative fraction, graphic scale, or written scale.

scarification (261) Human-induced mass movement of Earth materials, such as large-scale open-pit mining and strip mining.

scattering (diffuse radiation) (36) Deflection and redirection of insolation by atmospheric gases, dust, ice, and water vapor; the shorter the wavelength, the greater the scattering; thus, skies in the lower atmosphere are blue.

sea terrace (302) A wave-cut horizontal platform in the tidal zone, extending from a sea cliff out into the sea. Multiple sea level changes may cut multiple terraces.

seafloor spreading (224) As proposed by Hess and Dietz, the mechanism driving the movement of the continents; associated with upwelling flows of magma along the worldwide system of mid-ocean ridges. (*See* Mid-ocean ridge.)

secondary succession (365) Succession that occurs among plant species in an area where vestiges of a previously functioning community are present; an area where the natural community has been destroyed or disturbed, but where the underlying soil remains intact.

sediment (248) Fine-grained mineral matter that is transported and deposited by air, water, or ice.

sedimentary rock (219) One of the three basic rock types; formed from the compaction, cementation, and hardening of sediments derived from other rocks. (*Compare* Igneous rock, Metamorphic rock.)

seismic wave (230) The shock wave sent through the planet by an earthquake or underground nuclear test. Transmission varies according to temperature and the density of various layers within the planet; provides indirect diagnostic evidence of Earth's internal structure.

seismograph (230) A scientific instrument that records vibrations transmitted as waves of energy throughout Earth.

sensible heat (18, 34) Heat that can be measured with a thermometer; a measure of the concentration of kinetic energy from molecular motion.

September equinox (14) The time around September 22–23 when the Sun's declination crosses the equatorial parallel (0° latitude) and all places on Earth experience days and nights of equal length (also known as *autumnal equinox*). The Sun rises at the South Pole and sets at the North Pole. (*Compare* March equinox.).

sheetflow (270) Surface water that moves downslope in a thin film as overland flow; not concentrated in channels larger than rills.

sinkholes (255) Nearly circular depression created by the weathering of karst landscapes with subterranean drainage; also known as a *doline* in traditional studies; may collapse through the roof of an underground space. (*See* Karst topography.)

slab pulls (226) A force that occurs as a subducting tectonic plate, usually dense basalt, sinks into the hot mantle beneath it.

slopes (248) A curved, inclined surface that bounds a landform.

soil (352) A dynamic natural body made up of fine materials covering Earth's surface in which plants grow, composed of both mineral and organic matter.

soil creep (260) A persistent mass movement of surface soil where individual soil particles are lifted and disturbed by the expansion of soil moisture as it freezes or by grazing livestock or digging animals.

soil fertility (356) The ability of soil to support plant productivity when it contains organic substances and clay minerals that absorb water and certain elemental ions needed by plants through adsorption. (*See* Cation-exchange capacity.)

soil horizons (354) The various layers exposed in a pedon; roughly parallel to the surface and identified as O, A, E, B, C, and R (bedrock).

soil profile (354) A vertical section of soil extending from the surface to the deepest extent of plant roots or to regolith or bedrock.

Soil Taxonomy (384) A soil classification system based on observable soil properties actually seen in the field; published in 1975 by the U.S. Soil Conservation Service and revised in 1990 and 1998 by the Natural Resources Conservation Service to include 12 soil orders.

soil-moisture storage capacity (129) The ability of the soil to store moisture that percolates into the ground from precipitation and overland flow.

soil-water budget (128) A calculation that determines how the incoming precipitation "supply" is distributed to satisfy the output "demand" of plants, evaporation, and soil-moisture storage.

solar constant (10) The amount of insolation intercepted by Earth on a surface perpendicular to the Sun's rays when Earth is at its average distance from the Sun; a value of 1372 W/m^2 (1.968 calories/cm^2) per minute; averaged over the entire globe at the thermopause.

solar wind (6) Clouds of ionized (charged) gases emitted by the Sun and traveling in all directions from the Sun's surface. Effects on Earth include auroras, disturbance of radio signals, and possible influences on weather.

southeast trade winds (70) The prevailing southeast winds flowing from the subtropical high toward the ITCZ.

species (360) A population that reproduces sexually and can produce viable offspring.

specific heat (47, 88) The increase of temperature in a material when energy is absorbed; water has a higher specific heat (can store more heat) than a comparable volume of soil or rock.

specific humidity (92) The mass of water vapor (in grams) per unit mass of air (in kilograms) at any specified temperature. The maximum mass of water vapor that a kilogram of air can hold at any specified temperature is termed its maximum specific humidity. (*Compare* Vapor pressure, Relative humidity.)

speed of light (5) Specifically, 299,792 km (186,282 mi) per second, or more than 9.4 trillion km (5.9 trillion mi) per year—a distance known as a light-year; at light speed, Earth is 8 minutes and 20 seconds from the Sun.

spheroidal weathering (252) A chemical weathering process in which the sharp edges and corners of boulders and rocks are weathered in thin plates that create a rounded, spheroidal form.

Spodosols (388)　A soil order in the Soil Taxonomy. Occurs in northern coniferous forests; best developed in cold, moist, forested climates; lacks humus and clay in the A horizon, with high acidity associated with podzolization processes.

springs (134)　Natural features that occur when the water table intersects the surface. In some confined aquifers the water is under pressure, creating an artesian well where water may rise to the surface without pumping.

squall line (105)　A zone slightly ahead of a fast-advancing cold front where wind patterns are rapidly changing and blustery and precipitation is strong.

stability (94)　The condition of a parcel of air with regard to whether it remains where it is or changes its initial position. The parcel is stable if it resists displacement upward and unstable if it continues to rise.

steppe (166)　A regional term referring to the vast semiarid grassland biome of Eastern Europe and Asia; the equivalent biome in North America is shortgrass prairie, and in Africa, it is the savanna. Steppe in a climatic context is considered too dry to support forest, but too moist to be a desert.

storm surge (116)　A large quantity of seawater pushed inland by the strong winds associated with a tropical cyclone.

stratocumulus (97)　A lumpy, grayish, low-level cloud, patchy with sky visible, sometimes present at the end of the day.

stratosphere (19)　That portion of the homosphere that ranges from 20 to 50 km (12.5 to 30 mi) above Earth's surface, with temperatures ranging from –57°C (–70°F) at the tropopause to 0°C (32°F) at the stratopause. The functional ozonosphere is within the stratosphere.

stratus (97)　A stratiform (flat, horizontal) cloud generally below 2000 m (6500 ft).

stream discharge (276)　The volume of water moving past a point in a given unit of time, calculated using measurements of stream width, depth, and velocity of a given cross section of the channel. It is summarized in the equation

streamflow (132)　The flow or volume of water in a channel.

strike-slip (229)　Horizontal movement along a fault line—that is, movement in the same direction as the fault; also known as a *transcurrent* fault. Such movement is described as right lateral or left lateral, depending on the relative motion observed across the fault.

subduction zone (225)　An area where two plates of crust collide and the denser oceanic crust dives beneath the less dense continental plate, forming deep oceanic trenches and seismically active regions.

sublimation (89)　A process in which ice evaporates directly to water vapor or water vapor freezes directly to ice (deposition)　.

subpolar low-pressure cells (68)　A region of low pressure centered approximately at 60° latitude in the North Atlantic near Iceland and in the North Pacific near the Aleutians as well as in the Southern Hemisphere. Airflow is cyclonic; it weakens in summer and strengthens in winter. (*See* Cyclone.)

subsolar point (10)　The only point receiving perpendicular insolation at a given moment—that is, the Sun is directly overhead. (*See* Declination.)

subtropical high-pressure cells (68)　One of several dynamic high-pressure areas covering roughly the region from 20° to 35° N and S latitudes; responsible for the hot, dry areas of Earth's arid and semiarid deserts. (*See* Anticyclone.)

sulfur dioxide (SO₂) (24)　A colorless gas detected by its pungent odor; produced by the combustion of fossil fuels, especially coal, that contain sulfur as an impurity; can react in the atmosphere to form sulfuric acid, a component of acid deposition.

summer solstice (14)　The June solstice in the northern hemisphere, and the December solstice in the southern hemisphere. See June solstice and December solstice.

sunspots (6)　Magnetic disturbances on the surface of the Sun, occurring in an average 11-year cycle; related flares, prominences, and outbreaks produce surges in solar wind.

surface creep (308)　A form of eolian transport that involves particles too large for saltation; a process whereby individual grains are impacted by moving grains and slide and roll.

surge (325)　A brief but rapid advance of glacial ice at velocities up to 100 times faster than normal.

suspended load (278)　Fine particles held in suspension in a stream. The finest particles are not deposited until the stream velocity nears zero.

swell (300)　Regular patterns of smooth, rounded waves in open water; can range from small ripples to very large waves.

symbiosis (362)　An ecological relationship between organisms of two different species that live together in direct and close contact.

syncline (228)　A trough in folded strata, with beds that slope toward the axis of the downfold. (*Compare* Anticline.)

T

talus slopes (327)　A cone-shaped pile of irregular broken rocks at the base of a steep incline.

tarn (327)　A small mountain lake, especially one that collects in a cirque basin behind risers of rock material or in an ice-gouged depression.

temperate rain forest, (400)　A major biome of lush forests at middle and high latitudes; occurs along narrow margins of the Pacific Northwest in North America, among other locations; includes the tallest trees in the world.

temperature (44)　A measure of sensible heat energy present in the atmosphere and other media; indicates the average kinetic energy of individual molecules within a substance.

temperature inversion (23)　A reversal of the normal decrease of temperature with increasing altitude; can occur anywhere from ground level up to several thousand meters; functions to block atmospheric convection and thereby trap pollutants.

terminal moraine (328)　Eroded debris that is dropped at a glacier's farthest extent.

theory of plate tectonics (222)　The conceptual model and theory that encompass continental drift, seafloor spreading, and related aspects of crustal movement; accepted as the foundation of crustal tectonic processes.

thermal equator (52)　The isoline on an isothermal map that connects all points of highest mean temperature.

thermohaline circulation (77)   Deep-ocean currents produced by differences in temperature and salinity with depth; Earth's deep currents.

thermosphere (18)   A region of the heterosphere extending from 80 to 480 km (50 to 300 mi) in altitude; contains the functional ionosphere layer.

thrust (229)   A reverse fault where the fault plane forms a low angle relative to the horizontal; an overlying block moves over an underlying block.

thunder (110)   The violent expansion of suddenly heated air, created by lightning discharges, which sends out shock waves as an audible sonic bang.

tidal bore (300)   The leading edge of incoming tidal waters that resembles a large wave. Ocean basin characteristics (size, depth, and topography), latitude, and shoreline shape, all influence the size of the bore.

tidal flat (300)   A coastal wetland flooded and drained by tides. Occurs on level terrain wherever tidal influence exceeds wave action.

tide (300)   A pattern of twice-daily oscillations in sea level produced by astronomical relations among the Sun, the Moon, and Earth; experienced in varying degrees around the world.

till (328)   Direct ice deposits that appear unstratified and unsorted; a specific form of glacial drift. (*Compare* Stratified drift.)

tombolo (304)   A landform created when coastal sand deposits connect the shoreline with an offshore island outcrop or sea stack.

tornado (112)   An intense, destructive cyclonic rotation, developed in response to extremely low pressure; generally associated with mesocyclone formation.

topographic maps (i-19)   A map that portrays physical relief through the use of elevation contour lines that connect all points at the same elevation above or below a vertical datum, such as mean sea level.

traction (279)   A type of sediment transport that drags coarser materials along the bed of a stream. (*See* Bed load.)

trade winds (70)   Winds from the northeast and southeast that converge in the equatorial low-pressure trough, forming the intertropical convergence zone.

transparency (47)   The quality of a medium (air, water) that allows light to easily pass through it.

transpiration (128)   The movement of water vapor out through the pores in leaves; the water is drawn by the plant roots from soil-moisture storage.

trophic level (358)   The position of an organism in a food chain or web such as producer, consumer, or decomposer.

Tropic of Cancer (12)   The parallel that marks the farthest north the subsolar point migrates during the year; 23.5° N latitude. (*See* Tropic of Capricorn, June solstice.)

Tropic of Capricorn (12)   The parallel that marks the farthest south the subsolar point migrates during the year; 23.5° S latitude. (*See* Tropic of Cancer, December solstice.)

tropical climates (158)   Climates straddling the Equator from 20° N to 20° S, roughly between the Tropics of Cancer and Capricorn. Characterized by consistent day length and insolation, which produces steady warm temperatures.

tropical cyclone (114)   A cyclonic circulation originating in the tropics, with winds between 30 and 64 knots (39 and 73 mph); characterized by closed isobars, circular organization, and heavy rains. (*See* Hurricane, Typhoon.)

tropical rain forest (392)   A lush biome of tall broadleaf evergreen trees and diverse plants and animals, roughly between 23.5° N and 23.5° S latitude. The dense canopy of leaves is usually arranged in three levels.

tropical savanna (397)   A major biome containing large expanses of grassland interrupted by trees and shrubs; a transitional area between the humid rain forests and tropical seasonal forests and the drier, semi-arid tropical steppes and deserts.

tropical seasonal forest and scrub (396)   A variable biome on the margins of the rain forests, occupying regions of lesser and more erratic rainfall; the site of transitional communities between the rain forests and tropical grasslands.

troposphere (19)   The home of the biosphere; the lowest layer of the homosphere, containing approximately 90% of the total mass of the atmosphere; extends up to the tropopause; occurring at an altitude of 18 km (11 mi) at the equator, at 13 km (8 mi) in the middle latitudes, and at lower altitudes near the poles.

troughs (72)   An elongated region of low air pressure, usually oriented north-south.

true shape (i-18)   A map property showing the correct configuration of coastlines; a useful trait of conformality for navigational and aeronautical maps, although areal relationships are distorted. (See Map projection; compare Equal area.)

tsunami (301)   A seismic sea wave, traveling at high speeds across the ocean, formed by sudden motion in the seafloor, such as a seafloor earthquake, submarine landslide, or eruption of an undersea volcano.

tundra (164)   A vast and treeless Arctic region that spans Eurasia and North America, where the subsoil is permanently frozen.

typhoon (114)   A tropical cyclone with wind speeds in excess of 119 kmph (65 knots, or 74 mph) that occurs in the western Pacific; same as a hurricane except for location. (*Compare* Hurricane.)

U

Ultisols (388)   A soil order in the Soil Taxonomy. Features highly weathered forest soils, principally in the humid subtropical climatic classification. Increased weathering and exposure can degenerate an Alfisol into the reddish color and texture of these Ultisols. Fertility is quickly exhausted when Ultisols are cultivated.

undercut bank (280)   A steep bank formed along the outer portion of a meandering stream; produced by lateral erosive action of a stream; sometimes called a *cutbank*. (*Compare* Point bar.)

ungraded stream (274)   A stream that is actively eroding or depositing material in its channel so that its gradient is adjusted to carry the sediment load.

uniformitarianism (215)   An assumption that physical processes active in the environment today are operating at the same pace and intensity that have characterized them throughout geologic time; proposed by Hutton and Lyell.

upslope fog (98) Forms when moist air is forced to higher elevations along a hill or mountain and is thus cooled. (*Compare* Valley fog.)

urban heat island (50) An urban microclimate that is warmer on average than areas in the surrounding countryside because of the interaction of solar radiation and various surface characteristics.

U-shaped valley (327) The elongated troughs remain after glaciers recede. The valleys often display flat valley floors and steep, straight sides, greatly changed by erosion from their previous stream-cut V form.

V

valley fog (98) The settling of cooler, more dense air in low-lying areas; produces saturated conditions and fog. (*Compare* Upslope fog.)

valley glacier (322) A river of ice confined within a valley that originally formed by stream action. They grow in size as tributary cirque glaciers from higher elevations descend into valleys.

vapor pressure (92) That portion of total air pressure that results from water vapor molecules, expressed in millibars (mb). At a given dew-point temperature, the maximum capacity of the air is termed its saturation vapor pressure. (*Compare* Relative humidity, Specific humidity.)

ventifacts (307) A piece of rock etched and smoothed by eolian erosion—that is, abrasion by windblown particles.

vernal equinox (27) The spring or March equinox in the northern hemisphere, and the September equinox in the southern hemisphere. See March equinox and September equinox.

Vertisols (391) A soil order in the Soil Taxonomy. Features expandable clay soils; composed of more than 30% swelling clays. Occurs in regions that experience highly variable soil moisture balances through the seasons.

virga (88) Precipitation that falls from clouds, but evaporates before it reaches the ground.

volcano (234) A mountainous landform at the end of a magma conduit, which rises from below the crust and vents to the surface. Magma rises and collects in a magma chamber deep below, erupting effusively or explosively and forming composite, shield, or cinder-cone volcanoes.

W

warm front (104) The leading edge of an advancing warm air mass, which is unable to push cooler, passive air out of the way; tends to push the cooler, underlying air into a wedge shape; identified on a weather map as a line marked with semicircles pointing in the direction of frontal movement. (*Compare* Cold front.)

water table (134) The upper surface of groundwater; that contact point between the zone of saturation and the zone of aeration in an unconfined aquifer. (*See* Zone of aeration, Zone of saturation.)

waterfalls (281) A large stream nickpoint, or drop in the stream profile, resulting in free-falling water moves at high velocity under the acceleration of gravity.

waterspout (112) An elongated, funnel-shaped circulation formed when a tornado exists over water.

wave (300) An undulation of ocean water produced by the conversion of solar energy to wind energy and then to wave energy; energy produced in a generating region or a stormy area of the sea.

wave cyclone (106) *See* Midlatitude cyclone.

wave refraction (302) A bending process that concentrates wave energy on headlands and disperses it in coves and bays; the long-term result is coastal straightening.

wavelength (8) A measurement of a wave; the distance between the crests of successive waves. The number of waves passing a fixed point in 1 second is called the frequency of the wavelength.

weather (108) The short-term condition of the atmosphere, as compared to climate, which reflects long-term atmospheric conditions and extremes. Temperature, air pressure, relative humidity, wind speed and direction, daylength, and Sun angle are important measurable elements that contribute to the weather.

weathering (248) The processes by which surface and subsurface rocks disintegrate, or dissolve, or are broken down. Rocks at or near Earth's surface are exposed to physical and chemical weathering processes.

wells (134) An excavation into the ground to access groundwater in underground aquifers. The water is pumped or hand-drawn upward to the surface. In some confined artesian aquifers, the water is under pressure, and rises to the surface without pumping.

westerlies (70) The predominant surface and aloft wind-flow pattern from the subtropics to high latitudes in both hemispheres.

western intensification (76) The piling up of ocean water along the western margin of each ocean basin, to a height of about 15 cm (6 in.); produced by the trade winds that drive the oceans westward in a concentrated channel.

wetland (311) An area that is permanently or seasonally saturated with water and characterized by vegetation adapted to hydric soils; highly productive ecosystem with an ability to trap organic matter, nutrients, and sediment.

wind (65) The horizontal movement of air relative to Earth's surface; produced essentially by air pressure differences from place to place; turbulence, wind updrafts and downdrafts, adds a vertical component; its direction is influenced by the Coriolis force and surface friction.

wind vane (65) A weather instrument used to determine wind direction; winds are named for the direction from which they originate.

winter solstice (14) The December solstice in the northern hemisphere, and the June solstice in the southern hemisphere. See June solstice and December solstice.

Y

yardang (307) A streamlined rock structure formed by deflation and abrasion; appears elongated and aligned with the most effective wind direction.

Z

zone of aeration (134) A zone above the water table that has air in its pore spaces and may or may not have water.

zone of saturation (134) A groundwater zone below the water table in which all pore spaces are filled with water.

Credits

Photos

Introduction 1: (CO-Intro) All Canada Photos/Alamy. (I.2a) C1847 Erwin Patzelt Deutsch Presse Agentur/Newscom. (I.2b) Niranjan Shrestha/AP Images. (I.2c) Keith Thorpe/AP Images. (I.2d) - NASA. (I.2e) - Tang Chhin Sothy/AFP/Getty Images. (I.5a) USGS. (I.6a) AfriPics/Alamy. (I.6b) Danita Delimont/Gallo Images/Getty Images. (I.6c) Stephen Cunha. (I.6d) Ariel Skelley/Blend Images/Getty Images. (I.7b) NASA. (I.8) - Wasu Watcharadachaphong/Shutterstock. (I.9) TTstudio/Shutterstock. (I.10) Steve Humphreys/Getty Images. (I.11) - Offfstock/Fotolia. I.29B - Courtesy A. Chase and D. Chase, Caracol Archaeological Project, www.caracol.org (after A. Chase et al. 2012:Fig 4. (I.30ab) - USGS. (GLI.2) - atgc_01/Fotolia. (GLI.3) - Barry Winiker/Photolibrary/Getty Images. (HD-Ia) - NASA. (HD-Ib) - NASA. HD-ic - NASA.

Chapter 1: (CO-01) NASA. (1.1) Chintla/Shutterstock. (1.2) JSC Gateway to Astronaut Photography of Earth/NASA. (1.3a) Sdo/Steele Hill, 2012/NASA. (1.3b) Soho/EIT Consortium/NASA. (1.3b) Soho/EIT Consortium/NASA. (1.4a) Image Spacecraft GSFC/NASA. (1.4b) David Cartier, Sr., courtesy of GSFC/NASA.. (1.15) Jip Lambermont. (1.20) Red Bull Stratos/Zuma Press/Newscom. (1.21) NASA. (1.25a) Nasa Images. (1.25b) Ingolfur Juliusson/Reuters/Corbis. (1.26) David Gross/Corbis. (01.05UN) Nasa Images. (1.27) NASA. (1.28c) Bobbé Christopherson. (1.29) Nasa Images. (01-29 inset) Li Gang Xinhua News Agency/Newscom. (HD1a) Steve Bower/Shutterstock. (HD1c) Steve Parsons/AP Images. (HD1d) Per-Anders Pettersson/Getty Images News/Getty Images. **(R1.2)** Google, Inc.. (GL1.1) Chris Sattlberger/Blend Images/Corbis.

Chapter 2: (CO-02) Eric Lo/Getty Images. (2.7) John King/Alamy. (2.18) Stefan Christmann/Corbis. (2.23a) Kropic1/Shutterstock. (2.23b) Rgb Ventures/SuperStock/Alamy. (2.28) NASA. (R2.1b) Citizen of the Planet/Alamy. (HD 02-a) Justin Kase Zninez/Alamy. (HD 02-b) Ajit Solanki/AP Images. (HD 02-c) Joanne Hartley/Reuters. (HD 02-d) NASA. (2.4) Samo Trebizan/Shutterstock. (2.8a) Marc Shandro/Getty Images. (2.8b) Richard Whitcombe/whitcomberd/123RF. (2.11a) Patrick Pleul/dpa/Corbis. (2.11b) Iofoto/Fotolia I. (2.11C) - Jacques Descloitres, MODIS Rapid Response Team, NASA/GSFC. 2.16AB - NASA. Iofoto/Fotolia. (R2.1a) Steve Proehl/Proehl Studios/Corbis.

Chapter 3: (CO-03) Andrew Pielage/Zuma Press/Newscom. (3.1) Cameron Beccario/earth.nullschool.net. (3.4) Silvan Wick Water-Sports/Alamy. (HD-03A) Rick Bowmer/AP Images. (HD-03C) Adrees Latif/Reuters. (HD-03D) Gardel Bertrand/Hemis/Alamy. (3.2) Kirillica/Fotolia. (3.11) Marc Adamus/Aurora Photos/Corbis. **(R3.2)** NASA. (03-15b) NASA. (Figure R3.1) NASA's Earth Observatory.

Chapter 4: (CO-04) Mike Hollingshead/Corbis. (4.11) Bobbé Christopherson. (04-16a) Murat Subatli/Fotolia. (04-16b) Dmitry Knorre/Fotolia. (04-16e) David Wall/Alamy. (4.16f) Steve Austin/Papilio/Corbis. (4.16g) Pixel Memoirs/Fotolia. (4.16h) Bobbé Christopherson. (04-17) Paul Bruins Photography/Getty Images. (4.18b) NASA. (4.19) Manamana/Shutterstock. (4.22) Jeff Schmaltz/NASA. (4.24) NHPA/SuperStock. (04-24d) NASA. (04-30a) Douglas Peebles Photography/Alamy. (4.30b) Michele Falzone/Alamy. (4.50ab) USGS. **(04-16c)** PzAxe/Shutterstock. (4.21) National Park Service. (4.36) NOAA. (04-43d) Minerva Studio/Fotolia. **(04-16d)** Martchan/Shutterstock. (04-HDa(left)) Fabrice Coffrini/afp/Getty Images. (04-HDb) Yu Hyung-jae/Yonhap/AP Images. (04-HDc) Jason Clark/zumapress/Newscom. (04-HDc(inset)) David Mabe/Alamy. (04-Visual Analysis) U.S. Air Force photo by Master Sgt. Mark C. Olsen. (4.2) John Hyde/Perspectives/Getty Images. (4.29c) Zeb Andrews/Moment Open/Getty Images. (4.29d) Paul Gordon/Alamy. (4.32) Barcroft Media/Getty Images. (4.51) United States Department of Defense. (4.33) NASA/NOAA. (4.41) University Corporation for Atmospheric Research (UCAR). (4.47) Jeff Schmalt/LANCE/EOSDIS Rapid Response/NASA. (04-05) icarmen13/Shutterstock. (04-20) Michael Shake/Fotolia.

Chapter 5: (CO-05) Stringer Shanghai/Reuters/Corbis. (5.1) Stephen Cunha. (05-08) Sammy/Alamy. (HD 5a) Jeronimo Alba/Alamy. (HD 5d) Mike Goldwater/Alamy. (HD 5b) WitR/Shutterstock. (5.7a) Tim Gainey/Alamy. (05-07b) Stephen Cunha. (5.11) Stephen Cunha. (5.14a) USGS. (5.14b) USGS. (5.14c) NASA. (5.18b) Stephen Cunha. (5.20) U.S. Geological Survey. (5.21b) Stephen Cunha. (5.24a) Thomas Barrat/Shutterstock. (5.24b) Bobbé Christopherson. (05-24c) National Geographic Creative/Alamy. (5.25a) Thomas Barrat/Shutterstock. (5.25b) Bobbé Christopherson. (5.25c) National Geographic/Alamy. (5.26abc) Bobbé Christopherson. (5.27a) Rich Pedroncelli/AP Images. (5.28a) Rich Pedroncelli/AP Images. (5.29b) Stephen Cunha. (5.30) Beldesigne/Fotolia. (R5.1) Stephen Cunha. (GOOGLE-05) Google, Inc.. (GL05-04) John Rawlston/Chattanooga Times Free Press/AP Images.

Chapter 6: (CO-06) NPS Photo/Alamy. (6.2) Franck Monnot/Fotolia. (6.3) Bill Perry/Shutterstock. (6.5) Cpphotoimages/Shutterstock. (06-8b) Sue Cunningham/Alamy. (6.9b) J-F Perigois/Fotolia. (6.10b) Stephen Cunha. (06-12b) Tao Images Limited/Alamy. (06-13b) Science Source. (6.14b) Bobbé Christopherson. (6.16b) Bobbé Christopherson. (06-17b) Dave G. Houser/Documentary Value/Corbis. (6.18b) - Bobbé Christopherson. (06-19) Niebrugge Images/Alamy. (6.20) Stephen Cunha. (6.21) Andreanita/123RF. (6.23b) Thomas Pickard/Aurora Photos/Alamy. (6.25b) Stephen Cunha. (6.26b) Design Pics/Bilderbuch/Getty Images. (R6.1) Santiago Urquijo/Moment/Getty Images. (R6-03) Anna Bartosch Carlile/Alamy. (GL6-1) SuperStock/Glow Images. (GL6.2) Kevin Schafer/Alamy. (HD 6a) AlxYago/Shutterstock. (HD 6b) Nigel Cattlin/Alamy. (HD 6c) David Lomax/Robertharding/Alamy. (HD 6d) Scott Olson/Getty Images News/Getty Images.

Chapter 7: (CO-07) James Balog/Aurora. (7.1a) Cultura RM/Alamy. (7.1b) Hickey, D., Reich, C.D., DeLong, K.L., Poore, R.Z., Brock, J.C., 2013. Holocene core logs and site methods for modern reef and head-coral cores: Dry Tortugas National Park, Florida. Dept. of the Interior, U.S. Geological Survey, Washington DC., p. 27. (7.1c) International Ocean Discovery Program. (7.1d) Tim Burton/NEEM. (7.2) William Crawford/International Ocean Discovery Program. (7.3a) National Science Foundation. (7.3b) British Antarctic Survey. (7.5d) Peter von Bucher/Shutterstock. (07-07) Dietrich Rose/The Image Bank/Getty Images. (7.8a) Chris Howes/Wild Places Photography/Alamy. (7.8b) Ted Kinsman/Science Source. (07-15b) Arlan Naeg/Getty Images. (7.16) U.S. Army. (7.17) NOAA. (7.23b) NASA/Goddard Space Flight Center Scientific Visualization Studio. The Blue Marble data is courtesy of Reto Stockli (NASA/GSFC). (7.24b) Goddard Space Flight Center/NASA. (7.25b) Ashley Cooper pics/Alamy. (7.25c) National Geographic Image Collection/Alamy. (7.27) Gary Braasch-KPA/Zuma Press/Newscom. (7.28) Samantha Cristoforetti/ESA/NASA. (7.29) Eric Anderson/AP Images. (7.31) Jonathan Tennant/Alamy. (7.34a) Ulrich Doering/Alamy. (7.34b) Dave Bartruff/Corbis. (7.34c) Federico Rostagno/Fotolia. (7.34d) Parkerphotography/Alamy. (7.44) John G. Wilbanks/Alamy. (07-47) Arnaud Bouissou/Cop21/Anadolu Agency/Getty Images. (7.50) U.S. Department of Agriculture. (07-52a) William Perugini/Shutterstock. (07-52b) Jordan Tan/Shutterstock. (07-52c) Hgalina/Fotolia. (07-52d) Aerogondo/Fotolia. (07-52e) Esbobeldijk/Shutterstock. (07-52g) Steve Cukrov/Shutterstock. (07-52center) Nasa. (R7.2) Image Landsat copyright 2015 Digital Globe/Google Earth. (R7.1a) USGS. (R7.1b) USGS. (7.5b) GI0ck/Shutterstock. (7.5c) AuntSpray/Shutterstock. (7.5a) National Geographic Image Collection/Alamy. (HD 7d) Charles Sturge/Alamy. (HD 7c) David Hill/Photolibrary/Getty Images. (HD 7a) U.S. Department of Agriculture. (HD 7b) Pat Roque/AP Photo. (07-40) NASA. (7.52f) Martin Shields/Alamy.

Chapter 8: (CO-08) Patrick Taschler. (8.2a) Stephen Cunha. (8.3) Stephen Cunha. (8.8) Stephen Cunha. (8.9a) Bobbé Christopherson. (8.9c) - National Park Service/National Park Service. (8.9b) United States Geological Survey. (8.10a) Stephen Cunha. (8.10b) Stephen Cunha. (8.10c) Stephen Cunha. (8.11b) Stephen Cunha. (8.12c) Les Palenik/Shutterstock. (8.12d) Stephen Cunha. (8.12e) Sephirot17/Getty Images. (8.16c) Israel Hervas Bengochea/Shutterstock. (8.18b) Scott Dickerson/AP Images. (8.22c) Stephen Cunha. (8.22b) Juancat/Fotolia. (8.25d) Stephen Cunha. (8.25e) steve estvanik/123RF. (8.25f) Kevin Schafer/Alamy. (8.27c) Ali Hovaisi/Xinhua News Agency/Newscom. (8.29) Sipa/AP Images. (8.30b) Kimimasa Mayama/Corbis Wire/Corbis. (8.30) Aflo/Mainichi Newspaper/Epa/Newscom. (8.32b) Stephen Cunha. (8.34b) Stephen Cunha. (8.34c) Sergi Reboredo/Photoshot/Newscom. (8.34d) Tan Yilmaz/Getty Images. (8.34e) Lindsay Douglas/Shutterstock. (8.35) Bobbé Christopherson. (8.36) USGS. (8.37) Afp/Getty Images. (8.38a) Carlos Gutierrez/Reuters. (8.39b) David Cortes/Agenciauno Xinhua News Agency/Newscom. (8.39c) Felipe Trueba/Epa/Newscom. (8.39) Chaideer Mahyuddin/AFP/Getty Images. (8.41a) Dana Stephenson/Getty Images. (8.41b) Edwina Pickles/ZUMA Press/Newscom. (8.42b) Stephen Cunha. (8.43b) suronin/Shutterstock. (8.43c) Stephen Cunha. (GL8.2) John G. Wilbanks/Alamy. (HD 08b) inga spence/Alamy. (8.24b) Stephen Cunha. (8.43a) vdLee/Shutterstock. (8.24b) Stephen Cunha. (08-42b) Stephen Cunha. (8.18c) Stephen F. Cunha. (R8.1) Stephen F. Cunha. (R8.2) Google Earth.

Chapter 9: (CO-09) Bobbé Christopherson. (9.2b) Stephen Cunha. (9.3) Stephen Cunha. (09-4b) Dennis Frates/Alamy. (9.5b) Bobbé Christopherson. (9.6) Stephen Cunha. (9.7) Stephen Cunha. (9.8a) Bobbé Christopherson. (9.9) Jimmy Chin/Getty Images. (9.10) Damian Gil/Shutterstock. (9.11) Stephen Cunha. (9.12) Brenda Carson/Shutterstock. (9.13) Stephen Cunha. (9.14) Stephen Cunha. (9.16) Moises Castillo/AP Images. (9.17) SeanPavonePhoto/Fotolia. (9.18b) Westend61 Gmbh/Alamy. (9.19) Stephen Cunha. (9.20) Stephen Cunha. (9.21) NASA. (9.22a) DigitalGlobe/ScapeWare3d/Getty Images. (09-22b) Ted S. Warren/AP Images. (09-22c) Elaine Thompson/AP Images. (9.24) Stephen Cunha. (09-25) Rahmat Gul/AP Images. (09-28) Chris Hawley/AP Images. (09-29b) Heritage Image Partnership/Alamy. (9.30) Tatiana Grozetskaya/Shutterstock. (09-31a) Debbie Hill/Upi/Newscom. (HD 9B) Stephen Cunha. (HD 9c) Patrick Lin/Afp/Getty Images. (HD 9a) Schafer Hill/The Image Bank/Getty Images. (R9.1) kavram/Shutterstock. (R9.2) Aphotostory/Shutterstock. (GL9.3) Blaine Harrington III/Alamy. (R9.2) Google, Inc..

Chapter 10: (CO-10) Dirk Bleyer/ImageBROKER/Corbis. (10.1) Anatoliy Lukich/Fotolia. (10.3A) Mathess/Fotolia. (10.3C) Universal Images Group Limited/Alamy. (10.4) Bobbé Christopherson. (10.5A) Graham Eaton/Nature Picture Library/Alamy. (10.5B) Spacephotos/Age Fotostock. (10.5C) Stephen J. Krasemann/Science Source. (10.5D) H. Mark Weidman Photography/Alamy. (10.5E) Margus Muts/Getty Images. (10.5F) Corbis. (10.5G) Auscape/Getty Images. (10.7) National Geographic Image Collection/Alamy. (10.8b) Visage/Stockbyte/Getty Images. (10.9) Yevgen Timashov/Getty Images. (10.13) Pete Mcbride/Getty Images. (10.14) Ulana Switucha/Alamy. (10.15) Stephen Cunha. (10.16) Stephen Cunha. (10.17) Stephen Cunha. (10.18b) Stephen Cunha. (HD10b) Christopher Furlong/Getty Images News/Getty Images. (HD10a) USGS. (HD10d) USGS. (HD10c) Stringer Mexico/Reuters. (10.20) David Wall/Alamy. (10.23b) Stephen Cunha. (10.24B) Blickwinkel/Alamy. (10.26) Stephen Cunha. (10.28) United States Geological

Text

Chapter 4: (p. 117) L. Bengtsson, "Hurricane threats," Science 293 (July 20, 2001): (p. 121) © Google Earth. (04-01) From R. Christopherson, Geosystems: An introduction to Physical Geography, 9e (c) 2015 Pearson Education, Inc. (04-03) From R. Christopherson, Geosystems: An introduction to Physical Geography, 9e (c) 2015 Pearson Education, Inc. (04-04) From R. Christopherson, Geosystems: An introduction to Physical Geography, 9e (c) 2015 Pearson Education, Inc. (04-05a) From R. Christopherson, Geosystems: An Introduction to Physical Geography, 9e (c) 2015 Pearson Education, Inc. (04-06) From R. Christopherson, Geosystems: An Introduction to Physical Geography, 9e (c) 2015 Pearson Education, Inc. (04-07) From "Airs Total Precipitable Water Vapor, May 2009," NASA, available at http://www.jpl.nasa.gov/spaceimages/details.php?id=PIA12097. (04-08) From R. Christopherson, Geosystems: An Introduction to Physical Geography, 9e (c) 2015 Pearson Education, Inc. (04-09) From R. Christopherson, Geosystems: An Introduction to Physical Geography, 9e (c) 2015 Pearson Education, Inc. (04-10) Based on NOAA/ESRL Physical Sciences Division. (04-13) Tarbuck & Lutgens, The Atmosphere: An Introduction to Meteorology 12e, (c) 2013 Pearson Education Inc. (04-14) From R. Christopherson, Geosystems: An Introduction to Physical Geography, 9e (c) 2015 Pearson Education, Inc. (04-15) From R. Christopherson, Elemental Geosystems 7e, (c) 2013 Pearson Education, Inc. (04-17) From R. Christopherson, Geosystems: An Introduction to Physical Geography, 9e (c) 2015 Pearson Education, Inc. (04-23) From R. Christopherson, Elemental Geosystems 7e, (c) 2013 Pearson Education, Inc. (04-24) Part a: Based on Climatic Atlas of the United States, p. 53. (04-29) From R. Christopherson, Geosystems: An Introduction to Physical Geography, 9e (c) 2015 Pearson Education, Inc. and Tarbuck & Lutgens, The Atmosphere: An Introduction to Meteorology 12e, (c) 2013 Pearson Education Inc. (04-30) From R. Christopherson, Geosystems: An Introduction to Physical Geography, 9e (c) 2015 Pearson Education, Inc. and Tarbuck & Lutgens, The Atmosphere: An Introduction to Meteorology 12e, (c) 2013 Pearson Education Inc. (04-33) From R. Christopherson, Geosystems: An Introduction to Physical Geography, 9e (c) 2015 Pearson Education, Inc. (04-36) NCDC NEXRAD Data Inventory, NOAA, May 20, 2013. This image was created using GR2Analyst developed by Gibson Ridge Software. (04-37) Data from Daily Weather Maps, Februsry 21, 2014, NOAA/NWS/NCEP/HPC. (04-38) Data from NWS; Map Series 3; Climatic Atlas of Canada, Atmospheric Environment Service, Canada. (04-39) TRMM LIS-OTD image, NASA EOSDIS Global Hydrology Resource Center (GHRC) DAAC. (04-41) Based on image by G. Carbin, NOAA Storm Prediction Center. (04-42) From Tarbuck & Lutgens, The Atmosphere: An Introduction to Meteorology 12e, (c) 2013 Pearson Education Inc. and From R. Christopherson, Geosystems: An Introduction to Physical Geography, 9e (c) 2015 Pearson Education, Inc. (04-43a) From R. Christopherson, Geosystems: An Introduction to Physical Geography, 9e (c) 2015 Pearson Education, Inc. (04-43a) Data from the Storm Prediction Center, NWS, and NOAA sources. (04-44b) (b) R. Rohde/NASA/GSFC. (04-44c) Image reproduced courtesy of Unisys Corporation. (04-44b) Based on Adam Volland, "Earth's Disappearing Groundwater" Earth Observatory, November 5, 2014. earthobservatory.nasa.gov/blogs/earthmatters/2014/11/05/earths-disappearing-groundwater/. (04-44c) Data from the Storm Prediction Center, NWS, and NOAA sources. (04-45) Based on NOAA. (04-47a) (b) R. Rohde/NASA/GSFC. (04-47a) Based on NOAA. (GL 04.01) Based on Daily Weather Maps, October 8, 2012-October 14, 2012, NOAA/NWS/NCEP/HPC. (GL 04.02) From NOAA.

Chapter 5: (05-03) Source: http://sciencelearn.org.nz/Contexts/H2O-On-the-Go/Sci-Media/Images/Earth-s-water-distribution. (05-13) Source: Center for Environmental Systems Research, University of Kassel, April 2002- Water GAP 2.1D. (05-18) Source: http://www.health.state.mn.us/divs/eh/wells/images/spring 1.gif. (05-19b) Source: Adapted from "Water-level and sotrage changes in the high Planes aquifer, predevelopment to 2011 and 2009-2011" by V.L. McGuire, USGS Scientific Investigations Report 2012-5291, 2013,. 1. (05-19b) Source: Adapted from "Water-level and sotrage changes in the high Planes aquifer, predevelopment to 2011 and 2009-2011" by V.L. McGuire, USGS Scientific Investigations Report 2012-5291, 2013,. 1. (05-21a) Source: http://www.maryknollogc.org/article/water-and-keystone-xl-pipeline. (05-22) Source: Aguado and Burt, Understanding Weather and Climate, 4th Edition, Pearson Education, 2007. (05-24) Copyright © Pearson Education, Inc. (05-26b) Source: Based on "Instrumental Record of 'Naturalized' Colorado River Flow at Lees Ferry Near Page, Arizona" Institute of the Environment, The University of Arizona. http://www.southwestclimatechange.org/ures/lees_ferry_flow. (05-28b) Source: http://www.climatecentral.org/news/california-snowpack-obliterates-record-low-18847. (TBL05-01) Source: BGD, billion gallons per day. Population data from 2012 World Population Data Sheet (Washington, DC: Population Reference Bureau, 2012). CO2 data from PRB 2009.

Chapter 6: (06-23) Source: Based on http://www.south-georgia.climatemps.com/. (UN 06-02) (c) 2015 Google, Inc.

Chapter 7: (07-05) Source: Based on https://en.wikipedia.org/wiki/Paleoclimatology#mediaviewer/File:All_palaeotemps.png. (07-06) Source: http://www.knowqout.com/history-culture/finding-the-age-of-kings/. (07-09) Source: From R. B. Alley, "The Younger Dryas Cold Interval As Viewed from Central Greenland," Quaternary Science Reviews 19 (January 2000): 213–226; available at URL http://www.ncdc.noaa.gov/paleo/pubs/alley2000/. (07-10a) Source: http://www.climate4you.com/Sun.htm#Solar irradiance and sunspot number. (07-10b) Source: https://upload.wikimedia.org/wikipedia/commons/2/28/Sunspot_Numbers.png. (07-10c) Source: http://www.grida.no/publications/other/ipcc_tar/?src=/climate/ipcc_tar/wg1/2-20.htm. (07-11b) Source: https://upload.wikimedia.org/wikipedia/commons/7/7e/Milankovitch_Variations.png. (07-15a) Source: https://www.climatecommunication.org/climate/natural-factors/. (07-19) Source: Data from NOAA posted at URL http://www.esrl.noaa.gov/gmd/ccgg/trends/. (07-20) Source: Adapted from IPCC Fifth Assessment Report,

Climate Change 2013. The Physical Science Basis, Working Group I, FAQ 2.1, ure 1, p. 198. (07-22) Source: Based on data from NASA/GISS; available at URLhttp://climate.nasa.gov/key_indicators#co2. (07-23) Source: https://en.wikipedia.org/wiki/Instrumental_temperature_record#mediaviewer/File:Global_temperature_change_-_decadal_averages,_1880s-2000s_(NOAA).png. (07-24a) Source: http://www.skepticalscience.com/Accelerating-ice-loss-from-Antarctica-and-Greenland.html. (07-24b) Source: http://earthobservatory.nasa.gov/IOTD/view.php?id=8010. (07-24c) Source: http://www.skepticalscience.com/Accelerating-ice-loss-from-Antarctica-and-Greenland.html. (07-25a) Source: https://www.ccin.ca/home/ccw/permafrost/future. (07-27) Source: Based on Laboratory for Satellite Altimetry/NOAA. (07-30) Source: Adapted and updated from T. R. Karl, J. T. Melillo, and T. C. Peterson, 2009, Global Climate Change Impacts in the United States, Cambridge University Press, 189 pp. Updated map available at URL http://nca2014.globalchange.gov/report/our-changing-climate/heavy-downpours-increasing. (07-34) Source: From Greenhouse Gases Continue Climbing; 2012 a Record Year, NOAA, August 2013. (07-36) Source: Based on The NOAA Annual Greenhouse Gas Index (AGGI), NOAA, updated summer 2014. (07-38) Source: Based on ipcc 5ar wg1 report, ch5, p 393. (07-41) Source: IPCC Fifth Assessment Report, Climate Change 2013: The Physical Science Basis, Working Group I, ure TS-9, p. 60. (07-43) Source: Adapted by permission from J. Wiess, J. Overpeck, and B. Strauss, 2011, Implications of recent sea level rise science for low-elevation areas in coastal cities of the conterminous U.S.A., Climatic Change 105: 635–645. (07-45) Source: IPCC Fifth Assessment Report, Climate Change 2013: The Physical Science Basis, Working Group I, FAQ 12.1, ure 1, p. 1037. (07-48) Source: Taken from: http://www.infographicszone.com/environment/top-5-global-warming-infographics. (TBL07-01) Source: Data from NOAA. (TBL07-02) Source: The free Summary for Policy Makers for each AR5 Working Group is available at https://www.ipcc.ch/report/ar5/.

Chapter 8: (08-01) From R. Christopherson, Elemental Geosystems 7e, (c) 2013 Pearson Education, Inc. Data and update from Geological Society of America and Nature 429 (May 13, 2004): 124-125. (08-04) From R. Christopherson, Elemental Geosystems 7e, (c) 2013 Pearson Education, Inc. (08-07) From R. Christopherson, Elemental Geosystems 7e, (c) 2013 Pearson Education, Inc. (08-09) From R. Christopherson, Elemental Geosystems 7e, (c) 2013 Pearson Education, Inc. (08-17) Adapted from J. R. Heirtzler, S. Le Pichon, and J. G. Baron, Deep-Sea Research 13, © 1966, Pergamon Press, p. 247 and The Bedrock Geology of the World by R. L. Larson, et al., © 1985, W. H. Freeman and Company. (08-20) Source: U.S. Geodynamics, National Academy of Sciences and National Academy of Engineering. (08-29b) www.thegatewaypundit.com/2015/04/massive-7-8-earthquake-hitsnepal-avalanche-on-mount-everest-800-dead/. (TBL 08-02) Source: USGS Earthquake Hazards Program.

Chapter 9: (09-15) Source: Map adapted by Pam Schaus, after USGS sources; and D. C. Ford and P. Williams, Karst Geomorphology and Hydrology, p. 601. (09-18) Source: Based on http://illegaldumpingnwa.weebly.com/uploads/1/1/1/5/11159638/795752.jpg?1334018307. (09-31b) Source: http://www.dailyyonder.com/files/u2/MTRmap510.jpg. (TBL GL 09.01) Source: USGS.

Chapter 10: (10.01) Source: http://water.usgs.gov/edu/streamflow2.html. (10.02) Source: After U.S. Geological Survey; The National Atlas of Canada, 1985, "Energy, Mines, and Resources Canada"; and Environment Canada, Currents of Change—Inquiry on Federal Water Policy—Final Report 1986. (10.05) Republished with permission of American Association of Petroleum Geologists, from Drainage analysis in geologic interpretation; a summation, A. D. Howard,1967.permission conveyed through Copyright Clearance Center, Inc. (10.11) Source: http://www.whoi.edu/home/interactive/rivers/. (10.19) Source: Data from http://geoclasses.tamu.edu/ocean/wormuthwork/marinesediments/hjulstrom'scurve.gif. (GL 10.01) Source: http://geomorphologyresearch.com/2012/12/22/big-thompson-river-little-thompsonriver-drainage-divide-area-landform-origins-in-the-colorado-front-range-usa/. (GL10-03) Source: NOAA.

Chapter 11: (11-02) Source: NASA. (11-03) Source: http://wps.prenhall.com/esm_tarbuck_escience_11/32/8324/2131064.cw/content/index.html. (11-04) Source: Composite image of global ocean surface salinity from August 2011 to July 2012 using data from the Aquarius satellite, in orbit since 2011. [NASA.]. (11-05) Source: http://www.wri.org/resource/threat-coral-reefs-ocean-acidification-present-2030-and-2050. (11-13) Source: Based on http://www.hinchingbrookeschool.co.uk/geography/GCSERestlessEarth8.html. (11-29) Source: NOS/NOAA, 2011. (11-32b) Source: http://www1.american.edu/ted/ice/Bangladesh.html. (UN 11-01) Source: MapMaster. (UN 11-02) (c) 2015 Google. Data SIO, NOAA, U.S. Navy, NGA, GEBCO. Image (c) 2015 CNES/Astrium. (GL 11-02) Source: NOAA 2014. (GL 11-03) Source: NOAA 2014.

Chapter 12: (12-20) Source: NOAA arctic.noaa.gov/detect/detection-images/land-perm-dist-map-5.5.1.jog. (12-24) Source: http://imnh.isu.edu/digitalatlas/geo/basics/timeline.htm. (12-25) © Ron Blakey and Colorado Plateau Geosystems Inc. (12-26) Source: After USGS. (12-30) Source: Based on http://www.climatedata.info/Forcing/Forcing/milankovitchcycles.html. (GL 12.02) Based on USGS data.

Chapter 13: (13-04) Source Foley, J.A., I.C. Prentice, N. Ramankutty, S. Levis, D. Pollard, S. Sitch, and A. Haxeltine (1996) An Integrated Biosphere Model of Land Surface Processes, Terrestrial Carbon Balance and Vegetation Dynamics, Global Biogeochemical Cycles, 10, 603-628.: http://nelson.wisc.edu/sage/data-and-models/atlas/maps.php?datasetid=37&includerelatedlinks=1&dataset=37.Used by permission of The Center for Sustainability and the Global Environment, Nelson Institute for Environmental Studies, University of Wisconsin-Madison". (13-12) Source: Based on map prepared by USDA-NRCS, Soil Survey Division. (13-15) Source:

After USDA-NRCS, Soil Survey Manual, Agriculture Handbook No. 18, P.138 (1993). (13-19) Source: Based on http://instruct1.cit.cornell.edu/courses/ipm444/test/NotesEcologicalConcepts.htm. (13-20) Source: Based on http://www.youngtassiescientists.com/static/2005/bryant.html Diagram by McGonigal & Woodworth, 2001. (13-28a) Source: Based on http://www.npr.org/2012/08/23/159373770/the-new-normal-for-wildfires-forest-killing-megablazes. (13-28b) Source: Based on http://www.npr.org/2012/08/23/159373770/the-new-normal-for-wildfires-forest-killing-megablazes. (13-35) Source: http://cmsdata.iucn.org/downloads/species_susceptibility_to_climate_change_impacts.pdf. (13GE) Image Landsat (c) 2015 Google, Inc. Data SIO, NOAA, U.S. Navy, NGA, GEBCO. (GL 13-01a) Source: Atlas of the Biosphere, used by permission of The Center for Sustainability and the Global Environment, Nelson Institute for Environmental Studies, University of Wisconsin-Madison". (GL 13-02a) Source: Atlas of the Biosphere, "used by permission of The Center for Sustainability and the Global Environment, Nelson Institute for Environmental Studies, University of Wisconsin-Madison". (TBL13-01) Source: United Nations Environment Programme, Global Biodiversity Assessment (Cambridge, England: Cambridge University Press, 1995), Table 3.1–3.2, p. 118.

Chapter 14: (14-06) Source: http://www.shmoop.com/biogeography/island-biogeography.html. (14-08) Source: Adapted from Natural Resources Conservation Service maps, 1999, 2006. (14-18a) Source: Illustration adapted from N. C. Brady, The Nature and Properties of Soils, 10th ed., © 1990 by Macmillan Publishing Company. (14-44) Source: Aguado and Burt, Understanding Weather and Climate, 4th Edition, Pearson Education, 2007. (14GE) U.s. Dept of State Geographer. (c) 2015 Google. Image Landsat Data SIO, NOAA, U.S. Navy, NGA, GEBCO. (14HD) Source: Map courtesy of Erle Ellis, University of Maryland, Baltimore County, and Navin Ramankutty, McGill University/NASA; available at: http://earthobservatory.nasa.gov/IOTD/view.php?id=40554 Anthropogenic Biomes of the World (based on Ellis & Ramankutty, 2008). (IMA) Copyright (c) Google Earth.

Index

Note: Page numbers with "f" after them note an entry appears within a figure.

World – Physical

Great Basin	Land features
Caribbean Sea	Water bodies
Aleutian Trench	Underwater features

ARCTIC OCEAN

QUEEN ELIZABETH ISLANDS

GREENLA

Ellesmere Island

Beaufort Sea

Victoria Island

Baffin Island

Baffin Bay

Davis Strait

Great Bear Lake

Re

MACKENZIE MTS.

Mackenzie R.

Great Slave Lake

Hudson Bay

Labrador

Labrador Sea

Bering Strait

Yukon R.

△ Denali 20,310 ft (6,190 m)

ROCKY MOUNTAINS

NORTH AMERICA

Saskatchewan R.

Canadian Shield

Lake Winnipeg

Great Lakes

Island of Newfoundland

Bering Sea

Gulf of Alaska

Missouri R.

Aleutian Islands

Aleutian Trench

Vancouver I.

CASCADE RANGE

Columbia R.

GREAT PLAINS

Mississippi R.

Ohio R.

APPALACHIAN MTS.

Cape Cod

Sohm Plain

SIERRA NEVADA

Great Basin

Cape Hatteras

ATLANTIC OCEAN

Northeast

Mendocino Fracture Zone

Colorado R.

Rio Grande

Hatteras Plain

Bermuda Rise

Mid Atlantic Ridge

Murray Fracture Zone

Baja California

SIERRA MADRE

Mexican Plateau

Gulf of Mexico

Bahama Is.

Hawaiian Ridge

Tropic of Cancer

Molokai Fracture Zone

Cuba

Puerto Rico Trench

Hawaiian Is.

Pacific

Greater Antilles

West Indies

Johnston Atoll

Clarion Fracture Zone

CENTRAL AMERICA

Caribbean Sea

Demerara Plain

Central Pacific Basin

PACIFIC OCEAN

Middle America Trench

ANDES

Orinoco R.

Guiana Highlands

Line Islands

Clipperton Fracture Zone

Galápagos Is.

AMAZON

Cape S Roque

Equator

Basin

BASIN

Amazon R.

SOUTH AMERICA

Phoenix Is.

POLYNESIA

Marquesas Is.

East Pacific Rise

Brazilian Shield

Samoa Is.

Tuamotu Archipelago

Mato Grosso Plateau

Per

Tonga Is.

Cook Is.

Society Is.

Tahiti

Nazca Ridge

Atacama Desert

Gran Chaco

Tonga Trench

Austral Islands

Tropic of Capricorn

Pitcairn I.

Sala y Gómez Ridge

Peru-Chile Trench

Mt. Aconcagua 22,834 ft (6,960 m) △

Rio Gra Rise

Kermadec Tr.

Louisville Ridge

Southwest Pacific Basin

Easter I.

Challenger Fracture Zone

Juan Fernández Is.

Pampas

Rio de la Plata

Patagonia

Argentine Plain

40°S

Southeast Pacific Basin

Humboldt Plain

Falkland Is.

Sou Georg

Eltanin Fracture Zone

Strait of Magellan

South Georgia Ridge

Udintsev Fracture Zone

60

Cape Horn

Drake Passage

Pacific-Antarctic Ridge

Antarctic Circle

80°S

0	1,000	2,000 Miles
0	1,000	2,000 Kilometers

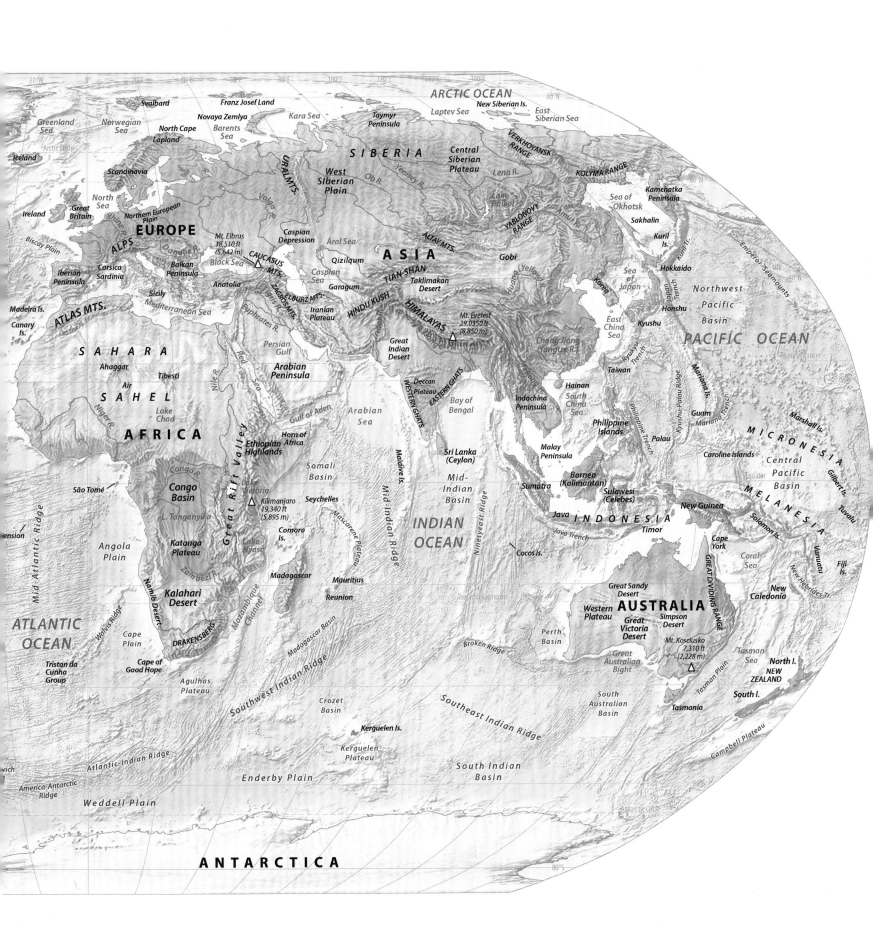